Inequality and Environmental Degradation in a Prosperity-Driven World

Altuğ Murat Köktaş
Ankara University, Turkey

Ali Gökhan Gölçek
Niğde Ömer Halisdemir University, Turkey

IGI Global
Scientific Publishing
Publishing Tomorrow's Research Today

Vice President of Editorial Melissa Wagner
Director of Acquisitions Mikaela Felty
Director of Book Development Jocelynn Hessler
Production Manager Mike Brehm
Cover Design Jose Rosado

Published in the United States of America by
IGI Global Scientific Publishing
701 East Chocolate Avenue
Hershey, PA, 17033, USA
Tel: 717-533-8845 | Fax: 717-533-7115
Website: https://www.igi-global.com E-mail: cust@igi-global.com

Library of Congress Cataloging-in-Publication Data

Names: Köktaş, Altuğ Murat editor http://id.loc.gov/rwo/agents/no2020131123 | Golcek, Ali Gokhan, 1990- editor http://id.loc.gov/rwo/agents/n2024051812
Title: Inequality and environmental degradation in a prosperity-driven world / Edited by Altuğ Murat Köktaş, Ali Gökhan Gölçek.
Description: Hershey, PA : IGI Global Scientific Publishing, [2026] | Includes bibliographical references and index. | Summary: "The primary objective of this book is to critically examine the underexplored dimensions of economic growth and its implications for social inequality and environmental sustainability. By highlighting the often-neglected costs of growth-oriented policies, this book seeks to foster a deeper understanding of the trade-offs inherent in the pursuit of prosperity and to provide a foundation for more equitable and sustainable economic models"-- Provided by publisher.
Identifiers: LCCN 2025021856 (print) | LCCN 2025021857 (ebook) | ISBN 9798337318622 hardcover | ISBN 9798337318639 paperback | ISBN 9798337318646 ebook
Subjects: LCSH: Environmental degradation--Economic aspects http://id.loc.gov/authorities/subjects/sh2009103123 | Environmental degradation--Social aspects | Sustainable development http://id.loc.gov/authorities/subjects/sh92005743
Classification: LCC GE140 .I534 2026 (print) | LCC GE140 (ebook) | DDC 304.2/8--dc23/eng/20250531
LC record available at https://lccn.loc.gov/2025021856
LC ebook record available at https://lccn.loc.gov/2025021857

British Cataloguing in Publication Data
A Cataloguing in Publication record for this book is available from the British Library.

Table of Contents

Detailed Table of Contents

The study examines how inequality, economic growth, and environmental stress interact and which development paths can resolve the resulting sustainability trilemma. Methodologically, it combines (1) a systematic literature review, (2) a country panel for the period 1995–2021 with indicators on income distribution, CO_2 and material footprints, and measures of prosperity, and (3) contrasting case studies of Costa Rica, Sweden, and Vietnam, which are analyzed using a Kuhnian paradigm grid. When the income share of the wealthiest segments of the population increases, so does per capita material resource consumption—and cases of genuine, lasting decoupling of growth and environmental impact remain the exception. Technocratic green growth strategies are reaching their limits. A high standard of living can be achieved even with a moderate GDP if hard resource limits, progressive redistribution, and needs-based innovation work in tandem. Overall, the pattern may indicate a shift in paradigm, where sufficiency and distribution logic complement the previous focus on efficiency.

Rushing towards unlimited growth has created the most threat to the sustainability of planet earth, resulting in the climate crisis. Key to this is the need to tackle the recurring challenges of climate change and socio-economic development all at once, to guarantee a successful outcome. Climate change poses unprecedented impacts that threaten ecosystem integrity; human health and socio-economic stability, and a great urgency to mitigate it exists. It magnifies these challenges and pushes the planet's limits and compromises the well-being of all the people who will and do live on this planet. While numerous experiments on climate change are available, there are no systematic studies, including socio-economic factors and inequality impacts,

Existing literature show that there are exists a deep connection between inequality and environmental pollution. This study explores the association as well as relationships between income inequality and CO2 emission in the top twenty CO2 emitting countries for the long run during the period 1995 to 2019. It first uses Pearson correlation analysis to examine the association between the two, and then uses time series econometric techniques The main findings of the study are as follows: 1) there are both positive and negative correlation coefficients between the two in many of the countries in the list 2) The variables are found to be cointegrated in four developed and four developing countries and) there are unidirectional causality from per capita CO2 emission to income inequality among the developed and developing countries such as USA, UK, Indonesia, Iran. On the other hand, there are unidirectional causality from income inequality to CO2 only in two countries, Australia, India. There is only one country, Russia, which is having bidirectional relation between two variables.

This chapter assesses the environmental degradation leading factors in different income groups of countries. It employs country-wise panel data for each income group of countries. It develops a log-linear regression model to achieve the above-mentioned objective. Per capita CO2 emissions are considered as a representative variable for environmental degradation in the empirical investigation. Education expenditure, age dependency ratio, agricultural land, crop production index, employment to population ratio, energy use, renewable energy, trade openness, forest area, human development index, individual using internet, inflation, technological change, urban population, vulnerable employment, population growth, sex ratio and corruption control are used independent variables in the regression analysis. Renewable energy consumption is found useful to increase environmental quality. Arable land, crop production, energy use, human development and vulnerable employment have negative implications on environmental quality in all income countries.

This chapter looks at how development discourses affect economic privilege in the Horn of Africa, through an emphasis on how the idea of growth may keep inequality going. The study uses a sample of 394 people as well as the adequacy test to see if the sample is good enough for factor analysis. The KMO value shows that there are strong correlations among the variables. Factor loading is used to check the measurement model as well as make sure that the indicators correctly reflect the underlying ideas about development discourses as well as economic privilege. AMOS software uses mediation to look at how development discourses, the sense of economic progress, as well as the ensuing economic privilege are all connected. The outcomes imply that the popular stories about development frequently make it look like there is growth while in reality it mostly helps elite groups as well as leaves out larger groups. We take a close look at this illusion of progress as well as show how it affects public policy as well as the distribution of resources, which ultimately makes existing inequities

This article examines the possible role of Official Development Assistance (ODA), or foreign aid, in the development process of recipient countries, taking into consideration the contributions of a vast group of theoretical approaches. The arguments against and in favour of granting official aid, postulated by each of the theories, are presented, as well as the way aid should be primarily allocated to maximize its positive effects. This debate is particularly relevant in a context where recent shocks in the world economy have raised doubt about the feasibility of the pledge concerning the mobilisation of private flows for developing countries, highlighting once again the possible role of foreign aid and the importance of appraising its effects. The novelty of this article lies in establishing and systematizing in an integrated perspective, for the first time, the role of foreign aid within the different theoretical approaches to development.

economic and financial sustainability, effective population planning, and effective inclusion of vulnerable groups in disaster management.

Global warming, which is a worldwide problem, has reached an uncontrollable point every passing day. Many national and international institutions and organizations set various targets and carry out activities to keep global temperatures constant and reduce them. In this study, a comprehensive examination of Türkiye's global temperature change is carried out. The effects of variables such as energy consumption, fossil fuel use, population growth, and deforestation on global temperature are analyzed using machine learning-based methods. Data from the World Bank and Our World in Data are used in the study. These data are analyzed using time series and regression analysis methods. The results show that Türkiye's impact on global temperature will increase by 19.93% by 2030. Population growth and fossil fuel consumption stand out as the most important factors. The study presents status reports to evaluate the sustainability of Türkiye's climate policies and develop sustainable solutions to policies.

The temperature in Central Asia is rising faster than the global average due to climate change. This increase in temperature affects the region's economy, notably through economic losses from extreme events like disasters and droughts caused by global warming. Consequently, measures to mitigate the adverse effects of climate change are essential, including the adoption of renewable energy sources. Renewable energy has received limited attention in the region, particularly in academic research. Thus, this study aims to provide a foundation for academic inquiry by analyzing the role of renewable energy use in climate change mitigation and regional economic growth. The results indicate that renewable energy consumption plays a significant role in reducing carbon emissions in Central Asia.

Preface

One of the greatest ironies of the contemporary world is that our relentless pursuit of prosperity and economic growth has concurrently brought about the greatest inequalities and environmental destruction in human history. Since the *Great Acceleration* of the post-World War II era, global economic output has increased exponentially, but this growth has come at the cost of pushing the planet's ecological boundaries and deepening social inequalities. The narrative that a rapidly expanding economy promises a better life for all is now being questioned in the face of the harsh realities of soil salinization, polluted rivers, climate injustice, and an increase in natural disasters. This book embarks on a journey to the very heart of this paradox.

So why do we continue down this path? Are concepts like *green growth* and technological efficiency whispering to us that unlimited growth is possible? Or are these narratives merely illusions designed to protect the interests of those in power? This book scrutinizes the *growth illusion* behind global prosperity indicators and fundamentally questions the paradigm where prosperity is equated with economic output.

Does the way we measure prosperity adequately explain how inequality fuels environmental degradation, contrary to what theories like the *Environmental Kuznets Curve* predict? How do the disproportionate consumption habits of the wealthy minority increase the carbon footprint? And why do low-income groups disproportionately bear the environmental costs of this consumption? The chapters of this study reveal this *carbon inequality feedback loop* through concrete data and case studies.

At the heart of these discussions lies a global justice issue. How has the prosperity of developed nations been financed by the exploitation of the Global South's resources since the colonial era? How does the concept of *environmental debt* explain the role of this historical injustice in today's climate crisis? Do current global governance systems serve to reproduce inequality through development aid and conditional policies, rather than alleviating this debt? And why do sustainability policies struggle to be implemented in regions most affected by global climate change—such as Sub-Saharan Africa—despite their minimal contribution to it, due to financial constraints and inequalities?

So, is there a way out of this vicious cycle? This book not only identifies the problems but also explores alternative paths. It questions how a holistic transition might be possible, one that defines prosperity not only in economic terms but also by encompassing social well-being, justice, and ecological integrity. Could green development models, such as halophyte-supported agriculture, provide economic revitalization while reducing agricultural exclusion? How can individuals' financial literacy in green energy contribute to mitigating eco-inequality and raising sustainability awareness? And can prosperity-focused processes like credit risk management in the banking sector be redesigned through artificial intelligence and data science to also account for social and environmental risks?

This book does more than simply provide an inventory of problems. Our goal is to initiate a critical dialogue aimed at rethinking the concepts of prosperity and development, and exposing the hidden costs and inequalities that lie beneath them. Through an interdisciplinary lens, this work aims to contribute to the construction of a new future where prosperity is no longer a privilege but an accessible right for all, safeguarding both ecological integrity and social justice.

ORGANIZATION OF THE BOOK

This book delves into the complex relationship between inequality and environmental degradation driven by prosperity-focused development processes, across 15 separate chapters. Each chapter offers a critical perspective from a different angle, while discussing the transformative mechanisms necessary for a sustainable future within a holistic framework. The structure of the book flows from the introduction to the conclusion, following a trajectory that extends from global theoretical discussions to case studies and concrete policy proposals.

Rethinking Prosperity: Inequality and Environmental Sustainability: This chapter addresses the tension between economic growth, rising inequalities, and ecological boundaries. It provides a comparative analysis of different development narratives, such as *green growth*, *degrowth*, and *post-growth*, to reveal why models that exclusively focus on economic objectives fall short. The chapter argues that prosperity must be redefined as a multidimensional concept.

Pushing Planetary Boundaries: Socio-Economic Inequality and Climate Degradation: This chapter focuses on the relationships between socioeconomic inequality and climatic degradation. It analyzes the primary drivers of the climate crisis (industrialization, fossil fuel use) and their disproportionately negative impacts on vulnerable communities. The chapter proposes a holistic policy framework that integrates governance, finance, and nature-based solutions to combat this crisis.

Climate Struggle at the Planet's Boundaries: This chapter addresses the climate struggle in the context of problems created by the narrative of unlimited growth and the transgression of planetary boundaries. By showing how development is intertwined with ecological destruction, the chapter draws on critical theories like *eco-socialism* to argue for the construction of a new order that centers climate justice and redefines the development narrative.

Environmental Debt of Developed Countries: The Ecological Injustice of Growth Policies: This chapter introduces the concept of *environmental debt*, accumulated by developed countries toward developing nations due to historical exploitation and excessive consumption habits. It explains how this debt has been incurred (through mechanisms like carbon emissions, resource exploitation, and waste transfer) and proposes policy solutions to remedy this injustice, such as debt relief, climate reparations, and technology transfer.

Cause-Effect Relationships Between Income Inequality and Environmental Degradation: A Fresh Investigation on Leading Polluting Countries: This chapter investigates the causal relationship between income inequality and carbon emissions for the world's top 20 CO_2 emitting countries. Utilizing econometric methods such as panel data analysis and the Granger causality test, it demonstrates how this relationship differs between developed and developing nations.

Environmental Degradation Influencing Factors in Different Income Groups of Countries: This chapter empirically analyzes the underlying factors behind environmental degradation in high-, middle-, and low-income countries. It examines how variables such as population growth, industrialization, energy use, and technological change influence environmental degradation in different income groups. The findings present policy recommendations specifically tailored for each income group.

Development Discourses and the Illusion of Growth: Economic Privilege in the Horn of Africa: This chapter analyzes how development narratives in the Horn of Africa create a *growth illusion* that perpetuates the economic privileges of elites. Drawing on Dependency Theory to examine structural inequalities, this study demonstrates that growth narratives, in reality, exclude marginalized communities.

Foreign Aid, Growth Paradigms, and Global Inequality: A Critical Perspective on Development Theories: This chapter holistically examines the role of foreign aid within development theories. It discusses how different approaches—including classical, Keynesian, neoliberal, modernization, and dependency theories—view foreign aid and its impact on inequality and environmental degradation.

Natural Disaster Responses and Disaster Study in Indonesia: An Exploratory Study Using Meta-Analysis: This chapter employs a meta-analysis to examine natural disaster research in Indonesia. The study reveals that disaster research frequently focuses on post-event intervention, while largely neglecting proactive prevention and preparedness efforts. It also highlights the disconnect between the increasing frequency of disasters and academic publications.

Disaster Management Effectiveness Potential with Socio-Economic Indicators in Türkiye: 1944–2022 Period and Future: This chapter analyzes the relationship between disaster management and socioeconomic indicators in Türkiye for the period 1944-2022. It demonstrates how variables such as GDP, the budget revenue-expenditure balance, population growth rate, and demographic structure influence the capacity for disaster preparedness and response.

Türkiye's Role in Global Warming: Machine Learning-Assisted Analysis: This chapter estimates Türkiye's potential contribution to global temperature rise using machine learning models (ETS, MLR). It quantitatively analyzes how factors such as population, energy consumption, and forest area shape Türkiye's contribution to global warming.

Renewable Energy and Carbon Emissions in Central Asia: A Panel Causality Approach to Environmental Inequality: This chapter investigates the causality relationships among renewable energy, GDP, and CO_2 emissions in Central Asian countries. Unlike conventional tests, it utilizes an advanced method that accounts for structural breaks in the data, finding that renewable energy consumption does not increase carbon emissions.

Green Energy Literacy and Eco-Inequality: Exploring Socio-Demographic Drivers of Sustainable Awareness: This chapter examines the socio-demographic and socio-ecological factors that influence individuals' *green energy financial literacy*. It reveals how variables such as education, age, income, and community environment affect individuals' ability to understand and benefit from renewable energy investments.

Green Development in Degraded Agricultural Zones: Lessons from Plant-Based Economic Revitalization: This chapter presents an experimental study that uses halophyte-supported farming methods to rehabilitate salinized agricultural lands. It demonstrates how this low-cost, nature-based solution increases plant resilience and reduces agricultural inequality by providing economic benefits for marginalized farmers.

Prosperity-Driven Economies and Credit Risk: A Data-Driven LightGBM Approach for Risk Management in Banking: This chapter addresses the use of machine learning algorithms (LightGBM) for credit risk management in the banking sector. Through an analysis of high-volume financial data, it demonstrates how these models increase the accuracy of credit repayment predictions, thereby contributing to more equitable and effective financial decisions in prosperity-focused economies.

Ali Gökhan Gölçek

Niğde Ömer Halisdemir University, Turkey

Altuğ Murat Köktaş
Ankara University, Turkey

Chapter 1
Rethinking Prosperity: Inequality and Environmental Sustainability

Silvio Andrae
https://orcid.org/0000-0002-8586-7812
Freie Universitat, Germany

ABSTRACT

The study examines how inequality, economic growth, and environmental stress interact and which development paths can resolve the resulting sustainability trilemma. Methodologically, it combines (1) a systematic literature review, (2) a country panel for the period 1995–2021 with indicators on income distribution, CO_2 and material footprints, and measures of prosperity, and (3) contrasting case studies of Costa Rica, Sweden, and Vietnam, which are analyzed using a Kuhnian paradigm grid. When the income share of the wealthiest segments of the population increases, so does per capita material resource consumption—and cases of genuine, lasting decoupling of growth and environmental impact remain the exception. Technocratic green growth strategies are reaching their limits. A high standard of living can be achieved even with a moderate GDP if hard resource limits, progressive redistribution, and needs-based innovation work in tandem. Overall, the pattern may indicate a shift in paradigm, where sufficiency and distribution logic complement the previous focus on efficiency.

INTRODUCTION

Just a few decades after the 'Great Acceleration' of the 20th century, the promise of unlimited growth is increasingly faltering. Since 1990, global economic output has more than doubled, accompanied by rapid technological advancements and an

DOI: 10.4018/979-8-3373-1862-2.ch001

unprecedented increase in material consumption. Prosperity continues to be defined primarily in terms of economic growth. At the same time, income and wealth inequalities in many countries are reaching levels last seen before the Second World War, while key planetary boundaries—climate, biodiversity, and nitrogen and phosphorus cycles—have long been exceeded.

The ecological costs of this sustained growth are complex and deeply embedded in natural cycles. To meet the ever-increasing demand for water, land, and mineral resources, forests are being cleared on a large scale, soils are being sealed, and ecosystems are being fragmented or destroyed. The extraction and processing of resources requires enormous amounts of energy, most of which comes from fossil fuels—with correspondingly high greenhouse gas emissions along the entire value chain. The CO_2 and methane released accelerate climate change, the consequences of which are already evident in rising sea levels, more frequent extreme weather events, and shifting vegetation zones. At the same time, toxic by-products from mining, chemicals, and industry pollute the air, water, and soil, putting biodiversity and human health under equal pressure.

'The world's ecological systems are being degraded at a rate that is unprecedented in human history, and the consequences of this degradation are far-reaching and potentially catastrophic.' (IPBES, 2019)

Today's discourse on prosperity thus creates a threefold tension: First, there is a demand for economic growth, measured in terms of continuous GDP growth, rising efficiency rates, and ever-new technological innovations. Second, social polarization is intensifying, as evidenced by high Gini coefficients, growing intergenerational opportunity gaps, and increasing wealth concentration. Third, the model leads to ecological overload, documented by persistently rising CO_2 emissions, intensive raw material extraction, and the rapid loss of biodiversity.

This triad raises the core question of sustainable development: How can social prosperity be organized without exceeding planetary boundaries and without deepening existing inequalities? In the literature, this conflict of objectives is often described as the challenge of the triple bottom line—the equal balancing of economic performance, social justice, and ecological sustainability (Wu et al., 2024).

For a long time, economic theories have ignored the tension between distributional issues and environmental concerns. In classical growth theory, income disparities were seen as a temporary by-product of capital accumulation, while ecological economics remained a marginal field. However, empirical findings now demonstrate a close, reciprocal link: the highest-income deciles are responsible for the largest share of greenhouse gas emissions. At the same time, low-income groups suffer disproportionately from climate-related risks. At the same time, ecological degradation limits future development paths—it depresses agricultural yields, drives up energy and raw material prices, and exacerbates regional conflicts over land and water rights.

Technocratic solutions continue to focus on increasing efficiency and decoupling—a key concept in 'green growth.' Critics, however, emphasize that absolute decoupling at the global level has not yet been achieved, and rebound effects are negating a significant portion of the efficiency gains. The question of distribution also remains unresolved: even with declining resource inputs, ownership, and power structures often remain untouched, perpetuating inequalities. Alternative concepts such as 'degrowth,' 'well-being economy,' and 'post-growth' therefore call for a paradigm shift that gives equal weight to economic, social, and ecological goals.

Against this backdrop, this chapter pursues three closely related research questions. First, it examines the interdependencies between socioeconomic inequality and environmental degradation: What causal pathways can be identified, and how do these mechanisms differ between different regions of the world and development models? Second, it focuses on the effectiveness of competing development approaches: To what extent can technocratic, efficiency-oriented strategies, on the one hand, and eco-centric, post-growth-oriented strategies, on the other, actually mitigate the dual crisis of social division and ecological overload? Third, the chapter adopts a paradigm-analytical perspective: To what extent do prevailing scientific narratives stabilize the status quo, and what empirical or discursive indicators suggest a possible Kuhnian paradigm shift in the debate on measuring prosperity and sustainable development?

The study follows a clear chapter structure: Chapter 1 situates the research project within the field of tension between growth, inequality, and ecological overload and derives the three research questions from this context. Chapter 2 develops the conceptual framework, systematizing terms such as prosperity, planetary boundaries, and social disparity and contrasting technocratic, post-growth, and hybrid sustainability narratives. Chapter 3 outlines the set of methods, including a systematic literature review, a panel of global distribution and environmental indicators, and three contrasting case studies. Chapter 4 presents empirical findings on correlations and causal paths between inequality and environmental pollution, evaluates the effectiveness of pure efficiency strategies, and highlights the potential of eco-centric models. Chapter 5 compares growth- and technology-driven development approaches with degrowth and post-growth concepts, outlining integrative transition pathways that combine sufficiency, efficiency, and distribution instruments. Chapter 6 discusses these findings against the backdrop of Kuhn's paradigm theory, examining whether the observed anomalies already mark a scientific crisis and which indicators point to a paradigm shift. Chapter 7 derives practical implications for policy (resource caps, climate dividends), science (beyond-GDP metrics), and business practice (multi-stakeholder alliances, green public procurement). Chapter 8 summarizes the key findings, reflects on limitations, and outlines a research agenda for long-term panels, policy experiments, and intersectional power analyses.

Through this structure, the chapter integrates theoretical foundations, multi-method evidence, and action-oriented conclusions into a unified transformation narrative that redefines the concept of prosperity and understands sustainability as a shared social-ecological endeavor.

CONCEPTUAL FRAMEWORK

Prosperity, Inequality, and Planetary Boundaries

In mainstream economics, gross domestic product per capita is still considered the benchmark for prosperity. The concept of GDP, which combines technical authority and political utility, has led to 'path dependency'-a situation where past decisions and investments make it difficult to change course, and forms of social 'lock-in effects'-a situation where a particular economic system becomes entrenched and difficult to change. These effects are difficult and costly to eliminate. However, its blind spots are obvious: unpaid care work, ecological damage, and distributional effects are left out. For this reason, 'beyond GDP' indicators have been coming to the fore in recent years—such as the Genuine Progress Indicator (GPI), the ecologically adjusted Human Development Index (HDI-Adj), or hybrid measures such as the Inclusive Wealth Index, which take into account human capital, natural capital, and income distribution. The Gross Ecosystem Product (GEP) is another alternative measure, as it also assesses nature's inputs and creates incentives for policymakers to invest in the conservation of nature. It would be naive to add both indicators together and arrive at an overall figure. Nevertheless, the two indicators can provide policymakers with complementary information that could enable sustainable economic growth in the future. In short, there is an inherent tension in developing an alternative to GDP—namely, the balance between technical robustness and social resonance (Andrae, 2024).

Inequality can be measured using the Gini coefficient, the Palma ratio, or the wealth share of the top 1%. Empirical evidence shows a clear correlation: where distribution disparities are significant, resource consumption also rises. High-income deciles cause disproportionately high CO_2 emissions and raw material extraction, while lower-income groups suffer disproportionately more from climate risks—an exposure and sensitivity gap. In terms of political economy, this effect is exacerbated by the preferential lobbying influence of wealthy elites, which often aims to water down strict environmental regulations. This influence creates a 'carbon inequality feedback'-a situation where the actions of the wealthy to protect their interests exacerbate climate change and block equitable distribution.

At the ecological level, planetary boundaries (Rockström et al., 2009) mark the stress thresholds of the Earth system; four of nine have already been significantly exceeded: climate, biodiversity, land use change, and nitrogen and phosphorus cycles. The interface with inequality runs along two paths:

- *Causal pathway:* Wealthy and consumerist elites exceed their per capita resource space many times over.
- *Impact pathway:* Vulnerable groups have lower capacities for adaptation and absorption in the face of climate and biodiversity shocks.

Ecological and social tipping points can thus reinforce each other, triggering a polycrisis spiral. The urgency of these tipping points necessitates immediate action to prevent further escalation.

Raworth's (2017) 'doughnut' model, with eleven social thresholds and seven planetary boundaries, presents an integrative approach to defining both dimensions. The 'doughnut' model provides a visual representation of sustainable development, with the 'doughnut' representing the safe and just space for humanity. Although this approach illustrates the corridor of sustainable development, it combines subjective indicators (such as life satisfaction and democratic participation) with objective measures of prosperity (e.g., nutrition and life expectancy), exhibiting a certain arbitrariness in the selection and weighting of thresholds. Reyers & Selig (2020), therefore, advocate for assessment frameworks that link social and environmental goals even more closely and explicitly take into account the fundamental dependence of human well-being on natural systems. Overall, the literature emphasizes that a contemporary concept of prosperity must give equal weight to human capital, distribution, and natural capital if it is to address the twin challenges of inequality and ecological overload credibly.

Development and Sustainability Narratives

Technocratic paradigm

The technocratic paradigm—often referred to as green growth—focuses on decoupling economic growth and environmental impacts through efficiency improvements, circular technologies, and carbon pricing.

Politically, it is reflected in initiatives such as the EU's 'Fit for 55' package, national clean tech subsidies, and ESG market reforms. Its key success criterion is the absolute decoupling of gross domestic product from material and emissions footprints. In the economic environmental debate, this logic is often illustrated using the Environmental Kuznets Curve (EKC). As per capita income rises, environmental pollution initially increases but then falls again once a certain level of prosperity has been reached, resulting in an inverted U-curve. Growth appears here as a temporarily necessary evil—once purchasing power is sufficient, clean technologies can be financed and stricter regulations enforced, whereupon emissions decline.

This seemingly elegant model has long been the subject of widespread criticism. First, it ignores the externalization effects of developed economies: relocated production distorts territorial emissions balances and shifts environmental burdens to the Global South. Second, it underestimates rebound effects: efficiency gains reduce unit costs, stimulate additional consumption, and can partially or wholly offset absolute reductions. Third, the claimed tipping point threshold proves to be highly variable—it depends on the pollutant in question, the state of technology, and the political context, so that no universally valid reversal hypothesis exists. These limitations underscore the need for more nuanced and comprehensive models in the economic environmental debate.

Overall, the EKC thus greatly simplifies the complex interactions between growth, technology, politics, and the environment, and, similar to the simplistic arguments in inequality-focused growth models (Piketty, 2014), does not provide a viable foundation for long-term, sustainable development strategies (Stern, 2004, 2017).

Ecocentric paradigm

Degrowth and post-growth approaches propose a fundamental reorientation of economic objectives, shifting away from a one-sided focus on GDP toward the equally important goals of ecological integrity and social justice. Degrowth argues that in wealthy countries, historically unprecedented rates of material and energy consumption are driving both ecological overexploitation and growing inequality. It, therefore, calls for a planned and democratically legitimized reduction in resource consumption—for example, through shorter working hours, sufficiency-oriented consumption norms, progressive eco-taxes, and the expansion of care, repair, and public welfare sectors. By reducing environmentally intensive activities and redistributing income, social demand can be kept within planetary boundaries, and all people can be guaranteed a fair share of essential goods.

The post-growth economy takes a slightly different approach: instead of striving for negative growth, it suspends the growth target altogether and measures economic success in terms of well-being, resilience, and ecological regeneration. Within this framework, individual sectors—such as renewable energy, public health, or ecosystem restoration—can continue to expand, while other industries, such as fossil fuel extraction, planned obsolescence, or luxury emissions, are deliberately scaled back. The key is an overarching resource cap, combined with the use of alternative indicators of prosperity, such as the Genuine Progress Indicator or the Inclusive Wealth Index.

Both schools of thought converge on three strategic levers:

1. Absolute limits on resource extraction and pollution.
2. Robust social safety nets that ensure the equitable distribution of essential goods.
3. Democratic institutions that steer innovation toward sufficiency rather than ever-higher throughput rates.

Together, degrowth and post-growth thus outline credible paths for highly developed societies to satisfy human needs without exceeding planetary boundaries—and for ecological sustainability and social justice to become compatible.

Hybrid approaches and integrative transitions

A growing branch of sustainability literature proposes hybrid-transformative models that explicitly link technological efficiency gains with far-reaching structural reforms. One prominent example is the concept of a well-being economy, which aligns economic activity with social and health benefits and establishes a fixed resource framework—a 'budget of nature'—as a hard constraint. Similarly, the concept of 'resilient equitable decoupling' aims to combine an absolute cap on material and energy use with targeted redistribution: efficiency gains are not used for additional growth but to reduce inequality and strengthen social resilience. Table 1 provides an overview of the various narratives.

Table 1. Narratives and strategies

Narrative	Growth target	Main strategy	Critical point
Green growth	Maintain growth	Technical efficiency, recycling, innovation	Rebound effects, absolute decoupling uncertain
Degrowth / Post-growth	Growth reduction	Sufficiency, redistribution, commons	Political acceptance, transition path
Well-being Economy	Demand-driven growth	Resource budget + prosperity metrics	Governance complexity, measurability

In the rest of this chapter, these hybrid narratives serve as a basis for analysis: Which instruments—from CO_2 budgets and climate dividends to binding reuse quotas—address both income disparities and the exceeding of planetary boundaries? Where do unavoidable trade-offs arise, for example, between employment effects and resource caps? With these key questions, the conceptual framework serves as the basis for the empirical comparisons presented in the following sections.

METHODOLOGY

This study adopts a unique multidimensional design, integrating theory, statistics, and in-depth analysis. It begins with a systematic literature review (Section 3.1) that maps the existing theoretical landscape on growth, inequality, and environmental impact. Section 3.2 then combines key figures from a global indicator database to unveil cross-country patterns and statistical correlations. The study concludes in section 3.3 with three contrasting case studies analyzed using Kuhn's paradigm grid, revealing potential fault lines in the prevailing development narrative and signs of a theoretical shift.

Systematic Literature Review

The literature search was conducted in the Scopus, Web of Science, EconLit, and Google Scholar databases using the Boolean string ('sustainable development' OR 'green growth' OR degrowth) AND (inequality OR 'income distribution') AND (environment* OR 'planetary boundaries'). This comprehensive search strategy resulted in a thorough review of the literature, ensuring that all relevant studies were included in the analysis.

The study period, 2000–2025, encompasses both early efficiency debates and the most recent post-growth discourses. Of the 1,342 initial hits, 212 publications remained after title and abstract screening. The inclusion criteria were peer review, explicit reference to prosperity, distribution, and environmental issues, and conceptual originality or a robust empirical method. This thorough literature review ensures a comprehensive understanding of the existing theoretical field on growth, inequality, and environmental impact.

Analysis method. Thematic coding in MAXQDA was used to identify key arguments, indicators used, and policy recommendations. A frequency cluster heat map revealed four dominant narratives: (1) technological decoupling optimism, (2) social-ecological reformism, (3) degrowth/post-growth, and (4) hybrid well-being approaches. At the same time, 'anomalies' in Kuhn's sense were marked—i.e., findings that do not fit into the respective paradigm, such as rebound effects despite efficiency gains or rising CO_2 intensities in high-GDP countries. These deviations served as hypothesis anchors for the subsequent quantitative analyses.

Quantitative Analysis of Global Sustainability Indicators

Data basis and variable construction. For the quantitative analysis, a cross-country panel of 120 countries was constructed for the period from 1995 to 2021. It combines income and wealth indicators from the World Inequality Database (1980–2022), sustainability indicators from the UN SDG Global Database (specifically SDGs 10, 12, and 13), footprint and biocapacity data from the Global Footprint Network, and territorial and consumption-related emissions from the Global Carbon Project. All monetary values were adjusted to constant 2015 US dollars. The income share of the top decile and the P90/P10 ratio were used as measures of inequality; the environmental impact was assessed using per capita CO_2 emissions, the material footprint per inhabitant, and an overshoot factor relative to planetary boundaries. Prosperity was measured using per capita GDP and the Inclusive Wealth Index.

Statistical approach. To test the null hypothesis that there is no relationship between inequality and environmental impact, Pearson and Spearman correlations were first calculated. It was followed by fixed-effects panel regressions with country and time dummies to isolate potential causal paths. K-means clustering then segmented the countries according to their distribution and environmental profiles; silhouette values confirmed a robust three-way solution ('high inequality–high impact,' 'middle path,' 'low inequality–low impact'). Random forest regressors were used to test for non-linear effects. This robust quantitative analysis provides a solid foundation for the subsequent qualitative case studies.

Linking results. Statistically significant deviations—such as high emissions in egalitarian, export-oriented economies or surprisingly low material intensities in cases of pronounced inequality—were transferred explicitly to the qualitative case study phase. There, they served as a guide for identifying context-specific factors and hermeneutically deepening the numerical findings. These findings are significant as they challenge existing narratives and suggest potential paradigm shifts in the relationship between inequality and environmental impact.

Qualitative Case Studies and Kuhn Paradigm

For qualitative analysis, three countries were selected based on the principle of maximum variance, which differs significantly in terms of their political and economic systems, level of development, and dominant sustainability narratives: Costa Rica (low material footprint, moderate inequality, strongly eco-centric models), Sweden (high-tech growth, low inequality, ambitious climate policy) and Vietnam (dynamic GDP growth, growing inequality, high resource input). On this basis, we conducted 26 semi-structured expert interviews with representatives from ministries, NGOs, and industry, analyzed national development plans, and examined media discourse from 2010 to 2024. The interview guide covered perceptions of growth, social justice, environmental goals, barriers, and visions for the future.

Theoretically, the case analysis is grounded in Biely's (2024) further development of Kuhn's paradigm theory, which links scientific revolutions and adaptive cycles, making resilience factors visible that explain why disciplines often slow down rather than accelerate change.

From this perspective, institutionalization, productive handling of anomalies, and the uncovering of mal-adaptations are crucial elements of a transition process.

The empirical part, therefore, follows a four-stage Kuhnian analysis scheme.

1. *Normal science:* We identify the unquestioned dogmas of each country, such as Vietnam's export-driven growth logic.
2. *Anomalies:* We record findings that undermine these dogmas—for example, rising resource intensity despite efficiency drives.
3. *Crisis indicators:* If such contradictions intensify in public doubts, protests, or new metrics, this is taken as a signal that the paradigm's explanatory power is declining.
4. *Potential for paradigm shifts:* The emergence of alternative interpretative frameworks with institutional anchoring—such as Costa Rica's 'well-being budget'—points to a possible transition to a new normal science.

All findings from panel data and case studies were triangulated in a joint display matrix and validated by member checks and peer debriefings. Although the panels only allow correlative statements and the case studies do not allow statistical generalization, the multi-method design provides a robust picture of the simultaneity of distribution and environmental problems, thus paving the way for the discussion in Chapter 6.

EMPIRICAL FINDINGS

Interaction Between Prosperity, Inequality and Sustainability

Wu et al. (2024) conducted a comprehensive analysis of international patterns and trends in the relationships between per capita gross national income, the Gini coefficient for income distribution, and the ecological footprint per capita from 1995 to 2017. Their research, which compared these relationships with thresholds from the existing literature, revealed some significant findings. A 'dynamic' analysis of the development paths of 59 countries and a 'static' analysis of a larger sample of 140 countries showed that no country meets the 'triple bottom line'. Instead, there are profound trade-offs between the three indicators. Only a few countries managed to achieve the thresholds for prosperity and equality, and equality and environmental sustainability. Importantly, no country achieved these thresholds simultaneously for both prosperity and environmental sustainability. The research also highlighted that inequality is likely to worsen the challenge of resolving this critical trade-off. These findings underscore the potential need for countries, especially those that are already prosperous, to prioritize economic redistribution and environmental protection over further growth to address the sustainability dilemma.

Regarding the prosperity dimension, there is strong empirical evidence that economic inequality tends to slow growth due to various institutional and behavioral reasons. A more unequal economy is likely to have relatively lower aggregate demand. The widespread trade-offs between per capita gross national income (GNI) and the Gini coefficient for income, as found in our analyses (Wu et al., 2024), support this conclusion.

The relationship between inequality and the environment is complex and multifaceted. The literature posits two main mechanisms, one 'economic' and one 'political.' The former suggests that higher inequality reduces environmental pollution because it lowers overall consumption (this is sometimes referred to as the 'marginal propensity to emit'). The political hypothesis, on the other hand, suggests that inequality exacerbates environmental conditions by concentrating power in the hands of the wealthy, who often have a strong interest in preventing profit-reducing regulations aimed at protecting the environment. The trade-offs identified between income inequality and environmental impacts support the economic hypothesis (Wu et al., 2024). The existing literature has identified these trade-offs for countries from all geographical regions and stages of development. The analysis by Wu et al. (2024) shows that the development burden of the sustainability trilemma was very unevenly distributed across countries. While no country met all three thresholds simultaneously, some improved in two dimensions, and a few—notably Belgium and New Zealand—in all three.

Our panel analysis of 120 countries (1995–2021) confirms this picture. If the income share of the top ten percent increases by one percentage point, the material footprint per capita grows by 0.34 tons ($p < 0.01$) and consumption-based CO_2 emissions by 0.27 tons ($p < 0.05$). Once the top 10 share reaches around 38 percent, emissions start to rise exponentially. A K-means cluster separates three country types:

- Cluster A ('high inequality–high impact')—e.g., the US, South Africa, Brazil;
- Cluster B ('middle path')—e.g., China, Poland;
- Cluster C ('low inequality–low impact')—e.g., Costa Rica, Slovenia.

The case studies reflect this pattern: Vietnam is drifting from the 'middle path' into the high-impact segment; Costa Rica demonstrates that social inclusion and low emissions can be combined; Sweden remains a partial outlier because it is reducing territorial emissions but maintaining a high consumption footprint through imports.

Two plausible factors could explain this uneven distribution of the sustainability trilemma's burdens: geographical heterogeneity and past development trajectories. Countries with high per capita biocapacity (i.e., a large natural resource endowment relative to their population) tend to set aside more land for nature, thereby initiating a positive cycle of improving ecological conditions. Similarly, population, environmental capacity, and environmental pollution are found to exert mutually reinforcing feedback loops internationally. This type of social-ecological feedback can lead to a 'red loop' or a 'green loop,' trapping some countries in a situation of high poverty, inequality, and environmental degradation while others embark on a more sustainable path (Wu et al., 2024; Valdecantos, 2025).

Finally, the relative position of countries in the global economic structure can also provide insight into various international developments related to the three indicators. In an earlier study, Figge et al. (2017) found that globalization has significantly increased the ecological footprint of countries, even after accounting for average income. Based on an analysis of the ecological footprint per capita and the Gini coefficient for income, Teixidó-Figueras and Duro (2015) conclude that reducing global environmental pollution requires a reduction in the significant inequalities between countries, including redistributive measures that prioritize international justice over the growth of individual economies. Therefore, in addition to domestic policies that place greater emphasis on equality and environmental integrity, structural reforms of international trade and financial networks may also be necessary to mitigate the sustainability trilemma.

Performance of Technocratic Efficiency Strategies

Empirical findings paint an ambivalent picture of technology-based sustainability transformation. A meta-analysis by Islam (2025) shows that economic and social drivers, as well as general technological advances, tend to favor the achievement of the SDGs; at the same time, however, it points to conflicting goals: 'Green' innovations correlate negatively with individual SDG aggregates as soon as they primarily target economic efficiency. Technology thus acts as a double-edged sword: it mitigates environmental impacts only when embedded in socially embedded institutions; otherwise, it shifts problems or creates new ones.

Our global panel analysis confirms the limits of the pure efficiency paradigm. Between 1995 and 2021, energy-related CO_2 intensity fell by 32 percent, while real global GDP grew by 145 percent)—yet absolute emissions rose by almost 60 percent. Only 27 countries)—almost exclusively advanced service economies)—show a temporary absolute decoupling. There are three paths behind this: technical substitution (Sweden), structural change (the UK), and CO_2 pricing (British Columbia). However, all three remain vulnerable to offshoring effects or rebound behavior: in 62 percent of cases, additional consumption offsets at least half of the efficiency gains.

Furthermore, the literature suggests that a purely technological energy transition)—such as replacing fossil fuel power plants with renewable ones)—can exacerbate socio-ecological problems if it overlooks underlying consumption patterns and resource limits (Ruggeri & Garrido, 2021; Biely & Chakori, 2025). Mealy and Teytelboym (2022) demonstrate that technological capabilities can still be a significant lever: countries with strong expertise in exporting complex green products exhibit significantly higher environmental patenting rates, lower CO_2 emissions, and stricter environmental policies—although they are also subject to strong path dependencies. Those who started early to build green capabilities are increasing their lead, while latecomers are struggling to catch up.

Overall, these findings underscore that technocratic strategies alone are insufficient: without binding resource caps, accompanying redistribution instruments, and a political corrective against rebound effects, even 'green' innovation dynamics risk cementing the sustainability trilemma rather than resolving it.

Potential of Ecocentric Models

Ecocentric or post-growth approaches have a noticeable impact as soon as they are institutionalized in policy. Since the 1990s, Costa Rica has linked payments for ecosystem services (Pagos por Servicios Ambientales) to social programs, thereby achieving a 'high' HDI level, while national CO_2 emissions are only around one-third of the OECD average. Slovenia, on the other hand, has reduced municipal waste by 59 percent with its 'Zero Waste Municipalities' program while creating around 0.1 additional jobs per 1,000 inhabitants)—an indication that sufficiency and circular economy strategies can generate local income.

A quantitative analysis of the Cluster C countries reveals three key mechanisms: First, legally enshrined resource ceilings set hard limits, such as Costa Rica's minimum forest area or Slovenia's residual waste cap. Second, progressive environmental taxes)—such as Canada's CO_2 tax with dividend payments)—link tariffs to consumption levels and feed into social funds. Third, alternative prosperity indicators, such as New Zealand's Well-being Budget, align national budget plans with goals beyond GDP, thereby shifting social priorities.

At the same time, the case studies point to the limitations of ecocentric models: their implementation requires high institutional capacities and broad social consensus; in export-oriented economies such as Costa Rica, hidden resource flows (e.g., through tourism) can cloud the balance sheet; external shocks such as the COVID-19 pandemic threaten to dry up funding sources.

The assessment of environmental change and resource management is becoming increasingly important (Troy & Wilson, 2006) despite the concept of natural capital being repeatedly criticized (Maechler & Boisvert, 2024). NC encompasses resources found in nature, such as minerals, oil, fish, and timber stocks in forests, water, and arable land, but also ecosystem services provided by ecosystems that are invisible to most people, such as air and water purification, protection against flooding or erosion, carbon sequestration, and storage, and the provision of habitats for species and others. The first step in accounting for natural assets and ecological services or functions for human activities in environmental-economic accounting is to define their spatial and functional dimensions. Measuring and evaluating these two dimensions and identifying the interrelationship between them and human activities, particularly in the context of the economy, enables the integration of ecosystems into an accounting structure (Obst et al., 2016). It is often difficult to represent both physical and monetary values. To solve these problems, Zhang et al. (2023) have integrated ecological footprint and ecosystem service methods into life cycle assessment. In the next step, a nature-based digital currency (NBDC) can be introduced, backed by fundamental values and serving as the foundation for a genuinely sustainable, nature-based economy. The NBDC and the underlying technologies offer a new perspective on how digital resilience can be built in ecological contexts (Andrae, 2025).

The synthesis shows that technocratic efficiency strategies can reduce environmental pressure, but they neither address structural income polarization nor prevent the externalization of environmental costs. Ecocentric approaches reduce both the ecological footprint and inequality; however, they have been limited so far to a few countries and municipalities. A viable path for the future must, therefore, combine efficiency with sufficiency and equitable redistribution, enshrine resource limits in law, and internalize rebound effects through progressive taxes. Only such an integrated strategy can achieve the dual goal of a sustainable standard of living that neither exceeds planetary boundaries nor deepens social divisions.

COMPARISON OF ECONOMIC DEVELOPMENT MODELS

Growth and Technology-Driven Approaches

The classic, growth-oriented development model equates social prosperity with a combination of GDP expansion, capital accumulation, and technological progress. Politically, this logic is reflected in 'green growth' strategies, which include CO_2

taxes, emissions trading, recycling technologies, and government R&D programs aimed at increasing efficiency gains faster than resource consumption. The success narrative often points to countries such as Sweden and Denmark, which have managed to reduce their territorial emissions despite growing GDP.

However, a closer examination of the panel data reveals that absolute decoupling is rare globally and typically only temporary. Vadén et al. (2020) do find evidence of decoupling for individual pollutants and, in a few countries, for land and water use. However, robust evidence for a comprehensive decoupling of economic growth and resource throughput at the international level remains scarce. Szigeti et al. (2017) do identify a positive trend: the ecological footprint intensity of growth has declined over the past two decades. However, it remains to be seen whether this trend is sufficient to resolve the trade-off between prosperity and the environment altogether.

In addition, national emissions reductions are often accompanied by the offshoring of heavy industry, which merely shifts part of the environmental burden to other countries. Rebound effects)—lower unit costs, higher consumption)—are estimated to neutralize up to one-third of efficiency gains. In terms of social policy, the green growth model addresses inequality only indirectly, for example, through the creation of new jobs in the clean tech sector. In highly polarized economies, such as the US, however, it is mainly high-income groups that benefit, with distributional effects remaining low.

In short, growth and technology paths are politically viable and enable the rapid dissemination of technical solutions. At the same time, they create structural blind spots—both in global material flows and distribution effects—that cannot be overcome without additional measures.

Post-Growth and Degrowth Models

Post-growth approaches radically question the fetishization of GDP. Although gross domestic product measures the monetary value of all goods and services produced, it completely ignores ecological and social costs. Degrowth authors, therefore, argue that in highly industrialized countries, the historically unprecedented throughput of materials and energy is directly responsible for environmental overexploitation and growing inequality. The answer is not efficiency but sufficiency and redistribution: shorter working hours, resource quotas, progressive eco-taxes, and the expansion of care, repair, and public welfare sectors should reduce absolute consumption while distributing income, time, and natural capital more equitably.

The theoretical roots range from Nicholas Georgescu-Roegen's bioeconomic thermodynamics argument (1971) to Serge Latouche's call for a cultural 'farewell to growth' (2010). More recent concepts, such as the post-growth economy (Kallis et al., 2025), build on this idea: growth does not become negative but instead becomes irrelevant. The success of an economy is measured by well-being, resilience, and ecological regeneration. Individual sectors—such as renewable energy or ecosystem restoration—may continue to expand while fossil extraction, planned obsolescence, and luxury emissions decline. Benchmarks are resource budgets and Beyond GDP indicators such as the Genuine Progress Indicator or the Inclusive Wealth Index.

The fact that consumption is the primary driver of the growth paradigm is evident from its side effects: resource and energy consumption are rising, waste streams are increasing, social exclusion is deepening, and rebound effects are eroding efficiency gains. A transformative path, therefore, leads through a genuine circular economy (design for repair, reuse, and recycling), sufficiency-oriented everyday practices)—such as sharing, second-hand, and frugal abundance (Plomteux, 2024)—and the regulation of excessive advertising and consumption incentives.

Conceptually, Corvallis, and Paulsson (2023) expand the discourse with the idea of resource shifting: degrowth is understood as a dynamic interplay between re-specification (turning things into resources) and de-respecification (decoupling resources from the market). This perspective opens up new socio-ecological paths for post-growth societies, where the focus shifts from exploiting resources for economic growth to redefining and repurposing resources for sustainable development.

Costa Rica provides empirical evidence of the potential: an electricity mix consisting of 98% renewable sources, payments for ecosystem services, and free basic services have propelled the country into the highest HDI category, while CO_2 emissions are only three tons per capita. However, the country's heavy reliance on tourism, which is a foreign exchange-dependent service, was exposed during the pandemic when global travel restrictions led to a significant drop in revenue. This highlights the vulnerability of such a model to external shocks.

The biggest hurdles remain political acceptance in affluent societies, fiscal sustainability in the face of shrinking tax revenues and scaling cultures of sufficiency across national borders. Nevertheless, studies show that post-growth combined with targeted redistribution—such as climate dividends—can increase basic security and drastically reduce environmental pressure. In short, degrowth and post-growth do not offer a backward-looking narrative of sacrifice but rather a path to a high quality of life within planetary boundaries.

Integrative Transitions: Socio-Technical and Socio-Ecological

Increasingly, authors are advocating for integrative transition models that combine technological efficiency programs with comprehensive structural reforms. The model is often referred to as the well-being economy, as Scotland and New Zealand are beginning to implement it. Such hybrid strategies work along three clear guidelines:

1. *Hard resource limits*: Absolute caps on CO_2 budgets, land use, or material throughput define the ecological framework within which economic activity must take place.
2. *Progressive redistribution & just transition funds*: Revenue from resource and CO_2 taxes are targeted at disadvantaged groups and regions. This cushions distribution risks and ensures social acceptance.
3. *Technology roadmaps with a sufficiency component:* Efficiency gains are accompanied by product-service systems, sharing models, or repair obligations to minimize rebound effects and reduce absolute consumption volumes.

Sweden provides evidence that such a hybrid model is effective: ambitious CO_2 taxes reduce territorial emissions, while the welfare state mitigates the social distribution effects. However, import leakage indicates that resource caps must be coordinated internationally.

The core idea is to link the performative power of technical progress with the normative agenda of sufficiency. Technical approaches (efficiency, consistency) optimize resource use and focus on renewables; sufficiency, in turn, aims at cultural and organizational innovations that change production and consumption patterns themselves. Value creation is understood as social-ecological co-design, where new technologies emerge along corridors of need that are shaped by planetary boundaries and equitable distribution.

With sufficiency as a strategy, social innovations are focused, and the absolute amount of raw materials and environmental impacts can be limited and reduced, allowing for operation within planetary boundaries. Sufficiency addresses both upper and lower limits. Consumption is graded into overconsumption, which exceeds planetary boundaries, and sustainable consumption, which enables a good life, meets basic needs and avoids scarcity. Sufficiency thus enables the simultaneous tackling of multiple problems, generating synergy effects and mitigating conflicting goals.

This brings us to the following interim conclusion:

- Growth and efficiency paths are politically viable but often fail due to rebound and distribution problems.
- Post-growth models offer significant ecological benefits and greater distributive justice but face limitations in terms of acceptance and scalability.
- Integrative transitions could combine both worlds but require hard resource caps, progressive redistribution, and innovative policies based on real needs rather than GDP.

This typology forms the basis for the following discussion (Chapter 6), in which we examine whether a paradigm shift from pure efficiency logic to sufficiency and distribution perspectives is already emerging.

DISCUSSION

Paradigm Analysis According to Kuhn—Status Quo vs. Paradigm Shift

According to Thomas S. Kuhn, every scientific community goes through cycles: a phase of 'normal science'—in which central concepts, methods, and objectives are rarely questioned—is followed by crises triggered by anomalies that the prevailing frame of reference can no longer explain; only when these contradictions become obvious does a paradigm shift occur. A systematic evaluation of our literature and indicator data suggests that growth- and technology-centered sustainability research still operates predominantly in the mode of normal science. Increasing efficiency is the go-to solution, GDP growth remains the unquestioned measure of prosperity, and technical progress is viewed as the primary means to solve problems. At the same time, questions about distribution and sufficiency are relegated to the sidelines.

However, the anomalies are piling up and shaking this paradigm on several fronts:

- *Persistent emissions*: Global CO_2 emissions continue to rise despite decades of efforts to improve efficiency.
- *Polarizing prosperity*: Material gains are concentrated in the top income and wealth deciles, which also have the most significant ecological footprint.
- *Exceeded planetary boundaries*: Climate, biodiversity, nitrogen, and phosphorus cycles exhibit persistent exceedances, fundamentally questioning the promise of 'green' decoupling.

Our case studies underscore this discrepancy. Costa Rica and Slovenia appear to be harbingers of a revolutionary phase in Kuhn's sense: both countries demonstrate that a high quality of life can be achieved even with stagnating or moderate GDP growth, provided that hard ecological limits)—such as forest protection or caps on residual waste)—are established and redistribution instruments are implemented. Vietnam, on the other hand, demonstrates the persistence of the old paradigm: rhetoric about efficiency masks ever-increasing resource intensities driven by export-oriented growth and a coal-heavy energy infrastructure.

Whether these anomalies already mark a fully-fledged crisis is debatable. However, several signals indicate a looming paradigm shift: the spread of alternative prosperity metrics, such as well-being budgets, the growing popularity of 'doughnut economics' as a policy-oriented framework, the expanding post-growth literature, and initial legislative initiatives on absolute resource caps. It indicates that the discourse is shifting from a pure efficiency logic to a focus on sufficiency and distribution)—even if this shift has not yet taken hold in the dominant policies of many countries. The old paradigm is faltering; the urgency of this shift is underscored by the fact that the longer it takes to institutionalize and translate the new guiding ideas into measurable progress, the more we risk exacerbating the crises we are already facing.

Global South and Climate Justice

The debate on prosperity within planetary boundaries gains analytical clarity when the asymmetrical starting conditions of the Global South are made explicit. Historically lower per capita emissions, combined with higher exposure to climate risks and structurally limited fiscal and financial scope for action, create a pressing need for fairness logics. These logics, which combine responsibility, capacity, and development needs, are crucial for addressing the unequal real burdens that result from identical nominal target values. It is recommended that national resource and emissions budgets be linked to per capita convergence paths that weight historical contributions and capacity indicators. At the same time, minimum social standards ('well-being') should be established as hard lower limits. As a result, measuring prosperity itself becomes an architecture of justice, a necessity in the current climate crisis. This approach is compatible with 'beyond GDP' approaches such as inclusive wealth or GEP, which provide an integrative representation of natural and human capital.

Second, it is important to address the structural mechanisms that shape the ecological paths in the South. Higher capital costs and currency risks make clean infrastructure more expensive. In addition, emissions from consumption-intensive economies are externalized through global value chains. Finally, 'informality', which refers to the lack of formal regulations or structures, weakens the ability to cushion price signals socially. The issue of territorial decoupling in conjunction with simultaneous import leakage discussed in this study illustrates the problem that national efficiency gains without adequate trade and financing regimes merely transfer the burdens along value chains rather than reducing them. This results in a discrepancy between aspiration and reality about the desired climate justice.

Third, a design logic for integrative transitions is crucial and can be derived from this. This logic synchronizes efficiency, sufficiency, and distribution, not as a normative bonus, but as a condition for effectiveness. The provision of hard resource budgets requires the establishment of an ecological framework. The use of progressive steering taxes in combination with climate protection measures shifts purchasing power in favor of the lower deciles and dampens rebounds. The use of just transition funds helps to reduce transition risks in carbon-intensive sectors and regions. For the South, it is also essential to make green financing structurally cheaper (e.g., through concessionality, reallocation of special drawing rights, or debt-linked resilience programs) so that identical technologies are not permanently unequal in price.

Fourth, measurability requires a systematic link between distribution and environmental indicators: panels that track income and wealth inequality together with CO2, material, and land use metrics should explicitly reflect consumption-based emissions and overshoot factors in order to make displacement effects visible. The present analysis shows that climate justice should therefore be viewed not as an appendix but as an evaluable component of the core results. This conclusion is consistent with the methodological design, which includes literature mapping, an indicator panel, and contrasting case studies.

Ultimately, climate justice is a question of the underlying worldview. The anomalies describe—persistent emissions despite efficiency, polarized wealth, exceeded limits—are exacerbated in regions where power and property structures block reforms. From the perspective of developing countries, it is thus clear that the emerging paradigm of sufficiency- and distribution-oriented pathways is not a choice but a necessary function to enable prosperity within ecological budgets in the first place. The policy implications are significant and encompass a broad spectrum of aspects, including the establishment of fair border adjustment systems with redistribution to supplier countries, the creation of technology corridors (local value creation, IP access), and the implementation of welfare-based budget processes. These elements are not just suggestions, but crucial steps toward a coherent governance architecture that is necessary for achieving climate justice.

Synthesis: Why Efficiency Alone Is Not Enough

The panel and case findings confirm that technical efficiency gains are indispensable but structurally insufficient. First, rebound effects partially offset savings, as falling unit costs stimulate additional consumption, and price advantages are passed on. Second, efficiency shifts environmental burdens spatially: territorially decoupled countries in Northern Europe import emissions via raw material- and energy-intensive supply chains. Third, the distribution mechanism remains unchanged: efficiency gains primarily increase the purchasing power of higher-income groups, while low-income households continue to suffer disproportionately from environmental damage.

Against this backdrop, Andreoni (2024) develops a framework for inclusive and sustainable structural change in which countries not only restructure their economies but also increase prosperity and well-being for ever-larger segments of the population—and do so in an increasingly environmentally friendly manner. The key is to link industrial upgrading with strict resource budgets and inclusive social policies.

Whittaker et al. (2020, 141) argue that

'the key challenge in the era of compressed development is not so much technological catch-up and upgrading per se, although this is difficult enough. Rather, it is to establish a virtuous loop between economic or technological development on the one hand, and equitable, inclusive growth and social development on the other.'

Creating positive feedback loops between economic and technological progress, on the one hand, and inclusive, sustainable structural change, on the other, is a complex process that extends far beyond mere efficiency gains. Integrative transitions, therefore, act on three interlinked levers. First, they set hard resource caps—absolute upper limits for CO_2 budgets, material throughput, or land use—thus creating the ecological framework within which all forms of value creation must take place. These hard ecological limits are crucial for ensuring that economic and technological progress does not exceed the planet's capacity to sustain it. Second, they anchor progressive redistribution mechanisms: revenues from resource and emissions taxes flow into just transition funds, climate dividends, or regional structural aid, thereby alleviating social hardship. Third, they pursue needs-based innovation paths in which research, corporate strategies, and public procurement are systematically geared toward sufficiency and circular solutions.

In this triad, efficiency assumes a new role: no longer a free pass for unlimited growth but an instrument for generating as much social welfare as possible within fixed ecological budgets. If all three levers are activated simultaneously, self-reinforcing cycles emerge: resource-efficient technologies reduce ecological burdens, redistribution ensures acceptance and demand, and innovation-friendly framework conditions accelerate the spread of post-growth business models. In this way, technical progress can become a driver of robust, fair, and planet-compatible prosperity.

The Role of Power and Distribution Structures

The analysis clearly shows that environmentally sustainable development is virtually impossible without intervention in power and ownership structures. High wealth concentration translates into significant political lobbying influence, which can water down or delay the implementation of strict climate and resource policies. At the same time, consumption by the highest income groups disproportionately contributes to environmental pollution, as luxury goods and resource-intensive lifestyles result in the most significant ecological footprint.

Breaking this power imbalance requires fiscal and institutional levers that combine ecological price signals with social compensation effects, such as progressive resource and wealth taxes, climate dividends, or just transition funds, which shift purchasing power to lower deciles and make sufficiency more economically attractive. Success stories demonstrate the effectiveness of such instruments: Costa Rica finances its forest protection through payments for ecosystem services, while Sweden combines an ambitious carbon dioxide tax with social redistribution. Both cases demonstrate that ecological and power issues must be considered in tandem.

Without this coupling, a dangerous dynamic threatens to emerge: ecological stress escalates, distributional injustices deepen, and cumulative anomalies (such as increasing CO_2 emissions, wealth concentration, and ecological exceedances) accelerate the collapse of the current paradigm (Winter, 2024).

It leads to the overall conclusion of the discussion: although efficiency and growth logic provide important technical building blocks, they do not permanently mitigate planetary boundaries or social polarization. An emerging paradigm oriented toward sufficiency and distribution closes these gaps but requires a conscious and deliberate restructuring of power, property, and value structures. This is not a task to be taken lightly, but it is a necessary step if we are to ensure that 'prosperity' will be understood in the future as an integrative measure of quality of life, fair participation, and ecological integrity.

Real Barriers to Implementation

The translation of welfare-oriented, ecologically sustainable strategies into effective policy, therefore, often fails less because of a lack of technology than because of political and economic obstacles. It can be observed that concentrated interests hurt measures aimed at devaluing assets (e.g., fossil fuels) or attacking pension systems (e.g., tax privileges, subsidies).

These actors have disproportionate agenda-setting power (lobbying, media access, litigation), while the winners are often diffuse and less well organized. This is a classic collective good problem that slows down ambitious redistribution and decarbonization. Veto players and institutional path dependencies create high barriers:

The implementation of multi-chamber systems, federal divisions of powers, and debt brakes contributes to the consolidation of the status quo. Even if majorities are in favor of redistribution, policy drift and regulatory inertia prevent indicators beyond GDP (e.g., inclusive wealth) from being bindingly incorporated into budget and investment decisions. The limitation of the ability to act is also a result of fiscal and financing conditions. Factors such as high public debt, procyclical ratings, volatile capital inflows, and risk premiums lead to higher costs for future investments that are crucial for a fair sustainability transition. This particularly affects investments in energy, mobility, and building renovation. Tax competition and aggressive profit shifting undermine the financing basis for progressive systems, making it difficult to levy socially equitable CO2 prices, climate dividends, and just transition funds.

The characteristics of the labor market and its structure have a negative impact, as a high level of informality or widespread solo self-employment makes it challenging to implement contribution-based social security systems and targeted transfers. In sectors affected by transitions (such as the coal industry, the basic materials industry, and energy-intensive logistics), there is an accumulation of regional path dependencies (one-sided economic structure, skill profiles) with high political sanction risks. In such cases, governments tend to water down measures or grant compensation without precise transformation requirements.

Another shortcoming stems from administrative and implementation deficits that prevent scaling. Incomplete registers, weak data infrastructures, fragmented IT systems, and limited evaluation capacities lead to leakage (free riding, misallocation) or slow rollouts. Measures aimed at sufficiency require detailed control (e.g., through spatial and temporal differentiation and social criteria). Without sufficient digital and organizational capacity, there is a risk of overburdening.

It should be noted that norms, narratives, and polarization influence political feasibility. For example, perceptions of 'performance' vs. 'need-based justice,' status quo losses, and identity conflicts can block objectively sensible reforms. Sufficiency is at odds with consumption norms and growth narratives unless positive semantics of prosperity (time prosperity, health and safety gains, local quality of life) are credibly communicated and institutionally anchored.

In addition, macroeconomic trade-offs, which are the competing economic choices that arise due to scarcity, are emerging: rigorous inflation control can narrow the scope for investment-driven transformation and social compensation. At the same time, too rapid an increase in spending exacerbates inflationary pressures. The timing of price signals (e.g., the price of CO2), initial investments (networks, public transport), and transfers is therefore crucial, as mistakes in the sequence can have significant political consequences.

Finally, the interlocking nature of international regimes restricts national room for maneuver. Trade and investment protection rules, heterogeneous product and data standards, and intellectual property rights determine whether local value creation in green chains succeeds or import dependencies are entrenched. Without co-design (e.g., border adjustments with redistribution revenues, IP releases/patent pools for key technologies), external economic headwinds will arise against domestic redistribution and transition.

From this perspective, redistribution and sustainability transformation can be seen as two sides of the same political economy. Without the reduction of extreme inequality, there is a lack of legitimacy and fiscal sustainability. Without ecological guard rails, redistribution fizzles out in rebounds and cascades of damage. In this context, a governance architecture that anticipates the aforementioned obstacles, addresses them institutionally, and chooses the sequence of reform steps in such a way that credible, visible gains offset short-term perceptions of loss is of crucial importance. The analysis clearly shows that environmentally sustainable development is virtually impossible without intervention in power and ownership structures. High wealth concentration translates into significant political lobbying influence, which can water down or delay the implementation of strict climate and resource policies. At the same time, consumption by the highest income groups disproportionately contributes to environmental pollution, as luxury goods and resource-intensive lifestyles result in the most significant ecological footprint.

IMPLICATIONS

Policy—Incentive Systems for Post-Growth Models

Governments can only truly open up the paradigm window described in Section 6 if they supplement traditional growth indicators with binding resource and welfare standards and shape the framework conditions in such a way that sufficient action becomes practically possible and socially attractive. These changes hold the potential to usher in a new era of sustainable prosperity. Four reform building blocks are central to this:

Binding resource budgets. CO_2, material, and land use caps should be enshrined in law, monitored transparently, and updated annually in line with fiscal deficit rules. Such budgets establish strict ecological guidelines within which all value creation must occur.

Progressive steering taxes plus climate dividends. An increasing tax on primary raw materials and fossil fuels internalizes environmental costs and creates incentives for sufficiency. The revenue is returned on a one-to-one basis as a per capita dividend. This dual mechanism rewards resource-efficient behavior, cushions social hardship, and reduces rebound effects.

Just transition funds. Industries transforming need reliable access to retraining, diversification, and conversion funds. Funds at the EU, federal, and state levels can help cushion the social impact of fossil fuel plant closures while supporting the development of regional industries of the future, such as repair and recycling clusters.

Well-being-oriented budgetary frameworks. Medium- and long-term government budgets should be reformed to adopt a well-being framework that gives equal weight to indicators such as life expectancy, quality of education, care work, and ecosystem services alongside fiscal indicators. It will systematically shift policy-making away from a focus on GDP to a focus on quality of life within planetary boundaries.

Currently, sufficiency is primarily a private matter: those who fly less, repair instead of buying, or use car sharing do so despite—not because of—the existing political framework. Without structural changes, such behavior will never move from niche to mainstream. Sufficiency policy, therefore, aims to design infrastructure, regulations, and incentive systems in such a way that resource-poor options become the obvious choice. Sufficiency does not arise solely from moral appeals or lifestyle trends but from politically designed opportunity structures that make sustainable behavior the easiest, cheapest, and most socially acceptable option. Only when resource caps, redistribution, and sufficiency-promoting infrastructures work together can the vision of more equitable prosperity within planetary boundaries become a reality.

Science—Alternative Metrics for Prosperity and Success

This results in a four-point priority list for research. First, robust indicators beyond GDP must be further developed and harmonized internationally. Future composite indices must meet three minimum requirements: (1) they explicitly reflect absolute resource limits, such as CO_2 budgets, material or land use limits; (2) they capture intra- and intergenerational distribution so that inequality is no longer a 'blind spot'; and (3) they are methodologically robust enough to be integrated into official statistical programs and national budget processes.

Second, empirical work should focus more on implementation pathways. Mixed-method designs—such as panel analyses combined with comparative case studies—can reveal which combination of resource caps, redistribution instruments, and technological innovations generates specific welfare outcomes. It enables the assessment of the effectiveness of individual policy packages under various contextual conditions.

There is a pressing need for an institutionalized form of 'anomaly tracking' in the sustainability landscape. This early warning system, which integrates scientific, official, and civil society data streams, is crucial for promptly identifying potential crises. For instance, it can signal when the efficiency paradigm is at risk of tipping into a crisis phase, such as in the case of sharp rises in import emissions or new waves of social polarization.

Fourth, research must become more open to practical application. Science, with its rigorous methods and commitment to truth, is a key enabler of a sustainable future. Public open data platforms that make CO_2 footprints, material flows, and distribution figures transparent enable start-ups, local authorities, and investors to align their business and development strategies with sufficiency and distribution goals. In this way, science moves from being a diagnostician to an active enabler of a concept of prosperity that respects ecological limits and ensures social participation.

Practice—Multi-Stakeholder Alliances and Governance

Ecological innovators now have a significantly greater scope for action once financial, structural, and market levers are systematically interlinked. At the financing level, more and more mission-oriented funds, impact loans, and revolving circular economy vehicles are emerging that specifically bridge the expensive proof-of-concept phase and share scaling risks. Public green bonds or sustainability-linked loans link interest rate advantages to measurable resource savings, thereby shortening the payback period for ecological technologies. At the operational level, path dependencies can be broken down through modular factory concepts and open data interfaces. Plug-and-play certifications standardize the verification of ecological benefits, reducing transaction costs. Digital twins and blockchain-based material passports provide real-time transparency into material flows, facilitating second-life or reuse models.

In the markets, scientifically based eco-labels and verifiable product passports ensure that sustainable performance is visible and reflected in premium prices, preferential market access, or lower financing costs. Binding minimum quotas for reuse or recycling creates predictable demand and lays the foundation for scalable take-back and repair networks. When these instruments are linked together, an 'enabling loop' is created: the more pioneering companies come on board, the more attractive the entire eco-ecosystem becomes)—which in turn motivates established players to convert their business models.

Companies, cities, and civil society can further expand this space for opportunity by forging alliances along the entire value chain. Repair clusters connect manufacturers, municipalities, and social enterprises, extend product life cycles, create local jobs, and reduce waste. Citizen energy cooperatives redistribute revenues from renewable sources locally and strengthen regional value chains. Cross-sector 'resource use pacts' commit competitors to ambitious caps on primary raw materials and binding reuse quotas, creating a level playing field and reducing first-mover risks. Such cooperative governance anchors public welfare criteria)—such as CO_2 budgets, material intensity, or distribution effects)—as hard benchmarks in procurement, investment, and innovation processes. Public procurers accelerate the momentum when they consistently align green public procurement with sufficiency, repair, and recycling requirements.

It creates a self-reinforcing cycle: each new cooperation reduces risks, attracts additional capital, and gradually transforms ecological entrepreneurship into a competitive, socially inclusive, and economically attractive path.

CONCLUSION

The analysis in this study confirms three key findings. First, there is a systemic nexus between economic growth, social inequality, and ecological overload: the more income and wealth are concentrated at the top, the higher per capita resource consumption is—and this correlation cannot be neutralized by efficiency gains alone. Second, technocratic green growth strategies are hitting structural limits. While they reduce emissions- and material-intensive domestic production, they often shift environmental burdens to global supply chains and largely ignore distribution issues. Third, the emergence of post-growth-oriented countries and municipalities offers a hopeful alternative, showing that a high standard of living can be achieved even with moderate or stagnating GDP, provided that hard resource limits, progressive redistribution, and needs-based innovation are interlinked. This potential of post-growth models indicates the beginning of a transition to a new paradigm in which sufficiency and distribution perspectives complement the previous focus on efficiency.

Nevertheless, the study has three limitations: First, despite the use of fixed-effects models, the panel analysis only allows for correlative statements and does not provide complete causal evidence. Second, the case studies on Costa Rica, Sweden, and Vietnam provide analytical insights but no statistical generalizations; path-dependent factors remain a possibility. Third, consumption-based emissions data are mainly available for well-documented economies, so leakage effects in emerging and developing countries may be underestimated.

This evidence suggests that the dilemma between economic prosperity and environmental sustainability will remain a key obstacle to global development in the future. Resolving this conflict of objectives is urgent and will require far-reaching changes in governance structures and cultural norms)—in particular, a shift away from the political primacy of economic growth in favor of social and environmental goals, especially in already prosperous countries (Wiedmann et al., 2020).

To break free from the narrow confines of reductionist transformation models, we must politicize sustainability, transition, and, ultimately, science itself—including its methods, institutions, and everyday practices. A genuine restructuring of the economic system can only succeed if worldviews and power dynamics shift in parallel: Who has the power to interpret? Who decides how resources are allocated? Who defines progress? Research should, therefore, increasingly examine the power constellations that act as levers or brakes on social change. Social ecology offers valuable starting points here. Applying ecological principles such as diversity, functional redundancy, and adaptive cycles to human communities provides a theoretical foundation for thinking about profound processes of change: resilient systems need pluralistic institutions; redundancies ensure the ability to act in crises; decentralized, cooperative networks strengthen collective intelligence. Power and social ecology-informed sustainability research could thus reveal the shifts in ownership, decision-making, and knowledge structures that are necessary to open up transformative scope for action—and how these changes can be initiated and supported in practice (Biely, K., & Chakori, 2025).

Against this backdrop, a five-point research agenda emerges: (1) Longitudinal panels should simultaneously record sufficiency, efficiency, and distribution indicators in order to reveal causal relationships over decades. (2) Randomized policy experiments—such as carbon dividends at the municipal level—can clarify which incentive systems are most effective in changing consumption and behavior patterns. (3) Intersectional inequality analyses must take gender, ethnicity, and region into account in order to map access to resources in a differentiated manner. (4) Company-related studies should investigate whether and how large corporations anchor sufficiency-oriented value creation internally or externalize it through the acquisition of circular start-ups. (5) Transdisciplinary modeling that links planetary boundaries, prosperity indicators, and power structures could simulate policy options in a 'digital twin earth' and make their distributional effects transparent.

Overall, the chapter suggests that prosperity should be understood as the result of integrated transformation: technological efficiency, social redistribution, and ecological sufficiency must be synchronized. The pursuit of equitable and ecologically sustainable prosperity is of utmost importance. Political guidelines, scientifically robust metrics, and practical alliances are equally necessary to achieve this goal. A social-ecological perspective reverses the classic logic of efficiency by specifically seeking conscious overlaps and 'productive redundancy' in order to increase the resilience of systems (Spangenberg, 2011). Niches and local experiments serve as strategic buffers against shocks, opening up space for solutions that the growth- and techno-centric paradigm has not yet offered. Only a pluralistic, interdisciplinary approach—one that takes alternative economic systems, diverse transition paths, and the close interconnection between humans and nature seriously—can do justice to the complexity of future sustainability processes and enable us to create a good life for all within planetary boundaries (Wiedmann et al., 2020).

REFERENCES

Andrae, S. (2024). The establishment of a Gross Ecosystem Product: A country comparison. In S. Boubaker & T. H. Le (Eds.), *Green finance and sustainable development goals* (pp. 369–407). World Scientific Publishing., 10.1142/9781800614482_0014

Andrae, S. (2025). The political economy of digital nature-based currencies: Opportunities for the BRICS countries? In A. Gedikli, S. Erdogan, & H. Ç. Terzioğlu (Eds.), *Changing the global political economy: BRICS countries and alternative relations strategies* (pp. 307–344). IGI Global Publishing., 10.4018/979-8-3693-7393-4.ch010

Andreoni, A. (2024). Reframing structural transformation towards sustainable and inclusive prosperity. In E. Scheja & K. B. Kim (Eds.), *Rethinking economic transformation for sustainable and inclusive development* (pp. 25–48). Edward Elgar., 10.4337/9781035348466.00006

Biely, K., & Chakori, S. (2025). Sustainability transition theories: Perpetuating or breaking with the status quo. *Sustainable Development, 33*(1), 52–68. 10.1002/sd.3101

Biely, K. (2024). Paradigm changes in research: Kuhn meets socio-ecological transition theory. *Journal of Cleaner Production, 480*, 144057. 10.1016/j.jclepro.2024.144057

Figge, L., Oebels, K., & Offermans, A. (2017). The effects of globalization on ecological footprints: An empirical analysis. *Environment, Development and Sustainability, 19*, 863–876. 10.1007/s10668-016-9769-8

Georgescu-Roegen, N. (1971). *The entropy law and the economic process*. Harvard University Press.

IPBES. (2019). Global assessment report on biodiversity and ecosystem services. https://www.ipbes.net/global-assessment

Islam, H. (2025). Nexus of economic, social, and environmental factors on sustainable development goals: The moderating role of technological advancement and green innovation. *Innovation and Green Development, 4*(1), 100183. 10.1016/j.igd.2024.100183

Kallis, G., Hickel, J., O'Neill, D. W., Jackson, T., Victor, P. A., Raworth, K., ... (2025). Post-growth: The science of wellbeing within planetary boundaries. *The Lancet. Planetary Health, 9*(1), e62–e78. 10.1016/S2542-5196(24)00310-3

Kuhn, T. (1970). The structure of scientific revolutions. 2nd, enlarged ed. University of Chicago Press.

Kuznets, S. (1965). *Modern economic growth.* Yale University Press.

Latouche, S. (2010). *Farewell to growth.* Polity Press.

Maechler, S., & Boisvert, V. (2024). Valuing nature to save it? The centrality of valuation in the new spirit of conservation. *Global Environmental Politics, 24*(1), 10–30. 10.1162/glep_a_00734

Mealy, P., & Teytelboym, A. (2022). Economic complexity and the green economy. *Research Policy, 51*(8), 103948. 10.1016/j.respol.2020.103948

Obst, C., Hein, L., & Edens, B. (2016). National accounting and the valuation of ecosystem assets and their services. *Environmental and Resource Economics, 64*(1), 1–23. 10.1007/s10640-015-9921-1

Piketty, T. (2014). *Capital in the twenty-first century.* Belknap Press.

Plomteux, A. (2024). Frugal abundance: Conceptualisation for degrowth. *Ecological Economics, 222*, 108223. 10.1016/j.ecolecon.2024.108223

Raworth, K. (2017). *Doughnut economics: seven ways to think like a 21st-century* economist. Chelsea Green Publishing.

Reyers, B., & Selig, E. R. (2020). Global targets that reveal the social–ecological interdependencies of sustainable development. *Nature Ecology & Evolution, 4*(8), 1011–1019. 10.1038/s41559-020-1230-6

Rockström, J., Steffen, W., Noone, K., Persson, A., Chapin, F. S., Lambin, E. F., ... (2009). A safe operating space for humanity. *Nature, 461*(7263), 472–475. 10.1038/461472a

Ruggeri, E., & Garrido, S. (2021). More renewable power, same old problems? Scope and limitations of renewable energy programs in Argentina. *Energy Research & Social Science, 79*, 102161. 10.1016/j.erss.2021.102161

Spangenberg, J. H. (2011). Sustainability science: A review, an analysis and some empirical lessons. *Environmental Conservation, 38*(3), 275–287. 10.1017/S0376892911000270

Stern, D. I. (2004). The rise and fall of the environmental Kuznets curve. *World Development, 32*(8), 1419–1439. 10.1016/j.worlddev.2004.03.004

Stern, D. I. (2017). The environmental Kuznets curve after 25 Years. *Journal of Bioeconomics, 19*, 7–28. http://link.springer.com/10.1007/s10818-017-9243-1

Szigeti, C., Toth, G., & Szabo, D. R. (2017). Decoupling – shifts in ecological footprint intensity of nations in the last decade. *Ecological Indicators, 72*, 111–117. 10.1016/j.ecolind.2016.07.034

Teixidó-Figueras, J., & Duro, J. A. (2015). The building blocks of international ecological footprint inequality: A regression-based decomposition. *Ecological Economics, 118*, 30–39. 10.1016/j.ecolecon.2015.07.014

Troy, A., & Wilson, M. A. (2006). Mapping ecosystem services: Practical challenges and opportunities in linking GIS and value transfer. *Ecological Economics, 60*(2), 435–449. 10.1016/j.ecolecon.2006.04.007

Vadén, T., Lähde, V., Majava, A., Järvensivu, P., Toivanen, T., Hakala, E., ... (2020). Decoupling for ecological sustainability: A categorisation and review of research literature. *Environmental Science & Policy, 112*, 236–244. 10.1016/j.envsci.2020.06.016

Valdecantos, S. (2025). The green transition dilemma: The impossible (?) quest for prosperity of South American economies. *Ecological Economics, 230*, 108508. 10.1016/j.ecolecon.2024.108508

Whittaker, D. H., Sturgeon, T. J., Okita, T., & Zhu, T. (2020). *Compressed development: Time and timing in economic and social development*. Oxford University Press.

Wiedmann, T., Lenzen, M., Keyßer, L. T., & Steinberger, J. K. (2020). Scientists' warning on affluence. *Nature Communications, 11*, 3107. 10.1038/s41467-020-16941-y

Winter, M. A. (2024). Towards a model for sociocultural transformation: Anthropocentric approaches to resilience, collapse, and resistance. *Journal of Ancient Near Eastern History, 11*(1), 83–103. 10.1515/janeh-2022-0012

Wu, T., Rocha, J. C., Berry, K., Chaigneau, T., Hamann, M., Lindkvist, E., ... (2024). Triple Bottom Line or Trilemma? Global Tradeoffs Between Prosperity, Inequality, and the Environment. *World Development, 178*, 106595. 10.1016/j.worlddev.2024.106595

Zhang, R., Li, P., Xu, L., & Zhong, S. (2023). Reconciling ecological footprint and ecosystem services in natural capital accounting: Applying a novel framework to the Silk Road Economic Belt in China. *Journal of Environmental Management, 330*, 117115. 10.1016/j.jenvman.2022.117115

KEY TERMS AND DEFINITIONS

Prosperity: Prosperity refers to a multidimensional concept that goes beyond GDP and integrates distribution, natural capital, and human development (e.g., inclusive wealth, GPI, GEP). Politically and economically, there is a tension between technical measurement robustness and social resonance; 'prosperity' thus becomes a balance between quality of life, participation, and ecological integrity within complex resource budgets.

Sufficiency: The term refers to the principle of reducing the absolute consumption of energy, materials, land, and emission budgets by adjusting demand, intensity of use, and levels of provision—while ensuring that basic needs are met. This study concludes that sufficiency plays a crucial role in complementing efficiency, enlightening the audience about the synergy between these concepts. By limiting the quantity, size, and pace of activities and consumption (e.g., less/smaller, less often, slower instead of just 'the same amount with less input'), we can achieve a more sustainable balance.

Green growth: Green growth stands for strategies for (ideally absolute) decoupling of GDP and environmental impacts through efficiency, circularity, and CO_2 pricing. Empirically, absolute decoupling and the environmental Kuznets curve remain the exception; rebound effects and offshoring shift burdens and limit the viability of purely technocratic approaches.

Inequality: Inequality (e.g., Gini, top 10/top 1 shares) correlates with higher per capita footprints: wealthy deciles cause disproportionate emissions and material consumption. At the same time, the political influence of concentrated wealth raises the barriers to strict environmental regulation—a 'carbon inequality feedback.'

Planetary boundaries: Planetary boundaries describe the limits of the Earth's systems; several (including climate, biodiversity) have already been exceeded. The interface with inequality is causal (overconsumption by the elite) and impact-related (higher exposure and lower resilience of vulnerable groups).

Sustainability transition: The sustainability transition, a socio-technical shift toward prosperity within ecological limits, is not just a theoretical concept. It's a pressing need that combines efficiency with sufficiency and fair distribution. The 'well-being economy' with hard resource limits, progressive steering taxes/dividends, and just transition funds offers a hopeful path for our future.

Paradigm shift: In Kuhn's sense, paradigm shifts are on the horizon when anomalies in the status quo accumulate: persistent emissions despite efficiency, growing inequality, and planetary boundaries being exceeded. Indicators such as 'well-being budgets' and resource ceilings are not just signals, they are crucial tools that signal a shift from pure efficiency logic to sufficiency and distribution logic as the new 'normal science.'

Rebound effect: The term refers to the partial or complete offsetting of expected energy or resource savings following an increase in efficiency. This phenomenon is caused by falling effective usage costs triggering additional demand (for example, a more efficient car leads to higher mileage). A distinction is made between direct rebound (more of the same service), indirect rebound (income effects that lead to the consumption of other resource-intensive goods), and macroeconomic/structural rebound (price, growth, and network effects). The term 'backfire paradox' (Jevons paradox) describes the situation where increased consumption exceeds efficiency gains.

Chapter 2
Pushing Planetary Boundaries:
Socio-Economic Inequality and Climate Degradation

Tanya Kapoor
Amity University, Noida, India

Prisha Agarwal
https://orcid.org/0009-0005-0871-1610
Amity University, Noida, India

Neeru Sidana
https://orcid.org/0000-0003-1087-4214
Amity University, Noida, India

ABSTRACT

Rushing towards unlimited growth has created the most threat to the sustainability of planet earth, resulting in the climate crisis. Key to this is the need to tackle the recurring challenges of climate change and socio-economic development all at once, to guarantee a successful outcome. Climate change poses unprecedented impacts that threaten ecosystem integrity; human health and socio-economic stability, and a great urgency to mitigate it exists. It magnifies these challenges and pushes the planet's limits and compromises the well-being of all the people who will and do live on this planet. While numerous experiments on climate change are available, there are no systematic studies, including socio-economic factors and inequality impacts, on environmental degradation. It impedes the development of a comprehensive set of strategies to stem the tide of the climate crisis.

DOI: 10.4018/979-8-3373-1862-2.ch002

INTRODUCTION

Climate change is widely recognized as one of the most critical problems of the modern day. It affects nearly every single ecosystem, economy, and society across the globe (Zhang et al., 2022). Global temperature rise, intensifying natural disasters, and sudden ecosystem shifts demand immediate and coordinated global action (Swaminathan et al., 2023). Research indicates that anthropogenic activities such as the burning of fossil fuels, logging, and the emission of gasses from factories greatly contribute to climate change (Boyd & Markandya, 2021). Global issues like climate change need to be addressed through a single integrated system that incorporates policies related to the environment, restructuring of the economy, and changes in governing systems (Trisolini, 2014).

Climate change is a cross-industry challenge requiring policy implementation, economic reform, and environmental action (Johnson et al., 2020). This study focuses on three core objectives: (i) Ensuring socio-economic resilience against climate change, (ii) analyzing key dimensions of the crisis, and (iii) proposing comprehensive mitigation policies (Zhang et al., 2022). These objectives aim to build climate legislation, improving climate resilience, and achieving sustainable development (Shaik & Abaranji, 2021).

In order to mitigate socio-economic instability, the socio-economic consequences of stranded assets in fossil fuel industries, climate change displacement, and the costs related to adaptation strategies must be managed (Frumkin et al., 2025). The U.S. withdrawal from the Paris Agreement further deepened the problems associated with international cooperation on climate change, but at the same time, underscored the steps taken by China as a key player driving global climate change action strategy (Ding et al., 2021). These challenges involving socio-economics are resolved by fostering both economic growth and regulations directed to environmental protection (Aldy & Pizer, 2015).

While the responsibility of climate change can be traced back to the factors of industrialization, urbanization, and agricultural development (Mercure et al., 2016). The need to fulfill the world's energy and food requirements has resulted in deforestation, soil degradation and greenhouse gas emissions, thus, worsening climate change (Khan et al., 2023). In addition, climate change has a disproportionately negative effect on the most vulnerable groups, especially in developing countries which depend on climate-sensitive sectors (Singh et al., 2023).

The socio-economic impacts are already under the strain from the exacerbation of extreme weather events like hurricanes, heatwaves, and wildfires (Mwangi et al., 2023). Adopting climate policies that are change responsive is of utmost importance as these natural calamities worsen food scarcity, economic volatility, and involuntary resettlement (Zhang et al., 2023). In high-altitude regions, climate impacts manifest through melting glaciers which threatens freshwater reserves, (Wang et al., 2023). Risks are made worse by anthropogenic factors like the changing of land use and airspace pollution (Li et al., 2023).

Economic, and sociological disadvantages should be dealt with in proportionate measures to fulfil adaptation and resilience in order to best rectify these issues (Chakraborty & Joshi, 2021). Key resilience strategies include early warning systems, climate-adaptive farming, and ecosystem restoration. These support long-term socio-economic stability (Maiti & Agrawal, 2021). Working towards having equal accessibility to resources and alleviating socio-economic discrimination in climate change measures is crucial for enduring these sustainability objectives (Silverstein, 2025).

Swaminathan and colleagues (2023) explain that the construction of comprehensive climate plans involves policy systems, economic activities, and conservation. The One Health strategy that synergistically looks at humans, animals, and the environment is gaining prominence as an effective climate resilience approach (Zhang et al., 2022). Governments can address climate and public health challenges simultaneously (Puig et al., 2025).

Accountability oversight combined with governance is essential to ensure the effective implementation and regular updating of climate policies (Trisolini, 2014). The British Columbia Climate Change Accountability Act is the paradigm of emission reduction through policy, showcasing challenges posed by independent climate advisory bodies to governance (COP29 Report, 2024). Such policies, however, need international collaboration, financial resources, and legal frameworks to be effective (Shaik & Abaranji, 2023).

Economic resilience is vital in managing the climate crisis because economic resources dictate the possibilities of undertaking adaptation measures (Boyd & Markandya, 2021). Transitioning to low-carbon economic development requires investment in sustainable infrastructure, climate finance, and green bonds (Investment Funds Report, 2025). The GCF and carbon pricing are equally important in aiding vulnerable economies to mitigate climate change (COP29 Report, 2024).

Environmental experts accept nature-based solutions (NBS) as a sustainable way to reduce greenhouse gases through their carbon capture function based on forests and protected regions and wetlands (Johnson et al., 2021). The coordination of conservation strategies within climate change policies provides governments potential to achieve biodiversity management and substantial cuts of greenhouse gases (Swaminathan et al., 2023). The Indian National Forest Policy demonstrates national initiative by emphasizing how forest conservation projects can contribute toward sustainable development (Corgo et al., 2024).

The implementation of flood and green infrastructure management systems and nature-based urban planning represents strategies for climate change mitigation at the city level according to How Cities Use NBS (Times,2025). Additionally coastal areas have used mangrove restoration and wetland protection to reduce storm surges and stop erosion (Corgo et al., 2024). These programs achieve high success rates when communities join forces with financial backing as well as well-defined policy measures (Swaminathan et al., 2023).

In summary, climate mitigation requires strategic alignment between governance, finance, and ecosystem preservation. The social and economic fields of climate mitigation require socio-economic susceptibilities together with proper policies and climate mitigation financial allocation (Boyd & Markandya, 2021; Trisolini, 2014). In addition, nature-based solutions Johnson et al., (2021) enables the sustainable reduction of emissions and enhancement of resilience to ecological disturbances. This research will review literature and frameworks to provide strategically appropriate gap analyses aimed at addressing the climate crisis incorporating resilience policies for future climate mitigation (Zhang et al., 2022).

However, this chapter extends current descriptive accounts by providing a synthesis of analytical perspectives, one that links socio-economic inequality, environmental degradation, and policies in response to it. The framework also helps in providing a more pragmatic comprehension of notions of climate governance, especially the ones based on high inequality and institutional weakness.

LITERATURE REVIEW

Addressing the climate crisis requires a multi-faceted approach incorporating political frameworks, economic motives, and environmental practices – everything from governance to the economy and conservation. Zhang et al. (2022) posited the framework's interdependence, arguing that sustainability guarantees a cohesive system to tackle climate change challenges. It includes conflict mitigation processes to streamline climate governance, which is critical in defining sectoral policies, setting limits, appointing accountability agents, and establishing baseline measurement frameworks to track mitigation undertakings (Trisolini, 2014). Climate policies anchored within strong governance structures have a greater chance of resilient long-term progression and enduring stress. Economically, climate risk poses exposure that escalates with inaction, necessitating addressed climate finance, investment, and economic policy (Boyd & Markandya, 2021). Climate mitigation policies complement the movement towards low-carbon economies through climate finance mechanisms, carbon pricing, and sustainable investment. Conservational efforts to augment carbon capture and biodiversity protective measures include the promotion of afforestation and ecosystem restores nature-based solutions (Johnson et al., 2020). These policies advocate for a comprehensive approach to tackling climate change via coherent policies, financing, and pro-active environmental preservation while reinforcing the need for integrated intervention. The focus of this literature review revolves around research gaps and underexamined areas amid the broader context of climate mitigation policies and provide a constructive delineation on where policies need financial commitment to establish enduring sustainability.

Socio-Economic Stability in Climate Mitigation Efforts

It is important to consider the intersection of climate change actions and sustainable development when developing policies. Integrated Assessment Models (IAMs) capture the dynamics of socio-economic systems and earth systems, but are often lacking in capturing the full scope of the development dimension (Pathak, Patel, & Some, 2024). Climate change policies from an economy-wide view have significant fiscal reverberations, contrainting government and market roles in mitigation processes (Bowen, Campiglio, &Tavoni, 2014). Adaptation to these changes requires decision-making by the insurance industry on climate risk management (Bowen et al., 2014).

Climate change impacts are multi-faceted including stranded costs in fossil fuel industries and social disruption due to climate events (Ausserladscheider, Ergen & Golka, 2025). Adaptive and mitigative financial plans are necessary to cope with these issues (Ausserladscheider et al., 2025). In the forestry industry, sustainable forest management in British Columbia, Canada, brings important socio-economic value as well as emission reduction (Xu, Smyth, Lemprière, Rampley & Kurz, 2018). These models demonstrate the advanced role of forestry in climate change counteractive actions beyond mere emission reduction (Xu et al. 2018).

Objectives;

- Ensuring socio-economic stability in climate mitigation efforts
- Identifying key drivers and consequences of the climate crisis
- Developing holistic strategies for climate crisis mitigation

Economic Implications of Climate Change Mitigation Policies

Fighting climate change comes at an expense which cuts across numerous industries, both directly and indirectly. The Organisation for Economic Co-operation and Development (OECD) examines the cost-effectiveness of aggressive policies aimed at emission reductions and highlights the need for comprehensive use of the least cost set of policy instruments on all emission sources (Burniaux et al., 2008). This strategy is essential in reducing the economic costs associated with mitigation activities while maximizing emission reductions. The OECD analysis makes it clear that additional technology-support policies are essential to lower emissions as well as future policy costs thus increasing the willingness of many countries to act against climate change (Burniaux et al., 2008).

Changes in technology and improvements in energy efficiency are essential in mitigating climate change. Gregory Casey's study focuses on the impacts of energy efficiency changes and technological innovations on energy use and climate policy (Casey, 2024). The implementation of suitable energy taxes would lead to the best possible welfare outcome and environmental success based on research and development and output advancement approaches (Casey, 2024). The government needs to create policies to foster private sector green technology and infrastructure investments because this drives a low-carbon transition of the economy (Casey, 2024).

Financial and macroeconomic policies play a critical role in enabling the transformation to a low-carbon economy according to the newly implemented policies. The review by Signe Krogstrup and William Oman examines different research to demonstrate how fiscal and monetary policies unite as a unified approach for preventing climate change (Krogstrup & Oman, 2019). The authors argue that climate change mitigation depends on carbon tax implementation and green infrastructure funding through public expenditures however requires complementary financial and monetary frameworks to handle economic impacts (Krogstrup & Oman, 2019). The economic transition demands a complete system of management to maintain social and economic stability.

The effects of mitigation policies on socio-economic stability require consideration when assessing climate change policy impacts on economic competitiveness together with emissions leakage. Joseph E. Aldy along with William A. Pizer evaluated these factors (Aldy & Pizer, 2011). The investigated data suggests the more energy-intensive industries will experience reduced productivity and increased imports because high energy prices from carbon pricing policies (Aldy & Pizer, 2011). Border carbon adjustments and subsidies for these industries require policies to minimize their detrimental effects on competitiveness (Aldy & Pizer, 2011).

Through a policy-oriented framework Alex Bowen and James Rydge conduct economic assessments of various mitigation options together with policy actions for environmental and economic goal evaluation (Bowen & Rydge, 2020). The authors support the notion that short-term economic costs from mitigation policies produce long-lasting advantages through climate impact reduction and cost-effective development of environmentally friendly technologies (Bowen & Rydge, 2020). Engineering solutions for lowering carbon emission levels prove crucial through strategic business analysis according to climate change mitigation economics principles (Bowen & Rydge, 2020).

The European Bank for Reconstruction and Development (EBRD) enhances their research on climate change mitigation economic effects through analysis of macroeconomic outcomes and essential systems reforms to establish decarbonization growth models (EBRD, 2020). The analysis executes a wide policy spectrum to reduce economic disruptions and social upheaval accompanying transformations (EBRD, 2020). According to the EBRD report meaningful climate change mitigation requires countries and world sectors to work actively together on all levels.

The study by J. F. Mercure together with other authors evaluates public perception and evaluates the positive potential of these climate change strategies by showing that traditional economic assessments fail to accurately represent policy benefits (Mercure et al., 2015). The authors emphasize the need for studying how the financial sector with its innovation and technology relates to non-equilibrium dynamics (Mercure et al., 2015). According to this view implementing effective mitigation policies increases economic development while driving technological progress and enhancing sustainability and resilience across the nation (Mercure et al., 2015).

Using appropriate tax and spending along with technological policies governments can support low-carbon economy transitions after optimizing socio-economic stability. Green technology and infrastructure development become key priorities among policy actions which aim to successfully reduce climate change effects according to the findings in these documents.

Socio-Economic Vulnerabilities and Adaptation Strategies

Examining the economic and social weaknesses along with climate adaptation methods has become essential research for areas setback by climate change such as seaside communities. Dolan & Walker (2006) examine socioeconomic vulnerability aspects affecting small rural coastal populations that result from their economic zone and location factors. Local communities need to implement self-generated responses based on their wisdom and resources for better adaptation according to Dolan and Walker (2006). The implementation of this approach provides essential means to create adaptation and resilience measures focused on vulnerable people which helps ensure successful implementation of these measures.

Social Vulnerability Indices (SVI) enable the identification of socio-economic factors increasing environmental risks therefore making these indices suitable for use in adapting strategy development. Damestoitet al.(2024) provide extensive research on how SVIs contribute to disaster and hazard analysis for vulnerability assessment and policy development. The analysis of Damestoitet al. (2024) indicates SVIs gain widespread use but more research is needed to create specific indicators that can derive from local understanding structures to explore vulnerability at its core (Damestoitet al., 2024). Both qualitative and ethnographic approaches should integrate into vulnerability assessments because they deliver better results when adapting to community-specific needs.

Hard engineering methods provide less socio-economic value than the adaptation solution known as Ecosystem-based Adaptation (EbA). Mar Riera-Spiegelhalder et al. (2023) study coastal zone adaptation planning by examining different strategies for their socio-economic outcomes. Severe challenges are addressed by EbA strategies according to Riera-Spiegelhalder et al. (2023) whose systematic review proves that wetland and mangrove restoration strategies benefit both nature and human communities and strengthen local populations' capacity to adapt to shifting conditions (Riera-Spiegelhalder et al., 2023). The study demonstrates the necessity of creating ecological integration between adaptation and socio-economic aspects of multidimensional sustainable development in vulnerable coastal zones.

Socio-economic factors require incorporation into climate change effect assessment for developing effective risk management strategies. The study from Dessai et al. (2014) demonstrates that socio-economic factors together with climatic projections both specify uncertain environmental conditions and determine necessary adjustments. The authors emphasized how integrating top-down and bottom-up scenario methods helps create a complete understanding about stakeholder involvement in the assessment process (Dessai et al., 2014). This well-rounded method provides better control of complex climate change uncertainties thus creating stronger bases for effective choices.

Sustainable adaptation strategies require socio-economic factors to become integral parts of vulnerability and impact assessments and climate change evaluations as per the United Nations Framework Convention on Climate Change (UNFCCC 2008). The background paper presents information about uniting socio-economic variables with analytical tools that incorporate socio-economic frameworks and conduct cost-benefit analyses. The integration provides necessary understanding about climate change effects while establishing fair and inclusive change measures.

The research paper by Raihan et al. (2023) delivers an extensive climate change overview that examines worldwide sector adjustments and mitigation methods. The authors demonstrate how biodiversity connects with environmental factors as well as the socio-political landscape and the socio-economic realm. The study stands for policy formulations which organize both adaptation and mitigation efforts to address climate change issues (Raihan et al., 2023). The stated method proves essential to preserve sustainable development with socio-economic equilibrium through regions vulnerable to climate change.

The adaptation strategies developed from the socio-economic approach need both local qualitative information and broad socio-economic systems. Policy developers obtain accurate engineered policies from the implementation of Social Vulnerability Indices along with ecosystem-based adaptation strategies. Incorporating socio-economic scenarios and pre-existing data into climate impact assessments fosters systematic informed decision-making while aiding in the crafting of fair sustainable adaptation strategies. These papers highlight the need for thorough and all-encompassing climate change adaptation approaches focusing on adequately equipping vulnerable groups to enable them build resilient and adaptive sustainable systems.

The Role of Governments and Markets in Climate Mitigation

With regard to driving the shift to a low-carbon economy, government and market roles are intertwined. In Newell's analysis, Richard G. Newell speaks on the necessity of bringing market forces into the picture as well as formulating policies that enable technological innovations to within the economy in order to reduce greenhouse gas emissions (Newell, 2010). Newell argues that effective policies for climate change should incorporate market solution elements such as carbon pricing which is fundamental in offering financial rewards to innovations aimed at development of cleaner technologies (Newell, 2010). Such an approach promotes investment by the private sector in green technologies while also making it possible for emission costs to be included, thereby fostering sustainable economic development (Newell, 2010).

Middle-income countries have particular problems of implementing market-based climate mitigation policies. The Center for Climate and Energy Solutions (C2ES) looks at how countries such as Brazil, China, India, South Africa and South Korea have been using market-based instruments like carbon pricing to tackle their climate challenges (C2ES, 2013). This report discusses the use of these policies for development together with their effectiveness in changing technological and behavioral practices and promotion of economic development (C2ES, 2013). The policies change the relative prices of goods and create new markets which motivate emission reduction to both producers and consumers(C2ES, 2013).

An alternative approach to the climate problem, Robert Mendelsohn integrates economic reasoning for consideration of the market and government roles in assisting the society adapt to a changing climatic condition. His paper focuses on efficient adaptation, arguing that it requires both a market system and government, where each has a different yet complementary role. Markets are capable of distributing private goods, while governments are needed to supply public goods and create frameworks that foster efficient adaptation within society's structural order (Mendelsohn, 2006). Such methods guarantee exceeding the costs of adaptation, which means maximizing societal welfare (Mendelsohn, 2006).

Adaptation to climate change and its mitigation has to be organized in a framework of governance. The World Bank (2010) analyzes the governance gaps and prospects for addressing climate change. The document comments on the need for well developed and well functioning institutions and governance bodies as a necessary condition for addressing the climate problem. It notes that all constituents have critical responsibilities that include, for the public sector, the provision of regulations and facilitative policies while for the private sector innovation and investment (World Bank, 2010). Good governance ensures that the implementation of climate policies is carried out in an efficient and just manner and captures the concerns of the most vulnerable populations while ensuring sustainable development (World Bank, 2010).

A low carbon economy transition relies on proper implementation of macro-economic and financial policy framework. The International Monetary Fund (IMF, 2019) surveys current studies linking fiscal and monetary and financial approaches to climate change and shows which structural policies can reduce climate changes effects. Fiscal policy focuses on implementing major carbon taxes alongside green infrastructure funding through public resources and developing additional financial and monetary elements to control broader economic impacts and structural policies (Krogstrup & Oman, 2019). A combination of policies must be applied because they help address economic transitions while raising socio economic outcomes (Krogstrup & Oman, 2019).

Extracting value from financial instruments together with market-based systems serves as a necessary method to fulfill the goals of Nationally Determined Contributions (NDCs) established through the Paris Agreement. The World Bank (2016) has studied how climate finance elements together with carbon market systems and blended financial solutions affect the improvement of climate mitigation efforts. The integration of blended climate finance and carbon markets produces opportunities to improve efficient project financing for low income countries and ultimately promotes better mitigation (Strand, 2016). This approach enhances mitigation outcomes by integrating diverse country situations while it builds international agreement and consortium (Strand, 2016).

Climate change mitigation through governments working with markets demands collective effort to achieve effectiveness because market incentives integrated with proper governance and macroeconomic policies help policymakers transition toward low-carbon economies. These research papers demonstrate through their findings that new technologies and international investments with global partnerships create climate mitigation success cases. Sustainable climate management requires public-private sector collaboration to properly handle the variety of challenges related to climate change and sustainable development goals.

The Drivers and Consequences of the Climate Crisis

Climate change has emerged as a critical global issue of the twenty-first century because natural phenomena unite with human activities to escalate dangers that damage entire ecological systems worldwide. Agricultural rural areas endure the worst effects of climate change due to their environmental-sensitive agriculture which must face poverty alongside inadequate infrastructure and lack of resources (Khan et al., 2023). Climate changes pose substantial threats to rural populations because their food supply security encounters risk alongside decreased income stability and weakened community power base from weather patterns alterations and strong environmental events and limited resource resources (Khan et al., 2023). Adaptation strategies implemented by these areas succeed differently in agricultural developments and resource diversification as well as improved climate data access while their outcomes depend on local funding allocation and environmental circumstances (Khan et al., 2023). The three main factors determine adaptation project success or failure through their influence on project alignment with local priorities and operational financial support and technical achievements combined with traditional knowledge integration and contemporary innovation delivery (Singh et al., 2023).

Project feasibility and sustainability potential are determined by climate patterns according to Singh et al. (2023). The effectiveness of adaptation efforts depends on resilience elements such as community solidarity combined with institutional backing and technology availability because it requires complete climate resilience strategies (Singh et al., 2023). The major group of farmers at risk from climate change adapts through crop spread methods combined with water saving methods and forest farming systems to defend their crops and keep productivity strong (Mwangi et al., 2023). The effectiveness of these adaptation methods faces barriers from inadequate policy support and insufficient resources along with deficient infrastructure so targeted resilience interventions are needed for smallholders (Mwangi et al., 2023). Bibliometric research into climate adaptation studies shows rising attention toward risk governance and policy reaction and worldwide extreme weather effects (Zhang et al., 2023) because these urgent matters need coordinated resolution. Climate extremes that grow more frequent and intensive throughout the Tianshan Mountains affect regional water supplies and natural systems and local human populations because of their vulnerable condition (Wang et al., 2023). Glacier mass balance shows direct links with climate extremes which produce a damaging chain reaction on water supplies along with farming output and emergency risks thus demanding adjustment measures to prevent these effects (Wang et al., 2023). The record-breaking wet bulb heat event in 2021 Southern China stood as a direct result of ongoing greenhouse gas emissions coupled with changing land use patterns and it created major negative effects for the public healthcare system and infrastructure systems and economic systems (Li et al., 2023). Addressing these challenges requires a multifaceted approach that integrates scientific research, policy innovation, and community engagement to build resilience and promote sustainable development in the face of an evolving climate crisis (Li et al., 2023).

Socio-Economic Vulnerabilities and Climate Change

The impact of climate change affects socio-economic disadvantaged communities the most since it intensifies financial inequalities while jeopardizing their job security (Zakharova et al., 2025). The dependency of developing rural communities on agriculture and fisheries creates extra challenges because these sectors react sensitively to climate changes (Chakraborty & Joshi, 2021). Socially marginalized people who lack both monetary resources and adaptive infrastructure find themselves at constant risk of climate-based extreme events (Maiti & Agrawal, 2021).

Poverty further continues to increase due to the variability of the climate that thwarts major livelihood systems, particularly food and water security. With warmer temperatures and changing rainfall come repeated crop losses among smallholder farmers, who are the main source of food in most countries. Not only does this deprive them of income, but it also leads to food insecurity on a large scale (Chakraborty & Joshi, 2021). That repeated stress results in the displacement of livelihoods coupled with forced migration and further economic vulnerability (Maiti & Agrawal, 2021).

The structural barriers are enhanced through the difference in adaptation between the industrialized countries and low-income countries. The rich nations have the means to fund mitigation measures whereas the poor countries may not even come up with the basic developmental demands (Silverstein, 2025). Most vulnerable states are not able to carry out sustainable systems of adaption they face bureaucratic obstacles, poor access to global funds, and lack of appropriate government at the local level (Associated Press, 2024).

There is also increased urban vulnerability. The impact of climate change on low-income urban neighborhoods is unbalanced since they are faced with more episodes of extreme heat, floods, and decreasing air quality (Silverstein, 2025). Environmental health crises are being fueled in the metropolitan Indian cities such as Delhi, where the shocking landscapes of sprawling informal settlements have no access to clean water and healthcare (Associated Press, 2024). An example that has succeeded is the Delhi Ridge Green Belt that can be used as the climate buffer and minimize urban heat islands and enhance the air quality in some of the densely inhabited districts. It shows that it is possible to utilize the ecosystem-based urban planning to achieve both environmental and social protection purposes (Kaur & Sharma, 2023).

Such multi-dimensional vulnerabilities need to be addressed through integrated programs such as the integration of climate adaptation and poverty alleviation and environmental justice. The concept of social resilience should be incorporated in rural development and urban governance process to guarantee that there are no backward groups in the unfolding climate emergency (Maiti & Agrawal, 2021).

Effectiveness of Climate Change Adaptation Strategies

Climate change adaptation strategies need to be judged effectively because they help reduce global warming impacts yet scientists and governments and regional communities must unite their understanding to succeed (Puig et al., 2025). Different adaptation measures such as early warning systems and resilient infrastructure and climate-smart agriculture are deployed worldwide because their results depend on local governance structures and economic capabilities and public approvals (Financial Times, 2025). The adaptation gap poses a severe challenge mainly in developing nations because scarce funding and technical limitations stop the establishment of strong climate resilience plans (The Guardian, 2025).

The implementation of urban greening and wetland restoration and reforestation has become prominent as effective climate risk mitigation strategies which simultaneously expand biodiversity along with their respective ecosystem services (Time, 2025). A growing number of cities across the world use green infrastructure methods to combat urban flooding and high temperatures which proves that environment-focused sustainability belongs in adaptation planning approaches (SpringerLink, 2025). These initiatives require continuous funding together with policy backing for success since political timeframes usually focus first on present financial requirements rather than climate resilience needs (Puig et al., 2025).

Business entities now include climate adaptation into their investment plans because they understand the financial risks linked to climate change (Financial Times, 2025). Businesses conducted by institutional investors and multinational corporations now focus on sustainable models that assess climate risks through their decision-making procedures to reduce economic disturbances (The Guardian, 2025). Clusters of industries work in a reactive manner over proactive measures because of insufficient climate adaptation preparations leading to insufficient disaster readiness (Time, 2025).

Adaptive governance stands essential for achieving enduring success of climate adaptation strategies because it demands joint involvement between governments businesses and local communities (SpringerLink, 2025). Adaptive policy frameworks with flexible regulations alongside stakeholder involvement combined with adaptive management principles produce better results when dealing with climate vulnerabilities (Puig et al., 2025). Institutional capacity strengthening and adaptive technology innovation will become mandatory to create resilience and address upcoming climate-related risks during the next decades (Financial Times, 2025).

Climate-Induced Disasters and Financial Consequences

Rising global temperatures convert extreme weather events into more severe occurrences that bring major risks for populations worldwide (The Guardian, 2025). Increased global warming generates stronger natural disasters such as hurricanes and wildfires as well as droughts and floods which create massive population displacement and destroy infrastructure and eliminate economic sources (Time, 2025). The effects of climate-related disasters intensify in urban areas because rapid urbanization together with poor infrastructure design makes them more susceptible (Associated Press, 2025).

To those ends, most cities are implementing nature-based methods of flood prevention, including restoring wetlands and controlling floodplains, which offer a form of ecological and economic fortitude (Time, 2025). The excessive intensity of storms and the increasing depth of the waters have resulted in the inability of the traditional means of flood control, which leads to changing it to transitional and ecosystem-based flood control (Financial Times, 2025). There are also droughts and heatwaves that are becoming more common, particularly in the Global South, which is endangering food and water security (Associated Press, 2024).

The snowball effect of such disasters goes deep down economically. Agriculture, energy, insurance and trade all enter into the realm of mounting risks due to clogged supply chains, crop failure, and devastating impacts to infrastructure (Financial Times, 2025). The losses due to climate are already in billions of dollars and this makes policy-makers develop climate risk assessment as part of financial and infrastructure planning (Time, 2025).

Existing models of financing in the real world provide an insight into the world of adaptations. The Green Climate Fund (GCF) has already financed early warning systems and better and more cyclone-resistant shelters in Bangladesh, with a direct impact in disaster-related deaths and destruction of infrastructure. The projects illustrate the role that international climate finance can play in enhancing resilience in most vulnerable areas. The same can be said about the Africa Adaptation Acceleration Program that helps to ensure climate-smart agriculture and resilient infrastructure in 30+ African countries, closing the gap between financing and actual production.

Nonetheless, the human impact is highest in places where there is low adaptive power, regardless of increased awareness. Increasing internal migration and displacement have been caused by drought, floods, and food shortage (The Guardian, 2025). This has increased pressures urgently calling upon the world to implement stronger policies, financial transfers, and use of technology to curtail vulnerability and sufferings of the human being (Financial Times, 2025).

The resilience begins with adaptive infrastructure and early warning systems, as well as, coordinated response to emergencies. However, this is only effective in case of proactive governance, ongoing financing, and cross-border collaboration (Associated Press, 2025). Unless major and sustained efforts in adaptation are made it will remain common that the financial and human costs of extreme weather events will increase.

Holistic Strategies for Climate Crisis Mitigation

For both the disaster faced by climate change globally, and its continued intensification, it is crucial to develop holistic strategies to mitigate climate crisis (Zhang et al., 2022). To enhance resilience and sustainability, a comprehensive approach that combines environmental, economic and policy perspectives is needed (Swaminathan et al., 2023). Including the climate conditional risks of the One Health approach, horizontal integration between health is underscored (Zhang et al. 2022). The increasing prevalence of the health crises caused in part by climate change justifies international cooperation for implementation of effective mitigation strategies such as that proposed by Zhang et al. (2022).

Climate accountability promotes that policies are in line with global sustainability aims and adaptive governance (Trisolini, 2014). Enforcement of mitigation policies tends to chalk stronger through the integration of legal frameworks and structures of climate governance (Trisolini, 2014).

Given the economic implications of climate change and rising costs of climate related disasters highly pressurize national economics (Boyd & Markandya, 2021). Stakeholders in Canada's economic sectors face significant financial costs if climate change remains unaddressed, and these costs can be minimized through policy-based interventions to manage long-term risk (Boyd & Markandya, 2021). Investing in the climate resilience and adaptation not only helps in reducing economic vulnerability but also promotes sustainable development (Boyd & Markandya, 2021). As a result of pricing carbon and green investment policy, market-based mechanisms have proved to be effective means of mitigating economic shocks caused by climate (Boyd & Markandya, 2021).

Nature based solutions including park and protected area carbon reduction programmes promise climate mitigation (Johnson et al., 2020). The integration of conservation and emission reduction strategies results in biodiversity preservation exercises and reducing climate change related phenomina (Johnson et al., 2020). However, these challenges remain for scaling up these programs because of funding issues and policy fragmentation (Johnson et al., 2020). Strengthening implementation of such initiatives must be coordinated efforts between governments, private sectors and environmental organizations (Johnson et al., 2020).

Holistic approach is taken by a climate governance approach that bring together mitigation and adaptation synergies and ensures that policy frameworks remain inclusive and dynamic (Trisolini, 2014). Nations can boost climate resilience leading simultaneously to the effective emission reduction of greenhouse gasses (Trisolini, 2014).

Integrated Climate Governance and Institutional Gaps

The complexity of the climate crisis requires the coherent architecture of governance which incorporates the financial, institutional, urban, and global policy devices. What Shaik and Abaranji (2021) emphasis is the fact that a robust policy-financial-communal aspect framework is requisite in order to achieve meaningful climate mitigation. Nonetheless, the present strategies tend to be disorganized in terms of sector integration due to which the implementation is disjointed. Governance of climate must be flexible and inclusive in its approach, focus moreover on ecosystem health, general health and social-economic resiliency.

This integration is essential in the urban governance. According to Gupta and Shukla (2024), cities are both the sources that cause high amounts of emissions and the source of innovation in solving problems related to fighting climate change. The well-known case is Heat Action Plan of Ahmedabad in the city of India, which is the first comprehensive approach to dealing with the urban heatwaves. It is a joint local government-non government organization development in collaboration with the public health organizations and consists of early warrants, education, and emergency responses. The plan presents the way in which localized governance in unison with the public health and climate science can diminish climate vulnerability in fast-growing cities (Kaur & Sharma, 2023).The incorporation of climate-sensitive measures into urban planning, including green infrastructure, low-emission transportation, and sustainable zoning processes allows to alleviate the vulnerability of cities, and also boost the economy (Kaur & Sharma, 2023). These area-specific plans should be subordinated within a larger national/international context to guarantee consistency and up-scaling.

One Health approach is a desirable governance model institutionally since it connects the health of the environment, animals, and humans when planning climate (Zhang et al., 2022). One of the provisions that reflect the urge to break down sectoral silos in mitigation is the inclusion of zoonotic risk, biodiversity conservation, and food security (Puig, 2025). Government should acknowledge such overlaps in order to develop holistic and fair climate responses.

At the financial level, green bonds, carbon pricing, and international funding modes require innovative instruments that should be underpinning the integrated governance (Kaur & Sharma, 2023). Such a shift towards renewable energy and sustainable methods needs financial reforms, particularly in underdeveloped economies, that will allow the necessary shift (Shaik & Abaranji, 2021). Clean energy infrastructure, including wind, solar, and hydrogen, is an essential factor determining the compatibility of a financial incentive to sustainability aimed at the long-term (Gupta & Shukla, 2024).

The international coordination system must be established under agreements such as the Paris agreement to promote the alignment of policies and to share knowledge and climate finances to achieve system responses (Zhang et al., 2022). International coordination not being there, national initiatives will pale in comparison with international goals.

Although institutional coherence can form the basis of effective governance, it needs to be aligned to specific financial instruments so that it has a real world application. The latter is discussed in terms of how these integrated strategies can be operationalized with green finance and macroeconomic instruments in the next section.

Green Finance and Economic Instruments

The funding of climate measures is the core of climate action implementation including resurrection of mitigation and adaption strategies in the Global South. Green finance has the financial backbone to support the transition to low-carbon and climate resilient economies systemically. According to Shaik and Abaranji (2021), financial flows should be aligned to address climate intentions using regulatory, institutional and investment-based mechanisms. It has had the current international financial system, however, in poor financial countries, fairness regarding sustainable development has been highly mismatched especially in regards to nations with feeble public monies, and high exposures.

An important climate finance leverage point is carbon pricing, a level playing field which establishes financial incentives to cut greenhouse gas (GHG) emissions. Market-based strategies like the use of trading emissions schemes etc. and carbon taxes can make the industries internalize the environmental costs and invest on cleaner technologies (Kaur & Sharma, 2023). But to work, carbon pricing needs to be either something that has substantial pricing incentives, that can be forecast, and that has regulation to avert manipulation and unfair dealing of the market.

Issuing green bonds is another important tool because this will leverage the participation of private capital in the governmental green projects, such as clean transportation, environmentally-friendly infrastructure, and restoration of the ecosystem. The article by Gupta and Shukla (2024) is based on the argument that the opening of green bond markets to the emerging economies can allow climate action without challenging their fiscal stability. But on the other hand transparency and accountability in such markets is a struggle committed they need to compile stronger frameworks of monitoring procedures and third party verification.

Other than the market tools, climate-related financial disclosures and tax incentives and sovereign green funds can be employed to redirect institutional investment portfolios. Institutions of public finance such as the central banks and development banks will be fundamental in de-risking climate investments and sparking off participation by the private sector (Shaik & Abaranji, 2021).

The investment on those renewable sources of energy, such as solar, wind, and hydrogen, can also not be overlooked. Such investments can not only result in the decrease of the reliance on fossil fuels but also bring green jobs and avenues of technological innovation. The lack of infrastructure and high initial investment, uncertainty of the policies, etc., are peculiar challenges of the developing countries that have to be overcome with the help of the concessional finance and the transfer of the technology (Kaur & Sharma, 2023).

A climate finance system also needs to take into consideration climate justice-how resources are used fairly and vulnerable groups are directly benefited. Unified macroeconomic and financial management is consequently not only a question of capital, but of transformation.

Even though finance is the muscle of climate action, nature-based strategies make its soul. Section 2 is an examination of the way ecosystem-based strategies can be used alongside financial and institutional reform.

Nature-Based Solutions and Social Resilience

Nature-Based Solutions (NBS) are an underutilized, but nonetheless important, avenue to tackle environmental and socio-economic frailty. Such methods exploit the functions of the ecosystem to help reduce climate effects as well as to successfully foster biodiversity, citizen health, and cultural robustness. Zhang et al. (2022) have stressed that NBS of improving wetlands, afforestation, and constructing green urban infrastructure offer cost-efficient mitigation instruments with co-benefits to disadvantaged groups in our societies.

NBS integration in the policy frameworks is poor because of institutional inertia, conflicts in land use and cross-sector coordination. Nonetheless, the fact is that, according to evidence provided by Riera-Spiegelhalder (2023) NBS can have a serious mitigation effect on floods and heatwaves in cities when integrated into the governance systems of local areas. Likewise, Damestoi (2024) illustrates how forest regeneration involving communities in South Asia has improved carbon sequestration and also improved the livelihoods of people living in the rural areas. These cases show how approaches nature-based have both ecological as well as social values.Scalability Urban NBS is shown in real-life scenarios. Mangrove restoration in Singapore has not only moderated the effects of storm surges due to its environment lowering effects on storm surges but intensify biodiversity and carbon sequestration. Such projects are a combination of ecological engineering and urban planning into projects that will serve more than one risk. Likewise, the Netherlands has adopted the concept of a sponge city through the permeable pavements, green roofs, urban wetlands to control excessive rainwater and avoid flash-floods. Such examples point to the fact that institutionalized planning that involves nature-based systems has the significant potential to provide ecological security and late climate vulnerability (Damestoi, 2024).

Of note, NBS should parallel the ideas of social resilience- not merely environmental functionality. The adoption of such tools as Social Vulnerability Index (SVI) enables the policymakers to focus on intervention of NBS in the most vulnerable communities (Puig, 2025). This supports justice-driven climate response in which responses are demographically specific and respond to historical disadvantages. In several situations, as Dessai (2024) observes, community ownership and the confidence in the institutions may prove to be the key factors to the success of the adaptation strategies and not necessarily on their technical feasibility.

Scaling NBS is challenging, however, with risks such as financial shortfalls, the under-representation of NBS in economic systems, and the absence of monitoring metrics. Long-term investment strategies, legal certainty on land rights, and inter-ministerial cooperation, are needed to integrate into national climate plans, and urban masterplans. In the absence of them, NBS risk being stuck as isolated pilot schemes as opposed to becoming transformative systems.

NBS should be an institutionalized component of mainstream development and climate agendas in order to maximize their potential. This needs a combination of ecological science, participatory planning, financial innovation and legal protections.

Natural-based solutions provide a representation that resilience is social and natural. When we combine the aspects of governance, finance, and ecosystems, one can notice that holistic climate strategies need to combine all these areas comprehensively. The conclusion below presents a template to bring these pillars together under one climate response framework.

Figure 1. Climate Crisis: Drivers, Impact, and Integrated Solutions

Climate Crisis: Drivers, Impacts, and Integrated Solutions

The following conceptual framework visualizes the core linkages between socio-economic inequality, governance limitations, and the climate crisis. It demonstrates how these drivers culminate in severe impacts and outlines integrated solution pathways.

■ Socioeconomic Inequality

↓

■ Governance Gaps → ■ Weak Climate Policy

↓

■ Climate Impacts (Displacement, Drought, Livelihood Loss)

↓

■ Integrated Solutions:

→ Nature-Based Solutions

→ Finance + Policy Reform

→ Global Cooperation

CONCLUSION

Possibly the most significant challenge of the 21st century is the climate crisis, which requires integrated strategies grounded in science and policies. The literature analyzed in this research study showcases the multifaceted relationships concerning climate, governance, economy, and environmental conservation (Zhang et al., 2022). Countries are presented with unprecedented socio-economic and environmental threats that come with great risk as global temperatures soar, demanding immediate and continuous action (Swaminathan et al., 2023).

This research focused on three primary goals: analyzing socio-economic balance concerning climate mitigation, assessing the causative and resultant aspects of the climate problem, and formulating integrated approaches to climate crisis mitigation. Evidence shows that multiple fields working together produce the best outcomes for dealing with complex climate emergency issues but single solutions cannot solve the problem (Johnson et al., 2021).

The sustainability problem requires response from all countries upholding sustainability commitments as Trisolini (2014) describes climate change as a distinctive issue affecting environment governance finance and societal structures. The present chapter amalgamates research from literature about policy enforcement alongside funding options and nature-based solutions to construct resilient and sustainable global futures (COP29 Report, 2024).

The effort to maintain balanced socio-economic operations during climate mitigation stands as a crucial obstacle during socio-ecological transformations (Boyd &

Markandya, 2021). A particular climate change policy leads to changes in industrial activity combined with energy use and labor participation along with its effects on economic expansion (Bowen et al., 2014). Today Integrated Assessment Models (IAMs) function as leading approaches to evaluate policy economic impacts yet they fail to address the intricate aspects of social inequality as well as time-based developmental processes (Ausserladscheider et al., 2025).

The economic challenge arises from fossil fuel companies who must remove their assets from markets because of climate change. Substantial losses can occur to capital assets in coal oil and gas infrastructure operations because decarbonization speeds up their obsolescence (Frumkin et al., 2025). The Paris Agreement withdrawal made the United States demonstrate that economic considerations alongside politics can force international climate cooperation into collapse. The reaction from China showed through DNG's (2021) analysis that developing nations can take advantage of the top position in global climate administration.

In order to maintain economic stability, climate policies need to consider the environmental objectives alongside employment, economic development, and industrial competition (Aldy & Pizer, 2015). Efforts to clean energy and green businesses must be comprehensive, actively protecting vulnerable groups from the worst consequences of climate policies (Xu et al., 2018).

Apart from economic factors, climate change fuels social vulnerabilities, especially those exacerbated by low income and marginalized groups (Chakraborty & Joshi, 2021). Impact of climate-change induced migration because of submerging coastlines, extreme weather conditions, and desertification is already affecting millions worldwide (Singh et al., 2023). Policies focused on climate adaptation and resilience need to address social equity, resource availability, and local development (Maiti & Agrawal, 2021).

These findings have stressed that achieving climate mitigation involves socio-economic balance which requires well-defined policies, adequate funding, and social services with a strong focus (Silverstein, 2025). This balance is only achievable through all-inclusive policies framed around addressing social equality and resilience to climate changes (COP29 Report, 2024).

Industrialization alongside deforestation, urban area expansion, and including agricultural activities is done on a larger scale are still the major drivers of the climate crisis (Mercure et al. 2016). These tendencies along with the use of fossil fuels for electricity tend to emit greenhouse gasses at a dangerously high rate compared to previously documented (Khan et al, 2023). Moreover, certain farming methods not only deplete the soil, but also reduce the variety of living organisms and add to carbon emissions (Singh et al., 2023).

The escalation of hurricanes, droughts, wildfires and other extreme weather conditions fuel economic and social vulnerability while needing full attention to adaption and harsher change mitigation plans (Mwangi et al., 2023).

The impacts of climate change are complex and multi-faceted, reaching global food systems, public health, economic performance (Wang et al., 2023). Regions with glaciers like the Tianshan Mountains have experienced extreme changes in climate, resulting in the rapid melting of glaciers which in turn disrupts the water resources available in the region, causing increased potential for destruction (Li et al., 2023).

There is also a growing threat posed by heatwaves, flooding, and the deterioration of air quality in urbanized areas, especially for lower income groups (How Cities Use NBS, Times, 2025). These issues require risk prioritization for proactive governance, community approach, and financial commitment to adaptive measures (Corgo et al. 2024).

An example of policy-based climate action integrating finance and governance is The Climate Change Accountability Act of British Columbia, which highlights the need for effective governance frameworks (COP29 Report, 2024). The existence of laws with legally binding emission targets, independent Committees, and advisory groups ensure growth of emissions sustainability in the long term (Trisolini, 2014). Still, the effectiveness of governance is reliant on international funding and cooperation (Shaik & Abaranji, 2023).

This means that climate mitigation should be supported by economic and financial mechanisms which enable the shift to sustainable energy and resilient infrastructure (Boyd and Markandya, 2021). According to the Investment Funds Report (2025), carbon pricing, climate financial and green investments are the necessary tools to reward emission reductions. On top of that, climate targets must be incorporated into the investment portfolios of financial institutions if these are to be mitigating systemic economic risks (COP29 Report, 2024).

Good results have been delivered by nature-based solutions (NBS) in carbon sequestration, disaster risk reduction and biodiversity conservation (Johnson et al., 2021). Forests, wetlands and protected areas are natural carbon sink and they help enhancing climate resilience while supporting ecological stability (Swaminathan et al., 2023). Urban planning with the involvement of green infrastructure, mangrove restoration, and sustainable land use, are also important to climate adaptation at various spatio-temporal scales (Corgo et al., 2024). The findings from this study emphasize that climate crisis mitigation requires a multidimensional approach integrating governance, economics, and environmental conservation (Trisolini, 2014; Zhang et al., 2022). Thus, according to the literature, enhancing policy accountability, expanding climate finance and leveraging nature based solutions are critical enablers for the long term sustainability (Johnson et al., 2021).

Future research should investigate new climate governance models, financial strategic for equity in climate adaptation, as well as in on nature based conservation (Swaminathan et al., 2023). The threats from the damaging impacts of climate change can only be combated with a collaborative global response that also leads to socio economic stability and environmental sustainability (COP29 Report, 2024).

The main analytical value of this chapter can also be found in the identification of four intertwining areas needing to be tackled in a concomitant manner to ensure sustainability of climate mitigation, namely: (1) Socio-economic drivers (inequality, poverty, marginalization); (2) Climate change impacts (disaster frequency, agricultural stress), (3) governance and institutional gaps, and (4) integrated mitigation pathways, such as Nature-Based Solutions and green finance. These spheres are mutually related and constitute a feedback loop: climate vulnerability associated with inequality intensifies, resulting in further inequality in terms of socio-economic stratifications. The current governance systems usually do not take an absolute view of this loop. The chapter also suggests a novel analytical framework, dubbed the "Quadruple Nexus of Climate Response" to help encompass all these four dimensions of climate policy planning. This framework is of special importance to the Global South, as climate impacts are overlapped by institutional vulnerability and long-standing inequalities.

Holistic approaches to integrating these strategies will then enable climate policies to lead transformation, reduce vulnerabilities, and build a resilient future for all (Boyd & Markandya, 2021).

REFERENCES

Abaranji, A., & Shaik, D. (2022). The Imperative Need for Climate Finance and Global Climate Mitigation Strategy–A Pragmatic Analysis. *National Journal of Environmental Law*, *5*(1).

Aldy, J. E., & Pizer, W. A. (2011). The competitiveness impacts of climate change mitigation policies. NBER Working Paper No. 17705. National Bureau of Economic Research.

Aldy, J. E., & Pizer, W. A. (2015). The competitiveness impacts of climate change mitigation policies. *Journal of the Association of Environmental and Resource Economists*, *2*(4), 565–595.

Associated Press. (2024, October 31). *UN report warns of famine, aggravated by conflicts and climate shocks.* AP News.

Associated Press. (2025, March 11). *US legal troubles for tycoon Adani expose shortcomings in India's booming solar sector.* AP News.

Ausserladscheider, V., Ergen, T., & Golka, P. (Eds.). (2025). *The socio-economics of loss and decline in the climate crisis* [Call for papers]. *Socio-Economic Review.* https://academic.oup.com/ser/pages/cfp-asset-stranding

Bowen, A., Campiglio, E., & Tavoni, M. (2014). A macroeconomic perspective on climate change mitigation: Meeting the financing challenge. *Climate Change Economics (Singapore)*, *5*(01), 1440005.

Boyd, R., & Markandya, A. (2021). Costs and benefits of climate change impacts and adaptation. In F. J. Warren & N. Lulham (Eds.), *Canada in a Changing Climate: National Issues Report* (Government of Canada). Government of Canada.

Burniaux, J. M., Chateau, J., Dellink, R., Duval, R., & Jamet, S. (2009). The economics of climate change mitigation. *Policies and options for the future.*

Casey, G. (2024). Energy efficiency and directed technical change: Implications for climate change mitigation. *The Review of Economic Studies*, *91*(1), 192–228.

Corgo, J., Cruz, S. S., & Conceição, P. (2024). Nature-based solutions in spatial planning and policies for climate change adaptation: A literature review. *Ambio*, *53*(11), 1599–1617.

Ding, Y. J., Li, C. Y., Wang, Y., Wang, S. X., Chang, Y. P., Qin, J., & Wang, Z. R. (2021). An overview of climate change impacts on the society in China. *Advances in Climate Change Research*, *12*(2), 210–223.

Dolan, A. H., & Walker, I. J. (2006). Understanding vulnerability of coastal communities to climate change related risks. *Journal of Coastal Research*, 1316–1323.

Donback, N. (2024, December 18). *How cities are using nature-based solutions to tackle floods. Time*. https://time.com/7202917/cities-nature-based-solutions-floods/

Dong-Qian, W. A. N. G., & Ying, S. U. N. (2022). Effects of anthropogenic forcing and atmospheric circulation on the record-breaking welt bulb heat event over southern China in September 2021. *Advances in Climate Change Research, 13*(6), 778–786.

European Bank for Reconstruction and Development (EBRD). (2020). The economic impacts of climate change mitigation policy. EBRD Report.

Financial Times. (2025, December 30). *The World Ahead 2025*. https://www.ft.com/content/deb33eae-37bc-4d50-98f6-716930438dca

Guja, M. M., & Bedeke, S. B. (2025). Smallholders' climate change adaptation strategies: Exploring effectiveness and opportunities to be capitalized. *Environment, Development and Sustainability*, *27*(8), 17927–17956.

Gupta, A., & Shukla, A. K. (2024). Optimal approaches in global warming mitigation and adaptation strategies at city scale. *Discover Sustainability*, *5*(1), 272.

Holland, D., & Young, G. (2020). The economic implications of climate change mitigation policies. *National Institute Economic Review*, *251*, R1–R2.

Investment Funds. (2025). *Investment funds report 2025: Transitioning to low-carbon economic development*. Global Climate Finance Initiative.

Johnson, D., Brune, S., Dagan, D. T., Meier, E., Wilkins, E. J., & Zhang, H. (2020, January). A holistic strategy for carbon reduction programs in parks and protected areas. In *Parks Stewardship Forum* (Vol. 36, No. 3, pp. 477-482). George Wright Society.

Kaur, A., & Sharma, V. (2024). Nexus between climate change and mitigation approaches for sustainable development: A bibliometric review. *Global Knowledge, Memory and Communication*.

Khan, N., Ma, J., Zhang, H., & Zhang, S. (2023). Climate change impact on sustainable agricultural growth: Insights from rural areas. *Atmosphere*, *14*(8), 1194.

Krogstrup, S., & Oman, W. (2019). Macroeconomic and financial policies for climate change mitigation: A review of the literature. Academic Press.

Mendelsohn, R., Dinar, A., & Williams, L. (2006). The distributional impact of climate change on rich and poor countries. *Environment and Development Economics, 11*(2), 159–178.

Mercure, J. F., Pollitt, H., Bassi, A. M., Viñuales, J. E., & Edwards, N. R. (2016). Modelling complex systems of heterogeneous agents to better design sustainability transitions policy. *Global Environmental Change, 37*, 102–115.

Mercure, J. F., Pollitt, H., Chewpreecha, U., Salas, P., Foley, A., Holden, P. B., & Edwards, N. R. (2013). Complexity, economic science and possible economic benefits of climate change mitigation policy. *arXiv preprint arXiv:1310.4403*.

Moarif, S., & Rastogi, N. P. (2012). *Market-based climate mitigation policies in emerging economies*. Center for Climate and Energy Solutions.

Newell, R. G. (2010). The role of markets and governments in climate change policy. *Energy Journal, 31*(1), 91–114. DOI: 10.5547/ISSN0195-6574-EJ-Vol31-No1-5

Ng, S. L. (2024). Bibliometric analysis of climate change and extreme weather adaptation based on Scopus. *Journal of Environmental Studies and Sciences*, 1–20.

Organisation for Economic Co-operation and Development (OECD). (2009). *The economics of climate change mitigation: Policies and options for global action beyond 2012*. OECD Publishing. DOI: 10.1787/9789264073616-

Pacillo, G., Nguyen, N., Paustyan, E., Cavatassi, R., Astralaga, M., & Läderach, P. (2024). What drives the success and failure of climate change adaptation projects? A qualitative comparative analysis. *Mitigation and Adaptation Strategies for Global Change, 29*(8), 89.

Painter, M. A., Shah, S. H., Damestoit, G. C., Khalid, F., Prudencio, W., Chisty, M. A., & Wilhelmi, O. (2024). A systematic scoping review of the Social Vulnerability Index as applied to natural hazards. *Natural Hazards, 120*(8), 7265–7356.

Pathak, M., Patel, S., & Some, S. (2024). Climate change mitigation and Sustainable Development Goals: Evidence and research gaps. *PLOS Climate, 3*(3), e0000366.

Puig, D., Adger, N. W., Barnett, J., Vanhala, L., & Boyd, E. (2025). Improving the effectiveness of climate change adaptation measures. *Climatic Change, 178*(1), 7.

Riera-Spiegelhalder, M., Campos-Rodrigues, L., Ensenado, E. M., Dekker-Arlain, J. D., Papadopoulou, O., Arampatzis, S., & Vervoort, K. (2023). Socio-economic assessment of ecosystem-based and other adaptation strategies in coastal areas: A systematic review. *Journal of Marine Science and Engineering, 11*(2), 319.

Sambrook, K., Harcourt, R., Dessai, S., & Taylor, A. (2024). New evidence of a strong public mandate for climate change adaptation in the UK.

Sarun, S., Ghermandi, A., Sheela, A. M., Justus, J., & Vineetha, P. (2018). Climate change vulnerability in a tropical region based on environmental and socio-economic factors. *Environmental Monitoring and Assessment, 190*(12), 727.

Silverstein, K. (2025, January 17). Commentary: Wealthy nations can't ignore climate effects on poorer countries. *My Journal Courier.* https://www.myjournalcourier.com/opinion/article/wealthy-nations-can-t-ignore-climate-effects-20027538.php

Strand, J. (2016). *Stranded assets: A climate risk challenge.* Inter-American Development Bank. https://publications.iadb.org/publications/english/document/Stranded-Assets-A-Climate-Risk-Challenge.pdf

Swaminathan, U., Reddy, G. B., Buot, I. E., & Ramamoorthy, S. (2023). Climate Change and Environment: Holistic Approaches Towards Climate Resilience. In *The Palgrave Handbook of Socio-ecological Resilience in the Face of Climate Change: Contexts from a Developing Country* (pp. 537–552). Springer Nature Singapore.

The Guardian. (2025, October 2). *Kemi Badenoch vows to repeal Climate Change Act.* https://www.theguardian.com/environment/2025/oct/02/kemi-badenoch-vows-to-repeal-climate-change-act

Trisolini, K. (2014). Holistic Climate Change Governance: Towards Mitigation and Adaptation Synthesis. *U. Colo. L. Rev., 85*, 615.

United Nations Framework Convention on Climate Change. (2024, November 22). *Summary of global climate action at COP 29.* https://unfccc.int/sites/default/files/resource/Summary_Global_Climate_Action_at_COP_29.pdf

United Nations Framework Convention on Climate Change (UNFCCC). (2008). *Integrating practices, tools and systems for climate risk assessment and management and strategies for disaster risk reduction into national policies and programmes* (Background Paper). https://unfccc.int/resource/docs/2008/tp/04.pdf

Vaughan, C., & Dessai, S. (2014). Climate services for society: Origins, institutional arrangements, and design elements for an evaluation framework. *Wiley Interdisciplinary Reviews: Climate Change, 5*(5), 587–603.

Venus, T. E., Bilgram, S., Sauer, J., & Khatri-Chettri, A. (2022). Livelihood vulnerability and climate change: A comparative analysis of smallholders in the Indo-Gangetic plains. *Environment, Development and Sustainability, 24*(2), 1981–2009.

World Bank. (2010). *World development report 2010: Development and climate change*. https://openknowledge.worldbank.org/handle/10986/4387

World Bank. (2016). *World Bank Group Climate Change Action Plan 2016–2020*. https://openknowledge.worldbank.org/handle/10986/24795

Xu, Z., Smyth, C. E., Lemprière, T. C., Rampley, G. J., & Kurz, W. A. (2018). Climate change mitigation strategies in the forest sector: Biophysical impacts and economic implications in British Columbia, Canada. *Mitigation and Adaptation Strategies for Global Change*, *23*(2), 257–290.

Zakharova, O., Karagulian, E., Viktorova, N., Gamukin, V., & Yablochkina, V. (2025). Risk assessment of socio-economic vulnerability to climate change: Case study of Tyumen region. *WORLD (Oakland, Calif.)*, *6*(1), 6. DOI: 10.3390/world6010006

Zhang, H., Wang, F. T., & Zhou, P. (2022a). Changes in climate extremes in a typical glacierized region in central Eastern Tianshan Mountains and their relationship with observed glacier mass balance. *Advances in Climate Change Research*, *13*(6), 909–922.

Zhang, H. B., Dai, H. C., Lai, H. X., & Wang, W. T. (2017). US withdrawal from the Paris Agreement: Reasons, impacts, and China's response. *Advances in Climate Change Research*, *8*(4), 220–225.

Zhang, R., Tang, X., Liu, J., Visbeck, M., Guo, H., Murray, V., & Zhou, L. (2022b). From concept to action: A united, holistic and One Health approach to respond to the climate change crisis. *Infectious Diseases of Poverty*, *11*(1), 17.

KEY TERMS AND DEFINITIONS

Adaptive Governance: A multi-tiered and de-centralized approach to environmental decision-making that allows scientific evidence, input of stakeholders, and an iterative policy process to address the uncertainty and complexity of the problem of climate.

Carbon Pricing: The pricing mechanism designed to impose a cost on the release of carbon dioxide and other greenhouse gasses in order to provide businesses and individuals with an incentive to minimize their emissions in a market-based mechanism.

Climate Crisis: An international crisis owing to rapidly worsening climatic change, in the shape of swelling temperatures, wild climatic conditions, and ecological imbalances, with socio-economic impacts, which jeopardize the sustainability of the world.

Climate Justice: Such a rule will entail a just sharing of burdens and benefits of climate change, pointing at the moral obligation to defend the vulnerable segments of the population and the future generations against environmental harm.

Climate Resilience: The ability of people, communities, organizations and natural systems to anticipate, adapt to, and recover not only after climate-related shocks but also support core livelihoods and change.

Chapter 3

Climate Struggle at the Planet's Boundaries

Kıvılcım Özge Kara
https://orcid.org/0000-0003-0883-1565
Ministry of Environment, Urbanisation, and Climate Change, Turkey

ABSTRACT

Climate uncertainty is the primary result of the disconnection of the existential connection with the planet. Environmental degradation caused by unlimited growth triggers this uncertainty and the climate crisis, and sustainability becomes a critical source of concern with its emphasis on equality and justice. The fact that this process is being discussed with an emphasis on equality and justice indicates that these values are in crisis at a global level. In this context, the study discusses that unlimited growth at the expense of pushing the boundaries of the planet causes urban costs such as environmental degradation, inequality, and decreasing efficiency. It is accepted that unlimited growth with traditional methods accelerates the transition beyond the boundaries of the planet and is the primary trigger of climate change, and it is concluded that the efficient and innovative use of limited resources can positively affect environmental performance and strengthen the climate struggle.

INTRODUCTION

One of the issues that cause concern in all societies is the relationship between humans and nature. This situation consists of related questions such as what kind of a life will be a common good biologically, how the use of limited resources will be effective and how the boundaries of the planet will be used fairly in distribution. These topics, which are perceived as the subject of pure economy at first glance, are now the subject of philosophical discussions in terms of perceiving nature and

DOI: 10.4018/979-8-3373-1862-2.ch003

living in harmony with it. In the relationship between humans and the environment on a global scale, the weight of an economic-centered market-based thought system is seen. In this approach, a common good perspective such as quality of life has been ignored, and while discussions on the functionality of economic processes continue, efforts and solutions on issues such as sustainable policies, efficient use of resources, pollution, level of damage to the environment, environmental performance and climate change remain inadequate in the long term.

In the historical process, environmental problems and those that have reached the risk dimension have been drawing attention since the 1990s. The already complex structure of environmental problems becomes even more complex when it comes to climate change. In this context, transformative changes and new orientations are needed in many socio-economic and environmental problems. Since it is understood that nature and society cannot be excluded from each other with the socio-ecological dimension; resilience and new ways of thinking about the boundaries of the planet are emerging. Indeed, environmental policies before the 20th century were shaped by the pressure created by the insufficiency of natural resources despite the increasing population and the subsequent ecological damage. This situation leads to the conclusion that environmental problems that can be seen when the world population is estimated to reach approximately 10 billion in 2050 will reach the dimension of competition and inequality and then the dimension of global risk. Addionally the urban population in underdeveloped countries is expected to increase by approximately two and a half times that of developed countries by 2070 (UN-Habitat, 2022). From this perspective the 21st century is getting acquainted with the causes and consequences of new generation inequalities as well as ongoing inequalities. One of these is climate change, which is the subject of this study and is an iconic problem that can be seen as the main paradigm shift that will shape the 21st century. The climate struggle; When considered as a welfare issue, a solution is sought with the sustainability dimension, it spreads to many variables in terms of causes, effects and results. The fact that the costs of climate change and unsustainable actions are not covered by those who carry them out creates negative effects between generations (Guida et al., 2025).

While climate change has been examined within environmental risks from the past to the present, today it is discussed together with social, ecological and economic risks and is positioned as a multidisciplinary research area due to its results. Climate change, the effects of which have become apparent in the last 50 years, is a subject of struggle due to the increasing competition effect on ecosystem services and scarce natural resources, and is associated with structural difficulties such as inequality in access to resources as it accelerates environmental degradation due to the growth trend of countries. In other words, the existential problem is managed with current conditions and future scenarios. In this context, the following research problems have been brought to the agenda in the study:

1. The growth and development concerns of economies and their motivation for unlimited growth despite the environment are critical constraints on the climate fight.
2. What are the socio-economic consequences of these restrictions and what effect do they have?

BACKGROUND

Literature Review

This study uses growth, limitless growth, climate change, eco-ecology approaches in the context of pushing the planet's boundaries. The study includes discussions of increasing exposure to climate change for the purpose of economic growth within the framework of equality and justice discourses.

Climate change and environmental concerns are the dominant scenarios of the future. The increase in extreme weather events and natural disasters intensifies climate adaptation and reduction scenarios especially in cities. Because cities are responsible for the vast majority of world carbon emissions. The transition to net zero emissions should be realized as much as possible in this context (UN-Habitat, 2022). Indeed, the United Nations Office for According to Disaster Risk Reduction (UNDRR), statements such as 'There are no natural disasters, human beings affect the creation of risks, the level of impact depends on their vulnerability, the most vulnerable groups are the most at risk, climate change worsens risks, and ultimately human actions and efforts can protect from disasters' indicate that climate change is human-induced and its consequences directly affect humans (UNDRR, 2022). For example, a study by the WEF found that the populations of East Asia are among those most exposed to climate hazards, while the populations of South Asia and Southern Africa are similarly highly exposed but also highly vulnerable (WEF, 2025).

Human-induced climate change projections are supply-side rather than demand-side. In other words, climate change, as one of the types of environmental degradation, affects the destruction and availability of natural resources on the supply side, while creating a decisive effect on costs. Accordingly, Maslin emphasized that adaptation measures that will reduce and slow down climate change should include low-cost steps (Maslin, 2011). Ignoring the effects of climate change that have no spatial boundaries creates a limitation in determining the costs exactly, and in 30%-60% of countries; it is seen that the total impact of climate change widely extends beyond the border (Kulmer et al., 2020).

According to Kruger et al., the use of natural resources by sectors such as energy and food causes ecological footprint and subsequently climate change (Krueger et al., 2022). Green innovation is attributed great importance in reducing the negative effects of such economic activities on the ecosystem (Alnafrah et al., 2023). By strengthening innovation infrastructure and supporting green technology ecosystems, carbon reduction and achieving zero carbon can reduce costs, and duplication in climate efforts can be greatly reduced. However, some studies have also reached conclusions that green innovation and technological progress cause an increase in emissions. On the other hand, environmental performance will be higher in regulations where environmentally friendly technologies and policies are adopted with high income levels and new energy technologies (Koengkan & Fuinhas, 2021). Alnafrah et al. found in their study that green innovation production focused on output rather than input is subject to the law of diminishing returns (Alnafrah et al., 2023). In this context, the yields of different ecosystems vary (Ponting, 2000). On the other hand, an economy with a high level of productivity is potentially an economy with a high tendency to pollute, and achieving economic growth with traditional means leads to more pollution (Eriksson & Persson, 2013; Hickel et al., 2022). Rakshit and Neog, in their study examining the effect of macroeconomic uncertainty on environmental degradation in India, concluded that higher economic growth and financial development can further increase environmental degradation (Persakis et al., 2024). On the other hand, in the study conducted by Isaksen, it was accepted that the use of chemicals in an economy is affected by the growth rate, capital inflow, export and rural-urban migration dynamics of that economy and it was concluded that a 1% increase in GDP per capita corresponds to a 1.1% increase in chemical use (Isaksen, 2020). In the study conducted by Ren et al., the relationship between economic growth target and environmental pollution was examined by selecting the local scale, and it was determined that the indirect pollution that emerged was limited by environmental localization. In addition, it was predicted that the improvement of the industrial structure, technological innovation, foreign direct investment and environmental regulations could reduce environmental pollution by removing the constraints on the economic growth target (Ren et al., 2023) (Figure 1). This approach will be possible especially if high-income countries contribute to climate justice by ensuring the ecological-economic balance.

Figure 1. Economic growth, pollution and environmental localization

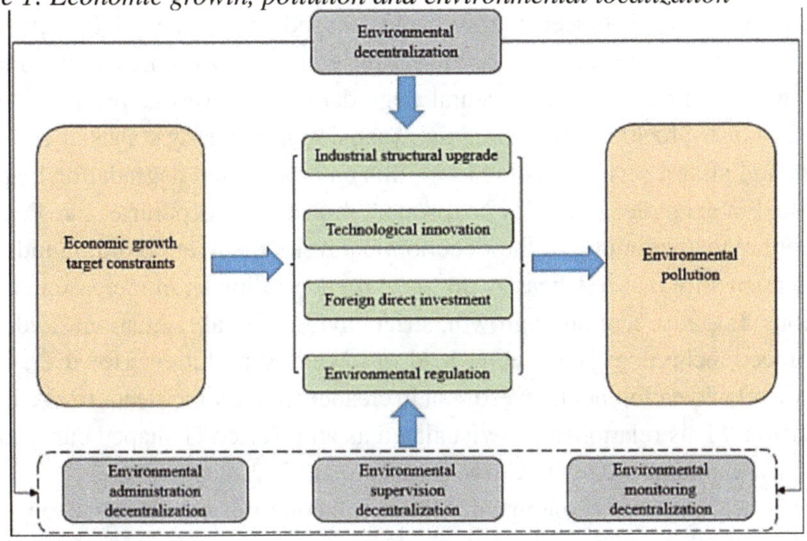

Source: Ren et al. 2023

On the other hand, in Freiswijk et al.'s study, a new behavioral-environmental model was proposed for the co-evolution of anthropogenic environmental impact and human behavior. In the model, individuals' behaviors were monitored in a social learning process and a reward system was designed to specifically encourage individuals to adopt or not adopt sustainable behaviors. However, it was determined that this process was affected by various factors, including sensitivity to environmental degradation, implementation of government policy interventions and higher costs associated with environmentally friendly behaviors (Frieswijk et al., 2025). In this context, the contribution of innovative and nature-based production methods to the reduction of emissions and the efficient use of natural resources with a more sustainable approach in the consumption of resources will be critical.

Theoretical Framework

Since the 1970s, discussions on the threat to humans and their environment have intensified within the scope of the status of socioeconomic structures and the attitudes of political actors. While the world is facing many ecosystem threats, our planet is progressing in a process beyond its borders and is facing the risk of causing irreversible damage to the world. Accordingly, instead of traditional environmental protection methods and discussions focused solely on the environment, dialectical discussions and an evolution towards biopolitics should be made (Lemke, 2013). In this context, adopting an economic growth approach without harming the environment in line with Foucault's biopolitics will serve the environmentally sustainable development approach. Indeed, the inverted U-shaped Environmental Kuznets, which explains the relationship between economic growth and environmental degradation, according to the curve, environmental degradation increases as priority is given to rapid industrialization and economic expansion in the early stages of economic growth, and after a certain income level, this environmental degradation begins to decrease. For example, a study by Naqvi et al. indicates that countries like Pakistan, which are at the beginning of their economic development, are building industries and infrastructure that rely heavily on fossil fuels, leading to an increase in carbon emissions. Initially, economic growth, strict environmental regulations, and a lack of advanced technology are seen to lead to increased pollution. However, after a certain level of development, investment in cleaner technologies leads to a reduction in emissions. This relationship is visualized as an inverted U-shaped curve known as the Environmental Kuznets Curve (Naqvi et al., 2025).

In the early stages of economic development, countries mostly strive to increase output despite the environment with production and consumption trends, and environmental regulations are often inadequate. The lack of advanced and renewable technology makes it difficult to control emissions. In the early stages of development, countries focus on GDP and development rather than environmental degradation, which worsens environmental degradation, but in the ongoing growth and development process, it is observed that a cleaner and more sustainable environment is established. Therefore, balancing economic development with environmental sustainability is important for a sustainable future. Naqvi et al., the existence of renewable resources and renewable technology, more economic growth, and increased environmental awareness contribute to the reduction of pollution, thus supporting improved environmental quality. Because as environmental awareness and consciousness increase, the government enacts stricter environmental regulations and contributes to the formation of a better environment with economic development reaching a certain level (Naqvi et al., 2025). In this context, it is understood that an approach based on sustainable development is essential. However, it is not possible to fully evaluate this curve within the framework of the climate change paradigm. Although emissions are high for developed countries with high environmental performance, over time, the transition to renewable energy, green technology and public interventions contribute to the climate fight by ensuring effectiveness in environmental policies.

However, since emissions have negative externalities, micro-level, i.e., country-based improvements cannot solve the global problem. Therefore, the course of the Environmental Kuznets Curve will not be concrete for climate change at this stage.

MAIN FOCUS OF THE CHAPTER

Climate Change and Its Major Interactions

One of the fundamental forces that shapes humanity is the climate. While the weather changes from year to year; long-term climate events (Ponting, 2000) are a determining factor on the existence of humanity by causing structural problems. It is in a determining position in all sectors and variables from the location of the continents to agriculture, energy and economy. Accordingly, ecosystems are not fixed, but develop over time through a series of changes. In this sense, one of the fundamental and most sensitive components of the environment is the climate system and one of the factors that increase the sensitivity here is the unnatural changes that occur in the climate system. The main reason for these is, as stated, the greenhouse effect that develops due to the use of fossil resources and the subsequent global warming relationship (Gautier, 2014). The climate system, which is not gentle but dynamic and mobile (Maslin, 2011), has been going beyond its natural boundaries, especially since the Industrial Revolution, with industrial activities, urbanization, migration etc. is increasing due to human activities, the results of this increase are being monitored through different parameters, especially temperature, and it is seen that exposure to climate hazards and inequality are increasing at the global level (Maslin, 2011). Structural changes such as green technology, with their eco-ecological structure, have recently had a reducing effect on scientists' concerns about sudden climate change.

As an environmental problem, climate change is one of the causes of environmental degradation. When the finiteness of the planet is accepted within the framework of the ecological economics approach, it would be meaningful to abandon the unlimited growth approach. Climate change has caused environmental degradation by affecting the healthy functioning of ecosystems in the last thirty years (Hamed et al., 2024). Therefore, understanding the complex cause-effect relationships between economic development, human health, climate change, and environmental quality is extremely important for applicable adaptation options within country dynamics (Hamed et al., 2024) . Climate change, which has bidirectional cause-effect relationships, is one of the leading factors in the re-creation of the environment today and has a wide-ranging impact on the economic, social and environmental systems of all continents as one of the most serious problems of humanity and the natural environment (Hashim et al., 2024). However, translating global climate policies into regional and local management practices is not an easy process, and there is a need to better understand how best to combine adaptation and mitigation policies (Dabaieh et al., 2024) .

According to the United Nations Framework Convention on Climate Change (UNFCCC), climate change is defined as "*A change in the climate resulting from human activities that directly or indirectly disrupt the composition of the global atmosphere in addition to natural climate changes observed in comparable time periods*" (UNFCCC, 1992); According to the IPCC, it is defined as "*Changes in the state of the climate that generally continue for decades or longer*" (IPCC/Glossary; IPCC/AR5, 2014). Accordingly, in its definition, UNFCCC distinguished between climate change due to natural causes and anthropogenic climate change. Rather than natural climate change, anthropogenic climate change closely affects almost all environmental policies with its complex and uncertain state globally and regionally. The deteriorating natural climate change effects due to uncontrollable human activities now exceed the boundaries of the planet in the short and long term and lead to environmental degradation (Hamed et al., 2024) .

Anthropogenic activities, especially industrialization and economic growth, reinforce the increase in emissions and exacerbate climate change. Because as the economy grows, industrial activities and therefore carbon emissions increase. Economic development based on fossil resources directly affects climate change due to emissions. Industrialization and urbanization are two basic factors of economic and social development that affect production and consumption, as well as emissions, and are also the main environmental problems (Naqvi et al., 2025). Increases in greenhouse gas emissions (GHG) that cause temperature increase and in the concentration of pollutants that cause the greenhouse effect increase the average temperature of the planet, and the negative externality of these increases leads to socio-economic consequences. While industrialization and economic growth efforts, which are the main causes of emission increases, cause environmental pollution; It leads to deterioration in all components of the environment, especially environmental values such as air, water, soil, and affects marginal costs. According to the conclusion to be reached here, an economy with a high level of productivity also has a great pollution potential and indirectly reinforces the climate crisis without eco-economic improvements. At this point, cities, which are the center of environmental problems; cannot be independent of the risks and dangers experienced by causing multiple crises. Indeed, rapid urbanization is at the center of economic opportunities and draws the limits of economic growth; and it causes inequality to emerge in places where there is no equal opportunity. Rapidly urbanizing places in developing countries show a profile that is more vulnerable and less compatible with climate change. Limited financial, human and technical resources, weak institutions and governance are the main factors limiting these economies. At the same time, global warming and its consequences are unbalanced in these places. The effects of climate change further increase the areas of struggle in cities and make it difficult to combat problems such as poverty, inequality and infrastructure deficits. This situation makes it difficult to achieve success in sustainable development goals (SDGs) (UN-Habitat, 2022) .

Figure 2. Resilient urban economies and productive urban futures

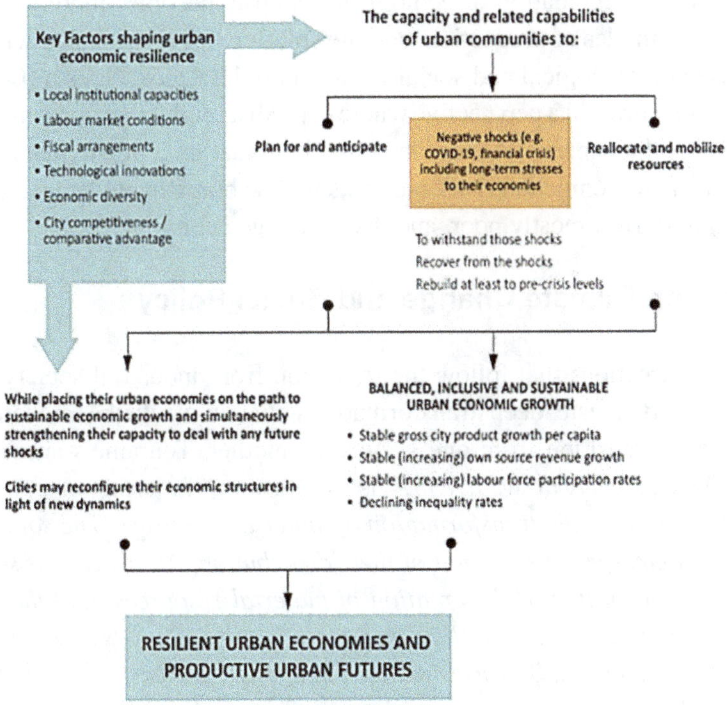

Source: (UN-Habitat, 2022)

Cities contribute to more than 70% of global emissions, leading to crisis-level climate change impacts. National climate strategies and action plans must reduce exposure of vulnerable and disadvantaged urban communities and strengthen the achievement of sustainable cities. Sustainable and resilient cities, as critical components of the climate struggle, also affect the resilience of cities in economic terms. Balanced, inclusive, sustainable economic growth is measured by low inequality rates, increasing labor force participation rates, growth based on equity, and an increase in GDP per capita, and urban welfare improves with the decrease in inequality (UN-Habitat, 2022) (Figure 2). Accordingly, the resilient economies of resilient cities also strengthen the climate change struggle. Considering that human behaviors and tendencies are at the origin of anthropogenic climate change, undesirable administrative problems again emerge from human-related results, and solutions can again be provided by making the climate struggle a development-oriented public policy. According to Amin, the huge polarization called development becomes evident due to the development of ecological destruction, which is a result of world capitalism (Amin, 2014:26). Accordingly, the social policy emphasis on how to adapt the built environment to climate change (Dabaieh et al., 2024) should be taken into consideration. Accordingly, climate change, one of the most critical consequences of environmental degradation or ecological collapse, becomes more pronounced under capitalist modes of production, and the solution lies in the construction of an order focused on ecological and social justice, based on ecosocialism (O'Connor, 1998). In other words, in a perspective where capitalist production and consumption trends come at the expense of environmental degradation, climate justice will be weakened, and environmental risks and costs will be borne by underdeveloped and developing countries, mostly poor, and disadvantaged communities.

Emphasis on Climate Change and Social Policy

The transformations that follow the transition from industrial society to post-industrial society create deep transformations that it would be reductionist to evaluate it as a transition from one society to another. Touraine's answer to the question "*Which society do we live in?*" is as follows; "*The principle of social life in industrial society is the transformation of material resources. The source of the main conflicts emerges at the point of their distribution. All aspects of social life are positioned according to the creation of material resources, and therefore by defining the actors and their conflicts as the production of goods and the distribution of existing resources. Because creation is no longer the case; the only situation observed is destruction and impoverishment. Financial capitalism accumulates, leading to financial collapse and social crises one after another. The danger of the world collapsing in this way with its unlimited search for profit is more than a symptom of a crisis*" (Touraine, 2016:21-22). When viewed from this perspective, a significant phase change is occurring, different social categories are emerging. Subsequently, with the statement "*All social categories demand certain rights*", it can be inferred that there will be different paradigm shifts (Touraine, 2016:88-90). The main one of these changes is social policy changes.

In this context, income and income redistribution, such as basic policies such as the environment, are fundamentally social policies and their implementations produce economic results. Leaving policies that stand out with their social dimension to the free-market economy has a negative impact on the effectiveness of all relevant policies. According to Harvey, the path to be followed regarding the widespread flaws of the market mechanism is a problem in itself (Harvey, 2015:55).

In this context, the main values of social policies include sustainability, spreading welfare to the base, and undertaking effective, personal, social and economic rights-based roles. In order for the equality and justice emphases that overlap with social policies to gain functionality, they must be placed on a legal basis and their implementation capability and motivation must be valid for the relevant economies. The holistic and inclusive nature of social policies in general; undertakes important roles regarding the establishment and continuation of social order.

The fundamentals of the changes and transformations that will form the basis of social policies all over the world today and in the future are shaped by climate change, poverty, unemployment, population growth, urbanization, migration, technology, etc. Climate change and environmental degradation are one of the most complex challenges that humanity faces in its current and future state, and one of the greatest threats to human well-being (Nguyen et al., 2023). The main reason for climate change, global warming, and the environmental disasters it triggers, are increasingly producing more intense and frequent social inequality, and the need for sustainable social policies is increasing. However, what is needed is to update the general components of conventional social policies and social policies that are not yet fully known; and to re-establish their tools, purposes, and practices.

Indeed, when each of the changes and transformations that will form the basis of social policies is evaluated as a crisis factor, Touraine's statement will be meaningful here: "*Crises accelerate the tendency of the economic system and social actors to separate from each other in the long term. While the sociologist is aware of these issues and questions the way to overcome the crisis; he brings a new idea without rejecting the technical solutions suggested by economists and politicians. The sociologist's thought works as follows; to re-establish social life and to end the dominance of the economy over society. What will ensure this is the increase in the importance given to human rights.*" Accordingly, only when economic systems are evaluated together with social systems can policies based on rights-based approaches be implemented (Touraine, 2016).

In fact, the extent and variety of inequalities increase with conjunctural changes, and this is followed by an increase in efforts to reduce inequality. Especially in underdeveloped countries, low environmental concerns, increasing environmental degradation and the inequalities that emerge subsequently strengthen conflicts and social divisions, and have a weakening effect on climate justice (Kara, 2024). In other words, considering that climate change can exacerbate pre-existing inequalities, communities with high levels of exposure are generally low-income and have previously been marginalized in some capacities (Vickery & Quinn, 2024). For example, the climate exposure of the Global South, which is predominantly underdeveloped, is high. The reasons for this are that the main source of income in these regions is agriculture and the sensitivity of the agricultural sector to climate and weather conditions is high; their low levels of welfare and low resilience; The lack of sufficient institutional capacity, low political stability and high security vulnerabilities

(Nguyen et al., 2023) are viewed as difficult and costly basic structural problems that strengthen inequality.

According to Creutzig et al., inequality and polarization reduce the ability to implement climate policies and erode the stability of the planet (Creutzig et al., 2022). Such situations significantly increase the dilemma between climate change and environmental degradation, security problems and cooperation problems (Nguyen et al., 2023). Since it leads to irreversible consequences, the effective sustainable solution to combat the climate crisis is collective action and global cooperation (Guida et al., 2025). At the global level, governments of all countries commit to strengthening environmental protection, equality, justice and prioritizing sustainable development, while accepting that this will be possible with interconnected goals (Alnafrah et al., 2023). Therefore, cooperation on a global scale can be expressed as the main administrative key to climate change. Otherwise, policy inconsistencies, capacity gaps and conflicts become apparent and the climate struggle weakens.

In general, while it is important to focus on a cost-effective goal by preferring integrated assessment models in the climate struggle, the principles of equality and justice should be taken as the basis in the models to be established, and the reproduction of injustice and inequality in social relations should be prevented. Therefore, climate justice discourses should be taken into account in all models for the climate struggle over time, and justice dimensions should be evaluated more clearly (Rivadeneira & Carton, 2022). For example, inequalities in the distribution of emissions disrupt social cohesion and the acceptability of reduction policies (IPCC, 2023).

This concern for governance, the concepts of equality and justice are rising values and have been accepted as the basic principles of the UNFCCC. On the other hand, the importance of democratic governance, which can be considered as supporting the viability of the concepts of equality and justice, is emphasized by Persakis. et al. 's study also emphasized the impact of accountable and transparent climate policies that ensure public participation on effectiveness (Persakis et al., 2024). Otherwise, the democratic order will not be an easy victory. A decrease in public participation, deregulation, transformation into economic liberalism, directing basic services only to the poor, not being seen as part of a general right, the collapse of the welfare state, and a decrease in public investments are the main effects that characterize post-democracy (Bauman & Bordoni, 2018:179).

In conclusion, effective climate policies are based on as stated Creutzig by et al., there are social systems (Creutzig et al., 2023). Considering the extent to which paradigm shifts have reached today, it is understood that it is necessary to go beyond traditional discourses in the understanding of social policy shaped in the context of green economy in socio-economic systems.

Within these systems, social justice is more than a goal; it is the basis of equality and resilient societies (UN-Habitat, 2025). In fact, climate-resilient development paths are progressing with difficulty due to temperature increases and the increase in socio-economic inequalities (IPCC, 2022). Policy makers want to reduce inequality, which is now discussed as a risk, ensure justice, protect the planet's borders by ensuring environmental sustainability, and at the same time increase economic growth and per capita income. Uncertainties in climate policy and administrative deficiencies and disruptions deepen environmental degradation and the climate crisis, and increase the need for comprehensive redistributive policies in this sense. Economic growth free from redistributive policies will not provide sufficient effectiveness in the climate fight. According to Van den Bergh, while growth can be abandoned in middle-income economies to achieve climate targets, this is not possible in low-income economies (Van den Bergh, 2023). This situation alone necessitates the monitoring of inequality in all parameters.

Climate Change with Its Socio-Economic Dimension

Some studies have found that integrated models of ecological degradation and climate change have failed, and therefore realistic approaches to the limits of growth are needed (Najafabady et al., 2021). In order to find universal solutions, as far as it is known from previous experiences, "*economic growth*" appears to be the main reason for the continuation and worsening of the problems encountered (Bauman, 2013).

Persakis et al. point out that macroeconomic uncertainty has an effect on environmental degradation, and that higher economic growth and financial development can worsen environmental degradation (Persakis et al., 2024). On the other hand, there are also results indicating that the effects of climate change decrease during periods of financial expansion and strong economic growth (Bettarelli et al., 2024). In this context, it is necessary to accept that the climate struggle is a socio-ecological struggle with a sustainability dimension and that the sustainability principle is the basic principle taken as reference. In fact, in order to overcome poverty and inequality in all geographies and economies, the need for sustainable development and not ignoring the consequences of environmental degradation in this process play a critical role. This situation is a kind of emphasis on eco-development and points to the importance of compatibility with the natural limits of growth.

In this context, climate change discussions establish an irreconcilable context with concerns about unlimited economic growth. While discussions continue on where we are on the environmental Kuznets Curve, more micro analyses and the inclusion of subheadings of environmental problems in the analysis are needed. Considering that multiple risks and successive crises make deficiencies and disruptions visible, it is important to turn these processes into opportunities; to establish cooperation and coordination structures at different scales and to strengthen existing ones. In this sense, considering the global spillover effects and limitlessness of crises, combinations of growth strategies aimed at local level decision-making and implementation should be considered.

Global North and Global South's Approach to the Environment and Its Impact on Climate Action

By the end of the 21st century, global warming will lead to very critical climate changes. Today, as global warming exceeds 1.5°C, our planet is entering a new period in which increasing climate risks are more intense and more frequent. In areas where climate change intersects with exposure and vulnerability, multidimensional crises inevitably turn into disasters. This situation is made more difficult by poverty, deprivation and subsequent inequality, and unsustainable development. Beyond the consequences of natural climate change, anthropogenic climate change is positioned before us as a process that needs to be managed in this sense. The need for systematic and structural transformations with its combined effects and chain consequences is therefore now imperative. Because global population growth causes an increase in the need for resources, and poverty and deprivation become a struggle with livelihood problems. In this struggle, a fair approach should be adopted and a stable future should be built by learning from the related problems that will arise in the absence of this approach.

The impact of these changes on the environment and economy is inevitable in this process. Since these effects will have direct effects on some sectors and indirect effects on others, estimating the exact effects is complex and uncertain at all administrative scales (Maslin, 2011:113-114). Climate change, which is at the intersection of the environment and economy, has been affecting both developing and developed countries severely in the last decade and has put many cities and communities at risk. This situation negatively affects people, their environments and all sectors in countries in various ways and levels, while directly affecting the cost of public intervention. Therefore, over the years, discussions have deepened that the consumption of natural and material resources should be made by observing environmental limits in order to prevent environmental degradation (Dabaieh et al., 2024; Kluza et al., 2024) . In this context, it is necessary to pay close attention to the transformations experienced in the world economic order. At the forefront of these transformations are those in large countries such as Russia, India and China. While these countries direct world trade; While they feed the production in the industrial and information sectors of the Northern countries with raw materials, energy and especially their exports, they gain economic growth potential (Touraine, 2016:37).

It is known that in the early stages of industrialization, energy use efficiency is low, output creates more pollution load, and after a certain level of welfare is reached, energy consumption efficiency increases with improvements and economic growth (Gautier, 2014:48). In their study on green innovation and environmental efficiency, Alnafrah et al. emphasized that China, South America and European Union countries are at the forefront in providing efficiency from green innovation for 2020; North American countries and India show relatively lower efficiency. On the other hand, it was determined that South Africa, the USA, India and Russia have the lowest efficiency scores for green innovation production. Factors such as lack of financing and weak financial regulations of South Africa, intensive and widespread industrialization and inadequate policies of China and the USA are the parameters explaining low efficiency. Again, it was understood in the study that Lithuania, Croatia and Slovenia are the most efficient countries in using inputs to produce greener innovations. In addition, it was determined that England and Chile have the highest environmental efficiency scores. This situation points to the importance of making progress together on SDGs and environmentally sustainable development goals ESDGs). Spain, Denmark, Mexico, France; have the best performance in using green innovation for ESDGs. However, it has also been stated that many countries lack progress in environmental efficiency by using green innovation to achieve ESDGs (Alnafrah et al., 2023).

When the relationship between green innovation and environmental efficiency is evaluated, it is seen that different dynamics emerge for each country. First of all, the production activities and tools of all countries are in different forms, and CO2 and GHG emissions are different. Especially in underdeveloped and developing countries, the need for energy and therefore fossil fuel consumption is high due to economic growth and development efforts, and their sectoral distribution and density are not evenly distributed. According to Ponting, the demand for resources has intensified due to the increasing population in villages and cities, and a demand for new and diverse products has emerged. However, while the need for heating and food has been constantly increasing, the forests have suffered the greatest damage, and as a result of local deforestation, increased soil erosion has caused major problems. For example; In the central regions of Jordan, as a result of the damage caused by soil erosion caused by deforestation in 6000 BC, there were decreases in the yield of products and insufficient food production (Ponting, 2000:63).

On the other hand, it is known that the countries responsible for the increase in CO2 and GHG in the atmosphere are mostly developed countries. While more than 90% of CO2 emissions in industry in the world originate from North America and Asia; South America, Asia and Africa are responsible for more than 90% of CO2 emissions due to land use (Maslin, 2011:22). According to the Global Carbon Budget, China, USA, EU and India are ranked globally in terms of GHG emissions

(Global Carbon Budget, 2021). China, USA and EU are followed as the three highest emitters on a global scale. The emissions of China and India are higher than the total emissions of the USA and EU. While China produces with resources dependent on pollutant emissions in order to grow economically, it has become the second largest economy in the world (Song et al., 2018). On the other hand, China is the world's largest primary aluminum producer and in the historical process, aluminum demand has increased with concerns about sustainable growth and development. However, the aluminum industry is important for the long-term carbon neutrality target due to its contribution to the green economy (Li et al., 2022). On the other hand, according to the World Meteorological Organization (WMO), Africa experienced its hottest decade. Due to extreme weather events and climate shocks, famine, forced migration, loss of natural resources and livelihoods were experienced. Kenya planned to create a climate-resilient structure by committing to a 32% reduction in emissions by 2030. With its updated national multi-statement (NDC), it contributes to the climate fight with development options suitable for its geographical characteristics. 21% of the required funding sources for this fight will be provided domestically and 79% from international support (UN- Habitat, 2025).

In terms of all these different results, different effects and different approaches, in order for environmental problems and especially climate change to be fair, there is a need for determining certain limits for GHG emissions and CO2 and a global structural change. Global warming, which is the main cause of climate change, is included in the sharing of the atmosphere beyond the conventional resource sharing. The main determinant of this sharing is the collective structure that can only be solved fairly through compromise. Thus, when an equality-based, sufficient and fair climate protection regime is built, the North-South divide will be far from deepening (Athanasıou & Baer, 2006). Indeed, the issue of sharing the atmosphere has become a search for justice beyond borders. Therefore, governments need to focus on providing better living conditions in underdeveloped and developing regions. Such efforts will deter unplanned rapid urbanization, reinforce green urban transformation, contribute to the reduction of emission problems and strengthen the SDG13: Climate Action target. Furthermore, green areas in urban support economic growth both directly and indirectly, while economic growth enables investment in these areas. While economic growth increases the demand for urban areas and causes a decrease in green areas, the pressure of intensive construction can lead to the loss of parks and natural areas, especially in developing countries. However, if proper planning is not done, unlimited growth will threaten green areas and make climate action difficult. The decrease in green areas in cities will negatively affect the quality of life and economic sustainability there.

The new global climate regime requires a green and carbon-neutral economy, in other words, the strengthening of the renewable energy capacity of the North, and the transition of the South to a production system free of fossil fuels. Because the equitable cooperation between the rich North, which produces fossil-based products and strengthens its industrialization, and the South, which aims for sustainable development, is now a global desire. Because the poorest and the least carbon-emitting countries but the most affected by climate change are the Southern countries, while the rich Northern countries, which are the biggest polluters, have more opportunities and tools to support the South, but they remain behind in terms of responsibility. In this case, making inclusive decisions is essential to find fair and collective solutions (UNU-EHS, 2025).

Otherwise, the results indicated by Harvey are seen, and the factors of wealth, population and migration are observed as the main reasons for the unbalanced distribution between regions, and an unequal geographical development mosaic emerges. Within this structure, some economies become richer when they are rich and poorer when they are poor. While developed economies can easily obtain new areas of activity thanks to their easy access to the production factors they need, the activity volumes of other economies are negatively affected by this (Harvey, 2015:157), and the geographical development mosaic is negatively affected. Considering such difficult results, the environmental approaches of countries have strategic importance in the climate struggle.

In this context, considering that the economies experiencing climate change at a crisis level are mostly underdeveloped and developing countries and that the economic and financial capacities of these countries are limited, it is not surprising that their exposure levels are high and their adaptation capacities are low. This situation points to the concept of "climate colonialism" in the literature, which leads to environmental and economic political inequalities. For example, it is known that the US and China are responsible for approximately 40% of the world's total CO2 emissions. These emissions exacerbate climate risks, particularly in low-income countries. Floods and droughts have been observed to increase in Pakistan and Bangladesh due to global warming stemming from the high emissions of these industrialized countries (UN Climate Report, 2023).

Relationship Between Economic Growth and Environmental Performance

A meaningful introduction can be made under this heading with the question: Is economic growth an indicator of prosperity? According to Dabaieh et al., economic growth increases welfare while causing increased consumption, deterioration in resource use balance and the environment. Prioritizing environmental values and their impacts is necessary for sustainable economic growth. In order to support this, it is important to integrate global climate change policies with national and local policies and to make adaptation and transformation processes functional (Dabaieh et al., 2024). Cities are important accelerators of economic growth. They contribute to more than 80% of global GDP. Due to both agglomeration effects and positive externalities, cities contribute to economic wealth by providing tangible benefits for their residents. The structure of cities, which are always open to economic shocks and crises, strengthens the discussions on the resilience of cities. Cities with strong economic resilience have high adaptive capacities since they have resources and institutional capacity. On the other hand, cities with high economic diversity capacities are expected to have high economic resilience However, future economic growth and resilience will be sustainable when the rural-urban disparity is reduced and infrastructure gaps are closed (UN-Habitat, 2022).

An unlimited economic growth policy approach does not eliminate the restrictive effects of climate change on production and consumption, while it has a negative impact on environmental performance with its negative environmental effects. The environmental performance index measures the policies implemented in environmental protection and improvement. Applications in the context of green economy increase environmental performance while creating a reducing effect on environmental costs (Silva et al., 2024). Therefore, sustainable economic growth will be possible with a sustainable environment. The need and obligation for countries to consider the potential synergies and trade-offs between these variables is increasing day by day. As a matter of fact, there are studies that conclude that environmental quality deteriorates with economic prosperity, and in this context, the need for policies that can balance economic and environmental sustainability with development is essential. Therefore, the close relationship between economic growth and environmental quality should be taken into account (Naqvi et al., 2025) .

In the study conducted by Rivas et al., it was determined that a high Gross Domestic Product (GDP) per capita is sufficient for a high environmental performance. While some authors concluded that a high income per capita would also mean high environmental performance; some determined that an increase in income would positively affect environmental protection. In this context, it can be concluded that economies with high income per capita will have better environmental performance through institutional environmental capacities, as stated by Neumayer, will be able to undertake environmental protection more easily and will have a positive effect in terms of environmental quality. On the other hand, in developed countries where individual rights and freedoms are protected; the existence of democratic participation, equality and justice, accountability, cooperation and sustainability will positively affect environmental performance (Rivas et al., 2022) and will contribute to the climate fight.

Thinking about Inequality: The Quest for Justice and Equality in the Growth Process

Climate exposure and increasing inequality levels are a global problem and vary across regions. According to Das, environmental problems and in particular climate change research results require a heterodox perspective due to the different and disproportionate impacts that different populations face (Das, 2024)s. Differential resource access, complex/conflicting governance patterns, institutional and cultural context point to the social aspects of disproportionate distribution, while comparable physical change levels support the explanation of differential exposure (Thomas et al., 2019) .

The most powerful way to overcome climate exposure can be expressed as increasing welfare. Welfare can be measured by the degree of equality and justice in a region. Although the most fundamental indicator of economic growth is monitored as GDP, it is accepted that it is not a sufficient determinant, although it seems to provide a guarantee for consumption (Zhao et al., 2025 . Therefore, unlimited growth will not be an indicator of the common good and common welfare. However, equality is an indispensable indicator for a safe, healthy and prosperous society. While the concepts of equality, justice and solidarity constitute the essence of political and legal systems, the unjust distribution of damage to the environment has played a role in the emergence of the environmental justice movement (Harvey, 2015). Both internal and external conditions are necessary for the existence of the isonomy of almost everything (Karatani, 2018:47). However, Danilo, who compiled the latest data on global inequality, According to Zolo, " It is dramatic that in the era of globalization, it does not require much information to confirm that the sun is setting on the 'Age of Rights' ". Indeed, inequality continues to deepen with its own logic and momentum today. As Miguel Çervantes de Saavedra pointed out 500 years ago, all types of social inequality derive from the distinction between the haves and the have-nots (Bauman, 2013:49). Accordingly, the call made all over the world is to aim for ways to achieve a collective life that includes justice and equality, where everyone has the right to live equally and well (Amin, 2014:157). Zhao et al. emphasized in their study that the inclusive development approach in China can be built on welfare, harmony and sustainability. These three dimensions show the basic components of the government's effort to ensure common welfare. Common welfare and development; It creates a balancing effect on the unlimited growth goal by raising general economic standards while reducing the disparity between the rich and the poor (Zhao et al., 2025).

Without these balancing effects, it would not be correct to say that economic growth will provide common prosperity. As Zhao et al. stated, ensuring common prosperity in climate change, which is a typical category of environmental degradation, can be built on efforts such as sustainability and adaptation. In the climate struggle, 'adaptation capacity' helps to reduce the possible effects of climate change, develop adaptation mechanisms, and is different for all countries (Dabaieh et al., 2024). Otherwise, exposure levels increase and turn into a struggle for rights. In the rich North-poor South discourse, the low adaptation capacity of the South and the high level of exposure are due to the lack of resources in the poverty struggle. Indeed, GHG emissions, which are the main cause of climate change, are affected by social aspects such as poverty. In countries where the poor segment in the total population is high; environmental degradation and energy poverty are higher. Poor people live in an unsustainable way and cause the destruction of natural resources in order to sustain their vital activities. Accordingly, possible environmental degradation and poverty caused by climate change, especially in underdeveloped countries, have a highly negative impact. On the other hand, the lack of green practices in these countries is high, which leads to more greenhouse gas emissions. Since poverty has also been shown to have an impact on the distribution of GHGs. In particular, for countries with high poverty levels, unsustainable consumption and consumption patterns that reinforce environmental degradation lead to poor living conditions (Kluza et al., 2024) and an increase in the ecological footprint.

Accordingly, the result of the destruction that humans have made to the natural ecosystem is the formation of artificial ecosystems and the subsequent emergence of a built environment. For example, it has been observed that trees on the island of New Guinea were cut down and fires were started 30,000 years ago. The main purpose of destroying the forest layer is to open up space for the planting of arable and consumable plants to meet food needs (Ponting, 2000:30) . As a result, a healthier ecosystem is an indicator of a healthier planet. Protecting the ecosystem contributes directly and indirectly to the reduction of poverty and climate change.

Pushing the Boundaries or Ensuring Sustainability?

Climate change continues to expose communities to unexpected risks (Dabaieh et al., 2024). This raises concerns about achieving sustainable socioeconomic development and the livelihood of the global population, reducing the gap between rich and poor countries, and sharing natural resources (Gautier, 2014).

Establishing economic, social and environmental resilience with a sustainable development approach requires appropriate governance approaches and institutional structures. Indeed, sustainability and management are hidden levers in the struggle to stay within planetary boundaries (UNCDD, 2024). These can be monitored within the framework of a new financial sustainability framework, greener investments and climate resilience, and stronger cooperation. A sustainable and healthy planet should also offer a just future. In order for no one to be left behind while moving towards climate neutrality, a just transition for vulnerable communities is also important. In this effort, priority should be given to equality and climate action should be progressed with its social dimension. For this, it is vital to socially and financially support low-income economies facing natural resource deprivation and poverty and to expand access to low-emission options in all sectors (EU Environment and Climate, 2025).

The concept of sustainable development is a process-based narrative that describes the change aimed at meeting everyone's basic rights and needs with the discourse of (Ponting, 2000) leaving no one behind. Because no human generation is aware that they are creating very big changes. On the other hand, the fact that the problems exceed spatial scales also creates a driving effect on political-economic powers. Dynamic and uncertain socio-ecological changes require reorientation and policy adjustments (Das, 2024). Indeed, the most revolutionary result of humankind's intervention in the ecosystem has been the unlimited use of resources in order to sustain life. So much so that this situation challenges the ecosystem's ability to withstand pressures beyond its limits. Throughout history, the most fundamental goal of humankind has been to find ways to obtain its livelihood resources from various ecosystems. This has weakened the ecosystem's endurance by intervening in ecosystems and has created a challenging effect on the planet's boundaries. These interventions in natural ecosystems are the main reason for the change in the environment. On the other hand, humanity's struggle to reduce its environmental footprint due to resource use is becoming more and more difficult day by day due to the effects of increasing population and economic growth all over the world. India, China and many Southern countries with high population density and natural resource potential are in the early and middle stages of development and it will take many years for them to achieve stability in economic growth. Indeed, on the one hand, countries struggling with poverty cause negative environmental impacts at the expense of growth. Therefore, there is a clear role for rich countries to take the lead in reducing the ecological footprint (Scherer et al., 2018). As stated by Alnafrah et al., achieving the SDGs in industrial economies is possible at the expense of negative environmental impacts (Alnafrah et al., 2023).

The first principle of the Rio Declaration on Environment and Development states that "*Human beings are at the center of sustainable development. They have the right to a healthy and productive life in harmony with nature.*", and that contexts beyond the dualistic structure between humans and nature are needed. The emphasis that this structure will only be possible with sustainable development and that it is an egalitarian right with intergenerational relationships is also included in the following articles. On the other hand, the fourth article also points out that sustainable development and environmental protection are complementary processes. For this purpose, it is underlined that the least developed countries and those with high levels of exposure to environmental degradation need to cooperate with the more powerful ones in the development process. Again, the importance of local knowledge is emphasized in the 22nd article, and the importance of starting efforts at the micro level in sustainable development is revealed (UN, 1992). Similarly, in the "principles" heading of the United Nations Framework Convention on Climate Change (UNFCCC), equity, equality, common but differentiated responsibility and relative capabilities, prudence, sustainable development, cooperation is emphasized (UNFCCC, 1992), and in many policy documents, it is indirectly expressed that the world needs these values and principles with the widespread emphasis on human, nature, sustainability, equality, etc. For this reason, planning activities have shifted from economic growth and development to sustainability, and have required the integration of socioeconomic systems with ecological systems.

Because one of the biggest challenges that countries face today is the unequal distribution of opportunities, resources, income and wealth and their inadequacy. In this sense, inclusive welfare and inclusive development are still difficult to achieve today and require a choice between many different policies and institutional arrangements. While the primary goal in the inclusive welfare and development discourse is economic growth and development, the aim is to achieve common welfare. In other words, the desired result is not only economic gains but also the continuation of unlimited growth despite environmental degradation and social inequality. Instead, a balance is sought between the environment, society and economy from a long-term perspective; an inclusive welfare approach with social justice and equality, where everyone has equal opportunities, comes to the fore (Rodrik, D., & Stantcheva, 2021). Especially as Kluza et al. have also pointed out, the depth of this impasse in underdeveloped countries is remarkable, and equality and justice are at the root of rights-based discussions. In places where there is no inclusive welfare, sustainability risk, or in other words, the risk of non-financial factors such as environmental, social and governance, is monitored as a type of risk that gains importance on the economy, in other words, on financial performance. The general view of the risks of all factors is reflected in the SDGs. SDG3 (good health and well-being), SDG13 (climate action), SDG12 (responsible consumption and production) etc. have a synergistic effect. Almost all of them are positively and directly related to welfare. The interactions between certain SDGs strengthen the climate fight. The direct relationship of SDG7 and SDG13 with emissions has the strongest effect on the development of climate policy (Kluza et al., 2024).

CONCLUSION

One of the biggest problems of the 21st century and beyond is climate change, and the main duties and responsibilities of states are changing and expanding within this scope due to their intergenerational consequences. The challenges created by the Anthropocene are widely reflected in economic and ecological disruptions. The World Economic Forum (WEF) emphasizes that the world is at a tipping point and most of the planetary boundaries have been violated. For this reason, in the Anthropocene, which has become a development issue, climate struggle and governance are now discussed as a necessity economy. The necessity economy concerns the entire economy of countries and constitutes the justification for adaptation actions. Because, it leads to environmental costs that make the sustainable development paradigm difficult against economic gains. As Gautier also stated, the issue of reducing emissions, which is at the origin of the climate struggle, is a public problem that requires innovative solutions. These innovative structural solutions serve as the most basic protective function against external disruptions with their environmental recovery power (Gautier, 2014:201). In other words, climate change, which will be understood as a crisis management process as long as structural changes are not made, cannot be defined in purely economic terms, and shows that new systemic transitions are needed. Touraine's statement " The aim is to recreate social life. There is a need to fight against the power of the strongest and the richest." supports the summary of the process (Touraine, 2016). Therefore, the ' ideological climate change' that began to be observed in the early 2000s (Gıntz, 2010). It is not surprising that it requires long-term discussions such as equality and justice.

The climate struggle, which reflects the shocking economic political process with its ideological dimension, leads to a tangible dimension of social inequality in this context, and requires that approaches that originate rights-based social policies be placed on a legal basis. Indeed, economies that have to struggle more frequently and more with environmental challenges, which are mostly underdeveloped and developing ones, have to strengthen their institutional contexts while growing economically. This situation creates a need for multiple strategies simultaneously. This approach will be possible by prioritizing a fair legal system, accountability mechanisms and sustainable perspectives. In this way, sustainable sectors can be actively subsidized and subsidies that strain the boundaries of the planet can be put to an end.

However, improving industrial innovations and technological solutions in the context of green development, revising the legislative infrastructure and supporting them with new ones contribute positively to climate policies. Prioritizing R&D expenditures and long-term investments in order to limit environmental degradation, regulating environmental taxes within the scope of equality and justice will reduce environmental degradation and have a positive effect on environmental quality and environmental performance. In this context, more cities should implement climate-friendly policies. Governments should develop and strengthen carbon taxes and international trading systems to encourage greener policy options and sustain unlimited growth. All these and similar policies support those aimed at both climate adaptation and mitigation. When environmental degradation, which started with the aim of unlimited growth, is faced with a crisis in the climate, integrated solutions will be needed. Success in the process will vary depending on how economies take into account the effects of climate change. Because it is important to understand climate change and its consequences well. However, the fact that climate change has local ecological, economic and social dimensions requires the same dimensions in global solutions, therefore the crisis of climate change in the Anthropocene can only be turned into an opportunity with global cooperation. Traditional growth concerns of underdeveloped and developing economies and production and consumption patterns at the expense of the environment make the transition to a green economy difficult, while causing them to fall behind in climate commitments and negatively affect the environmental performance index. In this context, social levers play a reinforcing role. Changing consumption patterns can create conditions for more permanent change.

Considering that the environmental performance index has developed cumulatively in the historical process, it can be concluded that the way in which humanity uses the planet's natural resources will shape this accumulation and will be a determinant of environmental quality. Accordingly, humanity's changing its own tendencies and behaviors in solving its own problems is a determinant of the process. In this sense, protecting and improving the boundaries of the ecosystem in which we live will have a positive effect on reducing climate uncertainty and will make future-oriented thinking mandatory.

In this context, future shocks significantly affect the urban economy, where climate change and environmental degradation are intensely experienced, while the difference in the resilience of cities is decisive for success in this process. On the other hand, in developing and underdeveloped countries where the informal economy is strong, the sectors that are valid in the urban economy should be supported to exit the informal economy. For this, cities that are well planned and managed economically and physically will have the transformative power for the inclusive growth and sustainability of future cities. However, current and potential crises reveal the importance of the resilience of cities every time. Especially developing countries with low resilience experience rapid urbanization without structural transformations. In moving from intention to implementation, the public and private sectors should increase resilience, identify strengths and weaknesses, exploit opportunities, and embrace collective action. Such integration will facilitate both strategic and systemic structural transformations.

This situation increases the vulnerability to climate change and related crises. In this context, cities should consider economic growth and economic diversification as critical factors for their resilience and should be supported for effective climate governance structures. On the other hand, United Nations University - Institute for Environment and Human Security (UNU-EHS, 2025) defines 5 paths to a better future; Rethink Waste, Realign with Nature, Reconsider Responsibility, Reimagine the Future, Redefine Value is underlined by The Theory of Deep Change The necessity of Change has been pointed out. In this way, with tools based on eco-ecological values that take welfare and ecological balance as the origin in public policies, the planet can be benefited and a less risky future can be achieved.

REFERENCES

Alnafrah, I., Okunlola, O., Sinha, A., Abbas, S., & Dagestani, A. A. (2023). Unveiling the environmental efficiency puzzle: Insights from global green innovations. *Journal of Environmental Management, 345*, 118865. DOI: 10.1016/j.jenvman.2023.118865 PMID: 37659369

Âmin, S. (2014). *Modernity, democracy and religion: critique of culturalisms.* Yordam.

Athanasıou, T., & Baer, P. (2006). *Deadly heat.* Bgst.

Bauman, Z. (2013). *Is the wealth of the few in the interest of us all?* Ayrıntı.

Bauman, Z., & Bordoni, C. (2018). *Crisis and the state.* İthaki.

Dabaieh, M., Maguid, D., El Mahdy, D., & Al Hegazi, S. (2024). Climate change adaptation at the margins. The case of Cairo, Egypt. *Cities (London, England), 154*, 105390. DOI: 10.1016/j.cities.2024.105390

Das, S. (2024). Perspectives on inequalities in social science research and environmental sustainability in India. *Environmental Science & Policy, 161*, 103891. DOI: 10.1016/j.envsci.2024.103891

EU Environment and Climate (2025). https://commission.europa.eu/strategy-and-policy/priorities-2019-2024/european-green-deal_en

Frieswijk, K., Zino, L., Morse, A. S., & Cao, M. (2025). A behavioural–environmental model to study the impact of climate change denial on environmental degradation. *Physica D: Nonlinear Phenomena, 476*(June 2024), 134648. https://doi.org/DOI: 10.1016/j.physd.2025.134648

Gautier, C. (2014). *Oil, water and climate.* Cambridge University Press.

Gıntz, C. (2010). *Elsewhere & Otherwise.* Dost.

Global Carbon Budget. (2021). https://www.globalcarbonproject.org/carbonbudget/

Hamed, L., Dhaouadi, L., Zehri, F., Tiba, S., Besser, H., & Karbout, N., & EMARA, E. I. R. (2024). Examining the relationship between the economic growth, energy use, CO2 emissions, and water resources: Evidence from selected MENA countries. *Journal of the Saudi Society of Agricultural Sciences, 23*(6), 415–423. DOI: 10.1016/j.jssas.2024.04.002

Harvey, D. (2015). *Seventeen contradictions and the end of capitalism.* Sel.

Hashim, B. M., Alnaemi, A. N. A., Hussain, B. A., Abduljabbar, S. A., Doost, Z. H., & Yaseen, Z. M. (2024). Statistical downscaling of future temperature and precipitation projections in Iraq under climate change scenarios. *Physics and Chemistry of the Earth Parts A/B/C*, *135*, 103647. https://doi.org/https://doi.org/10.1016/j.pce.2024.103647

Karatani, K. (2018). *Isonomia and the origins of philosophy*. Metis.

Kluza, K., Ziolo, M., & Postula, M. (2024). Climate policy development and implementation from the Sustainable Development Goals perspective. Evidence from the European Union countries. *Energy Strategy Reviews, 52*, 101321. https://doi.org/https://doi.org/10.1016/j.esr.2024.101321

Li, S., Niu, L., Yue, Q., & Zhang, T. (2022). Trajectory, driving forces, and mitigation potential of energy-related greenhouse gas (GHG) emissions in China's primary aluminum industry. *Energy*, *239*, 122114. https://doi.org/https://doi.org/10.1016/j.energy.2021.122114

Maslin, M. (2011). *Global warming*. Dost.

Naqvi, R. A., Almohsen, B., & Sohail, A. (2025). Modeling the environmental Kuznets Curve: A stochastic approach using economic and climate data. *Journal of Environmental Management*, *373*(January), 123108. DOI: 10.1016/j.jenvman.2024.123108 PMID: 39752958

Persakis, A., Tsakalos, I., Gkonis, V., & Nerantzidis, M. (2024). Climate policy uncertainty and environmental degradation: Does democracy moderate this relationship? *Cleaner Environmental Systems*, *15*, 100230. https://doi.org/https://doi.org/10.1016/j.cesys.2024.100230

Ponting, C. (2000). *A Green history of the world: The environment and the collapse of great civilizations* (Sander, A. B., Trans.). Sabancı University.

Rivas, S., Urraca, R., Palermo, V., & Bertoldi, P. (2022). Covenant of Mayors 2020: Drivers and barriers for monitoring climate action plans. *Journal of Cleaner Production*, *332*, 130029. https://doi.org/https://doi.org/10.1016/j.jclepro.2021.130029

Rodrik, D., & Stantcheva, S. (2021). A policy matrix for inclusive prosperity. *National Bureau of Economic Research.*, w28736.

Scherer, L., Behrens, P., de Koning, A., Heijungs, R., Sprecher, B., & Tukker, A. (2018). Trade-offs between social and environmentally Sustainable Development Goals. *Environmental Science & Policy*, *90*, 65–72. https://doi.org/https://doi.org/10.1016/j.envsci.2018.10.002

Silva, N., Fuinhas, J. A., Koengkan, M., & Kazemzadeh, E. (2024). What are the causal conditions that lead to high or low environmental performance? A worldwide assessment. *Environmental Impact Assessment Review, 104*, 107342. https://doi.org/https://doi.org/10.1016/j.eiar.2023.107342

Thomas, K., Hardy, R. D., Lazrus, H., Mendez, M., Orlove, B., Rivera-Collazo, I., Roberts, J. T., Rockman, M., Warner, B. P., & Winthrop, R. (2019). Explaining differential vulnerability to climate change: A social science review. *Wiley Interdisciplinary Reviews: Climate Change, 10*(2), 1–18. DOI: 10.1002/wcc.565 PMID: 31007726

Touraine, A. (2016). *After the crisis*. Yapı Kredi.

UN-Habitat. (2022). *Envisaging the future of cities*. World Cities Report.

UN-Habitat. (2025). https://www.linkedin.com/company/un-habitat-united-nation-human-settlements-programme-/posts/?feedView=all

UNCDD. (2024). https://www.unccd.int/resources/reports/stepping-back-precipice-transforming-land-management-stay-within-planetary

UNDRR. (2022). https://www.undrr.org/our-impact/campaigns/no-natural-disasters

United Nations University Institute for Environment and Human Security. (UNU-EHS) (2025). https://unu.edu/ehs/article/5-routes-better-future-turning-over-new-leaf

WEF. (2025). https://www.weforum.org/stories/2025/03/cities-global-trade-changing-seize-moment/

Zhao, C., Kong, L., Zhong, C., & Ying, N. (2025). Equality in the water: How does water ecological civilization city policy enhance inclusive development in China? *Cities (London, England), 156*, 105555. https://doi.org/https://doi.org/10.1016/j.cities.2024.105555

KEY TERMS AND DEFINITIONS

Anthropocene: Anthropocene is a new geological period and the Earth is being shaped and changed depending on people's behavior and tendencies.

Climate Change: The changes in the Earth's climate system caused by global warming resulting from greenhouse gas emissions and pollutant emissions.

Climate Justice: Climate justice is an environmental and socio-economic concept. It is the fair distribution of the costs and benefits of climate change for highly vulnerable communities.

Environmental Performance: Environmental performance refers to the extent to which economies manage the environment through sustainability, resource use, emissions, etc.

Socio-economics: Socio-economics is the dialectical study of social and economic events.

Sustainability: Sustainability is the fundamental principle required for a livable world in terms of environmental, economic and social aspects from an intergenerational perspective.

Chapter 4
Environmental Debt of Developed Countries:
The Ecological Injustice of Growth Policies

Muhammad Usman Tariq
https://orcid.org/0000-0002-7605-3040
Abu Dhabi University, UAE & University College Cork, Ireland

ABSTRACT

The concept of environmental debt highlights the disproportionate ecological impact that developed nations have on the planet, stemming from their centuries-long history of industrialization and unsustainable development. While affluent nations have gained economic advantages from resource extraction and carbon-intensive development, the environmental costs such as biodiversity loss and climate change are frequently shifted to poorer nations. This chapter examines the ecological injustices inherent in global growth policies, focusing on how historical emissions, land degradation, and excessive consumption in the global North contribute to the erosion of ecological resilience and hinder development prospects in the global South. Using environmental justice frameworks, the chapter examines how accountability and reparative actions can be achieved, such as through climate finance, technology transfers, and policy reforms that promote fairness in sustainability.

INTRODUCTION

The concept of environmental debt implies the total damage caused to the ecosystems and natural resources of the earth by the unsustainable actions of developed economies, leading to resource exploitation and environmental degradation,

DOI: 10.4018/979-8-3373-1862-2.ch004

mainly in developing countries. This concept emphasizes the inequity in which wealthy countries, because of centuries of overconsumption, unsustainable resource exploitation, pollution, and land grabbing, have exceeded their fair share of the earth's ecological ability to sustain life. Environmental debt can be traced back to the colonial era, when resource-wealthy parts of Asia, Africa, and the Americas were openly plundered of raw materials without much thought to environmental conservation and the long-term viability of usage. Such predatory practices continued even in the industrial age to the modern era of globalization, where the relatively affluent countries have the benefits of modernity and technology. Meanwhile, poorer nations are left to deal with the ecological impact. There is a wide disparity between developed and developing nations regarding the use of resources and their effects on the environment. Developed countries comprise only a small percentage of the total world population but contribute the largest percentage of world resource consumption, energy consumption, and all-time greenhouse gas emissions. For example, nations such as the United States, Canada, and most of Western Europe use much more resources per capita than a country in Africa, South Asia, or Latin America. These consumption patterns are supported by complicated global supply chains that offload environmental degradation into weaker countries, such as the mining of rare earth metals in the Congo, deforestation of soy and cattle in Brazil, or textile manufacture in arid parts of South Asia. At the same time, these emerging countries often have weak environmental policies and little political clout, making them more vulnerable to environmental exploitation. They also face the impact of climate change, including rising sea levels, droughts, floods, and loss of biodiversity, even though they are the least contributors to the causes of climate change (Ahmad et al., 2024). Environmental debt raises ethical questions of huge relevance in the joint global discourse on development issues. This runs counter to the popular perception that environmental stewardship is a shared responsibility, and there are differences in duties as well as historical responsibility. Environmental debt is not only a matter of science or economics; it is a matter of morality. Developed countries have acquired environmental benefits at the expense of people who have suffered negative effects, including contaminated water and air, evicted community members, and loss of life diversity. There is a need to redress this imbalance through meaningful actions: debt cancellation on the basis of environmental degradation, equitable climate financing, technology transfer, and the provision of a voice to local and indigenous communities in environmental decision-making. Considering environmental debt is important for any serious vision of sustainable development. It is imperative to shift from a growth model that wins at the expense of resources to one based on equity, environmental recovery, and respect for planetary boundaries.

HISTORICAL ROOTS OF ENVIRONMENTAL EXPLOITATION

The historical origins of environmental exploitation are deeply ingrained in the legacies of resource extraction from colonial and post-colonial times, which have significantly influenced the global economic systems and environmental disparities that persist to this day. During the colonial period, European imperial powers implemented exploitative systems in Asia, Africa, and the Americas, motivated by the desire for raw materials, agricultural goods, and labor to support the rapid industrialization of their home countries. These colonial ventures were not merely political conquests; they were economic endeavors driven by the conquest of the environment. Natural resources were systematically extracted, indigenous land management systems were dismantled, and ecosystems were altered to meet the needs of distant cities. This exploitation of resources and degradation of the environment in the Global South laid the foundation for what is now known as environmental debt, where the environmental damage and depletion in the Global South funded the economic growth of the Global North (Aladekomo, 2025).

The colonialists constructed infrastructure of extraction, such as railways, ports, and mono-plantations, whose sole purpose was to transfer resources such as rubber, cotton, and precious metals to Europe. This has transformed the entire ecology. In India, the cash-crop policy has substituted various growing systems, degraded soils, and increased the risk of famine. King Leopold II in the Congo slaughtered wildlife and caused the deaths of millions of people through forced labor on rubber and ivory. In Latin America, Spanish and Portuguese colonization disrupted indigenous farming methods and initiated large-scale silver and gold mining operations, resulting in long-lasting mercury pollution and extensive deforestation, which continue to affect the region today. The environmental burdens imposed on the Global South to support economic development in the Global North are not limited to the past (Dafermos, 2025). Post-colonial development, influenced by the institutions and trade systems established during colonization, continued to reinforce the role of many nations in the Global South as suppliers of raw materials within a global capitalist system. The economies that emerged after colonization have resulted in lasting environmental damage, where landscapes have been irreversibly changed, and local communities have been denied the advantages of the resources extracted from their lands. Nigeria's Niger Delta serves as a striking example, as multinational corporations have been extracting oil from the region since the 1950s, leading to severe environmental degradation (Tariq, 2025). The regular occurrence of oil spills, gas flaring, and water contamination has severely damaged biodiversity and public health. Despite the economic advantages, most benefits have gone to foreign corporations and national elites. Simultaneously, local communities receive minimal compensation and face repression for their environmental activism. Similarly, in the Amazon rainforest, extensive logging, mining, and agricultural expansion, much of it to meet the demand for soy and beef in global markets, have resulted in deforestation at an alarming rate in recent years. Despite the rhetoric surrounding rainforest conservation, global consumption patterns continue to exert pressure on this ecologically significant region. The negative consequences of these exploitative practices extend to marine ecosystems. The excessive fishing practices of colonial and post-colonial industrial fleets have resulted in the depletion of fish populations in West African waters, as these fleets from Europe and Asia continue to exploit resources. As the catches of local artisanal fishers decline, they struggle with food insecurity. Simultaneously, wealthier nations enjoy the affordability of seafood imports. This reflects a larger pattern of resource frontiers, where peripheral regions become hubs of intense extraction to sustain the consumption and growth patterns of the global North. It is particularly troubling that the Global South is being held responsible for environmental conservation under the guise of global climate and conservation agreements, despite the ongoing consequences of historical exploitation (Alenda and Kaufmann, 2024). Figure 1 illustrates how past exploitation continues to shape contemporary inequities.

Figure 1. Historical Roots and Continuity of Environmental Debt

MECHANISMS OF ENVIRONMENTAL DEBT CREATION

Environmental debt is caused by the economic actions of already developed countries, which have caused environmental pollution and depletion of the available resources, all of which have impacted mainly the welfare of developing nations. This conception underlines the fact that richer countries have prospered at the expense of carbon-emitting industries, resource extraction, and overconsumption, shifting the production cost to poorer countries, ecosystems, and future generations (Althouse and Svartzman, 2024). The build-up of environmental debt is an outcome of numerous intertwined processes, such as overconsumption and unsustainable resource production, the sweeping impacts of carbon, deforestation, biodiversity loss, and sheer inequalities in trade politics and global supply chains.

One of the leading causes of environmental debt is excessive consumption and unsustainable resource extraction in developed countries. The international consumption of global resources is excessive in affluent countries. Economies that focus on growth consume a lot of energy, water, minerals, and food. In North America and Europe, which account for less than 20 percent of the total world population, the consumption and production of materials and emissions are significantly higher than in all other regions (Alvarez-Veliz and Barton, 2024). Satisfying this demand requires massive raw material harvesting, frequently in vulnerable environments in the impoverished world. The relentless quest for inexpensive raw materials has led to extensive mining, deforestation, water diversion, and land degradation in Africa, Asia, and Latin America. These extractive practices harm ecosystems, deplete local resources, and yield minimal benefits to host communities in terms of sustainable development and environmental restoration.

The commercialization of nature, where forests, rivers, and minerals are primarily valued for their market potential, reinforces exploitative logic. Natural capital is often overlooked in global markets, leading to its overexploitation. The ecological costs of depletion, pollution, and habitat loss are not accounted for in the pricing of goods and services, resulting in what economists call negative externalities. These externalities disproportionately affect communities that depend directly on their environment for survival, such as Indigenous populations and smallholder farmers. By exceeding the planet's ecological limits, developed nations essentially "borrow" resources from other countries and future generations without any intention of repaying them, a defining characteristic of environmental debt (Auteri et al., 2024). Another mechanism is the buildup of carbon emissions, deforestation, and biodiversity loss, all of which contribute to global ecological imbalance. The Industrial Revolution triggered a rapid increase in greenhouse gas emissions, with industrialized nations such as the United Kingdom, Germany, and the United States being responsible for most of the historical carbon emissions. These emissions accumulate in the atmosphere and are primarily responsible for the current climate crisis. Although emerging economies have seen a rise in carbon emissions over the years, the per-capita carbon footprint of developed countries remains significantly greater. Additionally, a significant portion of emissions in the Global South can be traced back to the production of goods for consumption in the North, prompting discussions on how emissions should be allocated in a globally interconnected economy (Tariq, 2025). Most of the deforestation in tropical areas is caused by the global demand for commodities, such as beef, soy, palm oil, and timber products, which are primarily consumed in wealthy markets. When rainforests are cleared for farming or logging, local communities lose their means of income, global biodiversity declines, and carbon emissions increase. Biodiversity loss, often disregarded as a factor in environmental debt, exacerbizes ecosystem vulnerability and diminishes nature's ability to recover from climate-related challenges (Aydin and Degirmenci, 2024). These ecological processes are closely linked to trade policies and global supply chains, which contribute to the perpetuation of the environmental debt.

Globalization has caused the world to be highly interdependent in terms of the economy, with production being spread over various regions and, in some cases, continents away. It has given corporations the freedom to use cheap labor and lax environmental standards in the developing world and at the same time maintain the burden of environmental degradation in consumer countries. Deregulated trade and trade agreements that promote free market access and liberalization of investment and trade policies ensure that raw materials can be extracted and exported in developing countries with little environmental control policies. As a result of this structural imbalance, most countries possessing rich natural resources yet lagging in the economic development field are compelled to participate in activities that are environmentally degrading in international trade, supplying basic raw materials and cheap labor, and importing pollution, waste, and environmental degradation. Supply chains in any part of the world are usually full of secrets, making it difficult to know the full environmental impact of any product as it passes from its origin to the shop (Bakkar et al., 2024). Also, global trade agreements tend to include clauses that restrict government in the global South to impose environmental regulations because they fear adverse implications to foreign investment or breach of investor-state dispute settlements. These measures may produce impediments to safeguarding the natural environment of a nation, strengthening its dependency on foreign aid and debt servicing. Even the most benign technologies in terms of environmental impact, such as electric cars and solar panels, rely heavily on rare earth metals and minerals, which are mined in an environmentally and socially destructive manner in the Global South. This means that even a supposedly great green transition has the potential to reproduce exploitative processes unless managed justly. Environmental debt is also manifested in ecological dumping, whereby developed countries dump toxic waste, obsolete technologies, and carbon-intensive industries in less developed countries. Ecological risk transfer disproportionately affects communities with fewer resources to mitigate environmental damage and creates a system of economic-resource privilege for exposure. These actions undermine the sovereignty and sustainable development of developing countries, which continue to be subjected to a system in which the poorest and most vulnerable populations are the most unequally exposed to environmental hazards (Tariq, 2025). Only through a detailed reshaping of the current systems can one comprehend how environmental debt is created. Enforcement that promotes circular economies, diminishes dependence on long-haul supply chains, and includes environmental costs in pricing systems are essential elements for a sustainable future (Bohoslavsky et al., 2025). International cooperation must shift its organizational patterns of support from superiority to restructuring towards the principles of even-handedness, redressing, and environmental accountability. The real meaning of sustainability must entail the environment as well as the historical injustices that are ingrained in the economic systems worldwide. The prospect of redefining prosperity to include well-being, ecological balance, and intergenerational equity alongside rather than just GDP growth offers a way out of the situation that will consider the ecological debt owed to people and the planet.

INEQUITIES IN CLIMATE RESPONSIBILITY AND ACTION

The climate crisis is a topic that is always discussed in the context of collective responsibility for the earth. However, this view fails to acknowledge that the causes and effects of environmental damage are unevenly distributed. The reality is that climate change is a global problem, but its consequences have an unequal distribution of causes and outcomes. It is mostly poorer countries with historically little contribution to climate change that are most affected and barred by natural disasters, resource shortages, and failures in development (Corsi et al., 2024). At the same time, rich industrialized countries that have historically enjoyed the rewards of carbon-intensive development have and continue to exercise considerable leverage in shaping climate governance. This asymmetry is further compounded by other systems, such as carbon markets and offset programs, which, despite being intended to promote the negative emissions imperative, tend to reinforce these institutional inequities. To address climate change, the authors argue that technical solutions must also be accompanied by an in-depth analysis of the historical, economic, and geopolitical aspects that can also upturn environmental injustice (Tariq, 2025).

The SIDS, sub-Saharan African countries, and parts of South and Southeast Asia have been witnessing increased extreme weather conditions, creeping elevation, chronic droughts, and food insecurity to an extent that far surpasses their emissions rate (Dafermos, 2025). For example, Bangladesh, Vanuatu, and Mozambique have emitted an insignificant amount of greenhouse gases (GHG) worldwide. Nevertheless, they are at the receiving end of cyclones, floods, and other climate-induced disasters. These effects increase pre-existing socioeconomic weaknesses, such as poverty, poor infrastructure, health disparity, and political instability. Climate vulnerability can also be discussed at the social level of communities, in addition to the physical level of places. In different countries, disadvantaged communities, such as Indigenous peoples, women, people living in rural areas, and those with low incomes, experience higher levels of environmental risks because of systemic inequality and their exclusion from decision-making processes. Their means of sustenance, which are frequently reliant on natural resources and land, are significantly endangered by climate disruptions. However, they are seldom prioritized in national adaptation plans. This internal inequality reflects global disparities, perpetuating patterns of disenfranchisement that extend from the local to the international level (Elmhirst and Resurreccion, 2025; Gifford and Sauls, 2024).

In addition to these failures, the role of carbon markets and offsets is problematic, as they are intended to provide flexible ways to reduce emissions but often end up reinforcing existing inequalities (Hasan, 2025). Carbon markets enable countries or companies to purchase carbon credits from emissions-reducing projects in developing nations, often located outside their borders, to compensate for their emissions. In theory, this approach encourages the development of low-carbon solutions and helps achieve global targets while minimizing costs. Carbon offset schemes have sparked significant concerns regarding their environmental impact, transparency, and fairness towards marginalized communities. Many offset projects, such as forest conservation, biofuel plantations, and renewable energy installations, have been implemented without the informed consent of local communities, leading to land grabs, displacement, and violations of Indigenous rights. These projects often prioritize carbon accounting over ecosystem integrity and community well-being, resulting in what critics call "green colonialism." Additionally, by allowing polluters in the global North to continue emitting while paying for reductions elsewhere, offset mechanisms enable the externalization of environmental responsibility and delay systemic transitions toward decarbonization (Hay, 2025). One challenge is the accurate measurement and verification of the amount of carbon credits being traded. Many projects depend on speculative baselines and unverified assumptions about emissions that would have occurred without any intervention (Herrera, 2024). This results in inflated crediting and the issuance of "phantom offsets" that do not correspond to real reductions in emissions. Additionally, the emphasis on cost-effectiveness promotes investments in countries or sectors where cost reductions are most affordable rather than where they are most crucial for justice or for sustainability.

Therefore, carbon markets can potentially commoditize climate action without bringing the required systemic changes for a fair and equitable transition. In response to these critiques, carbon markets have been strengthened, and the employment of so-called high-integrity offsets with high levels of social and environmental standards has been encouraged. Unfortunately, these standards are not always implemented and do not address the inherent imbalance between emitters and offset hosts. Moreover, the persistence of offsetting as one of the key measures in terms of climate neutrality taken by many corporations and governments adopting the strategy of net zero raises some ethical considerations about responsibility and accountability (Hossen et al., 2025). To resolve these inequalities, climate policy should be based on a justifiable approach that emphasizes historical accountability, equity of delivery, and participatory decision-making. This means not only taking responsibility for the past emissions of industrialized countries but also appreciating the structural chains that prevent vulnerable nations from pursuing low-carbon growth strategies (Islam, 2024). International climate finance needs to be shifted to grants, focus on adaptation and resilience, and be easily accessible to marginalized communities without undue bureaucracy. The agenda should incorporate debt relief and reparation for climate loss and damage, which will not be charity but an issue of justice. In addition, climate action must be inclusive and equitable, with those most affected prioritized. Non-Western knowledge systems, feminist economics, and grassroots environmental movements offer an alternative representation of sustainability that criticizes technocratic and market-based approaches (Elmhirst & Resurrección, 2025). By including such perspectives in climate policy, we will be able to expand our understanding of resilience and create links between various countries and movements.

MEASURING ENVIRONMENTAL DEBT: TOOLS AND INDICATORS

Assessing environmental debt is crucial for evaluating the consequences of growth policies, determining fair obligations among nations, and formulating reparative and restorative approaches to address these debts. Nevertheless, creating trustworthy, all-encompassing, and morally sound measures to assess environmental debt poses substantial methodological and political challenges (Kaika et al., 2024). Despite these challenges, a combination of ecological footprint analysis, environmental input-output models, planetary boundary frameworks, and adjusted national accounting systems has provided the foundation for an emerging toolkit that can assess and compare environmental debt among nations. The ecological footprint is a widely recognized tool for measuring an individual's or a country's environmental impact. In the 1990s, Wackernagel and Rees created the ecological footprint, a tool that quantifies the amount of land and sea required to sustain a population's resource consumption and waste management, especially carbon emissions. This metric enables a comparison of countries' resource usage with respect to the global capacity of natural resources. Countries that consume more resources than they can sustainably produce are referred to as being in ecological deficit, relying on imports and bearing the environmental consequences of their actions (Kaklauskaite and Streimikienė, 2024). Wealthy nations consistently exhibit the largest ecological footprints per capita, underscoring their significant contribution to global environmental debt. However, many poor nations remain ecologically surplus. However, they always tend to suffer the negative effects of global environmental degradation.

The GPI incorporates negative externalities, such as pollution, declining resources, and income inequality, into the estimation of economic growth and provides more accurate measures of sustainable welfare. The World Bank considers natural resource depletion and environmental degradation essential to national savings estimations, which serve as an effective measure for assessing intergenerational equity and environmental liabilities (Kallis et al., 2025). Countries that regularly record negative values for their natural capital are losing their natural resources and creating environmental debt. Carbon accounting is vital for evaluating environmental accountability in the climate domain. The greenhouse gas (GHG) protocol identifies three scopes of emissions: direct emissions, indirect emissions caused by energy consumption, and all other indirect emissions, including those of the supply chain (Lang, 2024). Most developed countries show a lower level of emissions measured by the geographic area occupied by them but are much higher when measured based on the consumption pattern. Consumption-based accounting is a crucial tool for assessing environmental debt because it assigns emissions to the end consumer rather than the producer. This redistribution of responsibility emphasizes the environmental burden imposed by developed nations on the Global South through international trade.

Economists and environmental scientists have devised techniques for quantifying environmental debt in monetary terms through natural-capital accounting and environmental valuation. Examples include shadow pricing for ecosystem services, cost-benefit analysis of pollution reduction, and damage cost estimates for climate-related events (Liu et al., 2024). Although monetizing nature is a contentious topic, employing these methodologies can be politically advantageous for conveying the magnitude of ecological damage and shaping corrective fiscal measures. Nevertheless, the process of determining a company's value is constrained by cultural, ethical and scientific uncertainties. Not all natural environments can be easily converted into marketable commodities, and attempts to assign a monetary value to nature may undermine its inherent worth, beyond economic considerations. Despite the progress made in metrics, several limitations and challenges remain in evaluating environmental debt (Moyo, 2024). First, data accessibility and reliability are inconsistent across countries. Numerous developing countries face limited institutional capacity and financial resources, making it challenging to conduct comprehensive environmental monitoring and reporting. The lack of standardized methodologies, outdated datasets, and inadequate disaggregation impede the ability to compare assessments and diminish their reliability. Organizations such as the United Nations Environment Programme and Global Footprint Network play a vital role in promoting standardized reporting systems for national efforts; however, a lack of global cooperation and funding hinders their effectiveness. Environmental debt is complex and situation-specific in nature. Although it is possible to measure carbon emissions and resource extraction, it is challenging to quantify other factors, such as biodiversity loss, cultural displacement, or the spiritual connection people have with the land, using traditional methods. Indigenous knowledge systems and relational ontologies offer distinct perspectives on environmental harm, challenging the notion of quantification while being crucial for achieving environmental justice (Althouse & Svartzman, 2024). A metric-only approach to environmental debt could overlook the diverse knowledge systems and perspectives of Indigenous communities, thereby reinforcing existing power imbalances and colonial legacies.

Additionally, it is crucial to consider the political ramifications of environmental debt metric. Information on ecological footprints, material flows, and climate responsibility can reveal disparities that question the fairness of existing global governance systems. Nations burdened with significant environmental debt may reject or manipulate environmental assessments or data to evade responsibility. Conversely, developing nations may not possess the necessary political influence to advocate for reparations, even when they are supported by empirical evidence (Tariq, 2025). Consequently, translating environmental debt measurements into practical policy measures remains inconsistent and subject to debate. Fourth, the current indicators do not adequately capture intergenerational effects. Environmental debt encompasses not only the current ecological overshoot but also the responsibilities owed to future generations (Hasan, 2025). Over extended periods, such as decades or centuries, long-term issues such as soil degradation, ocean acidification, or glacial melting occur, making it challenging to assign responsibility using time-specific metrics. Incorporating foresight and precautionary principles into environmental accounting is crucial to ensure that debt evaluations are not overly focused on short-term factors (Álvarez-Véliz & Barton, 2024).

CASE STUDIES: ENVIRONMENTAL DEBT IN ACTION

Environmental debt becomes evident in real-life situations in which ecological damage in developing regions supports economic growth and consumption in developed countries (Rastegar et al., 2024). These case studies illustrate how the systematic extraction of natural resources and environmental degradation results in long-term ecological costs that are disproportionately borne by vulnerable populations. The first case study focuses on deforestation in Brazil for Agricultural Exports. The Amazon rainforest, commonly known as the 'lungs of the earth,' has experienced significant deforestation, primarily caused by the expansion of soybean farming and cattle ranching. These activities primarily target the export markets of

North America, Europe, and China. Between 2000 and 2020, a staggering number of hectares of forest were destroyed, resulting in the release of enormous quantities of stored carbon into the atmosphere. This change in land use not only significantly increases global carbon emissions, making Brazil one of the top emitters during deforestation peaks, but also disrupts regional rainfall patterns and biodiversity (Tunn et al., 2024). The carbon released from deforestation in Brazil is embedded in goods exported to other countries, illustrating how environmental costs are shifted to other regions of the world. Indigenous communities in the region face displacement, loss of their traditional ways of life, and heightened vulnerability to environmental hazards. Simultaneously, the benefits primarily accrue to agribusinesses and international markets. This is a clear example of environmental debt: the ecological costs of development and consumption in the global North are transferred to the ecosystems and populations of the global South.

The second case study examines the water scarcity and pollution resulting from mining activities in Africa. The extraction of valuable metals and rare earth elements, which are fundamental to the production of electronics, batteries, and sustainable technologies, is a significant contributor to water scarcity and pollution in some African regions (Okereke, 2024). Countries such as Ghana and Zambia have extensive mining operations for cobalt, copper, and gold, which are often controlled or influenced by foreign corporations. These industries employ vast amounts of water, release harmful substances into rivers and soils, and disrupt local water access for farming and drinking. For example, the DRC is responsible for supplying more than 60% of the world's cobalt, a crucial material for smartphones and electric vehicle batteries. However, mining communities in the DRC suffer from contaminated water sources, dangerous working conditions and economic exclusion (Islam, 2024). While the minerals used in digital and clean energy technologies drive advancements in developed countries, their environmental impact is concentrated and significant in these regions. This example illustrates how global economic systems generate environmental debt by extracting resources that benefit consumers who are not directly affected by environmental consequences.

The third case study focuses on oceanic degradation caused by industrialized fishing. Industrial-scale fishing, primarily driven by fleets from wealthy nations, has devastated marine ecosystems, particularly in the waters off Southeast Asia, West Africa and the South Pacific. Fishing has caused collapsing fish stocks and degraded marine biodiversity due to overfishing, bycatch, and other destructive practices such as bottom trawling. In Senegal, foreign fishermen overexploit fish stocks on which coastal communities depend to gain access to food and earn money (Pearse et al., 2025). These communities are enduring dwindling harvests, increasing food prices, and social displacement; thus, some of them must migrate. Most of the catch is sold to European or Asian markets. This is oceanic degradation and an example of environmental debt because marine resources are used in faraway markets. Meanwhile, local communities and ecosystems are damaged beyond repair. These examples explain why there is a need to ensure global accountability systems and fair environmental governance to address the unfair costs of environmental debt.

POLICY SOLUTIONS FOR ADDRESSING ENVIRONMENTAL DEBT

To address environmental debt, the unresolved ecological harm exposed to industrialized countries and, at a predominant level, to the Global South, requires policy interventions that are deliberate and holistically designed. Such efforts must not only be aimed at overcoming the damage of the present but also focus on trying to undo historical injustices. The ecological and economic imbalance is that the history of colonial exploitation, unfair trade practices, and unsustainable growth policies have favored developed countries at the cost of developing nations and increased their vulnerabilities. In relation to this, the fair sharing of environmental burdens, purposeful foreign development assistance, and executing green and restorative fiscal policies are the apt avenues towards the realization of ecological justice at the global stage (Pearse et al., 2025). Figure 2 provides an overview of the policy solutions framework.

Figure 2. Policy Solutions Framework

The first point in addressing environmental debt is to have a reasonable allocation of environmental responsibilities to developed and developing countries. This is the common but distinguished responsibilities and respective capabilities principle (CBDR-RC), which is incorporated in the United Nations Framework Convention on Climate Change (UNFCCC). This principle recognizes that, although every nation can do something to solve environmental problems, their past experiences and present capacities differ widely (Rastegar and Becken, 2024). Significant implementation of the CBDR-RC would involve industrialized nations making binding commitments to reduce emissions that are significantly higher than those required of developing nations, particularly the least-developed countries (LDCs). Furthermore, these responsibilities should not be confined to the reduction of emissions. However, they should also encompass other crucial aspects such as biodiversity conservation, ocean pollution control, and ecosystem restoration. Wealthy nations must also allocate resources for climate adaptation and resilience-building initiatives in developing countries. It encompasses assistance for early warning systems, coastal protection measures, reforestation efforts, sustainable farming practices and effective disaster management strategies (Wijburg, 2025). The focus should be on transitioning from charity-driven environmental diplomacy to legally binding climate reparations that accurately account for the unequal ecological harm inflicted by the wealthier nations.

To achieve a fair distribution, the environmental impact of global supply chains must be considered. Developed nations often shift their environmental burdens by importing resource-intensive products from developing countries while presenting themselves as environmentally conscious within their borders. Policy measures such as border carbon adjustments, environmental tariffs, and fair-trade certifications can play a role in internalizing these costs and encouraging more sustainable trading practices. Nevertheless, these tools must be precisely calibrated to prevent them from being used as instruments of protectionism that unfairly burden developing nations (Smith & Ely, 2025). Instead, agreements between multiple countries should focus on helping producers in low-income nations develop cleaner production technologies, obtain environmentally friendly certification, and gain fair access to green markets. The second significant aspect of tackling environmental debt is reconsidering the purpose and impact of international development assistance. Traditional forms of aid have frequently supported extractive practices, provided temporary economic aid, reinforced reliance, and caused environmental harm (Moyo, 2024). To promote ecological justice, aid distribution should be redesigned to prioritize sustainability, resilience, and self-determination in countries burdened by debt. Development aid should focus on allocating resources to renewable energy infrastructure, climate-resilient agriculture, reforestation efforts, water resource management, and waste reduction systems (Salah and Ament, 2025).

Climate finance systems, such as the Green Climate Fund (GCF), must be substantially increased and distributed effectively. Delays, bureaucratic hurdles, and donor-driven conditions have hindered the effectiveness of current climate finance (Herrera, 2024). Additionally, developed countries have consistently fallen short of their climate finance commitments, including the widely publicized $100 billion per year pledge made under the Paris Agreement. To ensure that international development aid effectively tackles environmental debt, it must be separated from geopolitical interests and redirected towards achieving long-term sustainability goals. Grant-based financing should be favored over loan-based approaches to prevent exacerbating the sovereign debt burden of developing nations. To ensure the success of green projects, blended finance models that involve private investment should be transparent, fair, and responsible to the communities they serve.

Debt-to-nature swaps or ties entwine the alleviation of debt and preservation. Writing off part of a country's debt is in exchange for investment in safeguarding the environment (Aladekomo, 2025). Examples may include marine programs in Ecuador and the Galapagos program or marine programs in Seychelles. These swaps do not eliminate the risks of debt, but they can keep ecosystems going. They flourish on open government, checks and balances, and the participation of locals. The successful implementation of these initiatives would largely require the establishment of clear governance frameworks, independent oversight, and community involvement. The third and final pillar of policy solutions is based on the incorporation of green fiscal measures and reparative policies to undo historical injustices and spur sustainable behaviors. Some green fiscal policies include environmental taxes and carbon pricing, subsidies for clean energy, and fossil fuel investment divestment (Liu et al., 2024). Such mechanisms can raise funds to finance climate action, shape consumer behavior, and increase corporate responsibility. However, their strategy must not harm low-income communities in any way; therefore, it must be fair and equal (Aydin & Degirmenci, 2024). For example, fuel taxes can harm rural and informal labor, which may require subsidies or investment in facilities to encourage the use of public transport.

By way of compensation, climate justice models are currently placing an increased focus on the need to formally admit to ecological harm and put in place a mechanism of compensation. This may take the form of climate reparations, where nations or corporations that have contributed significantly to environmental damage provide economic compensation to the impacted communities (Salah & Ament, 2025). Reparative justice includes restitution of land, support for communities affected by the project, and protection of environmental activists, who are usually met with violence and criminalization. Legal and institutional arrangements must be in place to incorporate ecological justice into policymaking. On the international front, many have advocated for the formation of an international environmental court to deal with incidents of transboundary environmental degradation and hold perpetrators of pollution accountable (Bohoslavsky et al., 2025). Internationally, other nations, including Ecuador, South Africa, and Norway, have used their constitutions as domestic methods for including environmental rights, which allows citizens and civil society to push policies towards more effective environmental protection.

It is also the task of realigning public finance. SWF, public investment banks, and multilateral development institutions must ensure that investment portfolios align with climate and sustainability goals. In addition, participatory budgets and fiscal decentralization can enable local communities to invest in projects that are in line with their environmental and social ambitions. These policy initiatives are vital to being open, responsible, and making fair decisions. Monitoring and evaluation programs should assess the environmental, social, and economic implications of assistance, fiscal policies, and conservation efforts (Kaklauskaite & Streimikiene, 2024). Civil society, academia, and the media are important in ensuring that environmental policies are beneficial to the masses at large and that they do not accelerate the social inequality that already exists. Using technology, it is possible to increase accountability in environmental finance and aid distribution by exploiting technology, including blockchain and open-data platforms. Ultimately, resolving environmental debt necessitates a paradigm shift from extractive economic models to regenerative systems that prioritize fairness, adaptability, and the limits of the planet. This necessitates moving beyond the mere use of sustainability as a buzzword and implementing policies that promote intergenerational solidarity and global collaboration. Although no single policy can fully address ecological injustice, a combination of fair burden sharing, sustainable aid approaches, and green fiscal tools offers a solid basis for transformative change. These endeavors require a transformation of development from an endless pursuit of growth to nurturing and sustaining all life within the boundaries of Earth's ecosystems. Only by addressing the underlying causes of environmental debt and promoting global solidarity can we create a future that is sustainable and equitable.

TRANSITIONING TOWARD SUSTAINABLE DEVELOPMENT MODELS

The current economic system has been pegged on the notion that constant economic growth, mainly in measured terms, GDP, is tantamount to progress, prosperity and development (Hossen et al., 2025). Unfortunately, the adverse environmental consequences of this growth-based approach have come to light in the long run. As a result of this re-examination, new frameworks have been established that emphasize the restoration of ecosystems, intergenerational fairness, and the general welfare of society over economic growth (Lang, 2024). Such novelties in thinking dispute the old-fashioned view that development is unlimited and promote the move toward sustainable development modes that consider the limits of the environment and human desires. The main opposition to growth-based models is the acknowledgment that permanent economic growth is not sustainable because the Earth has a finite amount of resources. The unchecked desire to achieve economic growth invariably ignores the adverse impacts on the natural environment, social injustices, and the loss of cultural heritage (Kallis et al., 2025). In addition, the benefits of this growth have not been shared equally, which has increased global inequalities. In wealthy communities, individuals consume more than the environment can support. Meanwhile, the poorest people are the most affected by environmental pollution, despite not being the primary source of such factors.

Alternative development models aim to reestablish the relationship between the economy, society, and nature. Another method is regenerative development, which exceeds the idea of sustainability, as it does not only cause little harm to the ecosystem but also takes action to recover and enhance its state. The focus on soil health, water cycles, reforestation, and the restoration of biodiversity is placed under the practices of regenerative farming, establishing a relationship between ecological renewal and economic and social resilience (Alenda et al., 2024). These models facilitate a shift out of resource-pulling industries and towards regenerative farming, renewable energy, and nature-based solutions to create systems that do not just sustain the well-being of life on earth but can also improve it. The aspect of fairness among and between generations is one of the essential postulates of alternatives to growth. The just transition framework is an example of how climate action must be linked with social justice, with the necessity of protecting workers and communities through this transition to fossil-free and sustainable economies (Bakkar et al., 2024). The strategy advocates policies that guarantee equality in employment, social security, and citizen participation, coupled with a decarbonized and rebuilt environment.

Among the primary alternatives proposed, one may speak about the concept of well-being economies is discussed. These frameworks prioritize the life and welfare of human beings and the environment over profit and production. New Zealand, Scotland, and Bhutan are among the countries that have started to measure well-being with indicators such as health, education, environmental quality, and mental health and have made it part of their national policy. These measures allow not only a shift in emphasis in measuring the quantity of the economy but also in measuring its quality, which adds value to a meaningful life within ecological limits. Alternative models also require localization and self-sufficiency. Local values and environmental sustainability can be ensured through communities that are resistant to the disruptive effects of the global economy and less harmful to the environment than global distribution chains. These can be seen in community-based projects, such as cooperatives, local currencies, and agro-ecological farms, which focus on participatory democracy, cultural continuity, and sustainable income. This territorial orientation resists the global capitalism of extractivism and contributes to place-based knowledge systems and ecosystems.

Crucially, these alternatives do not oppose technology; rather, they advocate for making technology accessible to all and using it in a manner that benefits both the environment and society. Technological advancements in energy, mobility, agriculture, and data can contribute to sustainable development if they are not driven by exploitative business models and are integrated into ethical and inclusive frameworks (Gifford & Sauls, 2024). Open-source platforms, decentralized renewable energy systems, and digital commons are examples of how technology can be used to benefit society rather than solely for corporate expansion. To facilitate this transition, policies that promote structural economic changes should be implemented. These include ecological taxation, which considers environmental costs, divesting from fossil fuels, reallocating subsidies to green infrastructure, and financial mechanisms that incentivize sustainability. In addition to individual actions, institutional changes are crucial, including shifting education towards sustainability, revising property laws to acknowledge the rights of nature, and establishing global governance frameworks that prioritize the well-being of the planet and its inhabitants.

Public imagination is of utmost importance in moving toward sustainable models. Shifting societal perspectives on success, worth, and advancement is crucial. Movements such as the Green New Deal, Doughnut Economics, and Buen Vivir have become influential catalysts for transforming consciousness and shaping policies that promote greater justice and sustainability in the future (Ahmad et al., 2024). These frameworks harmonize environmental preservation with social equity and are gaining traction among activists, scholars, and policymakers. Ultimately, moving beyond the growth paradigm necessitates a collective reimagining of human purpose and the role of economies in serving life, rather than exerting dominance over it. It requires bravery to confront established interests, ingenuity to develop innovative systems, and empathy to guarantee that no one is left behind during the transition (Okereke, 2024). By prioritizing ecological restoration and equity, alternative development models provide a hopeful path in a world burdened by the consequences of unsustainable growth.

As the flaws of the conventional growth-driven economic model become increasingly apparent, alternative frameworks such as degrowth, the circular economy, and community-driven development have gained popularity as potential solutions to ensure ecological sustainability and social fairness (Hay, 2025). These models aim to address the environmental damage caused by unrestricted economic growth while challenging the fundamental principles and structures that support current development approaches. Each offers a unique perspective on reconsidering progress in ways that prioritize well-being, ecological balance, and active citizen involvement. Degrowth is a radical critique of the conventional economic model, advocating the deliberate reduction of production and consumption in affluent nations to ensure that human activities remain within environmental boundaries. It is not just about negative growth or recession but about transforming economies to focus on well-being, fairness, and sustainability rather than profit and accumulation (Auteri et al., 2024). Degrowth challenges the notion that economic growth is always beneficial, asserting that beyond a certain threshold, additional growth in affluent societies results in diminishing returns in terms of overall well-being and causes irreversible harm to the environment. This would be accompanied by the expansion of public goods and services, such as healthcare, education, and public transportation, which enhance the quality of life without promoting excessive consumerism. Importantly, degrowth also emphasizes global justice: while wealthy nations reduce their economies to minimize environmental harm, developing countries can still pursue modest growth to meet basic human needs, provided it occurs within the framework of planetary boundaries.

Degrowth is inherently political, necessitating democratic discussions about the kind of society that individuals desire to inhabit (Corsi et al., 2024). It emphasizes having enough rather than striving for maximum efficiency, working together rather than competing against each other, and valuing care work over market labor. It also promotes the reconsideration of values such as success and prosperity, suggesting that flourishing should be measured by factors such as health, community unity, and ecological equilibrium rather than GDP. Although not as radical as other approaches, the circular economy provides a practical and widely accepted framework for integrating economic activity with environmental sustainability. Unlike the traditional linear "take-make-dispose" model, the circular economy recognizes waste as a design flaw. It aims to close the loop by implementing practices such as recycling, upcycling, reuse, and repair. Businesses and municipalities are increasingly adopting circular economic practices to minimize their environmental impact and remain competitive in the market. Urban areas are allocating resources to circular infrastructure, including composting facilities and materials recovery plants, to minimize landfill waste and foster sustainable urban environments (Smith and Ely, 2025). Nevertheless, the circular economy should not be limited to technological advancements and the efficient use of materials. However, it should encompass a broader approach to bring about significant changes. It is crucial to ensure that circular practices do not become mere greenwashing tactics for corporations but instead contribute to job creation, community empowerment, and affordable access to goods and services. A circular economy combines social and environmental objectives, guided by participatory governance and inclusive economic policies.

Community-driven development is a crucial component of envisioning sustainable futures. It focuses on empowering local communities as catalysts for change, highlighting participatory planning, local knowledge and democratic control over resources (Tunn et al., 2024). The CDD realizes that nobody can impose sustainable development on communities or nations and that sustainable development must be decided by those who are truly affected by ecological and economic decisions. Among the key strengths of community-driven development is that it is place-based, which tailors interventions to the specific ecological, cultural, and social context of a community. This contrasts with the development models that are universal to everybody without considering the situation and needs of each community. In many parts of the Global South, projects have enabled rural people to embrace agroecology, sustainable water management, and renewable energy systems that not only make them more resilient but also more self-determining.

CDD is also used to develop social capital networks of cooperation, mutual assistance, and trust, which are very important for long-term sustainability. It gives societies the power to confront governments and businesses and protect their rights to land, water, and livelihoods. Importantly, it has the capacity to be part of decolonizing development by reinstating local governance systems and opposing the imposition of externally defined development projects (Wartiainen et al., 2025). The community-driven development approach should be used to ensure that sustainability transitions are equitable and participatory, whereas the theoretical framework of degrowth and the practical scenario of the circular economy ensure that the necessity of change is understood and environmental damage is minimized. These three frameworks are not mutually exclusive; in fact, they complement each other because they deal with various facets of sustainability. The main idea of degrowth is a criticism of the current ideology of constant growth. Circular economy seeks to turn economic processes into restorative programs, whereas CDD brings attention to the significance of local agency and cultural meaning in development. By combining these frameworks, we can develop more comprehensive strategies for sustainable development. For example, a society that adopts degrowth principles might implement circular practices to minimize material consumption, while empowering communities to collectively manage local resources. In a society that values the commons, public investment would be directed towards infrastructure that benefits the environment and promotes social justice, such as community solar grids, cooperative housing, and decentralized food systems. However, substantial obstacles remain. Shifting to these alternative models necessitates political bravery, legal adjustments, and changes in societal norms. Education and public discourse need to adapt and change to challenge the idea of constantly buying things and help people understand the importance of caring for the environment, being kind to others, and feeling connected to the world (Wijburg, 2025).

Thus, global collaboration is vital. All countries in the Global South do not bear the cost of sustainability transitions. It is the obligation of rich countries to offer financial aid, technology transfer, and debt waivers to ease the transition process of developing countries, giving them the means they need to combat climate change. Moreover, the global government system must be reformed to better reflect the needs of all countries, including those regions that have become the most affected by environmental destruction and climate shifts (Ahmad et al., 2024). Making Indigenous knowledge and practices part of sustainable development is not symbolic but an essential plan for establishing ecological stability, social justice, and cultural survival. The natives of the world have coexisted with their environment in a coherent manner through successive generations using highly detailed systems of knowledge, religious beliefs, and refined management regimens.

Traditional ecological knowledge (TEK) is a general term used to describe indigenous knowledge that includes intimate knowledge of local biodiversity, climatic patterns, resource patterns, and ecological interdependencies (Aladekomo, 2025). In contrast to Western scientific paradigms, which tend to split knowledge into academic disciplines and treat humans as outside nature, TEK is unified, relational, and place-based. It is handed down generation by generation through stories and firsthand encounters, built on relationships between the natural world, and reinforced by the ability to share memories and adapt. For example, many native societies use rotational farming methods, polyculture, and agroforestry practices to maintain soil fertility, reduce erosion, and create biodiversity. Indigenous people in the Amazon rainforest have been found to be more successful in protecting forest cover and carbon sinks than governmental conservation programs. Indigenous hunters in the Arctic region have developed a unique skill to observe and understand the conditions of ice layers, the behavior of animals, and the seasons, which are sometimes even more effective than satellite data. Advanced methods are used by pastoralist societies in sub-Saharan Africa to respond to water shortages and grazing issues in arid areas. Such habits are not only a display of ecological sustainability but also an acknowledgement of the fact that all living things are interdependent (Alenda & Kaufmann, 2024).

Indigenous knowledge plays a significant role in developing environmental policies and climate adaptation strategies and sustaining them through integration. However, the idea of researching Indigenous territories must be regarded as highly respectful of their sovereignty, rights, and intellectual property. Often, Indigenous knowledge is stolen or used without consulting or asking the Indigenous populations, a form of epistemic injustice that resembles the colonial history of the exploitation of Indigenous peoples. Real inclusion also includes the co-creation of knowledge systems, equitable participation in the decision-making process, and acknowledgment of Indigenous people as right holders but not as stakeholders. Another important component of indigenous worldviews is the spiritual and ethical relationship that exists between them and the earth. In most indigenous cultures, people believe in the land, rivers, mountains, and other non-human beings as members of the family instead of things that can be used and exploited. This perspective creates a strong feeling of sacred duty and collective responsibility toward the planet's well-being, in sharp contrast to the current capitalist paradigm of nature as a commodity (Althouse and Svartzman, 2024). By recognizing and employing these values, we can turn sustainability into a moral and cultural change rather than a technical one. It causes societies to reconsider their notions of development, progress, and well-being, rather than merely focusing on GDP or resource extraction, adopting an alternative of harmony, reciprocity, and long-term sustainability. Indigenous governance systems often emphasize consensus, restorative justice, and community accountability, which can serve as the basis for democratic and inclusive sustainability practices. Such systems are dynamic, founded on oral transmissions, and often operate on different levels, including individual families and regional councils. Respecting such governance structures means giving a chance to Indigenous leadership in local, national, or international policymaking in environmental-related situations.

However, the implementation is random. Indigenous defenders globally are physically assaulted, forcibly dispossessed, and criminalized when they defend their territories against logging, mining, and agricultural businesses. Therefore, the necessity to use Indigenous knowledge in the context of sustainability cannot be separated from the context of Indigenous justice, decolonization, and human rights struggles (Auteri et al., 2024). The inclusion of Indigenous communities in national park management in Canada has resulted in better environmental management and conservation of cultural heritage. In Bolivia and Ecuador, constitutional changes have been made to recognize the rights of nature, inspired by indigenous thought, such as buen vivir and Pachamama. Such endeavors show that even in contemporary legal and governance frameworks, one can inculcate indigenous values that help people become responsible for the environment. Key also is education. The process of decolonizing sustainability education is based on acknowledging and valuing Indigenous and scientific knowledge, as well as training young people to value different types of knowledge and the opportunity to learn about Indigenous languages, stories, and ceremonies as repositories of ecological wisdom (Aydin & Degirmenci, 2024). Universities, non-governmental organizations, and governments need to invest in building the skills and knowledge of the Indigenous and non-Indigenous populations to enable cross-cultural learning and cooperative research. Indigenous ways of doing things provide examples of sustainable lifestyles that do not require taking something out of the environment but are based on sharing things, being satisfied with what is enough, and restoring and regenerating natural resources. These economies demonstrate that sustainability can be achieved by embracing cultural heritage and fostering a deep connection with the natural environment. To foster the growth of sustainable alternatives, it is crucial to provide fair trade, legal protection, and infrastructure investment to help combat global inequality and environmental degradation.

CONCLUSION

Environmental debt points out a major imbalance in the benefits and costs of economic growth worldwide. Advanced countries also have a huge ecological debt because of their historical industrialization and reliance on a perpetuated consumption-oriented economy, with the main burden resting on the ecosystem and populations of developing countries. This debt comes in many guises; it is the over-consumption of carbon leading to global warming, ecological erosion through the mining of resources that lead to lost biodiversity, and more tiresome, outsourcing, and dumping of waste that is exported through dumping. These activities have led to wealth creation in the Global North. However, they have also led to environmental issues and crippled sustainable development in the Global South. As disclosed in this chapter, the current situation of environmental indebtedness in developed countries is not a coincidental byproduct of economic development but a consequence of policies that focus on economic growth at the expense of ecological justice. Using past emissions data, trade volumes, and global supply chain studies, one can conclude that richer countries continue to transfer their ecological footprints to other countries, maintaining their political and economic power (Kaika et al., 2024). Global governance systems that do not consider the past and financial systems that do not consider the impact of their activities contribute to this imbalance. This injustice is only magnified by the lack of binding mechanisms to make nations responsible, as this enables rich countries to establish weak climate targets and postpone the necessary adoption of sustainability.

In the quest to close the gap between economic development and ecological justice, an essential step is to recast the concept of environmental debt as a legal, moral, and economic liability. It is the duty of advanced nations to lead in reparative payments, including offering fair climate finance, forgiving environmental-related debts, and providing equitable circulation of green technologies (Wartiainen et al., 2025). These measures should be incorporated into international agreements that acknowledge historical responsibility and the principle of common, but differentiated, responsibilities. Additionally, it is crucial to involve vulnerable and marginalized communities in the formulation of environmental policies to guarantee that justice frameworks are not just empty gestures but bring about real change. Filling this gap necessitates the involvement of both businesses and civil society in taking proactive measures to address the issue. Educational institutions and the media should also play a role in increasing awareness of ecological debt and its deeply rooted systemic causes.

REFERENCES

Ahmad, U. S., Rahman, Z. U., & Azam, M. (2024). Sustainability and Green Finance and its Relevance to Debt for Nature Swap Financing. In *The Palgrave Handbook of Green Finance for Sustainable Development* (pp. 645–671). Springer International Publishing.

Aladekomo, A. S. (2025). Trends in Gender, Feminism and Energy Transition in the Global South. *Journal of Sustainable Development Law and Policy (The), 16*(2), 53–75.

Alenda, J. M., & Kaufmann, M. (2024). From Marxist Development Theories to Their Translation in the Degrowth Discourse: Transforming Unequal International Structures for Environmental Sustainability.

Althouse, J., & Svartzman, R. (2024). Prospects and roadblocks to a" sustainable" international monetary and financial system. In *Understanding Green Finance* (pp. 182–198). Edward Elgar Publishing.

Álvarez-Véliz, R., & Barton, J. R. (2024). The historical geography of an idea: Sustainable development in Latin America, 1972–2022. *Journal of Historical Geography, 86*, 175–186.

Auteri, M., Mele, M., Ruble, I., & Magazzino, C. (2024). The double sustainability: The link between government debt and renewable energy. *Journal of Economic Asymmetries, 29*, e00356.

Aydin, M., & Degirmenci, T. (2024). The role of greenfield investment and investment freedom on environmental quality: Testing the EKC hypothesis for EU countries. *International Journal of Sustainable Development and World Ecology, 31*(6), 684–694.

Bakkar, Y., Jabeur, S. B., Mohammed, K. S., & Arfi, W. B. (2024). Environmental transition dynamics under external conflict risk: New evidence from European countries. *Journal of Cleaner Production, 472*, 143510.

Bohoslavsky, J. P., Laskaridis, C., & Prato, S. (2025). An International Debt Architecture and Debt Sustainability Assessments as if People and Planet Mattered. *Development*, •••, 1–4.

Corsi, G., Sapio, A., Guarino, R., Muñoz-Ulecia, E., & Franzese, P. P. (2024). *Uneven development and core-periphery dynamics: A journey into the perspective of ecologically unequal exchange* (No. ART-2024-138547).

Dafermos, Y. (2025). Climate finance and global justice. *Climate Policy*, •••, 1–17.

Elmhirst, R., & Resurrección, B. P. (2025). The coloniality of gender expertise in professional environment and development contexts. *Journal of Political Ecology*, *32*(1), 5960.

Gifford, L., & Sauls, L. A. (2024). Defining climate finance justice: Critical geographies of justice amid financialized climate action. *Geography Compass*, *18*(11), e70008.

Hasan, M. N. (2025). Unequal burdens and shared fates: Rethinking climate justice through reparative action and global solidarity. *Environmental Sociology*, •••, 1–9.

Hay, E. (2025). Towards an indigenous and traditional "pedagogy of repair": An ecological approach to environmental education. *South African Journal of Philosophy*, *44*(1), 17–29.

Herrera, H. (2024). Embedding Municipal Green Bonds in Mexico City's hydrosocial cycle:'Green'debt and climate action narratives. *Journal of political ecology: case studies in history and society.--*, *31*(1), 1-18.

Hossen, M. A., Benson, D., & Islam, M. (2025). Donor climate coloniality, ecological debt and vulnerabilities in coastal Bangladesh: Redesigning institutions for recognition justice. *Environmental Management*, •••, 1–15. PMID: 40159350

Islam, M. S. (2024). Rethinking Climate Justice: Insights from Environmental Sociology. *Climate (Basel)*, *12*(12), 203.

Kaika, M., Calvário, R., & Velegrakis, G. (2024). Austerity: An environmentally dangerous idea. *Journal of Political Ecology*, *31*(1), 67–81.

Kaklauskaite, U., & Streimikiene, D. (2024). Climate justice from the lens of sustainable development. *Sustainable Development*, *32*(6), 6788–6800.

Kallis, G., Hickel, J., O'Neill, D. W., Jackson, T., Victor, P. A., Raworth, K., & Ürge-Vorsatz, D. (2025). Post-growth: The science of wellbeing within planetary boundaries. *The Lancet. Planetary Health*, *9*(1), e62–e78. PMID: 39855235

Lang, M. (2024). Degrowth, global asymmetries, and ecosocial justice: Decolonial perspectives from Latin America. *Review of International Studies*, *50*(5), 921–931.

Liu, L., Chen, Z., Al-Hiyari, A., & Nassani, A. (2024). Sustainable growth in mineral rich BRI countries: Linking institutional performance, Fintech, and green finance to environmental impact. *Resources Policy*, *96*, 105159.

Moyo, G. (2024). Imperial Ecocide and the Bane of Global Climate Finance. In *Africa in the Global Economy: Capital Flight, Enablers, and Decolonial Responses* (pp. 103–123). Springer Nature Switzerland.

Okereke, C. (2024). Degrowth, green growth, and climate justice for Africa. *Review of International Studies*, *50*(5), 910–920.

Pearse, R., Schlosberg, D., Rickards, L., Della Bosca, H., Moraes, O., & de Kleyn, L. (2025). Compounding barriers to environmental justice. *Local Environment*, •••, 1–16.

Rastegar, R., & Becken, S. (2024). Embedding justice into climate policy and practice relevant to tourism. *Journal of Sustainable Tourism*, •••, 1–18.

Salah, M., & Ament, J. (2025). The ecological currency hierarchy: Empirical support for currency power driven asymmetries in environmental and social autonomy. *New Political Economy*, *30*(1), 77–99.

Smith, A., & Ely, A. (2025). From limits to growth to post-growth: The international politics of technology in historical perspective. *Science, Technology & Society*, •••, 09717218251326833.

Tariq, M. U. (2025). *AI-Driven Learning and Engagement in Higher Education*. IGI Global., DOI: 10.4018/979-8-3693-4074-5

Tariq, M. U. (2025). Strategic Place Marketing in Cross-Border Regions: Enhancing Territorial Cooperation and Economic Growth. In Sousa, B. (Ed.), *Cross-Border Regions Cooperation and Implications for Organizations* (pp. 107–136). IGI Global Scientific Publishing., DOI: 10.4018/979-8-3373-1912-4.ch004

Tariq, M. U. (2025). Digital Accountability: Revolutionizing Public Sector Transparency and Service Delivery. In Santos Ferreira, A., & Lourenço dos Santos, C. (Eds.), *Enhancing Public Sector Accountability and Services Through Digital Innovation* (pp. 129–154). IGI Global Scientific Publishing., DOI: 10.4018/979-8-3693-9251-5.ch004

Tariq, M. U. (2025). The Challenge of Artificial Intelligence in Environmental Sustainability: Opportunities, Risks, and Policy Implications. In Ahmadi, A., & Mna, A. (Eds.), *Nexus of Environmental Quality and Technology Innovation* (pp. 267–292). IGI Global Scientific Publishing., DOI: 10.4018/979-8-3373-1822-6.ch012

Tariq, M. U. (2025). Machine Learning Techniques for Personalized Rehabilitation: Enhancing Patient Outcomes Through Predictive Algorithms. In Korupalli, V., Ranasinghe, P., & Feldman, D. (Eds.), *Predictive Algorithms for Rehabilitation and Assistive Systems* (pp. 221–252). IGI Global Scientific Publishing., DOI: 10.4018/979-8-3373-0194-5.ch009

Tunn, J., Kalt, T., Müller, F., Simon, J., Hennig, J., Ituen, I., & Glatzer, N. (2024). Green hydrogen transitions deepen socioecological risks and extractivist patterns: Evidence from 28 prospective exporting countries in the Global South. *Energy Research & Social Science*, *117*, 103731.

Wartiainen, F., Boer, B., Pauw, P., & Romijn, H. (2025). Addressing the Twin Crises of Debt and Climate: Exploring Unconditional Debt Cancellation. *Development*, •••, 1–10.

Wijburg, G. (2025). On undevelopment and de-development: A geographical critique on perpetual growth and resource-based accumulation. *Area*, *57*(2), e12988.

KEY TERMS AND DEFINITIONS

Climate Accountability: Accountability for historical and current emissions.
Ecological Injustice: Unfair environmental burdens across global populations.
Environmental Debt: Unpaid ecological costs from economic growth.

Chapter 5
Cause–Effect Relationships Between Income Inequality and Environmental Degradation:
A Fresh Investigation on Leading Polluting Countries

Sumita Manna
Vidyasagar University, India

Ramesh Chandra Das
https://orcid.org/0000-0001-9305-7134
Vidyasagar University, India

ABSTRACT

Existing literature show that there are exists a deep connection between inequality and environmental pollution. This study explores the association as well as relationships between income inequality and CO2 emission in the top twenty CO2 emitting countries for the long run during the period 1995 to 2019. It first uses Pearson correlation analysis to examine the association between the two, and then uses time series econometric techniques The main findings of the study are as follows: 1) there are both positive and negative correlation coefficients between the two in many of the countries in the list 2) The variables are found to be cointegrated in four developed and four developing countries and) there are unidirectional causality from per capita CO2 emission to income inequality among the developed and

DOI: 10.4018/979-8-3373-1862-2.ch005

developing countries such as USA, UK, Indonesia, Iran. On the other hand, there are unidirectional causality from income inequality to CO2 only in two countries, Australia, India. There is only one country, Russia, which is having bidirectional relation between two variables.

INTRODUCTION

Environmental degradation and income inequality are among the most pressing global challenges today, requiring immediate and coordinated action. These issues are deeply interconnected with environmental harm often exacerbating social disparities, while inequality can hinder effective environmental governance.

In many economies, environmental degradation deepens the divide between the rich and the poor. Owner of an industry often accumulate wealth by exploiting natural resources and polluting the environment. While they live comfortably in cleaner, distant areas, far removed from the pollution their industries create. Other hand the poor who bear the brunt of this degradation. These vulnerable communities either work in these hazardous environments or live near the industries. As a result, they suffer from various health issues, increasing their medical expenses and reducing their quality of life. This added burden not only affects their well-being but also traps them in a cycle of poverty leading to inequality. Thus, environmental degradation is not just an ecological issue-it is a social one. According to Boyce (1994), many hazardous waste sites are others harmful environmental facilities in the United States are disproportionately located in neighborhood where low income residents, people of color, and immigrant communities live. Hajat et al. (2015) highlight that low income communities are generally more exposed to air pollution. As a result, they experienced poorer health outcomes, reduced ability to work, and limited income opportunities.

The level of environmental degradation caused by an activity is influenced by the power dynamics between those who benefit and those who are harmed. According to Boyce (1994), when the beneficiaries-those who gain from an activity- hold more power than those who suffer its negative effects, environmental damage tends to be greater. Conversely, when those affected by the harm have more influence or control, the extent of degradation is usually reduced. Dorling et al. (2007) examine the relationship of income inequality with personal consumption and generation of waste among affluent countries. Their findings indicate that wealthier countries with greater income inequality tend to use more resources and produce more waste per person. In contrast, countries with lower levels of inequality generally have lower per capita greenhouse gas (GHG) emissions even when their average income levels are similar to more unequal nations. Wang et al. (2021) demonstrate that in the presence of government corruption, the level of income inequality among the population influences how environmental protection regulations and policies are formulated. When official corruption is low, greater income inequality leads to stricter environmental regulations. But when corruption is high, income inequality causes environmental standards to decline-showing a threshold effect of corruption on this relationship. For a long list of 217 countries in the world with a long time series data Wan et al (2022) have shown that rising shares of low-income groups lead to higher CO2 emissions while rising shares of high-income groups contribute to lower CO2 emissions, and rising income inequality can decrease energy consumption and increase R&D expenditure, which in turn help to decrease CO2 emissions. With regard to the economic and environmental linkages in the developing and emerging countries Gao and Fan (2023) have shown that income inequality, economic growth, energy consumption, and agriculture significantly contribute to an increase in carbon emissions and a decrease in environmental quality.

Under the background, the present study analyzes the relationship between per capita CO2 emissions and income inequality across the top twenty CO2 emitting countries during the period 1995 to 2019. Panel data analysis can estimate the expected relationship between two variables in both the long run and the short run. In this study the countries are selected in such a way that the selected countries are emitting about 34% of total CO2 in the world.

LITERATURE REVIEW

The relationship between CO_2 emissions and income inequality has received growing attention in recent academic literature, reflecting concerns over the intersection of environmental sustainability and social equity. As global carbon emissions continue to rise, researchers have increasingly explored how income disparities within and between countries contribute to environmental degradation and the reverse causations also. This section reviews some key studies that examine the complex interplay between income distribution and carbon emissions to justify the present study through finding research gaps.

According to International Energy Agency (IEA, 2025), CO2 emissions in developed nations decreased by 1.1X109 ton, while emissions in developing countries rose by 6.1X109 tons, effectively nullifying the emission reduction progress made by the developed countries. In 2003, CO2 emissions from developing countries go beyond those of developed nations, contributing to nearly 50% of the world's total CO2 emissions (Zarzoso & Maruotti, 2011). According to Jiang et al. (2018) developing countries are becoming more influential in global climate change mitigation. While developed nations have contributed more to historical CO2 emissions, developing countries have seen a faster growth in emissions in recent decades and now account for a growing share of global emissions each year. In 2005, CO2 emissions from developing countries exceeded those from developed countries and now make up 63.4% of global emissions. From 1995 to 2009, CO2 emissions from coal use increased the most in developing countries, rising by 3.76 Gt. On the opposite side CO2 emissions from natural gas use rose the fastest in developed countries, increasing by 470 Mt between 1995 and 2009. Both developed and developing countries have experienced a steady increase in production-related CO2 emissions. However, between 1990 and 2014, developing countries saw a significantly higher annual growth rate of 3.2%, compared to just 0.4% in developed countries. In OECD countries, economic growth has reached a point where it could support environmental improvement, yet growing income inequality is resulting in greater environmental deterioration. That means as economic growth progresses and income inequality decreases, environmental pollution tends to be reduce (Kang, 2022). The reduction in co2 emissions, inequality is significantly smaller when taking into account the disparity in emissions between rich and poorer countries (emissions concentration), compared to when only the overall inequality in emissions is considered. Padilla & Serrano found that while the overall "simple" inequality in CO2 emissions has reduced, the disparity in emissions between residents of wealthy and poor countries has decreased to a lesser degree (Padilla & Serrano, 2005). Soares et al. (2017) analyze the relationship among CO2 emissions, Gini index and environmental efficiency in 2010 for 60 leading countries and their results finding that developed countries group (HH) demonstrated greater environmental efficiency compared to the other groups, whereas the low-income and middle-competitiveness countries group (ML) showed lower efficiency. In developing countries, a more equal distribution of income leads to lower CO2 emissions per person, while in most developed countries, income inequality has little impact on CO2 emissions (Chen et al. 2020). According to Uddin et al. (2020) there is a non-linear association between income inequality and CO_2 emissions. Between 1950 and 2000, income inequality significantly reduced CO_2 emissions. There was no significant relation found during the period 2000 to 2014. According to Liu et al. (2018) a growing income gap, particularly when income is unevenly distributed across different regions can be expected to worsen environmental quality and lead to higher CO_2 emissions. They are findings that a fairer distribution of income could have beneficial impacts on reducing CO_2 emissions in China. According to Grunewald et al. (2017) in low & middle-income countries, greater income inequality tends to lower carbon emissions whereas in upper-middle income and high-income countries, higher inequality is linked to increase per capita emissions. Basili et al. (2005) use pooled OLS and FE models to find the link between inequality and CO2 emissions. They find that both the models give opposite results. The Gini coefficient is positive but has low t value in FE model. On the other hand pooled OLS model gives negative and significant results. Belaid et al. (2020) used Westerlund's cointegration method, to study a long-term relationship between income inequality and carbon emissions. It finds that higher income inequality is linked to lower per capita CO_2 emissions in the long run, while in the short run, the relationship is positive. Additionally, a concave pattern is observed, indicating that both very high and very low levels of income inequality are associated with lower emissions. Dorsch et al. (2021) revisits the link between economic growth and environmental degradation by incorporating income inequality into the analysis. Using the ecological footprint as a measure, it finds that the Environmental Kuznets Curve (EKC) holds only under conditions of equal income distribution. When inequality is high, no clear turning point emerges, indicating that unequal growth leads to greater environmental harm compared to more equitable growth.

Demir et al. (2018) shows a negative relationship between CO_2 emissions and income inequality in Turkey, suggesting that higher inequality leads to lower emissions due to reduced overall consumption. This supports existing literature indicating that, in developing countries, environmental degradation may initially rise as income inequality decreases. Rufael et al (2016) use bounds testing approach to cointegration revealed the presence of a long-run relationship between income inequality and CO_2 emissions in both countries. However, this relationship was not found to be statistically significant in either the long-run or short-run estimates. While income and energy consumption remain the primary drivers of CO_2 emissions in both countries, the findings suggest that reducing income inequality may contribute to improved environmental quality in China, despite the lack of statistical significance. In contrast, income redistribution in India appears less likely to yield environmental benefits. Variance decomposition analysis revealed that income inequality played a minimal role in influencing CO_2 emissions, with no evidence of causality in either direction between the Gini coefficient and CO_2 emissions in both countries.

The lack of a significant effect from the Gini coefficient generally contradicts the marginal propensity to emit framework, which suggests that as income becomes more evenly distributed, lower-income individuals—now gaining greater purchasing power—will likely raise their consumption of energy and carbon-intensive goods as they transition toward middle-class lifestyles. Das et al. (2023) explores the long-term and short-term relationships among income, urbanization, energy consumption, and GHG emissions in the world's top 20 polluting countries from 1970 to 2018. Drawing on theoretical foundations, the study employs Johansen cointegration and VECM techniques to assess co-movements among the variables. Findings reveal significant long-run associations and short-run causal dynamics, particularly in developed nations. Wang (2024) finds that greater income equality helps reduce environmental degradation, particularly when renewable energy is well developed. This suggests that advancing renewable energy can ease the potential conflict between income equality and environmental goals, allowing both to progress together. Additionally, the positive environmental impact of renewable energy is amplified by fairer income distribution, with household consumption patterns and social norms playing a key role in this relationship. According to Khan et al. (2022) income inequality have a direct significant and positive effect on Carbon emission while trade openness and renewable energy significantly reduce carbon emission.

Boyce (1994) presents two key ideas. The First is that how much environmental harm takes place often depends on who holds more power-those who benefit from the harmful activity or those who suffer because of it. If the beneficiaries are more powerful, environmental damage is likely to be greater. This idea is described as a decision making process influenced by power, where stronger groups have more influence over outcomes. Pata et al. (2022) using data from 2002 to 2016 for four South Asian countries, their findings is that economic growth, income inequality, urbanization, and financial development raise the ecological footprint, while political stability and renewable energy reduce it. Causality tests reveal one-way effects from urbanization, renewable energy, income inequality, and growth to ecological impact. The results highlight the importance of political stability, income equality, and renewable energy in achieving sustainable development.

Based on the studies according to Wang et al. (2021) advancements in sustainable energy and reductions in income inequality are contributing to enhanced environmental quality. Additionally, natural resources appear to have a notably positive impact on environmental well-being. Political economic theory suggests that rising income inequality leads to higher consumption of polluting goods, increasing emissions. The competitiveness model also links inequality to greater consumption and longer working hours, both boosting energy use and CO2 emissions. Scruggs (1998) explores the link between political and economic equality and environmental degradation, drawing on Boyce (1994). It suggests that equality can lower degradation but isn't always essential—inequality may sometimes have a positive effect. Samani et al. (2024) investigates whether environmental degradation is a cause or consequence of income inequality. Their study shows that income inequality causes environmental degradation, but not vice versa. Income inequality, per capita GDP, and industrialization worsen degradation, while energy efficiency, education, and environmental spending reduce it. Meanwhile, environmental degradation, education, and higher GDP reduce inequality whereas environmental spending and taxation may increase it. Ching et al (2023) investigate the relationship between income inequalities, educational attainment and CO2 emissions by using panel data analysis for a group of 64 countries during the year 1990 to 2016. And their finding is that there has a negative association between unequal distribution of income and CO2 emissions. Also the study found that there are negative relation between educational attainment and environmental pollution. Zwickl (2015) study on US economy and the study finds significance regional difference in environmental inequality. African Americans face higher pollution exposure than whites in the Mid-West and South-central. For Hispanics, those in the live in the west or south both are more polluted cities and neighborhoods, on the other hand those live in the south central region they live in less polluted cities but more polluted neighborhoods within cities. That means a key question is that whether the regional differences seen here are mainly caused by where industries are built or where people choose to live. Using cross sectional augmented autoregressive distribution lags (CS-ARDL) Ogede et al. (2024) found that squared of income inequality has a negative impact on CO2 emissions, though the strength of this effect differs between the short term and the long term. Income disparity continue to have a positive impact on CO2 emissions, even when financial inclusion is taken into account both in the short run as well as in the long run. Coondoo and Dinda (2007) observed that for the country group of Europe and America, redistributing income more equally would lead to an increase in average emission. On the other hand significantly poorer country of Asia, such distribution would reduce average emission. While in Africa, the impact of mean emission is statistically insignificant. Ajide and Ibrahim (2021) using fully Modified Ordinary Least Squares (EMOLS) and Dynamics Ordinary Least Squares (DOLS) estimators indicate that income inequality has a positive and significant impact on environmental degradation in the G7 countries. And the country level analysis shows that the effect of income inequality on environmental degradation varies across individual in G7 countries, suggesting that a divergent pattern. The study found that there has a positive impact in case of Canada, Japan, & USA but negative for France & Germany. There has insignificant impact for Italy and UK. According to Sun (2023) over time the number of cities where income inequality inhibits carbon emission and green development has grown. In some cities, income inequality does not have much effect on green development. Barido and Marshall (2014) found that the elasticity may be depends on the strength of environmental policy of a country. Their study found that which countries has stronger environmental policy, urbanization is more beneficial of this countries that means there are less negative impact on emission. In which countries has weak environmental policies 1% increase in urbanization CO2 emissions will be increase 0.65% for higher income countries. On the other hand 1% increases in urbanization leads to increase 1.3% CO2 emission for weak environmental policy with lower income countries. In which countries has high income inequality, income inequality tend to decrease correspond with a reduction in per capita CO2 emission. Contrary, in societies with lower levels of income inequality, reduction of income inequality is leads to an increase in per capita CO2 emissions. Hailemariam et al. (2020) explore the impact of income inequality on carbon dioxide emissions across OECD countries. Utilizing advanced panel data estimation techniques the study finds that, when the richest people earn more compare to others, carbon emissions tend to increase. They also find that a rise in the Gini index, reflecting greater income inequality, is linked to lower carbon emissions, aligning with the marginal propensity to emit framework.

RESEARCH GAP

Some literature has focused on multiple countries, while other have studied only a single country. However this study examine top twenty CO2 emitting countries. Which addresses a gap in existing research. Moreover no existing literature has utilized data covering the period from 1996 to 2019, which this study incorporates to fill the gap.

OBJECTIVES

The objectives of the study are 1) to investigate whether there are any association between income inequality and CO2 emission of the selected countries, 2) to examine whether there is any long run relationship between the two variables and 3) to investigate whether there are any causal interplays between the two in the short run.

DATA AND METHODOLOGY

The study covers top twenty CO2 emitting countries of the world as of the year 2023. These countries are selected on the ground that they emit 32.81% of world CO2. The period of study is 1995-2019 since the post COVID 19 fluctuated data are not considered for the analysis. Since there are 25 year points in the study for both the series, there may have stochastic trends, hence, it is required to test for stationary of the two series for all selected groups and countries. Further, to get robust results, the countries are divided into two groups, namely, the developed and developing. The list of countries under the developed category contains USA, Japan, Germany, Canada, UK, Italy, South Korea, Saudi Arabia, Australia, Poland and the list for developing countries constitutes China, India, Russia, Indonesia, Iran, Mexico, Turkey, Brazil, South Africa, and Vietnam. In this study the measure for CO2 emission is considered in per capita CO2 emissions in ton. Gini coefficient is used for the data on income inequality. The data on emission and inequality are obtained from the World Bank database.

This study first investigates whether there is any association between income inequality and CO2 emission by conducting Pearson correlation analysis. In this study, the t value has been calculated to determine whether correlation between two variables is statistically significant or not. But it is a fact that correlation does not necessarily mean causality and interplay. To study the long run and short run linkages between environmental pollution and income inequality it is required to go for applying time series econometrics which involves studying the nature of the two series, whether stationary or not, and then study the cointegration for long run or equilibrium relations. Further, to justify whether such long run relation is stable it is required to model for error correction. Finally, the causal interplay in the short run between the two series the Granger causality is used.

Stationarity Checking

For testing stationarity of the two series, the study uses unit root test of the two series for all the selected countries. To test the null hypothesis of the existence of unit roots, the study uses Augmented Dickey-Fuller (ADF) (1979) technique. ADF test is based on the assumptions that the error term is serially independent and has a constant variance. For a data set (x_t, t=1, 2, ….., T), where t denotes time. Let us considered the following linear regression set up for unit root test for two versions of ADF regression, without time trend and with time trend -viz…

$$\Delta X_{t=} \alpha + \beta X_{t-1} + {}_{j=}^{p}\Sigma Y_j \Delta X_{t-j} + u_t \tag{1}$$

$$\Delta X_{t=} \alpha + \delta t + \beta X_{t-1} + {}_{j=}^{p}\Sigma Y_j \Delta X_{t-j} + u_t \tag{2}$$

If β = 0 (or ρ = 1) is rejected by ADF statistic in both the equations then it is said that the series is stationary. If this property works for the series of per capita CO2 emission and Gini coefficient then regression of one variable on the other can be done without the chance of getting spurious results. If not, it is required to test whether the series are integrated of order one (I (1)) or they are stationary at their first difference. If it is obtained that both series are stationary at first difference or non-stationary at levels, and their estimated error is stationary then it is said that both series have common natures over the time and is said that they are cointegrated and there is long run relationship between per capita CO2 emission and income inequality.

Cointegration Test

If two series are non stationary, i.e. having stochastic trends, then their errors might be non-stochastic and in that case a linear combination could be formed for the error term taking the two indicators. If that linear combination is stationary then it can be said that the two non stationary series might have a common relation for the long run. To test for cointegration between the variables the study uses Engel Granger cointegration test (1987). Again, to investigate the short run dynamics between the two the Error Correction Mechanism (ECM) is applied. For short run causal interplay between the variables, Granger Causality analysis (1969) is applied.

To execute the Engel Granger cointegration test, suppose, both dependent variable (Y_t) and independent variable (X_t) are integrated in order 1 i.e., I (1), there exists stationary. In that case we do not face the problem of spurious regressions. Then the regression equation to be estimated is as:

$$Y_t = \alpha + \beta X_t + u_t \tag{3}$$

u_t is the white noise error term, the linear combination of Y_t and X_t, i.e., Y_t- α- β X_t is stationary. Therefore, the estimated equation $\hat{Y}_t = \hat{\alpha} + \hat{\beta} X_t$ It represent long run relationship between Y_t and X_t. The estimated error ($\hat{u}_t$) connect Y_t, X_t and the estimated error $\hat{u}_t = \hat{Y}_t - \hat{\alpha} - \hat{\beta} X_t$

As mentioned, if the linear combination $u_t = Y_t - \alpha - \beta X_t$ is stationary, i.e., I (0) then both Y_t and X_t are said to be cointegrated. This implies that there is a meaningful long run equilibrium relations between the two non-stationary series. To determine this, we first estimate the regression equation (3). Then we obtain the residuals $\hat{u}_t = \hat{Y}_t$-α-$\hat{\beta} X_t$, and test whether these residuals are stationary (I (0)). If they are found to be stationary, then Y_t and X_t are cointegrated in the Engel-Granger sense. The estimated coefficients $\hat{\alpha}$ and $\hat{\beta}$ from this regression provide the long run equilibrium relation between X and Y. Thus the equilibrium equation for the i_{th} unit can be written as:

$$Y_{it} = \hat{\alpha} + \hat{\beta} X_{it} \tag{4}$$

To test the stationarity of the estimated error term $\hat{u}_t$, we can apply the Augmented Dickey-Fuller (ADF) test by estimating the following regression:

$$\Delta \hat{u}_{i,t} = \lambda\ \hat{u}_{i,t-1} + {}_{j=1}{}^{p}\Sigma \theta_i \Delta \hat{u}_{i,t-j} + \varepsilon_{i,t} \tag{5}$$

Null Hypothesis ($\lambda = 0$): the error term has a unit root, implying non-stationary.
Alternative Hypothesis ($\lambda < 0$): the error term is stationary.

If the null hypothesis is rejected, it suggests that Y and X are cointegrated. This implies that the existence of a stable, long run equilibrium relationship between the two variables.

Error Correction Mechanism (ECM)

Obtaining a long run relationship between the two series does not imply the absence of short run fluctuations around that equilibrium relation. These short run deviations are referred to as errors. It is important to examine whether these errors adjust to the equilibrium level over time. If they move back toward the equilibrium, the errors are said to be corrected. If they move away from the equilibrium, the errors are remain uncorrected. To capture this short run dynamics in relation to long run equilibrium, the Error Correction Mechanism (ECM) is used. The ECM is expressed as:

$$\Delta Y_{it} = a + \eta \Delta X_{it} + \gamma \hat{u}_{i,t-1} + e_{i,t} \tag{6}$$

In this equation, $\hat{u}_{t-1}$ represents the error correction term, and γ indicates the speed at which the system returns to equilibrium-reflecting convergence. If the coefficient γ is negative and statistically significant, then the variables are converging toward the long-run equilibrium. Any short run deviations are gradually corrected. Conversely, if γ is positive and significant, it indicates divergence, meaning the variables are moving away from equilibrium over time. Equation (6) captures the short run relationship between the variables. Here, η denotes the short run regression coefficient of the regression of Δ X on Δ Y.

Granger Causality Test

In the context of a bivariate non stationary model where both variables are integrated of order one, I(1), the Granger Causality (1969) test is conducted by estimating equations in their first differenced forms these equations incorporate error correction terms to capture both short term dynamics and long term equilibrium relationship between variables. Specifically for variables Y_t, X_t the Granger Causality test involves estimating the following equations:

$$\Delta Y_t = V_{yx} + {}_{j=1}^{T11}\Sigma \alpha_{1j} \Delta Y_{t-j} + {}_{j=1}^{T12}\Sigma \beta_{1j} \Delta X_{t-j} + \eta_{yx} ECY_{t-1} + u_{it} \tag{7}$$

$$\Delta X_t = V_{xy} + {}_{j=1}^{T21}\Sigma \alpha_{2j} \Delta Y_{t-j} + {}_{j=1}^{T22}\Sigma \beta_{2j} \Delta X_{t-j} + \eta_{xy} ECX_{t-1} + u_{it} \tag{8}$$

Let Δ represent the first difference operator. The term T_{tm}, where , m= 1, 2, 3, indicate the number of lagged values of ΔY and ΔX that influence the current values of these differenced variables. The parameters v, α, β, and η are regression coefficients, while u_{it} (for l=1, 2) are the disturbance terms, assumed to follow white noise properties.

The coefficients η_{yx} and η_{xy} are known as adjustment parameters. For the error correction mechanism to be meaningful, these parameter must be negative and statistically significant. The term ECY_{t-1} and ECX_{t-1} represent the error correction components derived from the residuals of long run regressions of y on x and x on y, respectively. Granger Causality is assesses by evaluating the F- statistics from the model. The direction and presence of causality are determined based on statistical significance of these F-values.

1. If all β_{1j}= 0 and η_{yx}=0, then X does not Granger cause Y.
2. If all α_{2j}=0 and η_{xy}=0, then Y does not Granger cause X.
3. If condition (1) is true but condition (2) is false, then there is unidirectional Granger Causality from Y to X
4. If condition (1) is false but condition (2) is true, then there is unidirectional Granger Causality from X to Y
5. If neither condition (1) nor (2) holds, then Granger causality is bidirectional
6. If both conditions (1) and (2) hold, then there is no Granger Causality between them.

By using the above methodologies we are calculate the results below.

RESULTS, ANALYSIS, AND DISCUSSION

Before exercising quantitative analysis, the study first presents the trends of the data on Gini coefficient and CO2 per capita across all the countries. Figure 1 presents the series for Gini coefficient and Figure 2 is for per capita carbon emission.

Figure 1. Trends of Gini coefficient for the countries

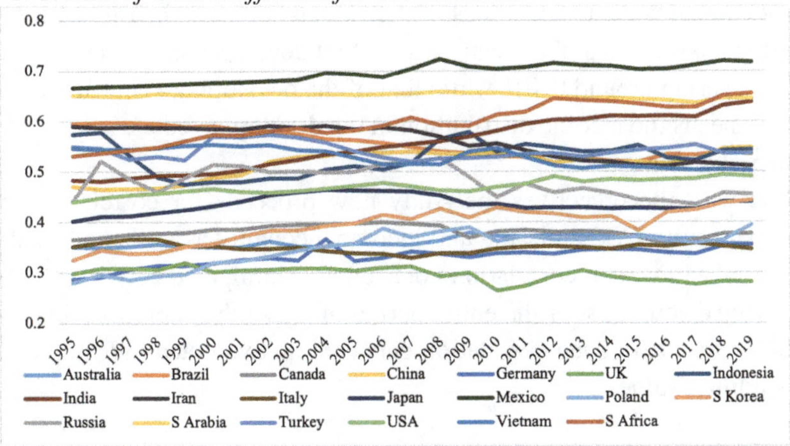

It is observed that the trends of the Gini coefficients are mainly rising in the listed countries but the average values of inequality are relatively larger than that of the developed countries. Mexico leads the group while UK follows the rest.

Figure 2. Trends of per capita CO2 emission (in ton) for the countries

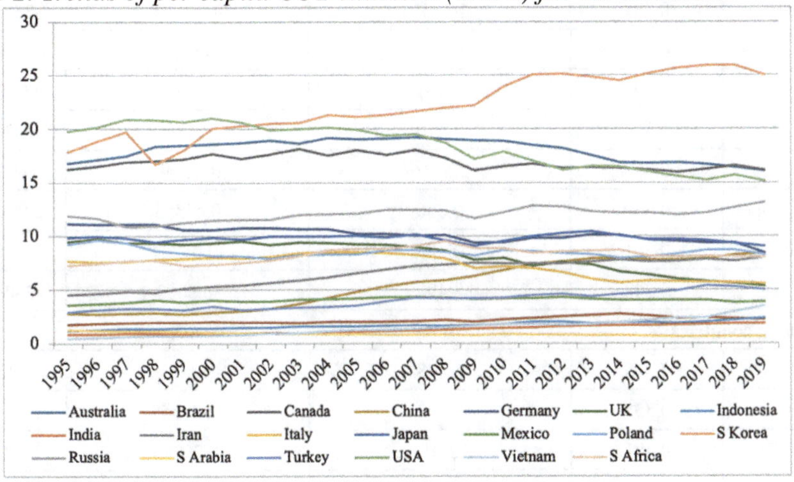

Figure 2 displays that the trends of carbon emission per capita are mostly rising for the developing countries while that of the developed countries, they are falling. The highest per capita emission is by South Korea all over the duration while that of Saudi Arabia, it is the lowest.

To have a clear idea on which country leads the emission and inequality indicators the study computes the mean values of the two and presents in Table 1 (column 2 and 3). It is observed that the mean values of all developed countries' Gini coefficient are hovering around 0.401 while that of the developing countries are around 0.556 that means the income distribution in the developing countries are relatively more disproportionate.

In line with the objectives the study now proceeds for correlation analysis between per capita CO2 emission and income inequality as measured by the Gini coefficient. The Pearson correlation coefficient is computed between the two for all the selected countries for the entire period of the study. Table 1 (column 4 & 5) presents the results of such correlation between the two noted indicators with the corresponding t values.

Table 1. Mean of Gini coefficient and per capita CO2 emission, and their correlation coefficients

	Mean of Gini	Mean of per capita CO2	Correlation coefficient	t values
Developed countries				
USA	0.471	18.415	-0.80648	-5.78699*
Japan	0.436	9.776	0.08338	0.354986
Germany	0.331	10.155	-0.68368	-3.97457*
Canada	0.381	16.909	0.8647668	7.3060267*
UK	0.297	8.151	0.75785	4.928209*
Italy	0.349	7.220	-0.2793	-1.23409
South Korea	0.393	22.116	0.84849	6.802413*
Saudi Arabia	0.650	0.856	0.44353	2.099579*
Australia	0.361	18.020	-0.07227	-0.30743
Poland	0.348	8.490	-0.5063	-2.49141*
Developing countries				
China	0.515	5.478	0.698633	4.142739*
India	0.555	1.272	0.978553	20.15422*
Russia	0.481	12.014	-0.20294	-0.87931
Indonesia	0.527	1.689	0.296478	1.317066
Iran	0.560	6.583	-0.8187	-6.0489*
Mexico	0.695	4.047	0.523616	2.607554*
Turkey	0.542	4.005	-0.29569	-1.31321
Brazil	0.557	2.158	-0.86492	-7.31107*

continued on following page

Table 1. Continued

	Mean of Gini	Mean of per capita CO2	Correlation coefficient	t values
South Africa	0.597	8.223	0.461926	2.209657*
Vietnam	0.529	1.478	-0.8209	-6.09877*

Source: Authors' calculations

Note: The * mark indicates 5% level of significance.

From the table it is found that out of ten developed countries only USA has highly negative and significant correlation in between per capita CO2 emissions and Gini coefficient. Out of the list of developed countries, the study finds only three countries having strongly positive and significant correlation between the two which are UK, Canada, and South Korea. On the other hand, there are only two countries, namely, Germany and Poland, in the list of the developed countries, which have negative and significant association between per capita CO2 and Gini coefficient.

Out of the list of developing countries three countries are having strongly negative as well as significant association between two variables namely, Iran, Brazil and Vietnam. India has highly positive and significant association between two variables. And there are three countries namely China, Mexico and South Africa which have positive and significant correlation between per capita CO2 emission and Gini coefficient. Only Mexico and South Africa have moderately positive and significant correlation between two.

Unit Root Test Results

To test the stationarity of the two series, per capita CO2 and income inequality as measured by the Gini coefficient, the study uses the ADF (1969) test for identifying the presence of the unit root in the series. The results of ADF test are given in Table 2.

Table 2. Unit root test results

	Per capita CO2(in ton)			Gini coefficient		
	ADF of CO2	Prob	Remarks	ADF of Gini	Prob	Remarks
	Developed Countries					
USA	-4.977197	0.0007	I(1)	-5.174284	0.0004	I(1)
Japan	-3.506311	0.0195	I(0)	-3.249710	0.0311	I(0)
Germany	-4.154550	0.0040	I(1)	-5.916956	0.0001	I(1)
Canada	-5.532982	0.0002	I(1)	-4.473410	0.0019	I(1)
UK	-6.372788	0.0000	I(1)	-6.049428	0.0001	I(1)

continued on following page

Table 2. Continued

	Per capita CO2(in ton)			Gini coefficient		
	ADF of CO2	Prob	Remarks	ADF of Gini	Prob	Remarks
Italy	-3.223041	0.0315	I(1)	-4.538264	0.0017	I(1)
South Korea	-4.926904	0.0007	I(1)	-7.969260	0.0000	I(1)
Saudi Arabia	-4.623298	0.0014	I(1)	-5.351386	0.0003	I(1)
Australia	-7.633582	0.0000	I(2)	-5.639128	0.0002	I(2)
Poland	-3.970981	0.0061	I(1)	-5.770005	0.0001	I(1)
	Developing countries					
China	-4.845720	0.0009	I(2)	-5.421409	0.0003	I(2)
India	-3.912578	0.0070	I(1)	-4.579909	0.0015	I(1)
Russia	-5.134395	0.0005	I(1)	-7.250954	0.0000	I(1)
Indonesia	-4.461021	0.0029	I(1)	-3.974241	0.0064	I(1)
Iran	-3.829148	0.0084	I(1)	-4.019533	0.0055	I(1)
Mexico	-5.019389	0.0005	I(1)	-6.156717	0.0000	I(1)
Turkey	-4.091030	0.0049	I(1)	-4.736800	0.0011	I(1)
Brazil	-4.186698	0.0038	I(1)	-4.059673	0.0050	I(1)
South Africa	-4.407682	0.0023	I(1)	-4.741055	0.0010	I(1)
Vietnam	-6.175075	0.0000	I(2)	-5.159660	0.0004	I(2)

Source: Authors' calculations

Note: I(1) and I(2) respectively means that the series are integrated of order 2. In other words, the series will be stationary if differencing is made twice.

To establish the presence of long run relationship, it is necessary that all series be integrated of order one i.e. I (1). Therefore, we test the stationarity of the first differences of the two series. The results are calculated by using ADF test. It is observed from Table 2 that, out of developed countries, both the series i.e. per capita CO2 emissions and Gini coefficient are stationary at their 1st differences. Only one country namely, Japan, having stationary at level and only one country having stationary at 2nd differences namely, Australia. Out of developing countries all having stationary at 1st difference. Only two countries having stationary at 2nd difference namely, China and Vietnam.

Cointegration Test Results

It has been mentioned in the methodology section that to investigate the existence of long run equilibrium relationship between the two variables, the study uses Engel Granger cointegration test. As explained earlier, the Engel Granger (EG) method tests whether the residuals of ADF are stationary. Table 3 shows the results for cointegration as well as error correction.

Table 3. Cointegration and error correction test results

	Regression Coefficient (Dependent variable: Gini coefficient)(Intercept)	Probability of Regression Coefficient	ADF test on Residuals	Probability of Residual ADF	Coefficient of Error Correction Test	Probability of Error Correction terms	Remarks (Cointegrated or not)
	Developed Countries						
USA	-5.447930(0.570961)	0.0000	-3.895509	0.0076	-0.422016	0.0046	Yes with significant error correction
Japan	4.350515(0.393836)	0.6919	-3.192459	0.0349	-0.159457	0.0519	Yes with significant error correction
Germany	-18.35516(0.517165)	0.0002	-4.236536	0.0032	-0.624256	0.0032	Yes with significant error correction
Canada	14.51479(0.135740)	0.0000	-2.759818	0.0804	-	-	
UK	7.308739(0.237384)	0.0000	-4.054726	0.0048	-0.819527	0.0008	Yes with significant error correction
Italy	-2.343665(0.365880)	0.1763	-1.703762	0.4167	-	-	
South Korea	10.30684(0.165484)	0.0000	-2.405775	0.1527	-	-	
Saudi Arabia	13.54322(0.638894)	0.0264	-1.720222	0.4089	-	-	
Australia	-0.684388(0.373334)	0.7314	-1.713180	0.4122	-	-	
Poland	-41.01779(0.696046)	0.0098	-1.351079	0.5885	-	-	
	Developing Countries						
China	8.661946(0.467997)	0.0002	-4.296328	0.0033	-0.249745	0.1252	Yes with insignificant error correction
India	139.2942(0.377891)	0.0000	-1.576029	0.4790	-	-	
Russia	-10.51889(0.607185)	0.3306	-2.756022	0.0797	-	-	
Indonesia	30.91834(0.474890)	0.1501	-2.488924	0.1304	-	-	
Iran	-22.12331(0.705374)	0.0000	-3.755738	0.0112	-0.171751	0.0185	Yes with significant error correction
Mexico	48.38930(0.499604)	0.0072	0.150455	0.9624	-	-	
Turkey	-6.299402(0.567062)	0.1513	-3.657257	0.0128	-0.343414	0.0528	Yes with significant error correction
Brazil	-89.48382(0.749927)	0.0000	-1.436588	0.5475	-	-	
South Africa	27.49611(0.371165)	0.0201	-0.424925	0.8896	-	-	
Vietnam	-20.04474(0.558594)	0.0000	-3.865938	0.0078	-0.313288	0.0624	Yes with insignificant error correction

Source: Authors' calculations

Table 3 shows the estimated long regression coefficient, the Augmented Dickey fuller test results on the residuals and the corresponding error correction term. Column (4) and (5) show the ADF test results to examine the stationarity of the residuals. Respective error correction (EC) terms are displayed in the columns 6 and 7. It is observed from the table that the two series in most of the countries are not cointegrated, that means, there is no relationship in the long run between the variables. Out of ten developed countries, CO2 and Gini are significantly cointegrated in the case of four countries namely, USA, Japan, Germany, UK. CO2 and Gini coefficient are significantly cointegrated in case of four out of ten developing countries namely, China, Iran, Turkey, Vietnam and have been error corrected of these countries.

Error Correction Results

It has been mentioned in the methodology section that to investigate the existence of long run equilibrium relationship between the two variables, the study uses Engel Granger cointegration test. As explained earlier, the Engel Granger (EG) method tests whether the residuals of ADF are stationary. Table 3 shows the results for cointegration as well as error correction.

Granger Causality Test Results

Absence or presence of long term equilibrium relation does not mean that there do not have any relation, rather interplay, between the two variable in the short run. The Granger Causality test is conducted to examine the short term interplay between per capita CO2 emission and income inequality, as measured by Gini coefficient across selected developed and developing countries. As mentioned in the methodology that the null hypotheses (H_0) represent no way of causal interplay between the two variables and alternative hypotheses (H_(1)) represent there have unidirectional or bidirectional causal relation between two. Table 4 represents the results of Granger Causality tests between CO2 emission and income inequality in the selected countries.

Table 4. Causality test results

Null hypothesis: No way of causality Alternative hypothesis: There exists causality (unidirectional or bidirectional)				
Null hypothesis	**Optimum Lag**	**F-statistics**	**Probability**	**Directions of Causality**
Developed countries				
USA	3	4.92479 1.06303	0.0141 0.3941	CO2 is the cause of Gini Gini does not Granger cause of CO2
Japan	1	0.00145 0.02866	0.9700 0.8672	No way of causality
Germany	1	5.44547 3.97444	0.0296 0.0593	CO2 is the cause of Gini Gini does not cause of CO2
Canada	1	3.27040 0.52201	0.0849 0.4780	No way of causality
UK	2	6.52362 1.86942	0.0074 0.1829	CO2 is the cause of Gini Gini does not Granger cause of CO2
Italy	5	6.59149 1.91541	0.0076 0.1871	CO2 is the cause of Gini Gini does not Granger cause of CO2
South Korea	2	0.03268 0.92002	0.9679 0.4165	No way of causality
Saudi Arabia	1	0.49895 0.60350	0.4877 0.4459	No way of causality
Australia	1	1.93600 17.5271	0.1787 0.0004	CO2 is not Granger cause of Gini Gini is the cause of CO2
Poland	1	0.55125 0.02811	0.4660 0.8685	No way of causality
Developing countries				
China	1	0.30178 38.0556	0.5886 0.0000	No way of causality
India	1	0.11806 6.46716	0.7346 0.0189	CO2 is not granger cause of Gini Gini is the cause of CO2
Russia	2	7.96016 6.87770	0.0033 0.0060	Causality for both
Indonesia	2	3.93634 0.30288	0.0382 0.7424	CO2 is the cause of Gini Gini does not Granger cause of CO2
Iran	1	7.96673 2.29942	0.0102 0.1443	CO2 is the cause of Gini Gini does not Granger cause of CO2
Mexico	1	2.96551 1.35865	0.0998 0.2568	No way of causality
Turkey	1	0.43662 0.66787	0.5159 0.4230	No way of causality
Brazil	1	0.75576 2.22322	0.3945 0.1508	No way of causality

continued on following page

Table 4. Continued

Null hypothesis: No way of causality Alternative hypothesis: There exists causality (unidirectional or bidirectional)				
Null hypothesis	**Optimum Lag**	**F-statistics**	**Probability**	**Directions of Causality**
South Africa	1	0.04091 0.01470	0.8417 0.9046	No way of causality
Vietnam	1	4.58450 0.18388	0.0441 0.6724	CO2 is the cause of Gini Gini does not Granger cause of CO2

Source: Authors' calculations

By estimating the equation (7, 8))), the study observes among the developed countries that a unidirectional causality runs from per capita CO2 emission to the Gini coefficient in four namely, USA, Germany, UK, and Italy. It means the change in carbon emission significantly influence income inequality in these economies, or, environmental pollution makes a causal influence to the income inequality in these four countries. In these countries, carbon emissions aggravate income inequality through consumption patterns, access to green technologies, and political influence. Richer individuals tend to consume more carbon-intensive goods and services, driving up overall emissions and disproportionately impacting lower-income groups. Unequal access to green technologies and environmental awareness also contributes to this disparity, as lower-income households may be forced to rely on more polluting options (Bergougui et al, 2025). Conversely, in case of Australia, causality runs from Gini coefficient to per capita CO2 emission, that means inequality may drive change in the CO2 emission in the country. Income inequality in Australia, particularly the consumption habits of the wealthiest, is a significant driver of increased carbon emissions. The wealthy tend to have larger carbon footprints due to higher consumption of energy-intensive goods and services, as well as increased travel, leading to a greater overall environmental impact (Cheng et al, 2023). Out of all the developed countries, there are five countries, namely, Japan, Canada, South Korea, Saudi Arabia, Poland, where there is no way of causality i.e., we accept H_0: No way of causality, for these countries.

From the results of the developing countries, the study finds that per capita CO2 emission makes a cause to the Gini coefficient in unidirectional way in three developing countries out of ten namely, Indonesia, Iran, Vietnam. Unequal access to green technologies and environmental awareness in these countries might contribute to this disparity, as lower-income households may be forced to rely on more polluting options (Bergougui et al, 2025). Only in case of India CO2 is the Granger cause of Gini coefficient. The result resembles Australia (Cheng et al, 2023). Out of ten developing countries, only Russia is having bidirectional causality. There are five countries out of the developing countries which are namely, China, Mexico,

Turkey, Brazil, South Africa where the study does not find any direction of causality between two.

Table 5 is the summary table consisting of correlation, cointegration, error correction and causality results for the two sets of the countries.

Table 5. Summary results of correlation, cointegration and causality

Subjects	Developed countries	Developing countries
Positive and significant correlation	Canada, UK, South Korea, Saudi Arabia	China, India, Mexico, South Africa
Negative and significant correlation	USA, Germany, Poland	Iran, Brazil, Vietnam
Significant cointegration	USA, Japan, Germany, UK	China, Iran, Turkey, Vietnam
Significant cointegration with error correction	USA, Japan, Germany, UK	China, Iran, Turkey, Vietnam
Causal influence from inequality to CO_2	Australia	India
Causal influence from CO_2 to inequality	USA, Germany, UK, Italy,	Indonesia, Iran, Vietnam
Both way causality	Germany	Russia

Source: Authors' own construction

Out of the twenty countries eight countries produce the results of significant cointegration with error correction while carbon emission makes a cause to inequality for seven countries with two countries having bidirectional causal interplays. The results for the developing countries go in line with the studies such as Gao & Fan (2023), and for the results of the developed countries and the developing countries together, they mostly resemble to the study of Wan et al (2022), Boyace (1994), Belaid et al (2020), among the others.

CONCLUSION

This study aimed to explore the association as well as relationships between income inequality and CO2 emission in the world's top twenty CO2 emitting countries for the long run during the period 1995-2019. Using correlation analysis, it finds both positive and negative correlation coefficients between the two in many of the countries in the list. The cointegration test finds that CO2 and Gini values maintain long run and stable relationships in four developed and four developing countries such as USA, UK, China, Iran. Further, the Granger causality test shows a mixture of results in either of the directions of causality. The relationship between CO2 emission and income inequality is justified because wealthier nations disproportionately contribute to pollution while suffering fewer negative consequences, and low income countries are more vulnerable to the adverse health and economic impacts of pollution. This creates a cycle where high inequality leads to more pollution, and in turn, pollution deepens existing inequalities by disproportionately harming marginalized groups and hindering their recovery from environmental destruction.

The study thus recommends that controlling income inequality and carbon emission might lead to the betterment of the countries in particular and the world as a whole in general as these twenty countries hold the carbon emission share of more than 60 per cent globally. The countries should adopt the policies for managing the two curse to human society and existence leading to attain sustainable development. The recommended policies are categorized as-

- To maintain proper distribution of working activities and income across the society
- To minimize the corruption and illegal financial flows to reduce the differences between the upper and lower income people
- To increase spending upon green energy developments
- R&D activities are to be strengthened

DISCLOSURE STATEMENT

The authors do hereby disclose that they do not have any conflict of interest and have no funding support behind the work.

REFERENCES

Ajide, K. B., & Ibrahim, R. L. (2021). Environmental impacts of income inequality: Evidence from G7 countries. *Environmental Science and Pollution Research International, 29*(2), 1887–1908. PMID: 34363162

Ansari Samani, H., Rouzbahani, M., Dalvandi, H., & Sadeghi, R. (2024). Income inequality and environmental degradation in the provinces of Iran. Environmental Health Enginnering and Management Journal, 11 (1), 25-30. http://ehemj.com/article-1-1301-en.html

Basili, M., Franzini, M., & Vercelli, A. (2005). *Environment, Inequality and Collective Action*. Routledge., DOI: 10.4324/9780203481509

Belaid, F., Boubaker, S., & Kafrouni, R. (2020). Carbon emissions, income inequality and environmental degradation: The case of Mediterranean countries. *The European Journal of Comparative Economics, 17*(1), 73–102.

Bergougui, B., Ben-Salha, O., & Murshed, S. M. (2025). How income inequality drives climate inequality in the world's largest emitting nations: Novel evidence from multivariate quantile-on-quantile and wavelet quantile methods. *Journal of Cleaner Production, 513*. Advance online publication. DOI: 10.1016/j.jclepro.2025.145666

Boyce, J. K. (1994). Inequality as a cause of environmental degradation. *Ecological Economics, 11*(3), 169–178. DOI: 10.1016/0921-8009(94)90198-8

Chen, J., Xian, Q., Zhou, J., & Li, D. (2020). Impact of income inequality on CO2 emissions in G20 countries. *Journal of Environmental Management, 271*, 110987. DOI: 10.1016/j.jenvman.2020.110987 PMID: 32579533

Cheng, C., Suhui, L., Qi, Y., Qing, L., Xin, G. & Huixian, Z. (2023). Does income inequality have a heterogeneous effect on carbon emissions between developed and developing countries? Evidence from simultaneous quantile regression, Frontiers in Environmental Science, 11, DOI=DOI: 10.3389/fenvs.2023.1271457

Ching, S. L., Lau, L. S., & Choong, C. K. (2023). Income inequality, educational attainment & environmental degradation: Evidence from global panel. *Environmental Science and Pollution Research International, 30*(15), 43056–43067. PMID: 35508851

Coondoo, D., & Dinda, S. (2007). Carbon dioxide emission and income: A temporal analysis of cross-country distributional patterns. *Ecological Economics, 65*(2), 375–385. DOI: 10.1016/j.ecolecon.2007.07.001

Das, R. C., Chatterjee, T., & Ivaldi, E. (2023). Co-movements of income and urbanization through energy use and pollution: An investigation for world's leading polluting countries. *Ecological Indicators, 153*, 110381. DOI: 10.1016/j.ecolind.2023.110381

De Leon Barido, D. P., & Marshall, J. D. (2014). Relationship between Urbanization and CO2 Emissions Depends on Income Level and Policy. *Environmental Science & Technology, 48*(7), 3632–3639. DOI: 10.1021/es405117n PMID: 24422489

Demir, C., Cergibozan, R., & Gök, A. (2018). Income inequality and CO2 emissions: Empirical evidence from Turkey. *Energy & Environment, 30*(3), 444–461. DOI: 10.1177/0958305x18793109

Dickey, D. A., & Fuller, W. A. (1979). Distribution of the estimators for autoregressive timeseries with a unit root. *Journal of the American Statistical Association*, •••, 74.

Dorsch, M. T., & Kirkpatrick, B. (2021). Economic growth, inequality and environmental degradation. *International Journal of Sustainable Development, 24*(2), 124. DOI: 10.1504/ijsd.2021.118844

Granger, C. W. J. (1969). Investigating causal relations by econometric models and cross spectral methods. *Econometrica, 37*(3).

Grunewald, N., Klasen, S., Martínez-Zarzoso, I., & Muris, C. (2017). The trade-off between income inequality and carbon dioxide emissions. *Ecological Economics, 142*, 249–256. DOI: 10.1016/j.ecolecon.2017.06.034

Hailemariam, A., Dzhumashev, R., & Shahbaz, M. (2020). Carbon emissions, income inequality and economic development. *Empirical Economics, 59*(3), 1139–1159.

IEA. (2025), Global Energy Review 2025, IEA, Paris https://www.iea.org/reports/global-energy-review-2025

Jiang, J., Ye, B., & Liu, J. (2018). Research on the peak of CO2 emissions in the developing world: Current progress and future prospect. *Applied Energy, 235*, 186–203. DOI: 10.1016/j.apenergy.2018.10.089

Kang, H. (2022). Impacts of income inequality and economic growth on CO_2 emissions: Comparing the Gini coefficient and the top income share in OECD countries. *Energies, 15*(19), 6954. DOI: 10.3390/en15196954

Khan, H., Weili, L., Khan, I., & Han, L. (2022). The effect of income inequality and energy consumption on environmental degradation: The role of institutions and financial development in 180 countries of the world. *Environmental Science and Pollution Research International, 29*(14), 20632–20649. PMID: 34741747

Liu, Q., Wang, S., Zhang, W., Li, J., & Kong, Y. (2018). Examining the effects of income inequality on CO_2 emissions: Evidence from non-spatial and spatial perspectives. *Applied Energy, 236*, 163–171. DOI: 10.1016/j.apenergy.2018.11.082

Ogede, J. S., Oduola, M. O., & Tiamiyu, H. O. (2024). Income inequality and carbon dioxide (CO_2) in sub Saharan Africa countries: The moderating role of financial inclusion and institutional quality. *Environment, Development and Sustainability, 26*(7), 18385–18409.

Padilla, E., & Serrano, A. (2005). Inequality in CO_2 emissions across countries and its relationship with income inequality: A distributive approach. *Energy Policy, 34*(14), 1762–1772. DOI: 10.1016/j.enpol.2004.12.014

Pata, U. K., Yilanci, V., Hussain, B., & Naqvi, S. A. (2022). Analyzing the role of income inequality and political stability in environmental degradation: Evidence from South Asia. *Gondwana Research, 107*, 13–29. DOI: 10.1016/j.gr.2022.02.009

Scruggs, L. A. (1998). Political and economic inequality and the environment. *Ecological Economics, 26*(3), 259–275. DOI: 10.1016/s0921-8009(97)00118-3

Soares, T. C., Fernandes, E. A., & Toyoshima, S. H. (2017). The CO_2 emission Gini index and the environmental efficiency: An analysis for 60 leading world economies. *Economía, 19*(2), 266–277. DOI: 10.1016/j.econ.2017.06.001

Sun, Y. (2023). Income inequality, carbon emissions and green development efficiency. *Environmental Science and Pollution Research International, 30*(8), 21081–21091. PMID: 36264465

Uddin, M. M., Mishra, V., & Smyth, R. (2020). Income inequality and CO_2 emissions in the G7, 1870–2014: Evidence from non-parametric modelling. *Energy Economics, 88*, 104780. DOI: 10.1016/j.eneco.2020.104780

Wang, J. (2024). Renewable energy, inequality and environmental degradation. *Journal of Environmental Management, 356*, 120563. DOI: 10.1016/j.jenvman.2024.120563 PMID: 38479288

Wang, Y., Uddin, I., & Gong, Y. (2021). Nexus between Natural Resources and Environmental Degradation: Analysing the Role of Income Inequality and Renewable Energy. *Sustainability*, *13*(15), 8364. DOI: 10.3390/su13158364

Wolde-Rufael, Y., & Idowu, S. (2016). Income distribution and CO2 emission: A comparative analysis for China and India. *Renewable & Sustainable Energy Reviews*, *74*, 1336–1345. DOI: 10.1016/j.rser.2016.11.149

KEY TERMS AND DEFINITIONS

Causal interplay: It gives the direction of causal interplays between the pair of variables that happen in the short run. It is developed by Granger (1969).

CO_2 emission: It is the discharge of carbon dioxide gas into the atmosphere, primarily from the burning of fossil fuels, the manufacture of cement, and other human activities which disrupts the natural carbon cycle, adding a significant amount of a powerful greenhouse gas that contributes to global warming and climate change.

Cointegration: It explains whether two time series variables do have any long run relationships. It is first developed by Engel & Granger (1987).

Correlation: It measures how the two variables are associated. It is the ratio of covariance to variances of the two variables. It ranges from -1 to +1.

Environment: It is a sum total of all the living and non-living elements and their effects that influence human life.

Gini coefficient: It is an important measure of income, wealth and consumption inequality. It is a statistical measure whose values range from 0 to 1, where 0 indicates perfect equality (everyone has the same income) and 1 indicates perfect inequality.

Inequality: It is the uneven distribution of a population's annual earnings, wages, profits, and transfers, while wealth inequality is the uneven distribution of a population's net worth. Gini index/coefficient is the popular measure of income inequality.

Pollution: It is the introduction of harmful or contaminating substances into the environment, causing adverse changes and potentially harming living organisms and ecosystems. It may be of air source, water source, soil source etc.

Chapter 6
Environmental Degradation Influencing Factors in Different Income Groups Countries

Ajay Kumar Singh
https://orcid.org/0000-0003-0429-0925
Department of Humanities and Social Sciences, Graphic Era University, Dehradun, India

S. Srinivasan
https://orcid.org/0009-0002-0179-9849
Department of Humanities and Social Sciences, Graphic Era University, Dehradun, India

ABSTRACT

This chapter assesses the environmental degradation leading factors in different income groups of countries. It employs country-wise panel data for each income group of countries. It develops a log-linear regression model to achieve the above-mentioned objective. Per capita CO2 emissions are considered as a representative variable for environmental degradation in the empirical investigation. Education expenditure, age dependency ratio, agricultural land, crop production index, employment to population ratio, energy use, renewable energy, trade openness, forest area, human development index, individual using internet, inflation, technological change, urban population, vulnerable employment, population growth, sex ratio and corruption control are used independent variables in the regression analysis.

DOI: 10.4018/979-8-3373-1862-2.ch006

Renewable energy consumption is found useful to increase environmental quality. Arable land, crop production, energy use, human development and vulnerable employment have negative implications on environmental quality in all income countries.

INTRODUCTION

Environmental degradation is continuously increasing due to several reasons like urbanization, population growth, industrialization, consumption of non-renewable energy, agricultural intensification, and unsustainable use of natural resources in production activities across countries (Alvarado et al., 2018; Zhang et al., 2019; Uzar & Eyuboglu, 2023). The high economic growth is found to be a significant variable which is creating environmental issues in the world. However, high economic growth and development cannot be achieved without use of energies and natural resources in the production units (Wang, 2024). While extensive consumption of non-renewable energy and natural resources are also creating a negative impact on the environment (Zhang et al., 2019). Further both are diminishing the common properties of natural resources and sources of energies. Alvarado et al. (2018) argued that energy consumption is increasing CO_2 emissions in all income groups of countries. Environmental quality is a key aspect of sustainable development. Therefore, environmental degradation is creating multiple issues for global countries to achieve sustainable development and SDGs. Environmental degradation brings multiple health issues, and it has several negative impacts on social and economic sectors in the long-run (Osman et al., 2024). For example, poor quality of air and environment are caused to increase multiple health issues (Shetty et al., 2023). Thus, environmental degradation has an adverse impact on human health. Agricultural production produces a negative impact on ecosystem services (Jatav et al., 2023; Ali et al., 2023). Osman et al. (2024) noted a negative impact of agricultural expansion on the environment in Somalia. Moreover, extensive infrastructural development is helping to increase economic development, while it is producing a negative impact on the environment (Awad et al., 2023). Most developed countries have the appropriate resources to maintain the pace in sustainable development and environmental sustainability. Low-income and larger agrarian countries are also facing multiple problems due to climate change, population growth, urbanization and industrialization. These countries have to achieve high economic growth through extensive industrialization and intensification of production sectors to be globally competitive countries. Thus, low-income countries cannot implement strict environmental policies to reduce environmental degradation. Low-income countries also do not appropriate physical infrastructure and technical resources to increase environmental sustainability. Industrial production is based on eco and environmentally friendly ecosystems in developed countries (Rahman et al., 2023). Thus, developed countries have the ability to maintain the pace in high economic growth and environmental sustainability. While most developing nations wish to attain high economic growth, thus, they are paying relatively low attention towards environmental sustainability.

There are other reasons like technological development, green innovation, green technologies, green products, multiple sources of renewable energy and low pressure of population of natural resources are providing benefits to increase protection of natural resources in developed countries (Eweade et al., 2024). While economic growth also increases load capacity factor (LCF) in the developing countries (Onwe et al., 2024). LCF is essential to increase economic development (Hongjie & Muzi, 2025). Economic growth appears positive to increase environmental quality in high income countries (Luo et al., 2023). Conversely, many developing countries appear unable to manage the industrial wastage for further environmental development. Therefore, many developed countries have less priority towards environmental development. Hence, it is necessary for low-income countries to reduce CO_2 emissions (Luo et al., 2023). Reduction in global carbon emissions is the only solution to increase environmental quality (Zhang et al., 2019). The high competition to achieve high economic growth would lead to increased fossil fuel consumption and environmental degradation across emerging countries (Nain et al., 2015; Li, 2022; Chen et al., 2023).

The transition from fossil fuels to renewable energy would be a crucial initiative to increase the environmental quality (Ssekibaala et al., 2022). High economic growth also increased the imbalance in the environmental sector in developing countries (Pujiati et al., 2023). Hence, these countries have more environmental degradation as compared to developed countries. There is a significant diversity in environmental degradation across countries. Previous studies could not examine the environmental degradation influencing factors in different income groups of countries using robust empirical models. Other dimensions of demographic patterns like urbanization and population density are increasing environmental degradation in different forms (Bello et al., 2018; Alvarado et al., 2018; Hanif, 2018; Wang & Dong, 2019; Nathaniel et al., 2020; Osman et al., 2024; Kumar & Datta, 2024). Urbanization led to increased demand for energy and enhanced environmental degradation (Wang & Dong, 2019). Overwhelming population growth is increasing enormous pressure on ecosystem services, natural resources and environment (Cropper & Griffiths, 1994). Population is accountable to increase air, water and solid-waste pollution (Cropper & Griffiths, 1994; Osman et al., 2024).

Natural resources are crucial inputs for agricultural development. Agricultural sector, therefore, is also facing multiple problems due to scarcity of natural resources. Application of natural resources are depleting due to change in food habits and consumption patterns of people. Farm management practices and agricultural intensification are highly essential to increase farm population (Jatav et al., 2023). While advanced farm management is creating multiple issues for the environmental sector. Furthermore, the industrial sector is using an extreme quantity of natural resources to sustain their production scale and meet the demand of a growing population. Technological change and innovation are helping industrial and manufacturing sectors to produce innovative goods and services to the consumers. These are valuable for extensive industrial development. Wastages from industrial, agricultural and food sectors are also a challenge for environmental Infrastructural development is crucial to create jobs and physical infrastructure.

The above-mentioned activities are contributing to increase GHGs and CO2 emissions in the environment. Moreover, global countries are highly keen to achieve high economic growth to maintain their global competition. Globalization could be beneficial for global countries to increase foreign trade. Many developing countries are enjoying trade surplus due to international trade (Zhang et al., 2019). Trade may be positive to open the possibilities of sustainable development (Barkat et al., 2024). However, foreign trade is accountable for increasing environmental degradation in many developing and emerging countries. Trade is a significant driver to increase energy-intensive production processes (Barkat et al., 2024). Trade openness is caused to increase deforestation (Ssekibaala et al., 2022). Trade openness is highly accountable for environmental degradation. Although, FDI is also found harmful for environmental development (Nguyen et al., 2024). Udemba (2021) noticed the FDI showed a negative impact on the environment in Chile. High economic growth and development are not possible without using natural resources (i.e., water, air, and energy) in the production of goods in all sectors including agriculture, industrial, transport, and business. Hence, economic development is significantly linked with natural resources (Coronado et al., 2024). While high economic growth hampers the environmental quality (Rakshit & Neog, 2021).

Further, economic development is positive to increase social and human development. Conversely, these activities are harmful for environmental development. For instance, Gu et al. (2017) explored that rapid socioeconomic progress is increasing pollution in China (Gu et al., 2017). Moreover, income inequity, poverty, low literacy and gender inequality create multiple environmental challenges (Wang et al., 2021). Mahalik et al. (2018) recognized a positive impact of income inequality on environmental degradation in Brazil, India and China. Climate change and global warming are being more sensitive due to environmental degradation (Singh et al., 2019; Zhang et al., 2022; Li, 2022; Nguyen et al., 2024; Coronado et al., 2024). It has been a global concern due to its implications on all sectors associated with human livelihoods (Ahakwa et al., 2023). Further, climate change produces a negative impact on the agricultural and health sector (Singh, 2020). Environmental degradation and quality are becoming more sensitive due to GHGs and CO2 emissions worldwide (Wei et al., 2022).

BACKGROUND

Earlier studies observed the impact of economic and social development on environmental sustainability and performance using various indexes in the empirical investigations (Onwe et al., 2024). Renewable energy, green technology, green entrepreneurship, green innovation, green business, green jobs, green fund and green GDP and green projects would be conducive to reduce environmental degradation (Ahmad et al., 2020; Udemba et al., 2021; Zhang et al., 2022; Rahman et al., 2023; Kumar & Datta, 2024). Earlier studies have used CO_2 emissions as a representative variable for environmental degradation (Vidyarthi, 2013; Wang, 2024). Accordingly, previous research develops an empirical association between energy consumption, economic growth and environmental degradation in different income groups of countries (Vidyarthi, 2013; Wang, 2024). Furthermore, it is observed a diverse association of the environmental sector with renewable energy and urbanization worldwide (Bello et al., 2018; Acheampong, 2018; Zhang et al., 2019; Uzar & Eyuboglu 2023; Kumar & Datta, 2024). EKC theory explores that initially economic growth is caused to increase environmental degradation (Acheampong, 2018). Environmental Kuznets Curve (EKC) is the oldest in environmental economics (Rakshit & Neog, 2021; Ssekibaala et al., 2022; Osuntuyi & Lean, 2022; Naseem et al., 2022; Kumar & Datta, 2024). Environmental degradation and economic growth do not have a uniform association across countries under EKC theory (Kumar & Datta, 2024; Wang, 2024). Past evidence infers that environmental degradation and economic growth have a U-shaped and inverted U-shaped relationship in different countries under EKC hypothesis (Hanif, 2018).

Existing literature could detect an ambiguous relationship between environmental sustainability and digital technologies. As different income groups, countries have a diverse ecosystem with regards to above-mentioned variables. Thus, these countries have diversity in environmental degradation. However, existing researchers could not investigate the environmental degradation increasing factors in different income groups of countries. This chapter, thus, desire to assess the answer for following research questions:

1. What is the impact of macro level indicators on environmental degradation in different income groups of countries?
2. How environmental degradation is creating barriers for SDGs?
3. How can global countries reduce environmental degradation?

This chapter is attaining following research objectives:

- To estimate the environmental degradation influencing factors in different income groups of countries.
- To provide policy suggestions to reduce environmental degradation and achieve SDGs.

LITERATURE REVIEW

Past literature examines the factors affecting environmental degradation across countries (Zhang et al., 2019; Rakshit & Neog, 2021; Osuntuyi & Lean, 2022; Luo et al., 2023; Uzar & Eyuboglu 2023). Environmental development has multifaceted association with different factors. Researchers and scientists accepted that rising GHGs and CO2 emissions are increasing environmental degradation. CO2, N2O, SO2 and CH4 are highly harmful for ecology and environment (Munir & Ameer, 2018; Kumar & Datta, 2024). Most studies, therefore, applied diverse indicators for environmental degradation and ecological footprint (EF) for assessing its inter-linkages with social and economic promoting indicators (Bello et al., 2018; Wang & Dong, 2019; Ahmad et al., 2020; Ünal & Aktuğ, 2022; Haq et al., 2023; Uzar & Eyuboglu 2023; Eregha et al., 2023; Hongjie & Muzi, 2025).

Many researchers assess the impact of various indicators in the domain of economic growth, income inequality, poverty, foreign trade, FDI, globalization, natural resource rent, green technology, green innovation, environmentally-friendly technologies, inflation and financial inclusion on environmental performance (Xiao et al., 2022; Ssekibaala et al., 2022; Zhang et al., 2022; Grolleau & Weber, 2024; Onwe et al., 2024; Hondroyiannis et al., 2025). Moreover, previous literature explored the implications of ecological footprint on human health, total fertility rate, economic and social development, financial inclusion, foreign trade, etc. (Coronado et al., 2024). Wei et al. (2022) found that GHGs emissions are accountable for human health disaster risk. Haq et al. (2023) studied the association of ecological footprint with total fertility rate in 31 cross countries.

Economic Growth and Development: Economic growth and development depend on production of products in a nation. A country can increase economic development by increasing extensive production. While production requires more natural resources. Economic growth, thus, is highly accountable for environmental degradation (Hanif, 2018; Uzar & Eyuboglu 2023). Vidyarthi (2013) investigated the impact of economic growth on CO_2 emissions in India. Alvarado et al. (2018)

explored the association between per capita output and CO_2 emissions. Pradhan et al. (2024) noticed that per capita GDP has a positive contribution in environmental pollution in G7 and Asian countries. For instance, high per capita GDP requires more consumption of energy (Pradhan et al., 2024). While energy consumption has a detrimental impact on the environment. Acheampong (2018) investigated the causal relationship between CO_2 emissions and economic growth in 116 countries. It noted that carbon emissions have a positive impact on economic growth. Pradhan et al. (2024) also found a positive impact of CO_2 emissions on economic growth in G7 and South Asian countries. Economic growth is a challenge for environmental protection in global countries (Osman et al., 2024; González-Álvarez & Montañés, 2023; Ganda & Ruza, 2025). However, economic growth could be helpful to increase environmental quality in many developed countries.

Agricultural Growth and Development: The agricultural sector requires more natural resources like land and water to increase its production. Without irrigation and land agricultural production is not possible. Agricultural sector has a significant share in CO_2 and GHGs emission. Moreover, at present agriculture depends heavily on chemical fertilizers, pesticides and mechanization, while these are emitting high quantities of GHGs emissions. Thus, it causes increased environmental degradation and this sector also bears its negative impacts.

Technological Change and Green Innovation: Technological change and green innovation may be positive and negative for the environmental sector (Naseem et al., 2022). Green innovation helps to develop and discover green products. Thus, green innovation creates a healthy and clean environment (Topuz et al., 2025). Ahmad et al. (2020) noted the positive impact of technological innovation on the environmental sector. Environmentally friendly technologies would also enhance sustainability of the environmental sector (Rakshit & Neog, 2021). Chen et al. (2023) suggested that environment-related technologies would be imperative to reduce environmental degradation. Green economy would nurture a sustainable and active path to increase environmental quality (Zhang et al., 2022).

Uncertainty in Macroeconomic Indicators: Uncertainty in macroeconomic indicators is also valuable to increase environmental degradation. The association of macroeconomic uncertainty with environmental degradation is observed by Rakshit and Neog (2021) in India. It noted that macroeconomic uncertainty leads to increased environmental degradation. Sahu et al. (2023) found that macroeconomic stability is negatively impacted due to environmental degradation in 22 emerging countries.

Income Inequality: Income inequality is caused to increase EF (Uzar & Eyuboglu, 2023). Further, income inequality may be harmful for the environment (Wang et al., 2021).

Education Quality: Education is impactful to increase the awareness among the people towards environmental issues. Thus, it may drive environmental quality.

Education is found suitable to increase environmental sustainability (Osuntuyi & Lean, 2022). Further education would be favourable to increase energy-efficient products and green technology investment.

Demographic Change: Demographic changes like population growth, urbanization, population density are also harmful for the environment (Alvarado et al., 2018; Xiao et al., 2022; Naseem et al., 2022; Pujiati et al., 2023). Alvarado et al. (2018) noted an increasing trend in CO_2 emissions due to urbanization in 151 countries. Osman et al. (2024) noticed that population is associated with environmental degradation in Somalia. However, the impact of urbanization on the environment is also reported positive in some countries (Bello et al., 2018). Raghutla and Chittedi (2021) claimed an increase in environmental quality due to increase in urban population in BRICS countries. Hanif (2018) found that urban areas are caused to increase environmental degradation in Sub-Saharan African countries. Pujiati et al. (2023) determine the factors affecting environmental quality in Indonesia. Environmental quality is significantly determined by HDI, FDI and population growth. Kumar and Datta (2024) investigated the association of economic growth, and energy consumption with air pollution (i.e., CO2, N2O, SO2 and CH4) across 55 countries using a EKC theory. Alvarado et al. (2018) observed a strong U-shaped relationship between real per capita output and CO_2 emissions in middle-high and low-income countries. Zhang et al. (2019) found an inverted U-shaped EKC in few countries. González-Álvarez and Montañés (2023) examined the association between CO_2 emissions, energy consumption and economic growth in 31 countries. Bello et al. (2018) found an inverted U-shaped association between GDP and environmental degradation in Malaysia. Hongjie and Muzi (2025) investigated the load capacity factor (LCF) promoting variables in France.

Industrial Development: Rahman et al. (2023) examined the impact of industrial production and energy consumption on environmental degradation in G8 and southeast Asian countries. It noticed that industrial production is environmentally friendly in developed countries. Ahakwa et al. (2023) noticed that industrial development accelerates environmental degradation in the long-run.

Financial Development (FD): Raghutla and Chittedi (2021) claimed that financial development could be helpful to increase environmental quality in BRICS nations. Yang et al. (2021) also noted a negative impact of FD on environmental quality in GCC countries. Ganda and Ruza (2025) argued that financial development helps to reduce CO2 emissions after a certain extent.

Energy Security and Efficiency: Energy security is viable to increase environmental quality (Rahman et al., 2023). As economic development has an extensive dependence on energy consumption (Coronado et al., 2024). However, extensive energy growth may produce a negative impact on the environmental sector (Singhal et al., 2022). Yang et al. (2021) noticed that energy utilization could be harmful for environmental sustainability in GCC countries. Management of energy demand, thus, would be beneficial to improve environmental quality (Naseem et al., 2022). Sources of renewable energy would lead to improved air quality and environmental development. Furthermore, clean and renewable energy would also promote environmental quality (Nathaniel et al., 2020; Raghutla & Chittedi, 2021; Zhang, 2023). Carbon-free energy sources will enhance environmental quality (Vidyarthi, 2013). Renewable electricity consumption helps to minimize CO_2 emissions (Pradhan et al., 2024). Energy transition is also required to minimize climate change impact and environmental degradation (Patel et al., 2024). Further, energy transition helps to reduce carbon inequality across countries (Patel et al., 2024). Promotion of energy efficiency would be positive to increase environmental quality (Vidyarthi, 2013). Eregha et al. (2023) also noticed that energy consumption is creating environmental degradation. While Rakshit and Neog (2021) noticed that renewable consumption intensity is positive to increase environmental degradation in India. Li (2022) examined the factors affecting energy productivity in 7 emerging countries. Energy productivity may be an effective solution to resolve the challenge of environmental degradation. Energy development is essential to increase protection of natural resources. While energy consumption also seems negative for environmental quality. (Uzar & Eyuboglu, 2023). Efficient energy technologies would play an active role to increase environmental quality (Pradhan et al., 2024). Thereupon, energy inflation is also found a primary driver of CO2 emissions in high income countries (Hondroyiannis et al., 2025).

Renewable Energy Consumption (REC): Extensive energy consumption is caused to increase environmental degradation in terms of increasing CO_2 emissions (Zhang et al., 2019). REC is helpful to reduce CO_2 and GHGs emissions. Thus, REC would be positive to reduce environmental degradation worldwide. However, CO_2 emissions are causing more energy consumption. Acheampong (2018) analyzes the impact of energy consumption on CO_2 emissions in 116 countries. Energy consumption is also an important driver of economic growth (Pradhan et al., 2024).

Fossil Fuels: Application of fossil fuels are increasing due to expansion of economy and technological change (Wang, 2024). Fossil fuels are contributing GHGs and other gases in the environment (Hanif, 2018; Ssekibaala et al., 2022; Wang, 2024). Many countries apply fossil fuels to meet their energy requirements. Hence, it is accountable for increasing water footprint and environmental degradation (Bello et al., 2018; Udemba et al., 2021). While hydroelectricity consumption is found suitable to decrease environmental degradation (Bello et al., 2018).

Foreign Trade and Trade Openness: Foreign trade is a valuable determinant of economic growth (Ssekibaala et al., 2022). Foreign trade, thus, is significantly associated with the environment (Nathaniel et al., 2020). Vale et al. (2017) examined the impact of international trade on CO2 emissions. Ssekibaala et al. (2022) noted that international trade appears harmful for the environmental sector in Sub-Saharan countries. Topuz et al. (2025) claimed that export diversification is decreasing environmental quality in 21 European countries.

Globalization: The global countries are also bound to maintain the competition in production activities due to globalization. Globalization is increasing deforestation (Zhang et al., 2022). Chen et al. (2023) also recognized that globalization is increasing environmental degradation in 17 emerging countries. Shayanmehr et al. (2023) claimed that economic globalization has an adverse impact on ecological footprint. Yang et al. (2021) found deterioration in the environment due to globalization in GCC countries.

Environmental Policy and Environmental Tax: Environmental taxes can be conducive to reduce environmental degradation (Wolde-Rufael & Mulat-Weldemeskel, 2021). Topuz et al. (2025) noted that environmental taxes could be harmful for environmental quality in 21 European countries. Nain et al. (2015) also suggested that environmental policies should be implemented to sustain the link between energy consumption and GDP in India. Shayanmehr et al. (2023) reported that environmental tax is useful to minimize environmental degradation. Environmental regulations may be beneficial to increase environmental protection. The government policies must be implemented to increase the protection and conservation of forest area, rivers and lakes. While adoption of strict environmental policies depends on good governance and political regime. Hence, political regime and governance are also important to increase environmental performance. Mignamissi et al. (2024) analyze the role of political regime and governance in environmental degradation in Africa. It found a negative association between better institutions and pollution. Therefore, the government should make it mandatory for the production sector to apply environmentally friendly technologies to increase environmental quality (Alvarado et al., 2018).

Natural Resources: Extensive application of natural resources also reduces environmental quality (Ahakwa et al., 2023). It proposed that high-income countries should improve natural resource rent policies to reduce environmental degradation. Ahakwa et al. (2023) explored the impact of natural resources on CO2 emissions in Ghana. Table 1 highlights the environmental degradation affecting factors as per the existing studies.

Climate Change and Natural Disaster: Climate change is also a significant challenge for environmental sustainability. Climate change is creating multiple issues for the environment (Acheampong, 2018). While environmental degradation is also caused to increase the fluctuation in climatic factors. Thereupon, climate change is creating obstacles to increase sustainable development and sustainable development goals (González-Álvarez & Montañés, 2023).

Table 1. Environmental degradation affecting factors

Author(s)	Research Area	Objectives	DVs/IVs	Main findings
Bello et al. (2018)	Malaysia	Examine the environmental degradation due to hydroelectricity consumption.	Hydroelectricity consumption, various forms of environmental degradation, GDP and urbanization.	Significant association between CO_2 emissions and GDP.
Hanif (2018)	Sub-Saharan African countries	Examine the factors affecting environmental degradation.	CO_2 emissions, economic growth, renewable energy, fossil fuels	Non-existence of U-shaped association between economic growth and environmental degradation.
Munir and Ameer (2018)	Emerging Asian counties	Examine the long-run and short run association of SO2 emissions with economic growth and trade openness	SO2 emissions, trade openness, urbanization, economic growth, technology	Trade openness and technology is caused to enhance SO2 emissions.
Wang and Dong (2019)	Sub-Saharan countries	Determine the factors affecting environmental degradation	CO_2 emissions, economic growth, and urbanization	EF is significantly increase due to economic growth.
Rakshit and Neog (2021)	India	Association of macroeconomic uncertainty with CO_2 emissions	CO_2 emissions, and financial development	Environmental quality is negatively impacted due to financial development.
Raghutla and Chittedi (2021)	BRICS	Impact of FD, and technology on CO2 emissions	CO_2 emissions, FD, technology, energy consumption, urbanization	FD, urban population and technology are found significant for environmental quality.
Ahmad et al. (2020)	Emerging countries	Impact of technological innovation and economic growth on EF	EF, technological innovation and economic growth	Economic growth is lead to increase EF. While technological innovation is appeared negative for EF.
Nathaniel et al. (2020)	CIVETS	Identify the EF enhancing factors	EF, renewable energy, urbanization, economic growth and trade openness	Environmental degradation is increased due to urbanization.
Osuntuyi and Lean (2022)	92 countries	Impact of education on environmental degradation	CO_2 emissions, education, energy consumption and economic growth	Energy consumption is caused to increase environmental degradation.
Naseem et al. (2022)	BRICS countries	Examine the association of environmental quality with economic growth	Economic growth, decoupling index	Economic expansion showed a negative impact on environmental quality.

continued on following page

Table 1. Continued

Author(s)	Research Area	Objectives	DVs/IVs	Main findings
Agozie et al. (2022)	BRICS countries	Assess the interconnection of CO2 emissions with FDI, ECI, renewable energy and urbanization	CO2 emissions, FDI, urbanization, renewable energy, economic complexity index (ECI)	Important association between ECI and environmental degradation.
Ssekibaala et al. (2022)	41 Sub-Saharan countries	Observe the impact of international trade on CO_2 emissions	CO2 emissions, international trade, economic growth	International trade is caused to reduce deforestation.
Rahman et al. (2023)	G8 and southeast Asian countries	Impact of industrial and energy consumption on environmental degradation.	Environmental degradation, industrial production, energy consumption	Developed countries are applying environmentally friendly ecosystem in industrial production.
Chen et al. (2023)	17 emerging countries	Explore the impact of material productivity on environmental degradation	CO2 emissions, globalization, economic growth, material productivity	FD, economic growth, and globalization are increasing CO_2 emissions.
Uzar and Eyuboglu (2023)	USA	Determinants of EF	EF, and cropland	Income inequality, and economic growth has a positive impact on EF.
Luo et al. (2023)	Asian countries	Impact of natural resource rent and economic growth on CO_2 emissions	CO_2 emissions, natural resource rent, economic growth	Economic growth has a positive impact on CO2 emissions.
Kumar and Datta (2024)	55 countries	Relationship of ecological footprint with air pollution	CO_2, N_2O, SO_2 CH_4, economic growth, FDI, energy consumption, urbanization	Diverse association of air pollution with economic growth.
Mignamissi et al. (2024)	Africa	Examine the impact of political regime and government on CO_2 emissions.	CO_2 emissions, political regime, industrialization, natural resources	Political regime drives environmental quality.
Onwe et al. (2024)	India	Impact of income inequality, renewable energy and economic growth on load capacity factor	Environmental sustainability, and economic growth	Renewable energy help to increase environmental quality.
Nguyen et al. (2024)	Developed and developing countries	Examine the impact of natural resource rent, FDI and economic growth environmental performance.	Environmental performance, natural capital, regional development, FDI, green technologies	Environmental performance improves as green innovation increase.

continued on following page

Table 1. Continued

Author(s)	Research Area	Objectives	DVs/IVs	Main findings
Topuz et al. (2025)	21 European countries	Examine the impact of environmental taxes and green innovation on CO2 emissions.	CO2 emissions, green innovation and economic growth.	CO2 emissions could be increased due to increase in environmental taxes and economic growth.
Hongjie and Muzi (2025)	France	Impact of eco innovation on LCF	Renewable energy, supply chain, eco-innovation.	Eco-innovation should be promoted.

Source: Author's compilation as per the previous literature.

Infrastructural Development: Infrastructural development may be positive to increase overall productivity of a nation (Awad et al., 2023). However, infrastructural development may be accountable for environmental degradation (Awad et al., 2023). Building and road construction, and transportation facilities are the basic activities under infrastructural development. Infrastructural development is increasing deforestation and increasing CO_2 and GHGs emissions.

Sustainable Development: Environmental quality and sustainability are crucial determinants of sustainable development (Barkat et al., 2024). Environmental quality is helpful to increase various SDGs as well. Thus, Environmental degradation is also creating barriers to attain sustainable development.

THEORETICAL FRAMEWORK

EKC theory is applied by existing researchers to examine the association of environmental degradation with economic growth (Bello et al., 2018; Hanif, 2018; Acheampong, 2018; Osuntuyi & Lean, 2022; Xiao et al., 2022; Wang, 2024; Ganda & Ruza, 2025). Mignamissi et al. (2024) examine the association of political regimes with environmental degradation in Africa using EKC theory. Acheampong (2018) also noted evidence of EKC globally. EKC theory can be used to observe the economic growth and income gap (Wang, 2024). Several researchers also adopted this theory to explore the association of environmental degradation, environmental quality and environmental performance with economic growth and development, and their enhancing variables like FDI, foreign trade, financial inclusion, employment rate, green technology, renewable energy etc. However, it found a diverse association between these variables with environmental degradation in different countries. Accordingly, the different forms of empirical models are formulated by the research communities in the area of environmental economics. The theoretical and empirical review of related studies is provided in Table 2.

Table 2. Theorical and empirical review

Author(s)	Form of data (time period)	Theory	Empirical model	Main suggestions
Gu et al. (2017)	Cross-sectional data (2013)		Logit multilevel analysis	It is noted a complex association between environmental degradation and socioeconomic development.
Bello et al. (2018)	Time series data (1971-2016)	EKC hypothesis	VECM, sensitivity analysis	Hydroelectricity consumption is positive to reduce environmental degradation.
Hanif (2018)	Panel Data (1995-2015)	EKC hypothesis	Panel data regression analysis	Renewable energy abates air pollution and enhance environmental quality.
Munir and Ameer (2018)	Panel data (1980-2014)	EKC hypothesis	STRIPAT model	R&D sector should develop new technologies to reduce SO_2 emissions.
Wang and Dong (2019)	Panel Data (1990-2014)		Penal data regression analysis using AMG estimator	Renewable energy consumption would reduce EF.
Rakshit and Neog (2021)	Time series data (1971-2016)		PGARC heteroskedasticity model	Use of environmental-friendly technologies would reduce pollution.
Raghutla and Chittedi (2021)	Panel data (1998-2016)		Fully modified OLS estimation	Technological development help to reduce CO_2 emissions.
Wang et al. (2021)	Panel data (1995-2017)		Panel regression model	-
Ahmad et al. (2020)	Panel data (1984-2016)	EKC hypothesis	CS-ARDL model	Protection of natural resources and technological innovation help to increase sustainable development.
Nathaniel et al. (2020)	Panel data		Panel cointegration and causality test	Use of renewable energy would enhance environmental sustainability.
Naseem et al. (2022)	Imbalanced panel	EKC hypothesis	ARDL model	Use of green technologies would reduce environmental degradation.
Osuntuyi and Lean (2022)	Panel data (1985-2018)	EKC hypothesis	ARDL and panel causality test	Education is positive to reduce environmental degradation. Education promotes energy-efficient product and investment in green technologies.

continued on following page

Table 2. Continued

Author(s)	Form of data (time period)	Theory	Empirical model	Main suggestions
Agozie et al. (2022)	Panel data (1990-2019)	EKC theory	MOLS estimation model	Healthier manufacturing activity would be energetic in environmental sector.
Ssekibaala et al. (2022)	Panel data (1990-2017)	EKC theory	LSDVC model	Environmental policies are proactive to increase protection of environmental sector.
Rahman et al. (2023)	Panel data (1971-2011)	Toda Yamamoto technique		Energy security and efficiency are significant to reduce environmental degradation.
Luo et al. (2023)	Panel data (1990-2018)		Mean group regression	Adoption of environment-friendly economic growth would be positive for environmental sustainability in middle and low-income countries.
Uzar and Eyuboglu (2023)	Time series (1965-2017)		Fourier ARDL and Fourier ADL models	
Chen et al. (2023)	Panel data (1995-2019)		Driscoll Kraay standard estimation model	Environmental related technologies would minimize environmental degradation.
Kumar and Datta (2024)	Panel data (1990-2018)	EKC theory	Random and fixed effect models	
Mignamissi et al. (2024)	Panel data (1196-2020)	EKC theory	Dynamic panel data regression model	Digitalization is appeared positive to reduce pollution.
Onwe et al. (2024)	Time series data (1970-2019)		Novel Fourier ARDL model	Massive investment in renewable energy would enhance environmental sustainability.
Nguyen et al. (2024)	Panel data (2002-2022)		AMG estimator	Green innovation helps to increase environmental performance.
Topuz et al. (2025)	Panel data (1995-2020)		Panel Co-integration test	Green innovation is significant to reduce CO2 emissions.
Ganda and Ruza (2025)	Panel data (1990-2019)	EKC Theory	Quantile regression model	FD is supportive to increase environmental quality.

Source: Author's compilation as per the previous literature.

RESEARCH METHODOLOGY

Study Area and Time Period

This chapter includes 117 countries with different income groups. These are selected as per the availability of statistical data for prescribed variables related to this study during 2000 – 2022. These countries are divided in four groups i.e., high income (HI) (43), upper middle-income (UMI) (36), lower-middle income (LMI) (32) and low income (LI) (6) (Table 3). Income-wise distribution of countries is based on the World Bank's definition. This chapter develops a separate imbalanced panel for HI, UMI, LMI and LI to attain the objectives.

Table 3. List of selected countries

Income Group	Countries
High income (HI)	1. Australia, 2. Austria, 3. Belgium, 4. Brunei Darussalam, 5. Canada, 6. Chile, 7. Croatia, 8. Cyprus, 9. Czech Republic, 10. Denmark, 11. Estonia, 12. Finland, 13. France, 14. Germany, 15. Greece, 16. Hungary, 17. Ireland, 18. Israel, 19. Italy, 20. Japan, 21. Korea Rep., 22. Kuwait, 23. Latvia, 24. Lithuania, 25. Luxembourg, 26. Malta, 27. Netherlands, 28. New Zealand, 29. Norway, 30. Poland, 31. Portugal, 32. Qatar, 33. Saudi Arabia, 34. Singapore, 35. Slovak Republic, 36. Slovenia, 37. Spain, 38. Sweden, 39. Switzerland, 40. United Arab Emirates, 41. United Kingdom, 42. United States, 43. Uruguay
Low income (LI)	1. Congo Dem. Rep., 2. Haiti, 3. Niger, 4. Syrian Arab Republic, 5. Togo, 6. Yemen Rep.
Lower middle income (LMI)	1. Algeria, 2. Angola, 3. Bangladesh, 4. Benin, 5. Bolivia, 6. Cambodia, 7. Cameroon, 8. Congo Rep., 9. Cote d'Ivoire, 10. Egypt Arab Rep., 11. El Salvador, 12. Ghana, 13. Honduras, 14. India, 15. Iran (Islamic Republic of), 16. Kenya, 17. Kyrgyz Republic, 18. Mongolia, 19. Morocco, 20. Nepal, 21. Nicaragua, 22. Pakistan, 23. Philippines, 24. Senegal, 25. Sri Lanka, 26. Tajikistan, 27. Tanzania, 28. Tunisia, 29. Ukraine, 30. Vietnam, 31. Zambia, 32. Zimbabwe
Upper middle income	1. Albania, 2. Argentina, 3. Armenia, 4. Azerbaijan, 5. Belarus, 6. Botswana, 7. Brazil, 8. Bulgaria, 9. China, 10. Colombia, 11. Costa Rica, 12. Dominican Republic, 13. Ecuador, 14. Gabon, 15. Guatemala, 16. Indonesia, 17. Iraq, 18. Jamaica, 19. Jordan, 20. Kazakhstan, 21. Libya, 22. Malaysia, 23. Mauritius, 24. Mexico, 25. Moldova (Republic of), 26. Namibia, 27. North Macedonia, 28. Panama, 29. Paraguay, 30. Peru, 31. Romania, 32. Russian Federation, 33. South Africa, 34. Thailand, 35. Turkey, 36. Venezuela RB

Source: Selected by Author's as per the availability of statistical data of relevant variables.

Sources of Data and Explanation of Variables

The required data is taken from world development indicators and OECD. Although, air pollution is increasing due to CO_2, N_2O, SO_2 CH_4 emissions which are also enhancing environmental degradation (Kumar & Datta, 2024). Existing researchers applied different measures like environmental quality, ecological footprint, carbon footprint, water footprint, environmental performance and environmental sustainability to explore the progress of the environmental sector (Shayanmehr et al., 2023; Pujiati et al., 2023). Few studies adopt an index-based approach to examine the environmental performance across countries. Despite that, most studies prefer per capita CO_2 emissions to analyze the progress of the environmental sector (Wang, 2024). Per capita CO_2 emissions ($PCCO_2E$), therefore, are considered as representative variables for environmental degradation in this chapter (Osuntuyi & Lean, 2022; Thai, 2023; Kumar & Datta, 2024).

Education is a significant determinant to increase the process of economic growth (Awad et al., 2023). Further, education is a significant driver to increase the understanding of the people towards environmental concern (Osuntuyi & Lean, 2022). Educated people have better knowledge towards environmental concern and its negative implications on their life. Education, thus, is a significant determinant to increase environmental sustainability (Osuntuyi & Lean, 2022). It is positive to increase the behaviour of people towards the environment and increase efficient use of energy (Osuntuyi & Lean, 2022).

Hence, it would be imperative to reduce environmental degradation. Age dependency ratio (ADR) signifies the role of people to reduce environmental degradation. Agricultural sector is using advanced technologies, machine, chemical fertilizer, pesticides and modern irrigation facilities which are helping to improve its intensification. However, CO2 and GHGs emissions from this sector are increasing due to above-mentioned initiations. Thus, agricultural intensification is harmful for ecosystem services and the environmental sector (Wang & Dong, 2019; Zhang et al., 2022). Thus, agricultural land (AL) and crop production index (CPI) are applied to capture the impact of agricultural intensification on environmental degradation.

Employment to population ratio (EPR) is significant to increase the income of people. Accordingly, it may increase per capita income and GDP. While per capita GDP may produce a negative impact on environment (Acheampong, 2018; Alvarado et al., 2018). Extensive energy use (EU) is caused to increase environmental degradation. While renewable energy consumption (REC) helps to minimize CO_2 emissions (Hanif, 2018; Osuntuyi & Lean, 2022; Uzar & Eyuboglu 2023; Kumar & Datta, 2024). Renewable green investment would also increase the environmental performance (Zhang, 2023). Xuan and Hung (2024) also infer that green investment is useful to reduce environmental degradation in Vietnam. Thus, REC is considered as an important determinant to increase environmental sustainability in this study.

Trade openness (TO) helps to increase economic growth (Barkat et al., 2024). Thus, TO shows a complex association with environmental degradation (Barkat et al., 2024). TO may be accountable for extensive use of natural resources in the production. TO is a useful variable that measures the progress of international trade of a nation with other countries. Previous studies provided the controversial relationship between international trade and environmental development. Thus, exports of goods and services are included as explanatory variables to explore association of trade openness with environmental degradation in this research.

Forest area (FA) is a prime driver to maintain sustainability in ecosystem services (Wang et al., 2025). Forest area minimizes the negative impact of maximum temperature on earth. Forest areas absorb CO_2 emissions and release O_2 in the environment. Accordingly, forest area is positive to increase purification of the environment. Thus, FA is supportive to increase climate and ecosystem sustainability (Wang et al., 2025). Further, FA can be positive to diminish the negative impact of climate change and socio-economic and demographic changes on environmental quality.

Gross domestic product (GDP) is caused by increased environmental degradation (Osuntuyi & Lean, 2022). Life expectancy rate also determines environmental quality. Human development index (HDI) is applied as an independent variable to capture the integrated impact of per capita real GDP, life expectancy and literacy rate on environmental degradation in this chapter (Haq et al., 2023; Pujiati et al., 2023).

Individuals using the internet (IUI), thus, are used to explore the impact of digitalization on environmental degradation. Accessibility of the internet maintains the appropriate communication of different stakeholders like the common people and government. Accordingly, stakeholders can take effective actions to increase environmental sustainability. Hence, IUI is used to assess the implications of digitalization on environmental degradation in this research (Mignamissi et al., 2024).

Inflation (IGDPD) has a negative impact on production of all sectors. Hence, inflation also shows a worthy impact on environmental degradation (Hondroyiannis et al., 2025). Hondroyiannis et al. (2025) noted a significant implication of inflation on carbon emission in high-income countries. While Grolleau and Weber (2024) noted a negative relationship between inflation and CO_2 emissions.

Technological change is valuable to discover and develop new technologies in different dimensions. Technological change is helping to develop green innovation, environmental technologies, green products, and green business in many countries (Naseem et al., 2022). Conversely, technological change may be worthy and beneficial to increase industrial development that has a negative impact on environmental development. Hence, technological change provides a complex impact on environmental development. Therefore, scientific and technical journal articles (STJA) are included as explanatory variables to examine the impact of technological change on environmental degradation in this research.

Urbanization is caused to increase more utilization of arable land and forest to meet the demand of the population in urban areas. Further it also enhances infrastructural development in urban areas. Hence, urban population (UP) is used to assess the impact of urbanization on environmental degradation (Alvarado et al., 2018; Wang & Dong, 2019; Xiao et al., 2022; Naseem et al., 2022; Pujiati et al., 2023). While green urbanization would be positive to increase environmental sustainability.

Vulnerable employment (VET) signifies poor economic conditions of workers due to their low productivity. Therefore, VET may have a negative impact on the environmental sector.

Population growth (PGA) always produces a negative impact on environmental quality (Cropper & Griffiths, 1994; Pujiati et al., 2023). Sex ratio (SRB) is also valuable to increase the equality among the male and females. Equal sex ratio the equal opportunities for male and females in the social and economic sector. Environmental degradation, thus, has dependence on sex ratio. Corruption is caused to reduce the effective implementation of government policies in social, economic, health and environmental sectors. While control of corruption (CCNS) may be supportive to increase environmental quality. Accordingly, these variables are included as core, mediator and control variables in the empirical investigations (Table 4).

Table 4. Explanation of the variables

Variables	Code	Unit
CO_2 emissions (metric tons per capita)	$PCCO_2E$	MT/PC
Adjusted savings: education expenditure (% of GNI)	*ASEE*	%
Age dependency ratio (% of working-age population)	*ADR*	%
Agricultural land (% of land area)	*AL*	%
Crop production index (2014-2016 = 100)	*CPI*	Number
Employment to population ratio, 15+, total (%)	*EPR*	%
Energy use (kg of oil equivalent per capita)	*EU*	Kg./PC
Renewable energy consumption (% of total final energy consumption)	*REC*	%
Exports of goods and services (% of GDP)	*TO*	%
Forest area (% of land area)	*FA*	%
HDI-Value	*HDI*	Number
Individuals using the Internet (% of population)	*IUI*	%
Inflation, GDP deflator (annual %)	*IGDPD*	%
Scientific and technical journal articles	*STJA*	Number
Urban population (% of total population)	*UP*	%
Vulnerable employment, total (% of total employment)	*VET*	%
Population growth (annual %)	*PGA*	%
Sex ratio at birth (male births per female births)	*SRB*	Number
Control of Corruption: Number of Sources	*CCNS*	Number

Source: Finalized by Author's as per existing theoretical and empirical literature.

Statistical and Empirical Analysis

Log-linear regression equations are developed to examine the impact of included core, mediator and control variables. Osuntuyi and Lean (2022); Osman et al. (2024); Ganda and Ruza (2025) also applied log-form of variables in the regression analysis. This, chapter assumes that per capita CO_2 emissions ($PCCO_2E$) is a function of the mentioned variables in table 4. The mathematical functional relationship of $PCCO_2E$ with these variables are explained as:

$$(PCCO_2E) = f\{(ASEE), (ADR), (AL), (CPI), (EPR), (EU), (REC), (TO), (FA), (HDI), (IUI), (IGDPD), (STJA), (UP), (VET), (PGA), (SRB), (CCNS)\} \quad (1)$$

The clarification of the variables mentioned in equation (1) is with Table 1. The equation (1) is transformed in following form after adopting log-linear regression model:

$$ln(PCCO_2E)_{ct} = \beta_0 + \beta_1 \, ln(ASEE)_{ct} + \beta_2 \, ln(ADR)_{ct} + \beta_3 \, ln(AL)_{ct} + \beta_4 \, ln(CPI)_{ct} + \beta_5 \, ln(EPR)_{ct} + \beta_6 \, ln(EU)_{ct} + \beta_7 \, ln(REC)_{ct} + \beta_8 \, ln(TO)_{ct} + \beta_9 \, ln(FA)_{ct} + \beta_{10} \, ln(HDI)_{ct} + \beta_{11} \, ln(IUI)_{ct} + \beta_{12} \, ln(IGDPD)_{ct} + \beta_{13} \, ln(STJA)_{ct} + \beta_{14} \, ln(UP)_{ct} + \beta_{15} \, ln(VET)_{ct} + \beta_{16} \, ln(PGA)_{ct} + \beta_{17} \, ln(SRB)_{ct} + \beta_{18} \, ln(CCNS)_{ct} + \phi_{ct} \quad (2)$$

Here, β_0 is constant coefficient; β_1, β_2, ..., β_{18} are the egression coefficients of associated variables; ln is natural logarithms of corresponding variables; *c* is cross-sectional country in a specific income group; *t* is time period; ϕ_{ct} is error-term in equation (2).

Measurement of Elasticities of $PCCO_2E$ with Respect to *IEVs*

Elasticities of environmental degradation with the respect to individual explanatory variables (*IEVs*) are estimated through simple mathematical technique. Following formula is used to examine the elasticities of $PCCO_2E$ with respect to *IEV*:

Elasticity of $PCCO_2E$ with respect to *IEV* = Regression coefficient of *IEV* × [Mean of *IEV*/Mean of $PCCO_2E$] (3)

Equation (3) measure the variation in $PCCO_2E$ due to 1% change in respective explanatory variables. While, the regression coefficients of *IEVs* are estimated through equation (2).

Robustness Check and Application of Statistical Software

The regression coefficients are estimated through general feasible least square estimation due to the existence of autocorrelation and serial correlation in the panel data. The robustness of regression coefficients is estimated through correlation coefficients of various lags of error-terms as per the respective regression equations. SPSS and STATA statistical software are used to analyse the data and estimate the coefficients of *IEVs* with $PCCO_2E$.

FINDINGS

The coefficients of explanatory variables (*EVs*) with $PCCO_2E$ are provided in Table 5. The statistical values of *F*-tests are reported significant at 1% significance level for HI, LI, LMI and UMI. The estimates, therefore, suggested that undertaken *EVs* had an independent impact on $PCCO_2E$. The R^2 values are found 0.8486, 0.9874, 0.887 and 0.9095 for HI, LI, LMI and UMI, respectively. The results provide a clarification that around 85%, 99%, 89% and 91% variation in $PCCO_2E$ can be determined by included EVs in HI, LI, LMI and UMI, respectively. The estimates also infer the different income groups countries have diverse ecosystems and infrastructure to deal with environmental issues.

The coefficient of education expenditure with $PCCO_2E$ are appeared negative in HI, LMI and UMI. Hence, education expenditure would help to reduce environmental degradation in these income group countries. Osuntuyi and Lean (2022) also noted that education is found positive to increase environmental sustainability. While education expenditure is noted insignificant to reduce environmental degradation in LI. LI have a poor quality of education. Thus, poor quality of education may not be significant to reduce environmental degradation. The coefficients of *ADR* with $PCCO_2E$ are detected negative in HI, LI, LMI and UMI. The estimates suggested that environmental degradation is predicted to decrease as age dependency ratio increases across countries.

Table 5. Impact of IVs on per capita CO2 emissions

Income Groups	**HI**		**LI**		**LMI**		**UMI**	
Number of obs.	**988**		**138**		**735**		**828**	
Number of groups	**43**		**6**		**32**		**36**	
F(18, 22)	**141107.59***		**20736.35***		**109587.26***		**45818.13***	
R-squared	**0.8486**		**0.9874**		**0.8870**		**0.9095**	
Mean VIF	**3.13**		**35.21**		**3.29**		**2.52**	
log($PCCO_2E$) **=[DV]**	*Coef.*	*Std. Err.*	*Coef.*	*Std. Err.*	*Coef.*	*Std. Err.*	*Coef.*	*Std. Err.*
log(ASEE)	-0.1918*	0.049	0.1480	0.126	-0.0065	0.053	-0.1918*	0.034
log(ADR)	-0.4254*	0.032	-0.794**	0.376	-1.0720*	0.267	-0.4800*	0.023
log(AL)	0.0893*	0.005	0.6850*	0.093	0.0144	0.024	0.0923*	0.014
log(CPI)	0.0322	0.030	0.0055	0.116	0.2399**	0.089	0.1260*	0.035
log(EPR)	-0.164**	0.062	-0.1902	0.264	-0.169***	0.094	-0.3045*	0.056
log(EU)	0.9376*	0.030	0.9071*	0.177	0.5364*	0.036	0.8415*	0.048
log(REC)	-0.0458*	0.010	-0.4271*	0.086	-0.0792*	0.013	-0.2040*	0.014
log(TO)	-0.1167*	0.022	-0.0111	0.038	0.2090*	0.020	0.0261	0.019
log(FA)	-0.0566*	0.013	-0.2222*	0.076	-0.0687*	0.006	0.0209*	0.005
log(HDI)	2.4473*	0.321	1.5989*	0.436	0.2928	0.438	0.1858	0.230
log(IUI)	-0.1246*	0.036	-0.0829*	0.023	0.0344	0.025	-0.0641*	0.019
log(IGDPD)	-0.0076	0.008	-0.0048	0.008	0.0145	0.015	-0.027**	0.011
log(STJA)	-0.030**	0.012	-0.10***	0.052	-0.0028	0.004	0.0266*	0.003
log(UP)	-0.4102*	0.067	-0.8023*	0.264	0.6760*	0.135	-0.0379	0.054
log(VET)	0.0559*	0.008	0.1292	0.404	0.1673*	0.028	0.0225	0.030
log(PGA)	-0.0049	0.013	0.0100	0.046	0.058***	0.029	0.0056	0.008
log(SRB)	1.7917	2.091	-26.59**	10.20	7.2840*	0.854	-3.7258*	0.149
log(CCNS)	-0.0511	0.035	0.0824	0.063	-0.2121*	0.038	-0.0010	0.033
Cons. Coef.	-0.0071	0.639	1.4474	2.966	-2.7159*	0.399	-1.6870*	0.520

Source: Author's estimation. Note: *, ** and *** infer that coefficients are statistically significant at 1%, 5% and 10% significance level.

Agricultural intensification is accountable for increasing environmental degradation. Accordingly, coefficients of *AL* and *CPI* with $PCCO_2E$ are noted positive. The results reveal that arable and crop production has a negative impact on environmental performance. Wang and Dong (2019); Zhang et al. (2022) also recognized the negative impact of the agricultural sector on environmental quality. The coefficients of *EPR* with $PCCO_2E$ are observed negatively in all countries. The results, thus, indicate that employment to population ratio would assist to minimize environmental degradation in HI, LI, LMI and UMI. Conversely, vulnerable employment is observed harmful for environmental quality in these countries.

Eregha et al. (2023); Uzar and Eyuboglu (2023) highlighted that ecological footprint increases due to energy consumption in 11 emerging countries. Singhal et al. (2022) proposed that extensive energy growth would increase environmental degradation. Conversely, renewable energy consumption would reduce pressure on the environmental sector. Thus, coefficients of *REC* with $PCCO_2E$ are found negative in HI, LI, LMI and UMI. The results showed that renewable energy would be useful to reduce environmental degradation. Previous scholars like Hanif (2018); Wang and Dong (2019); Nathaniel et al. (2020); Raghutla and Chittedi (2021); Ssekibaala et al. (2022); Ahakwa et al. (2023); Onwe et al. (2024) claimed that environmental sustainability and environmental quality improved due to increase in renewable energy consumption. Ganda and Ruza (2025) also noticed a similar association between renewable energy consumption and CO_2 emissions.

Table 6. Validity of the regression results

HI								
No. of lags	***Lag* (1)**	***Lag* (2)**	***Lag* (3)**	***Lag* (4)**	***Lag* (5)**	***Lag* (6)**	***Lag* (7)**	***Lag* (8)**
Lag (1)	1.0000							
Lag (2)	0.9951*	1.0000						
Lag (3)	0.9912*	0.9950*	1.0000					
Lag (4)	0.9875*	0.9908*	0.9948*	1.0000				
Lag (5)	0.9830*	0.9870*	0.9905*	0.9945*	1.0000			
Lag (6)	0.9775*	0.9824*	0.9865*	0.9900*	0.9942*	1.0000		
Lag (7)	0.9710*	0.9766*	0.9817*	0.9858*	0.9895*	0.9939*	1.0000	
Lag (8)	0.9638*	0.9698*	0.9756*	0.9808*	0.9851*	0.9889*	0.9936*	1.0000
LI								
No. of lags	***Lag* (1)**	***Lag* (2)**	***Lag* (3)**	***Lag* (4)**	***Lag* (5)**	***Lag* (6)**	***Lag* (7)**	***Lag* (8)**
Lag (1)	1.0000							
Lag (2)	0.9937*	1.0000						
Lag (3)	0.9868*	0.9935*	1.0000					
Lag (4)	0.9819*	0.9864*	0.9933*	1.0000				
Lag (5)	0.9766*	0.9812*	0.9860*	0.9931*	1.0000			

continued on following page

Table 6. Continued

HI								
No. of lags	*Lag* **(1)**	*Lag* **(2)**	*Lag* **(3)**	*Lag* **(4)**	*Lag* **(5)**	*Lag* **(6)**	*Lag* **(7)**	*Lag* **(8)**
Lag (6)	0.9708*	0.9758*	0.9807*	0.9855*	0.9929*	1.0000		
Lag (7)	0.9672*	0.9700*	0.9753*	0.9802*	0.9851*	0.9927*	1.0000	
Lag (8)	0.9634*	0.9663*	0.9695*	0.9747*	0.9796*	0.9849*	0.9928*	1.0000
LMI								
No. of lags	***Lag* (1)**	***Lag* (2)**	***Lag* (3)**	***Lag* (4)**	***Lag* (5)**	***Lag* (6)**	***Lag* (7)**	***Lag* (8)**
Lag (1)	1.0000							
Lag (2)	0.9980*	1.0000						
Lag (3)	0.9957*	0.9980*	1.0000					
Lag (4)	0.9930*	0.9956*	0.9980*	1.0000				
Lag (5)	0.9896*	0.9929*	0.9955*	0.9979*	1.0000			
Lag (6)	0.9856*	0.9893*	0.9926*	0.9954*	0.9979*	1.0000		
Lag (7)	0.9816*	0.9852*	0.9890*	0.9924*	0.9952*	0.9978*	1.0000	
Lag (8)	0.9773*	0.9809*	0.9847*	0.9886*	0.9921*	0.9951*	0.9978*	1.0000
UMI								
No. of lags	***Lag* (1)**	***Lag* (2)**	***Lag* (3)**	***Lag* (4)**	***Lag* (5)**	***Lag* (6)**	***Lag* (7)**	***Lag* (8)**
Lag (1)	1.0000							
Lag (2)	0.9940*	1.0000						
Lag (3)	0.9885*	0.9939*	1.0000					
Lag (4)	0.9831*	0.9883*	0.9937*	1.0000				
Lag (5)	0.9776*	0.9827*	0.9879*	0.9936*	1.0000			
Lag (6)	0.9711*	0.9769*	0.9821*	0.9875*	0.9934*	1.0000		
Lag (7)	0.9654*	0.9703*	0.9763*	0.9816*	0.9871*	0.9931*	1.0000	
Lag (8)	0.9596*	0.9640*	0.9692*	0.9754*	0.9808*	0.9866*	0.9928*	1.0000

Source: Author's estimation. Note: * Infer that coefficient is statistically significant at 1% significance level.

Trade openness is found effective to reduce environmental degradation in HI and LI. While trade openness is causing environmental degradation in LMI and UMI. Findings of Munir and Ameer (2018); Ganda and Ruza (2025) also proposed that trade openness is caused to increase air pollution. However, Barkat et al. (2024) claimed that TO may be positive to increase sustainable development as it promotes technology transfer across countries. Forest area helps to minimize maximum temperature, and sustain the common properties of other ecosystem services (Wang et al., 2025). Forest area releases O_2 and absorbs harmful emissions. Thus, forest area is an important source to purify the ecosystem services directly and indirectly. Forest area is found positive to reduce environmental degradation in HI, LI and LMI. Forest area seems harmful for the environmental sector in UMI. UMI should increase the appropriate management of forest area for increasing its positive implications on environmental sustainability.

The coefficients of *HDI* with $PCCO_2E$ are reported positive. The results imply that human development is leading to increased environmental degradation in HI, LI, LMI and UMI. Pujiati et al. (2023) also observed a positive association of HDI with per capita CO_2. Availability of the internet creates a favourable digital ecosystem and digitalization. Digitalization may be an essential tool to minimize environmental degradation (Mignamissi et al., 2024). Hence, it is expected that digitalization would be effective to maintain environmental quality. The coefficients of IUI with $PCCO_2E$ are detected negative in HI, LI and UMI. It suggested that individuals using the internet would be valuable to increase environmental quality in these income groups of countries. While digitalization showed a negative impact on environmental quality in LMI. Thereupon, inflation appeared positive to reduce environmental degradation in HI, LI and UMI. Similar associations between inflation and carbon emission are reported in high income countries (Hondroyiannis et al., 2025).

The coefficients of *STJA* with $PCCO_2E$ are perceived as negative and statistically significant in HI, LI and LMI. The results reveal that technological change could be beneficial for improving environmental quality in these income groups of countries. Numerous scholars have already claimed that technological change brings innovation for increasing environmental sustainability (Raghutla & Chittedi, 2021).

Table 7. Elasticities of environmental degradation with respect to individual IVs

	HI		LI		LMI		UMI	
	Reg. Coef.	*Elasticity*	*Reg. Coef.*	*Elasticity*	*Reg. Coef.*	*Elasticity*	*Reg. Coef.*	*Elasticity*
ASEE	-0.192	-0.090	0.148	0.698	-0.007	-0.017	-0.192	-0.202
ADR	-0.425	-2.062	-0.794	-114.818	-1.072	-50.153	-0.480	-6.685
AL	0.089	0.323	0.685	59.167	0.014	0.433	0.092	0.975
CPI	0.032	0.312	0.006	0.863	0.240	14.769	0.126	3.009
EPR	-0.164	-0.946	-0.190	-17.534	-0.169	-6.569	-0.305	-4.316
EU	0.938	433.122	0.907	625.526	0.536	262.427	0.842	359.519
REC	-0.046	-0.074	-0.427	-41.019	-0.079	-2.425	-0.204	-1.130
TO	-0.117	-0.675	-0.011	-0.409	0.209	4.766	0.026	0.259
FA	-0.057	-0.178	-0.222	-6.353	-0.069	-1.383	0.020	0.179
HDI	2.447	0.216	1.599	1.293	0.293	0.121	0.186	0.035
IUI	-0.125	-0.845	-0.083	-1.626	0.034	0.461	-0.064	-0.609

continued on following page

Table 7. Continued

	HI		LI		LMI		UMI	
	Reg. Coef.	*Elasticity*	*Reg. Coef.*	*Elasticity*	*Reg. Coef.*	*Elasticity*	*Reg. Coef.*	*Elasticity*
IGDPD	-0.008	-0.003	-0.005	-0.261	0.015	0.117	-0.027	-0.066
STJA	-0.030	-89.170	-0.100	-14.857	-0.003	-8.778	0.027	101.084
UP	-0.410	-3.241	-0.802	-52.553	0.676	21.418	-0.038	-0.642
VET	0.056	0.061	0.129	15.112	0.167	6.441	0.023	0.192
PGA	-0.005	-0.001	0.010	0.048	0.058	0.074	0.006	0.002
SRB	1.792	0.190	-26.590	-47.205	7.284	5.269	-3.726	-1.023
CCNS	-0.051	-0.047	0.082	1.260	-0.212	-1.593	-0.001	-0.003

Source: Author's estimation.

Urban population and population growth produce a positive impact on the environmental sector in HI. Urban population also showed a negative impact on environmental degradation in LI and UMI. Osman et al. (2024) also reported that population growth is increasing environmental degradation in Somalia. Similarly, sex ratio and corruption control do not produce a uniform impact on environmental degradation HI, LI, LMI and UMI. Sex ratio is found suitable to reduce environmental degradation in LI and UMI. Corruption control has appeared positively to increase environmental quality in HI, LMI and UMI.

Many studies also pursue similar techniques to check the validity of regression coefficients in a model. The correlation coefficients among the 8 lags of error-terms are estimated for further validation of regression coefficients (Table 6). The estimates suggested the error-terms have statistically significant correlation with their various respective lags in HI, LI, LMI and UMI. Hence, the regression coefficients estimated through equation (2) has rationality.

The elasticities of $PCCO_2E$ with individual *IV*s are given in Table 7. Age dependency ratio (*ADR*), energy use (*EU*) and technological change (*STJA*) has a greater impact on environmental degradation in HI. The variation in environmental degradation varies due to change age dependency ratio (*ADR*), arable land (*AL*), employment to population ratio (*EPR*), energy use (*EU*), renewable energy consumption (*REC*), forest area (*FA*), technological change (*STJA*), urban population (*UP*), vulnerable employment (*VET*) and sex ratio (*SRB*) in LI. While the variation in environmental degradation exists due to change in *ADR*, *CPI*, *EPR*, *EU*, *REC*, *TO*, *FA*, *STJA*, *UP*, *VET*, *SRB* in LMI. Finally, *ADR*, *CPI*, *EPR*, *EU*, *REC*, *STJA* and *SRB* are accountable for increasing variation in environmental degradation in UMI.

CONCLUSION

The results provide a clarity that environmental degradation is a global issue in all income groups of countries. Thus, increasing environmental degradation would hamper all development related activities including sustainable development. Economic, social and technological development are accountable for environmental degradation. High income groups of countries are able to cope with environmental degradation due to their green and innovative capabilities, and adoption of technological advancement. Green energy transition could be helpful to reduce environmental degradation in high income groups of countries. While it is seemed difficult for lower and low-income groups of countries to implement environmental policies to reduce environmental degradation. Low technological green innovation adoption capacity is increasing more environmental degradation in these countries. These countries have low resource efficiency and productivity in production sectors. Hence, lower- and low-income groups of countries should increase resource efficiency for minimizing environmental degradation.

Agricultural intensification, energy use, human development and vulnerable employment are reported as prime determinants of environmental degradation (Jatav et al., 2023). While age dependency ratio, renewable energy consumption, employment to population ratio, forest area, digitalization and technological change are determined as a positive variable to reduce environmental degradation (Hanif, 2018). Renewable energy consumption would be positive to increase sustainable development (Hanif, 2018). Renewable sources of energy would be proactive to reduce environmental degradation (Munir & Ameer, 2018; Luo et al., 2023). Corruption control appears favourable to increase environmental quality in HI, LMI and UMI. Thereupon, education expenditure produces a negative implication on environmental degradation in HI, LMI and UMI. Trade openness could be harmful for the environmental sector in LMI and UMI. Each income group of countries are required to implement separate environmental policies to reduce environmental degradation. Highly agrarian countries should pursue sustainable agricultural management policies to increase their sustainability and reduce environmental degradation. Renewable energy consumption, age dependency ratio, technological change are found to be the most significant variables to reduce CO2 emissions in global countries. Thus, the empirical findings of this chapter are helpful for each income group of countries to reduce environmental degradation and attain sustainable development.

POLICY IMPLICATIONS

Lower and low-income groups of countries should give high priority to control population growth and urbanization to reduce environmental degradation (Wang & Dong, 2019). Low and lower-income countries must give high priority to control over population growth to increase environmental quality (Cropper & Griffiths, 1994). Larger agrarian countries should pursue sustainable agricultural management practices to protect environmental development. Water conservation and water resource sustainability are essential to increase environmental sustainability. Resource efficiency must be promoted to increase protection of natural resources (Singhal et al., 2022). Food wastage must be reduced to increase environmental sustainability. Traditional industrialization must be transformed as green industrialization.

Green innovation and green technologies would help to increase green growth. Investment in technologies is also required to reduce destruction in the environmental sector (Osman et al., 2024). Use of environmentally friendly technologies in the financial sector must be promoted to increase environmental quality (Rakshit & Neog, 2021). Energy policies would be favourable to increase environmental sustainability (Mahalik et al., 2018). Initiation of efficient renewable energy technologies would be positive to increase environmental quality (Pradhan et al., 2024). Accordingly, industrial and business sectors should develop green products for increasing environmental sustainability. Green urbanization must be effective and conducive initiatives towards sustainability. Green business opportunities in rural areas would be positive to increase environmental sustainability. Application of renewable energy would minimize GHGs emissions and enhance environmental sustainability. Appropriate integration of economic and environmental policies is required to overcome the challenge of environmental degradation world-wide (Kumar & Datta, 2024).

INNOVATIONS AND VALUE ADDITION OF THIS CHAPTER

A comprehensive review of literature with regards to environmental degradation, environmental sustainability and environmental quality are provided in this chapter. The comprehensive literature review related to environmental degradation and its influencing factors are also explained. The comparative progress of environmental degradation and its influencing factors in different income groups of countries are given. It examines percentage change in environmental degradation due to one unit in explanatory variables using elasticities measurement. Consequently, this chapter suggests several policy proposals to increase environmental development in different income groups of countries.

LIMITATIONS AND SCOPE OF FURTHER RESEARCH

This chapter has multiple limitations which can be considered in further studies. For instance, this chapter applied per capita CO2 emissions as a dummy variable for environmental degradation. While environmental degradation is increasing due to many GHGs emissions (Kumar & Datta, 2024). EF, carbon footprint and water footprint are also causing environmental degradation (Bello et al., 2018). Hence, these emissions can be included in future studies. Climate change is also causing environmental degradation (Singh et al., 2019). While this chapter could not observe the implications of climate change on environmental degradation. Further research can examine the significance of green technologies, eco-friendly technologies, green innovation and green product development in the environmental sector (Ssekibaala et al., 2022; Nguyen et al., 2024). Green urbanization would enhance the protection of natural resources. Strict environmental policies are proactive to increase environmental quality. While agricultural intensification is leading to environmental degradation. The above-mentioned relationship can be observed in further research. Sustainable energy consumption (SEC) would also be positive to increase environmental quality. Hence, further research can assess the impact of SEC on environmental sustainability.

REFERENCES

Acheampong, A. O. (2018). Economic growth, CO_2 emissions and energy consumption: What causes what and where? *Energy Economics*, *74*(1), 677–692. DOI: 10.1016/j.eneco.2018.07.022

Agozie, D. Q., Akwasi, G. B., Victor, B. F., Ozturk, I., & Taha, A. (2022). Environmental Kuznets Curve hypothesis from lens of economic complexity index for BRICS: Evidence from second generation panel analysis. *Sustainable Energy Technologies and Assessments*, *53*(1), 10–25. DOI: 10.1016/j.seta.2022.102597

Ahakwa, I., Xu, Y., Tackie, E. A., & Mangudhla, T. (2023). How crucial are natural resources in descending environmental degradation in Ghana? A novel dynamic ARDL simulation approach. *Journal of Cleaner Production*, *420*(1), 1–13. DOI: 10.1016/j.jclepro.2023.138427

Ahmad, M., Jiang, P., Majeed, A., Umar, M., Khan, Z., & Muhammad, S. (2020). The dynamic impact of natural resources, technological innovations and economic growth on ecological footprint: An advanced panel data estimation. *Resources Policy*, *69*(1), 1–10. DOI: 10.1016/j.resourpol.2020.101817

Ali, E. B., Gyamfi, B. A., Bekun, F. V., Ozturk, I., & Nketiah, P. (2023). An empirical assessment of the tripartite nexus between environmental pollution, economic growth, and agricultural production in Sub-Saharan African countries. *Environmental Science and Pollution Research International*, *30*(1), 71007–71024. DOI: 10.1007/s11356-023-27307-4 PMID: 37160515

Alvarado, R., Ponce, P., Criollo, A., Córdova, K., & Khan, M. K. (2018). Environmental degradation and real per capita output: New evidence at the global level grouping countries by income levels. *Journal of Cleaner Production*, *189*(1), 13–20. DOI: 10.1016/j.jclepro.2018.04.064

Awad, A., Mallek, R. S., Ozturk, I., & Abdalla, Y. A. (2023). Infrastructure development's role in environmental degradation in sub-Saharan Africa: Impacts and transmission channels. *Journal of Cleaner Production*, *414*(1), 1–10. DOI: 10.1016/j.jclepro.2023.137622

Barkat, K., Alsamara, M., Kwifi, O. S. A., & Jarallah, S. (2024). Does trade openness mitigate environmental degradation in organization for Economic Co-operation and Development (OECD) countries? Implication for achieving sustainable development. *Natural Resources Forum*, *49*(1), 677–698. DOI: 10.1111/1477-8947.12412

Bello, M. O., Solarin, S. A., & Yen, Y. Y. (2018). The impact of electricity consumption on CO2 emission, carbon footprint, water footprint and ecological footprint: The role of hydropower in an emerging economy. *Journal of Environmental Management, 219*(1), 218–230. DOI: 10.1016/j.jenvman.2018.04.101 PMID: 29747103

Chen, F., Ali, S., Ma, J., Arshad, S., & Ahmad, S. (2023). Material productivity and environmental degradation: Moderating role of environment-related technologies in achieving carbon neutrality. *Gondwana Research, 117*(1), 155–168. DOI: 10.1016/j.gr.2023.01.010

Coronado, V. M. H., Valderrama, C. J. M., Valencia-Arias, A., Huamán, H. I. M., Valencia, J., & Cardona-Acevedo, S. (2024). Analysis of research trends on economic growth and environmental degradation: A bibliometric study. *Sustainable Environment, 10*(1), 1–10. DOI: 10.1080/27658511.2024.2345445

Cropper, M., & Griffiths, C. (1994). The interaction of population growth and environmental quality. *The American Economic Review, 84*(2), 250–254. https://www.jstor.org/stable/2117838

Eregha, P. B., Nathaniel, S. P., & Vo, X. V. (2023). Economic growth, environmental regulations, energy use, and ecological footprint linkage in the Next-11 countries: Implications for environmental sustainability. *Energy & Environment, 34*(5), 1327–1347. DOI: 10.1177/0958305X221084293

Eweade, B. S., Akadiri, A. C., Olusoga, K. O., & Bamidele, R. O. (2024). The symbiotic effects of energy consumption, globalization, and combustible renewables and waste on ecological footprint in the United Kingdom. *Natural Resources Forum, 48*(1), 274–291. DOI: 10.1111/1477-8947.12392

Ganda, F., & Ruza, C. (2025). Unveiling the non-linear dynamics: Quantile regression of financial development's impact on environmental degradation in BRICS and G7 nations. *Discover Sustainability, 6*(60), 1–26. DOI: 10.1007/s43621-025-00792-y

González-Álvarez, M. A., & Montañés, A. (2023). CO_2 emissions, energy consumption, and economic growth: Determining the stability of the 3E relationship. *Economic Modelling, 121*(1), 1–10. DOI: 10.1016/j.econmod.2023.106195

Grolleau, G., & Weber, C. (2024). The effect of inflation on CO2 emissions: An analysis over the period 1970-2020. *Ecological Economics, 217*(1), 1–10. DOI: 10.1016/j.ecolecon.2023.108029

Gu, L., Rosenberg, M. W., & Zeng, J. (2017). Competing forces of socioeconomic development and environmental degradation on health and happiness for different income groups in China. *International Journal of Health Services*, *47*(4), 752–777. DOI: 10.1177/0020731417725470 PMID: 28847235

Hanif, I. (2018). Impact of economic growth, nonrenewable and renewable energy consumption, and urbanization on carbon emissions in Sub-Saharan Africa. *Environmental Science and Pollution Research International*, *25*(15), 15057–15067. DOI: 10.1007/s11356-018-1753-4 PMID: 29552722

Haq, S. M. A., Chowdhury, M. A. F., Ahmed, K. J., & Chowdhury, M. T. A. (2023). Environmental quality and its impact on total fertility rate: An econometric analysis from a new perspective. *BMC Public Health*, *23*(1), 2397. DOI: 10.1186/s12889-023-17305-z PMID: 38042784

Hondroyiannis, G., Papapetrou, E., & Tsalaporta, P. (2025). Environmental degradation and inflation in high-income countries. *Economic Change and Restructuring*, *58*(1), 1–28. DOI: 10.1007/s10644-024-09840-5

Hongjie, D., & Muzi, F. (2025). How supply chain disruptions, renewable energy consumption, and eco-innovation mitigate environmental degradation? a path towards sustainable development in France. *Energy Economics*, *145*(1), 1–5. DOI: 10.1016/j.eneco.2025.108346

Jatav, S. S., Nathoo, B., Mubeena, Md., Arakhita, B., Pooja, R., & Rukmani, D. (2023). Development of agricultural sustainability index for different regions of Uttar Pradesh, India. *Environment and Ecology*, *41*(4C), 2781–2789. https://www.environmentandecology.com/volume-41-2023

Kumar, P., & Datta, S. K. (2024). Environmental degradation and economic growth: A longitudinal analysis across different income group countries. *The Indian Economic Journal*, *72*(1), 72–89. DOI: 10.1177/00194662231199238

Li, T. (2022). Energy productivity and financial inclusion: Evidence from non-parametric novel panel methods. *Energy Efficiency*, *15*(7), 51–63. DOI: 10.1007/s12053-022-10057-6 PMID: 36000077

Luo, J., Ali, S. A., Aziz, B., Aljarba, A., Akeel, H., & Hanif, I. (2023). Impact of natural resource rents and economic growth on environmental degradation in the context of COP-26: Evidence from low-income, middle-income, and high-income Asian countries. *Resources Policy*, *80*(C), 1–13. DOI: 10.1016/j.resourpol.2022.103269

Mahalik, M. K., Mallick, H., Padhan, H., & Sahoo, B. (2018). Is skewed income distribution good for environmental quality? A comparative analysis among selected BRICS countries. *Environmental Science and Pollution Research International*, *25*(23), 23170–23194. DOI: 10.1007/s11356-018-2401-8 PMID: 29862476

Mignamissi, D., Mougnol, A., Ekoula, H. W., & Thioune, T. (2024). Institutions' quality and environmental pollution in Africa. *Environment and Development Economics*, *29*(3), 206–233. DOI: 10.1017/S1355770X24000044

Munir, K., & Ameer, A. (2018). Effect of economic growth, trade openness, urbanization, and technology on environment of Asian emerging economies. *Management of Environmental Quality*, *29*(6), 1123–1134. DOI: 10.1108/MEQ-05-2018-0087

Nain, M. Z., Ahmad, W., & Kamaiah, B. (2015). Economic growth, energy consumption and CO_2 emissions in India: A disaggregated causal analysis. *International Journal of Sustainable Energy*, *36*(8), 807–824. DOI: 10.1080/14786451.2015.1109512

Naseem, S., Mohsin, M., Zia-UR-Rehman, M., Baig, S. A., & Sarfraz, M. (2022). The influence of energy consumption and economic growth on environmental degradation in BRICS countries: An application of the ARDL model and decoupling index. *Environmental Science and Pollution Research International*, *29*(9), 13042–13055. DOI: 10.1007/s11356-021-16533-3 PMID: 34564817

Nathaniel, S., Nwodo, O., Sharma, G., & Shah, M. (2020). Renewable energy, urbanization, and ecological footprint linkage in CIVETS. *Environmental Science and Pollution Research International*, *27*(16), 19616–19629. DOI: 10.1007/s11356-020-08466-0 PMID: 32219652

Nguyen, T. H., Deng, H., Abbas, Z. Z., Lam, T. T., & Abbas, H. R. (2024). The effect of natural capital, regional development, FDI, and natural resource rent on environmental performance: The Mediating role of green innovation. *Resources Policy*, *91*(1), 1–12. DOI: 10.1016/j.resourpol.2024.104923

Onwe, J. C., Uche, E., Dhayal, K. S., Uwazie, I. U., & Ashibogwu, K. N. (2024). Advocating green economy in India: The tug of war among income inequality, export diversification, and environmental quality. *Sustainable Development*, *32*(5), 4879–4894. DOI: 10.1002/sd.2927

Osman, B. M., Shire, S. A., Ali, F. H., & Mohamed, A. A. (2024). Examining the drivers of environmental degradation in Somalia: The role of agriculture, economic and population growth. *Discover Sustainability*, *6*(150), 1–13. DOI: 10.1007/s43621-024-00786-2

Osuntuyi, B. V., & Lean, H. H. (2022). Economic growth, energy consumption and environmental degradation nexus in heterogeneous countries: Does education matter? *Environmental Sciences Europe*, *34*(48), 1–14. DOI: 10.1186/s12302-022-00624-0

Patel, G., Pal, S., & Sahu, A. K. (2024). Does energy transition reduce carbon inequality? A global analysis. *Environmental Science and Pollution Research International*, *31*(1), 34689–34708. DOI: 10.1007/s11356-024-33542-0 PMID: 38713356

Pradhan, K. C., Mishra, B., & Mohapatra, S. M. (2024). Investigating the relationship between economic growth, energy consumption, and carbon dioxide (CO_2) emissions: A comparative analysis of South Asian nations and G-7 countries. *Clean Technologies and Environmental Policy*, *26*(1), 3349–3367. DOI: 10.1007/s10098-024-02802-5

Pujiati, A., Nurbaeti, T., & Damayanti, N. (2023). What are the factors that determine differing levels of environmental quality? Evidence from Java and other islands in Indonesia. *Management of Environmental Quality*, *34*(2), 290–307. DOI: 10.1108/MEQ-02-2022-0034

Raghutla, C., & Chittedi, K. R. (2021). Financial development, energy consumption, technology, urbanization, economic output and carbon emissions nexus in BRICS countries: An empirical analysis. *Management of Environmental Quality*, *32*(2), 290–307. DOI: 10.1108/MEQ-02-2020-0035

Rahman, M. S., Noman, A. H. M., Shahari, F., Aslam, M., Gee, C. S., Isa, C. R., & Pervin, S. (2023). Efficient energy consumption in industrial sectors and its effect on environment: A comparative analysis between G8 and Southeast Asian emerging economies. *Energy*, *97*(1), 82–89. DOI: 10.1016/j.energy.2015.12.109

Rakshit, B., & Neog, Y. (2021). Do macroeconomic uncertainty and financial development cause environmental degradation? Evidence from an emerging economy. *International Journal of Social Economics*, *48*(9), 1264–1289. DOI: 10.1108/IJSE-10-2020-0690

Sahu, A. K., Mahalik, M. K., Patel, G., & Pal, S. (2023). The role of environmental degradation in macroeconomic instability: Panel evidence from emerging economies. *Environmental Science and Pollution Research International*, *30*(37), 86879–86891. DOI: 10.1007/s11356-023-28389-w PMID: 37410330

Shayanmehr, S., Radmehr, R., Ali, E. B., Ofori, E. K., Adebayo, T. S., & Gyamfi, B. A. (2023). How do environmental tax and renewable energy contribute to ecological sustainability? New evidence from top renewable energy countries. *International Journal of Sustainable Development and World Ecology*, *30*(6), 650–670. DOI: 10.1080/13504509.2023.2186961

Shetty, S. S., Deepthi, D., Harshitha, S., Sonkusare, S., Naik, P. B., Kumari, S. N., & Madhyastha, H. (2023). Environmental pollutants and their effects on human health. *Heliyon*, *9*(9), 1–12. DOI: 10.1016/j.heliyon.2023.e19496 PMID: 37662771

Singh, N. P., Bhawna, A., Singh, S., & Arshad, Kh. (2019). Mainstreaming climate adaptation in Indian rural developmental agenda: A micro-macro convergence. *Climate Risk Management*, *29*(1), 30–41. https://www.sciencedirect.com/science/article/pii/S221209631830130X

Singh, S. (2020). Farmers' perception of climate change and adaptation decisions: A micro-level evidence from Bundelkhand Region, India. *Ecological Indicators*, *116*(1), 1–10. https://www.sciencedirect.com/science/article/abs/pii/S1470160X2030412X

Singhal, S., Thapar, S., Kumar, M., & Jain, S. (2022). Impacts of sustainable consumption and production initiatives in energy and waste management sectors: Examples from India. *Environment, Development and Sustainability*, *24*(12), 14184–14209. DOI: 10.1007/s10668-021-02026-3

Ssekibaala, D. S., Ariffin, M. I., & Duasa, J. (2022). Economic growth, international trade, and environmental degradation in Sub-Saharan Africa. *Journal of Economic Development*, *24*(4), 293–308. DOI: 10.1108/JED-05-2021-0072

Thai, H. N. (2023). Causal relationship between globalization, economic growth and CO2 emissions in Vietnam using Wavelet analysis. *Energy & Environment*, *34*(7), 2386–2412. DOI: 10.1177/0958305X221108498

Topuz, H., Kazak, H., Rahman, M. M., Kılıç, C., Akcan, A. T., & Özekicioğlu, H. (2025). Is export diversification detrimental to environmental quality? An examination of the roles of green innovation and environmental taxation. *Environmental Research Communications*, *7*(1), 1–10. DOI: 10.1088/2515-7620/adac36

Udemba, E. N. (2021). Mitigating environmental degradation with institutional quality and foreign direct investment (FDI): New evidence from asymmetric approach. *Environmental Science and Pollution Research International*, *28*(1), 43669–43683. DOI: 10.1007/s11356-021-13805-w PMID: 33837941

Ünal, H., & Aktuğ, M. (2022). The impact of human capital and bio-capacity on the environmental quality: Evidence from G20 countries. *Environmental Science and Pollution Research International*, *29*(30), 45635–45645. DOI: 10.1007/s11356-022-19122-0 PMID: 35149945

Uzar, U., & Eyuboglu, K. (2023). Does income inequality increase the ecological footprint in the US: Evidence from FARDL test? *Environmental Science and Pollution Research International*, *30*(1), 9514–9529. DOI: 10.1007/s11356-022-22844-w PMID: 36057056

Vale, V. A., Perobelli, F. S., & Chimeli, A. B. (2017). International trade, pollution, and economic structure: Evidence on CO emissions for the North and the South. *Economic Systems Research*, *30*(1), 1–17. DOI: 10.1080/09535314.2017.1361907

Vidyarthi, H. (2013). Energy consumption, carbon emissions and economic growth in India. *World Journal of Science. Technology and Sustainable Development*, *10*(4), 278–287. DOI: 10.1108/WJSTSD-07-2013-0024

Wang, J., & Dong, K. (2019). What drives environmental degradation? Evidence from 14 Sub-Saharan African countries. *The Science of the Total Environment*, *656*(1), 165–173. DOI: 10.1016/j.scitotenv.2018.11.354 PMID: 30504018

Wang, L., Wei, F., Fang, Z., & Svenning, J. C. (2025). Transforming forest management through rewilding: Enhancing biodiversity, resilience, and biosphere sustainability under global change. *One Earth*, *8*(3), 1–10. DOI: 10.1016/j.oneear.2025.101195

Wang, S., Zhang, W., Wang, H., Wang, J., & Jiang, M. J. (2021). How does income inequality influence environmental regulation in the context of corruption? a panel threshold analysis based on chinese provincial data. *International Journal of Environmental Research and Public Health*, *8*(15), 1–21. DOI: 10.3390/ijerph18158050 PMID: 34360341

Wang, X. (2024). Correlation analysis of energy consumption, carbon emissions and economic growth. *Energy Informatics*, *7*(45), 1–12. DOI: 10.1186/s42162-024-00349-9

Wei, J., Rahim, S., & Wang, S. (2022). Role of environmental degradation, institutional quality, and government health expenditures for human health: Evidence from emerging seven countries. *Frontiers in Public Health*, *10*(1), 1–20. DOI: 10.3389/fpubh.2022.870767 PMID: 35400071

Wolde-Rufael, Y., & Mulat-Weldemeskel, E. (2021). Do environmental taxes and environmental stringency policies reduce CO2 emissions? Evidence from 7 emerging economies. *Environmental Science and Pollution Research International*, *28*(18), 22392–22408. DOI: 10.1007/s11356-020-11475-8 PMID: 33417133

Xiao, D., Yu, F., & Yang, H. (2022). The impact of urban-rural income inequality on environmental quality in China. *Complexity*, *2022*(1), 1–10. DOI: 10.1155/2022/4604467

Xuan, H. T., & Hung, N. T. (2024). Does green investment mitigate environmental degradation in Vietnam: The time-frequency effect of nonrenewable energy investment and globalization? *Management of Environmental Quality*, *35*(5), 1005–1027. DOI: 10.1108/MEQ-09-2023-0332

Yang, B., Jahanger, A., Usman, M., & Khan, M. A. (2021). The dynamic linkage between globalization, financial development, energy utilization, and environmental sustainability in GCC countries. *Environmental Science and Pollution Research International*, *28*(1), 16568–16588. DOI: 10.1007/s11356-020-11576-4 PMID: 33387307

Zhang, J. (2023). Assessing the effect of the improvement of environmental damage compensation legal system and green finance project on the re-establishment of the ecological environment. *Environmental Science and Pollution Research International*, *30*(25), 67662–67675. DOI: 10.1007/s11356-023-26877-7 PMID: 37118386

Zhang, L., Xu, M., Chen, H., & Li, Y. (2022). Globalization, green economy and environmental challenges: State of the art review for practical implications. *Frontiers in Environmental Science*, *10*(1), 1–7. DOI: 10.3389/fenvs.2022.870271

Zhang, X., Zhang, H., & Yuan, J. (2019). Economic growth, energy consumption, and carbon emission nexus: Fresh evidence from developing countries. *Environmental Science and Pollution Research International*, *26*(1), 26367–26380. DOI: 10.1007/s11356-019-05878-5 PMID: 31290044

KEY TERMS AND DEFINITIONS

Ecological Footprint (EF): EF measures the degradation of ecosystem services due to human driven activities.

Economic Development: It is an integrated component of the economy that increases social welfare, human development, living standard and well-being of people.

Economic Growth: It measures the percentage change in gross domestic product (GDP) in current year as compared to previous year or base year.

Environmental Degradation: It measures the quantity and quality of ecosystem services and natural resources are decreasing due to increasing CO2 and GHGs emissions.

Environmental Policy: It is the precautionary action taken by a specific country to abate GHGs and CO2 emissions, and provide the protection of ecosystem services and natural resources.

Environmental Sustainability: It ensures the protection and conservation of ecosystem services to increase their productivity and efficiency to meet the needs of current and future generations.

Green Innovation: It is environmentally supportive technology to create green products and enhance environmental quality.

Human Development Index (HDI): HDI is the integrated components of real GDP, life expectancy at birth and literacy rate. HDI reflects the overall economic, social and health progress of people.

Technological Change: It is applications of innovation and scientific mechanisms in production sectors to increase their efficiency and productivity.

Chapter 7
Development Discourses and the Illusion of Growth:
Economic Privilege in the Horn of Africa

Wako Jio
https://orcid.org/0009-0003-6192-459X
Bule Hora University, Ethiopia

Shashi Kant
https://orcid.org/0000-0003-4722-5736
Lovely Professional University, India

ABSTRACT

This chapter looks at how development discourses affect economic privilege in the Horn of Africa, through an emphasis on how the idea of growth may keep inequality going. The study uses a sample of 394 people as well as the adequacy test to see if the sample is good enough for factor analysis. The KMO value shows that there are strong correlations among the variables. Factor loading is used to check the measurement model as well as make sure that the indicators correctly reflect the underlying ideas about development discourses as well as economic privilege. AMOS software uses mediation to look at how development discourses, the sense of economic progress, as well as the ensuing economic privilege are all connected. The outcomes imply that the popular stories about development frequently make it look like there is growth while in reality it mostly helps elite groups as well as

DOI: 10.4018/979-8-3373-1862-2.ch007

leaves out larger groups. We take a close look at this illusion of progress as well as show how it affects public policy as well as the distribution of resources, which ultimately makes existing inequities

INTRODUCTION

The Horn of Africa is an area through huge differences in wealth as well as power, as well as political instability (Bannister & Carver, 2025; Ward & Lucas, 2024). It is a good a case of how development ideologies as well as economic privilege may work together (Mausethagen, 2025; Costa et al., 2024). This issue is important on a global scale because of the region's vital geopolitical position as well as its ability to affect economic patterns across Africa as well as beyond (Amevor, 2025). As countries deal through problems through development, it is very important for policymakers as well as international organizations to understand how growth stories affect economic reality (Rac et al, 2024). Post-colonial theories that criticize traditional economic models as well as focus on the stories that shape development practices are where the idea of development discourses came from (Azkar, 2025). These conversations typically mirror the interests of powerful people, which keeps cycles of privilege as well as inequity going (Akudugu & Ogwu, 2024). Research has shown that there is a big difference among the reported economic progress as well as the real-life experiences of marginalized populations (Hadar & Ergas, 2025). This shows that development projects often don't fix the root causes of structural inequality (Jio et al., 2025).

There are several real-world problems in this area, such as not having enough data, ongoing corruption, and not having governing mechanisms that include everyone. These problems make it harder to put good development plans into action, and there are still big gaps in our understanding of how development stories affect economic privilege. This chapter tries to fill in these gaps in the research by utilizing a strong technique that comprises a sample size of 394, KMO analysis, CFA, and SEM through AMOS software. It does this in order to better understand the complicated connections among development discourses and economic privilege, and to provide information that can help through future study and policy decisions. The main issue that this research tries to answer is: How do the prevailing development stories in the Horn of Africa make it seem like there is growth, which keeps elite people economically privileged while pushing larger groups to the edges? To establish development plans that are fairer and more successful, we need to answer this issue. These policies should really help disadvantaged groups and encourage long-term growth in the region.

BACKGROUND

The Horn of Africa, which includes Ethiopia, Somalia, Kenya, and Djibouti, has always had a complicated social and economic environment full through both opportunities and problems (Amevor, 2025). Stories on the region's development generally focus on how quickly the economy is growing, especially in areas like agriculture, telecommunications, and infrastructure (Akudugu & Ogwu, 2024). But these stories often hide deeper problems and don't lead to real gains for people who are already at a disadvantage (Hadar & Ergas, 2025). For example, in case of Ethiopia research shows that Ethiopia's Growth and Transformation Plan (GTP), which aspired to make the country a middle-income nation, is a real-world a case of these problems. The plan showed resulting GDP growth rates, but some people have criticized it since the benefits were not evenly distributed (Rac et al, 2024). For a case, rural populations and pastoralists did not see much of the economic expansion (Azkar, 2025). For example, in case of Somalia, too, international development efforts have relied mainly on donor-driven projects that focus on short-term stabilization rather than long-term structural improvements (Ward & Lucas, 2024). The Somali Compact was intended to solve political and economic problems, but it has been hard to put into action because of problems like corruption and a lack of inclusive government (Mausethagen, 2025). Because of this, many people still can't see the benefits of growth, and local communities are routinely left out of decision-making processes (Costa et al., 2024).

For example, in case of Kenya research shows that The Vision 2030 program in Kenya aspires for long-term growth and development; and it has run into a lot of problems because of corruption and inequality (Alghamdy, 2024). The advantages of modernizing and building new infrastructure have mostly gone to urban elites, leaving people in rural areas throughout enough access to services and opportunities (Kant et al., 2025). These case studies show a big discrepancy among what people in the Horn of Africa say about development and what they really experience. They also show how hard it is to put good ideas into action (Jio et al., 2025). This background makes it even more important to look more closely at how development stories may generate the appearance of progress that keeps economic privilege going. To come up through plans that not just boost economic growth but also make sure that all groups of people in the Horn of Africa are included and treated fairly, you need to understand how these things work.

Statement of Problem

The Horn of Africa has a complicated challenge through how development narratives affect economic privilege (Rac et al, 2024). Development projects should ideally encourage fair growth that helps all parts of society and fixes problems that already exist (Azkar, 2025). And, research that goes against this shows that many of these programs give the impression of development while really benefiting elite groups more than the general population (Akudugu & Ogwu, 2024). Even if a lot of money has been put into development, studies show that communities are still poor and don't have a voice, which raises important issues about how well existing policies are working (Mausethagen, 2025). There are big holes in the research of how development stories affect economic privilege (Costa et al., 2024). A lot of studies don't give clear answers about the real advantages of stated growth. They also don't typically talk about how these stories affect people's views and policy choices (Amevor, 2025). There are also gaps in theory as well; current frameworks may not fully take into consideration how cultural and socio-political settings affect development outcomes (Kant et al., 2025). We don't fully comprehend how local realities affect global discourses, which frequently leads to development initiatives that don't meet the demands of local people (Alghamdy, 2024).

Corruption, weak governance, and not enough data is just a few of the real-world problems that make it much harder to put successful development strategies into action (Butson & Spronken-Smith, 2024). These problems typically make people in communities less trusting of one other, which makes development initiatives less effective (Ward & Lucas, 2024). This study tries to fill up these gaps by using a strong methodological approach that combines a thorough look at development discourses, economic privilege, and real-world data from a sample of 394 respondents. The research will use KMO analysis, CFA, and SEM using AMOS software to give a more detailed picture of these linkages, giving insights that are both based on theory and applicable to the situation. In the end, our study aims to help future academics and politicians come up through more inclusive and successful development plans that really help marginalized people in the Horn of Africa, making sure that the promise of prosperity leads to real benefits for everyone.

Theoretical Foundation

Dependency Theory: Dependency Theory came forth in the late 20th century as a response to modernization ideas that said all countries could progress in a straight line. It says that developing countries are still economically dependent on industrialized countries because of past exploitation as well as uneven power relationships. This theory shows how outside economic factors as well as internal structural disparities keep people poor as well as underdeveloped. Dependency Theory is useful in the Horn of Africa because it helps explain how global economic systems keep inequality going, creating the appearance of prosperity that only serves a small few while leaving the bulk behind (Rac et al, 2024).

Theory of World Systems: Immanuel Wallerstein came up through World Systems Theory, which builds on Dependency Theory by putting countries into three groups: core, semi-periphery, as well as peripheral. This concept focuses on how the economies of the world are linked as well as how core nations take use of periphery nations for labor as well as resources. It gives you a way to look at how development talks in the Horn of Africa may strengthen economic privilege by making some people rely on markets as well as decisions made outside of the region. This idea is very important for figuring out what structural problems are getting in the way of fair growth in the area (Alghamdy, 2024).

Theory after Colonialism: Post-Colonial Theory looks at how colonialism has affected former colonies in the long term, through a focus on themes of identity, power, as well as cultural dominance. It criticizes how stories about development frequently show colonial views by calling other civilizations "backward" or "in need of development." This idea helps us understand how past wrongs as well as cultural misrepresentations affect current development strategies in the Horn of Africa. It gives important information about how local people see as well as interact through development projects, which shows how complicated the idea of growth really is (Ward & Lucas, 2024).

Constructivism in Society: According to social constructivism, social interactions as well as processes create knowledge as well as meaning. This theory is especially useful for figuring out how development discourses affect how people in countries see progress as well as privilege. In the Horn of Africa, stories about progress may help people agree on things that either support or question the way authority is set up now. Researchers can learn more about the elements that create the illusion of growth as well as its effects on economic privilege by looking at how these stories are made as well as spread (Amevor, 2025).

Dependency Theory: This theory is the main theory used in this study. It gives us a strong way to think about the structural disparities that still exist in the Horn of Africa as well as how global economic trends keep these inequities going. This theory will help us comprehend the complicated link among development discourses as well as economic privilege by looking at the dependence ties that have formed throughout time. It will also help us understand the problems that marginalized groups in the region confront.

Definition and Origins of Concept

Development Discourses: Development discourses are the stories as well as frameworks that people use to talk about as well as understand development policies as well as practices. These ideas come from modernization theories that came forth after World War II. They impact how people think about development, as well as they typically represent the objectives of powerful organizations. They have an impact on public policy, financing priorities, as well as the execution of development projects, which in turn affects how resources as well as opportunities are distributed through in societies (Alghamdy, 2024).

Economic Advantage: Economic privilege is the perks as well as advantages that some people or organizations get because of their social as well as economic standing. This idea comes from theories of social stratification as well as inequality that look at how money, access to resources, as well as power dynamics work together to make differences. In the Horn of Africa, economic privilege typically shows up as a concentration of wealth among the elites, which keeps poor populations in a cycle of poverty (Ward & Lucas, 2024).

Illusion of Growth: The illusion of growth is the false belief that an economy is doing well when, in fact, the advantages of expansion are not evenly shared. This phrase draws attention to the differences among official economic data as well as the real-life realities of regular people. It came from criticisms of standard economic measures that don't show how progress affects people in different ways, especially in places where growth doesn't mean better living conditions for everyone (Costa et al., 2024).

EMPIRICAL LITERATURE REVIEW

Development Discourses and Economic Privilege

A lot of research has been done on the link among development discourses and economic privilege, especially in the Global South, which includes the Horn of Africa (Amevor, 2025). Many academics say that the stories about progress frequently make things worse by reinforcing existing inequities instead of making them better (Alghamdy, 2024). For a case, research shows that development discourses are built from a Western point of view, making non-Western cultures seem "backward" or in need of help (Costa et al., 2024). This kind of view can lead to policies that put the needs of the elite foremost while seeming to promote national progress (Rac et al, 2024). This keeps a small group of people rich. Studies that looked at the Horn of Africa show how these things work in real life (Akudugu & Ogwu, 2024). For example, in case of Ethiopia research shows that Ethiopia criticizes the government's development and Transformation Plan because it promised high development rates but only benefited urban elites, leaving rural areas behind (Smith, 2024). This discovery supports the idea that development projects frequently don't take into account the unique demands of different groups of people and the local environment, which gives the impression of progress that doesn't lead to fair economic possibilities (Ward & Lucas, 2024).

Also, research shows how important it is to know the historical as well as social-political settings in which development discourses take place (Hadar & Ergas, 2025). He says that the effects of colonialism as well as neocolonialism are still felt in today's development strategies, which create systems that favor outside interests as well as keep people dependent on them (Mausethagen, 2025. This point of view shows how development stories may hide the truth about economic privilege by leaving out the perspectives as well as experiences of those who are already excluded. A comparative research shows how complicated development discourses may be (Azkar, 2025). It looks at how different stories about gender as well as development affect women's job prospects in different places. Sadat-Razavi et al. (2025) research shows that some development projects try to give women more power, but they frequently just strengthen the power systems that already exist as well as give some groups more authority than others. This shows how important it is to carefully think about how development narratives affect different categories of people in society (Butson & Spronken-Smith, 2024). Also, the work on the capacity approach gives us a good way to look at how well development discourses operate (Jio & Kant, S2025). Sadat-Razavi et al. (2025) believes that real development should be about more than just looking at economic growth metrics; it should also be about giving people more freedom as well as talents. This approach says that we need to rethink how we define success in development projects, putting more emphasis on fairness as well as making sure everyone has access to resources (Bannister & Carver, 2025).

H1: Development Discourses and Economic Privilege having empirical association

Development Discourses and Illusion of Growth

A lot of research has been done on the idea of the "illusion of growth" in development talks. This research shows how stories of success may hide the truth about inequality and underdevelopment (Amevor, 2025). Researchers say that development talks typically focus on quantitative economic metrics, such GDP growth, and ignore qualitative factors that affect people's well-being and social fairness (Azkar, 2025). For a case, some studies say that using GDP as a metric of success might give people the wrong idea about how well things are doing (Akudugu & Ogwu, 2024). In many circumstances, big economic development doesn't mean that most people would have a better quality of life, especially in emerging areas like the Horn of Africa. Several studies show how government and international stories of progress may keep the illusion of success alive in the Horn of Africa (Bannister &

Carver, 2025). For a case, a research on Ethiopia's fast economic growth points out that even though the government has produced good GDP statistics, the advantages have mostly gone to urban elites, leaving rural communities poor and on the sidelines (Alghamdy, 2024). This difference shows how stories about growth may make things look successful when they are really just hiding the ongoing disparities that exist in many cultures (Costa et al., 2024).

The book also looks at how global neoliberal policies affect local development discourses, typically putting market-driven growth ahead of social equality (Rac et al, 2024). He says that these kinds of policies can bring in money in the short term, but they don't think about how they will affect the economy in the long run or how they would affect everyone (Ward & Lucas, 2024). This criticism is especially important in the Horn of Africa, where reforms that focus on the market have sometimes made existing problems worse instead of fixing them completely. The study also criticizes the role of international aid in making the illusion of progress stronger (Butson & Spronken-Smith, 2024). Sadat-Razavi et al. (2025) says that help frequently makes people dependent as well as doesn't fix the problems that produce poverty. Aid can make it seem like growth is happening throughout making life better for the poor, instead than encouraging self-sufficiency as well as long-term progress (Hadar & Ergas, 2025). This observation shows how important it is to carefully look at how development discourses affect how people think about growth as well as the success of assistance efforts (Mausethagen, 2025). As we spoke about, social constructivism gives us a new way to look at the idea of the illusion of growth. He says that growth stories are not only reflections of factual truths, but are made via social interactions as well as power relations (Kant et al., 2025). This point of view shows how the way development is framed may affect how people think about it as well as what policies are made, which can frequently lead to a gap among reported progress as well as the real-life realities of regular people (Lemi et al., 2025).

H2: Development Discourses and Illusion of Growth having empirical association

Illusion of Growth and Economic Privilege

More and more empirical research is looking at the link among the illusion of development and economic privilege (Alghamdy, 2024). This shows how false stories about economic progress may hide deep-seated disparities (Azkar, 2025). Researchers say that even while certain areas may claim high GDP growth rates, these numbers don't often show what life is really like for excluded groups (Butson & Spronken-Smith, 2024). For a case, study looks at how economic development in undeveloped countries might make it look like things are getting better, while in fact it only lets the rich become richer and the poor stay poor (Ward & Lucas, 2024). For example, in case of Ethiopia research shows that gap among growth measures and real-life social and economic conditions is especially clear in places like the Horn of Africa, where development stories typically leave out the experiences of those who are not included (Jio & Kant, 2025). Many experts have looked at the story of Ethiopia's remarkable economic rise (Hadar & Ergas, 2025). A research concluded that Ethiopia has one of the fastest-growing economies in Africa, but the gains have not been shared equally (Rac et al, 2024). The outcomes show that government policies have mostly helped the urban elite, leaving rural populations and pastoralists out of the economic benefits (Bannister & Carver, 2025). This circumstance shows the illusion of growth, where economic measures that just show the surface hide the fact that many people are still poor and unequal (Amevor, 2025).

For example, in case of Kenya research shows that The Vision 2030 program in Kenya also attempted to change the economy and raise the level of living (Mausethagen, 2025). And, data shows that even if the country's economy has improved, these improvements have mostly helped cities and rich people. In this case, the illusion of progress has caused social instability and unhappiness among marginalized groups who feel left behind by policies that put the needs of the privileged foremost (Costa et al., 2024). This real-world data shows how important it is to think critically about how growth stories may keep economic privilege and current power structures in place (Butson & Spronken-Smith, 2024). Another important area of study is the part that international financial institutions play in keeping the illusion of development alive (Kant et al., 2025). Researchers say that the conditions that come through loans from groups like the World Bank and the IMF generally put macroeconomic stability ahead of social equality (Azkar, 2025). These kinds of measures can lead to short-term economic development, but they don't fix the underlying problems that cause poverty and inequality (Akudugu & Ogwu, 2024). So, even though governments may say they are growing, the truth is that this development does not lead to better living conditions for most people. This shows a big gap among policy goals and social consequences (Hadar & Ergas, 2025). Also, as we spoke about, the idea of the illusion of growth is closely related to social stratification and economic privilege. His ideas about capital—social, cultural, and economic—suggest that the advantages of growth typically go to people who already have different kinds of capital. This concentration makes existing disparities worse since those through privilege are better able to take advantage of economic opportunities, while others who are excluded are stuck in cycles of poverty (Kant & Adula, 2025).

H3: Illusion of Growth and Economic Privilege having empirical association

Development Discourses, Economic Privilege, Illusion of Growth

The relationship among development discourses, economic privilege, and the illusion of progress has been a major focus of empirical literature (Rac et al, 2024). These studies show how stories about development may hide structural disparities while also reinforcing privilege (Akudugu & Ogwu, 2024). Globalization and neoliberal economic policies, which put market-driven growth at the top of the list of what counts as success, is commonly used to shape development discourses (Bannister & Carver, 2025). Scholars like these say that these discourses not only impact how people think about development, but they also make policies that help the rich seem more legitimate (Azkar, 2025). This framing can make it look like the economy is growing while, in fact, poverty and disenfranchisement is still common, especially in groups that are already on the margins. In the Horn of Africa, stories of development typically use fast economic growth as proof that things are getting better. But studies show that even while GDP numbers may show a flourishing economy, the advantages are often only felt by a limited group of people (Alghamdy, 2024). For example, in case of Kenya research shows that their research shows that development strategies in countries like Ethiopia and Kenya mostly benefit those who live in cities and are wealthy, which economic privilege keeps going and puts rural areas at a disadvantage (Mausethagen, 2025). This shows how development discourses may make it look like things are getting better when they really aren't, hiding the structural impediments that disadvantaged groups confront and creating an appearance of growth that doesn't lead to fair outcomes (Hadar & Ergas, 2025).

In addition, the illusion of growth is mediated by development discourses, which are further investigated by someone who stresses the need to understand the social environment in which these stories are made (Bannister & Carver, 2025). Sadat-Razavi et al. (2025) says that development plans don't often deal through the core causes of inequality since they are based on a limited view of progress that puts economic metrics ahead of social well-being (Hadar & Ergas, 2025). This narrow view can make current power structures stronger, as people who already have economic power are better able to take advantage of growth, which keeps inequality going. It's also important for international organizations to play a part in these processes (Butson & Spronken-Smith, 2024). Research shows that the World Bank as well as IMF's policies, which are based on neoliberal ideas, generally put macroeconomic stability ahead of social equality. This method can bring short-term economic benefits that hide the long-term structural problems that cause poverty as well as inequality (Azkar, 2025). This keeps the illusion of growth alive, which lets the rich get the advantages while the poor see little change in their living situations (Amevor, 2025). The idea of social capital, which we spoke about earlier, is also a helpful way to think about how development discourses might affect the link among economic privilege as well as the false sense of growth (Costa et al., 2024). Bourdieu says that those who have access to social networks as well as cultural resources are more inclined to take advantage of economic chances (Szedlak et al., 2025). This point of view shows how development stories may keep privilege going by favoring people who already have social capital, which keeps inequities in place (Ward & Lucas, 2024).

H4: Development Discourses. Economic Privilege and Illusion of Growth Mediation having empirical association

Figure 1. Conceptual framework

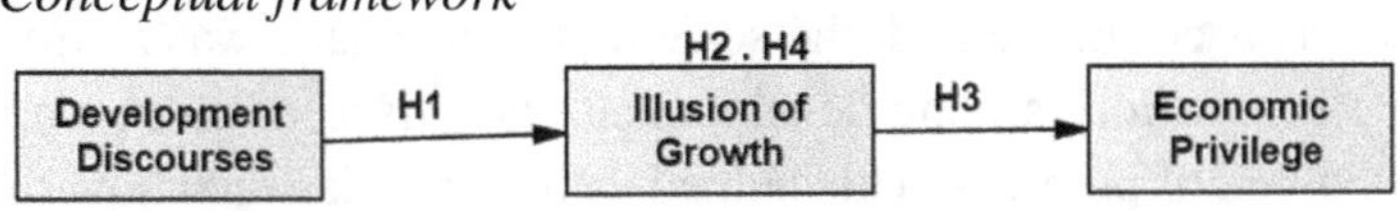

RESEARCH METHODOLOGY

The study used a quantitative research method to look at the links among development narratives, economic privilege, as well as the false sense of growth in the Horn of Africa. Stratified random sampling was used to choose a sample size of 394 respondents. This was done to make sure that the sample was representative of people of different ages, genders, as well as social as well as economic backgrounds. The goal of this sampling method was to get a wide range of opinions from the public so that a full study of the problems could be done. We used the adequacy of data to see if the sample was good enough for factor analysis. The KMO value was above the permitted level of 0.6. This meant that the sample could be looked at more closely. After that, reduction of factor was used to find the basic structure of the factors that had to do through economic privilege, development discourses, as well as the illusion of growth. EFA helped figure out how many factors there are as well as how much each item contributes to each component. This made it easier to see how these constructs are related to each other. After EFA, we used AMOS software to do factor loadings to check the measurement model. This stage made sure that the observed variables properly reflected the constructs, which proved that the scales were reliable as well as valid. We next used mediation to look at how the constructs were related to each other. This gave us information on both direct as well as indirect impacts through in the proposed framework.

A thorough examination of related literature provided the basis for the scale development for this work. We looked at existing scales that deal through development discourses, economic privilege, as well as the illusion of growth to help us come up through new items. The research showed that there were important themes as well as features related to these notions, such as how people see growth, how easy it is for them to get resources, as well as how unequal their socio-economic status is. Items were made to represent these themes, making sure they conveyed the spirit of each build as well as were in line through existing metrics in the field. Each item was carefully phrased to make it clearer as well as more relevant, so that responders could accurately describe their experiences as well as thoughts. A small group of people tested the scale before it was used to make sure the items were clear as well as to improve them. To make the scale more reliable, feedback from this pre-test was used. In the end, the scale that was created was a strong tool for assessing the connections among development discourses, economic privilege, as well as the illusion of progress. This helped us better understand how these things work in the Horn of Africa.

DATA ANALYSIS

Table 1 shows the study's data adequacy metrics, through an emphasis on the findings of the sample extent test as well as the adequacy. The KMO score of .924 shows that the sampling is very good, which means that the data may be used for factor analysis. A KMO rating higher than 0.8 is usually seen to be "magnificent," which makes the sample utilized in this study more reliable. The Bartlett's Test of Sphericity also gives a Chi-Square value of 19,484.673 through 780 degrees of freedom. The p-value of .000 shows that the correlation matrix is quite distinct from an identity matrix, which means that there are enough correlations among the variables. This conclusion is important because it shows that component analysis is appropriate as well as that the variables are connected enough to each other to merit further investigation of their underlying structure.

Table 1. Data Sample Extent

Meyer Kaiser Olkin tof Adequate.		**.924**
Sphericity Test of Bartlett	Chi-Square (Relative)	19484.673
	Degree of freedom	780
	p-value	.000

Source: Authors

Table 2. Variance Sum

Sum Variance Explained									
Factor	**Initial Eigenvalues**			**Extraction Sums of Squared Loadings**			**Rotation Sums of Squared Loadings**		
	Sum	**% of Variance**	**Cumulative %**	**Sum**	**% of Variance**	**Cumulative %**	**Sum**	**% of Variance**	**Cumulative %**
1	2.7	26.7	26.76	2.7	26.7	26.7	8.7	21.9	21.9
2	1.9	14.9	41.71	1.9	14.9	41.7	6.0	15.0	37.0
3	1.3	10.9	52.68	1.3	10.9	52.6	4.6	11.6	48.6

Abstraction: PCA.

Table 2 shows the outcomes of the factor analysis, through an emphasis on the eigenvalues as well as the amount of variation that each component explains. The foremost eigenvalue shows that the foremost component has an eigenvalue of 2.7, which means it explains 26.7% of the total variance. This means that the foremost factor explains a lot of the differences in the data set, which makes it an important part of comprehending the constructs being assessed. The second component has an eigenvalue of 1.9, which accounts for 14.9% of the variance. This brings the total proportion to 41.7%. This makes the foremost two components even more important for interpreting the overall data structure, as they account for about 42% of the variance. The third element has an eigenvalue of 1.3 as well as adds 10.9% to the total, bringing the total to 52.6%. The Extraction Sums of Squared Loadings show that the components stay the same after extraction as well as explain the same amount of variance. The Rotation Sums of Squared Loadings show a sharper separation of the variables, which is important. After rotation, the foremost component explains 21.9% of the variation, the second factor 15.0%, as well as the third factor 11.6%. This shows how rotation makes it easier to understand the factors by spreading out the variance as much as possible.

Table 3. Path Validation

Path Validation			Estimate	S.E.	C.R.	P
Illusion of Growth	**<-->**	**Development Discourses**	**.563**	**.094**	**5.989**	**.00**
Illusion of Growth	<-->	Economic Privilege	.493	.056	8.803	.00
Development Discourses	<-->	Economic Privilege	.652	.105	6.209	.00

Source: Authors

Table 3 shows the covariance estimates among the constructs of Illusion of Growth, Development Discourses, as well as Economic Privilege. The correlation among Illusion of Growth as well as Development Discourses is thought to be .563, through a standard error (S.E.) of .094. The critical ratio (C.R.) of 5.989 as well as a p-value of 0.00 show that there is a very strong association. This means that while people think that development discourses are becoming better, the illusion of progress likewise gets stronger. This large correlation backs through the idea (H2) that stories about development have a big effect on how people see growth. The correlation among Illusion of Growth as well as Economic Privilege is also thought to be .493, through a S.E. of .056. The C.R. of 8.803 as well as the p-value of .00 show a strong connection, which supports the idea that people's ideas about growth are strongly linked to the economic advantages that some groups have (H3). This research implies that when people think they are experiencing an illusion of progress, they may also see or feel an increase in their economic advantages. The last thing to note is that the correlation among Development Discourses as well as Economic Privilege is .652, through a standard error of .105. The C.R. of 6.209 as well as the p-value of.00 show that there is a strong association (H1). This significant correlation shows that the stories we tell about development not only change how people think about growth, but they also change how wealth is spread out among the people.

Table 4. Validity Concern

	CR	AVE	MSV	MaxR(H)	DD	IG	EP
DD	0.732	0.643	0.219	0.852	**0.654**		
IG	0.758	0.657	0.132	0.763	0.171	**0.665**	
EP	0.793	0.603	0.274	0.797	0.383	0.193	**0.684**

Note: DD = Development Discourses; IG= Illusion of Growth; EP = Economic Privilege

Table 4 shows the validity issues through the constructs of Development Discourses (DD), Illusion of Growth (IG), as well as Economic Privilege (EP). DD is 0.732, IG is 0.758, as well as EP is 0.793. These numbers are higher than the usual cutoff of 0.7, which means that the items in each construct are good at measuring the ideas behind them. The AVE numbers back up this dependability even further. DD has an AVE of 0.643, IG has an AVE of 0.657, as well as EP has an AVE of 0.603. All of these numbers are higher than the suggested minimum of 0.5, which means that their constructions capture a large part of the difference among the observed variables. The MSV values show how much variation is shared among constructs. For a case, DD has a value of 0.219, IG has a value of 0.132, as well as EP has a value of 0.274. These numbers are rather low compared to the AVE values, which mean that each construct is different as well as doesn't overlap too much through the others. The MaxR(H) values also support this. They are 0.852 for DD, 0.763 for IG, as well as 0.797 for EP, all of which show that the data is quite reliable.

Figure 2. Structure equation model

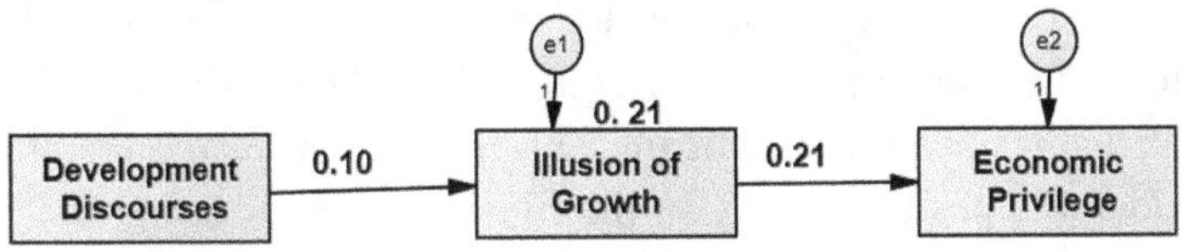

Table 5. Model Fitness

Sig.	Chi-Sq	RMR	GFI	PNFI	TLI	RMSEA
0.003	2.163	.037	0.929	0.925	0.921	.029

Source: Authors

Table 5 shows the model fitness statistics for the structural equation model used in the study. The acceptance level of p is 0.003, which means that the chi-square statistic is significant. This means that the model fits the data well. A high chi-square value usually means that the model fits well, as well as it can also change depending on the size of the sample. The Chi- Sq is 2.163, which is rather low. This adds to the idea that the model is a good fit for the data. The RMR value is 0.037, which means that the fit is satisfactory because values below.05 is usually seen as acceptable. A number of fit indices give the model further proof. The GFI is 0.929, which is higher than the usual cutoff of 0.90. This means that the fit is strong. The PNFI is 0.925, as well as the TLI is 0.921, which both show that the model fits well. These indices show how well the model can handle the data while still being simple. The RMSEA value is .029, which is far lower than the suggested threshold of 0.05. This makes the models fit even better. A low RMSEA score means that the model is a good approximation of the population covariance matrix.

Table 6. Critical ratio

Comparative			Approx.	S.E.	C.R.	P	Ass.
Illusion of Growth	**<---**	**Development Discourses**	**.716**	**.405**	**1.767**	**.06**	**H2**
Economic Privilege	<---	Illusion of Growth	.651	.197	3.304	.00	H3
Economic Privilege	<---	Development Discourses	.563	.105	5.361	.00	H1

Source: Authors

Table 6 shows the outcomes of the regression test, which shows how the ideas of Development Discourses, Illusion of Growth, as well as Economic Privilege are related to each other. The computed regression coefficient for the link among Development Discourses as well as Illusion of Growth is 0.716, through a standard error of 0.405. The crucial ratio of 1.767 as well as the p-value of 0.06 show that there is a link that is not quite significant. This means that Development Discourses have a considerable positive effect on the Illusion of Growth, but the evidence is slightly below the standard significance level of 0.05. This data partially supports hypothesis H2, which says that development narratives have a big effect on how people think about growth. As well as, more research may be needed. On the other has well as, the association among Illusion of Growth as well as Economic Privilege has a regression coefficient of 0.651 as well as a S.E. of 0.197. A C.R. of 3.304 as well as a p-value of 0.00 show a very strong association, which strongly supports hypothesis H3. This means that as people get more aware of the illusion of development, they also become more aware of economic privilege. This shows that there is a clear link among these two ideas. Lastly, the regression route from Development Discourses to Economic Privilege has a coefficient of 0.563 as well as a S.E. of 0.105. The C.R. of 5.361 as well as the p-value of.00 show that there is a strong link, which strongly supports hypothesis H1. This data supports the idea that development stories not only change how people think about growth, but they also directly affect how economic advantage is spread among the population.

Table 7. Mediation

	Worth
Development Discourses → Economic Privilege	.10
Development Discourses → Illusion of Growth → Economic Privilege	.93*.23=.21
	.21

Source: Authors

Table 7 shows the mediating effects that were seen in the interactions among Development Discourses, Illusion of Growth, as well as Economic Privilege. The foremost row shows the direct effect of Development Discourses on Economic Privilege, which has a route influence of 0.10. This number implies that the direct association is not important, which means that Development Discourses by them don't have a big effect on Economic Privilege when other factors are not taken into account. The second row, on the other has well as, shows the indirect effect, where Development Discourses change Economic Privilege through the Illusion of Growth. To get the calculated route influence, you multiply the coefficients: 0.93 (the effect of Development Discourses on Illusion of Growth) as well as 0.23 (the effect of Illusion of Growth on Economic Privilege). This gives you an indirect effect of 0.21. This outcome shows that the Illusion of Growth is what connects Development

Discourses through Economic Privilege. The term "full mediation" for this indirect impact makes it clear that the Illusion of Growth completely explains how Development Discourses as well as Economic Privilege are connected. To put it another way, the Illusion of Growth shapes how people see Development Discourses, which then affects Economic Privilege. This study shows how important perceptions are in turning development stories into real economic outcomes. It also suggests that policymakers should focus on changing the stories about growth in order to have a bigger impact on social as well as economic situations.

DISCUSSION

The outcomes of this study shed light on the complex connections among development discourses, the illusion of progress, and economic privilege in the Horn of Africa. These constructs are strongly linked to each other, which show how important stories about progress are in creating economic reality. It's interesting to note that development discourses have a big effect on the illusion of growth, but they don't seem to have much of an effect on economic privilege. This means that just telling stories about growth isn't enough; it's important to think about how diverse groups of people see and feel these stories. The study's outcomes show that the illusion of progress is a strong link among discourses about development and economic privilege. This mediation demonstrates that the way people think about and talk about growth may have a big impact on who benefits from economic progress. When people think there is an illusion of growth, it becomes linked to the economic advantages that elite groups have, which makes existing disparities even worse. This outcome is in line through what other researchers have said: that development projects generally put the demands of the rich ahead of those of the poor and don't help the people who need it most. Also, the evidence of complete mediation shows how important it is to think about the stories that people tell about progress when designing development strategies. Policymakers shouldn't only look at economic statistics; they need also think about how this information are shared and understood. It could be feasible to make the distribution of economic rewards fairer by encouraging a more open discussion that includes the experiences of all parts of society. This study adds to the expanding body of research that stresses the need of critically looking at development stories. By understanding how perceived growth and economic privilege interact, we can come up through better ways to help underrepresented groups. It can be observed that during the debate on the environmental implications within the context of the development discourses in the Horn of Africa, there has been a critical intersection of inequality with environmental degradation. Growth discourses are seen to favor top sections of society at the expense of the poor, as it usually serves the interest of making economic gains at the cost of ecological well-being. Unsustainable activities that consist of industrial agriculture and deforestation are often presented as a way of progress, but they result in massive environmental degradation, which is found in soil degradation, water shortage and the loss of biodiversity. These two foci bring out the point that environmental-unfriendly policies have a role to play in institutionalizing unequities, where marginalized groups are at a higher risk of environmental degradation and therefore unable to adapt or recover. Hence, the policymakers need to reconsider their approach to the development process, with the key priorities being the inclusion of resources must be distributed devoid of any bias and true sustainability that would allow all communities to become more resilient. Managing these complicated problems, we will be able to pave a more equal and sustainable path into the future along the Horn of Africa. In the end, this study shows how complicated development is in the Horn of Africa and calls for more sophisticated approaches that connect talk through real-world social and economic outcomes.

CONCLUSION

This study has helped us understand how development discourses, the illusion of progress, as well as economic privilege all work together in the Horn of Africa. The outcomes show that development stories have a big impact on how people think about growth, but they don't have a big impact on economic privilege. Instead, the illusion of progress is a key mediator that shows how the way development is framed may hide underlying disparities as well as keep privileged groups' advantages going. This shows how important it is to carefully look at development initiatives to make sure they are inclusive as well as represent the different realities that excluded population's experience. This study calls for a change in the way development projects are planned as well as talked about by focusing on how perception affects economic outcomes. Policymakers should make it a priority to tell stories that not only celebrate progress but also recognize as well as work to fix the problems that still exist in society. In the end, encouraging a more fair conversation about growth is necessary to turn economic progress into real advantages for everyone. This will help the area thrive in a way that is both long-lasting as well as open to everyone. This analysis sets the stage for more research as well as policy work to break down the obstacles to fair growth so that the promise of development may be realized for all people in the Horn of Africa.

MANAGERIAL IMPLICATIONS

The outcomes of this study have important significance for managers as well as policymakers who are working on development projects in the Horn of Africa. Foremost as well as foremost, managers need to understand how development discourses affect how people think as well as how the economy works. By knowing that stories about development might hide real inequities, managers can use more subtle ways to communicate that clearly show both triumphs as well as problems. When creating development policies, companies should also make sure to include all stakeholders. Managers may make sure that development projects meet the needs as well as goals of all parts of society, not just the privileged, by actively incorporating disadvantaged groups in the decision-making process. This interactive method can help people trust each other as well as work together, which will lead to better as well as longer-lasting outcomes. Also, managers should put money into programs that help vulnerable groups get involved in as well as benefit from development activities. This might include training programs, access to resources, as well as support for local businesses, which can help close the gap among stories of progress as well as real economic advantages. Lastly, it is very important to keep an eye on as well as evaluate development projects all the time. Managers may find differences as well as make changes to make sure that growth gains are fairly shared by looking at how policies affect different demographic groups in the actual world. In short, using a broad as well as inclusive approach to development may make managers more successful as well as help the Horn of Africa's economy flourish in a fairer way.

PRACTICAL IMPLICATIONS

The outcomes of this study have a number of real-world effects on how development projects may be carried out in the Horn of Africa. Foremost, it's important for professionals to understand how important it is to write development stories that truly reflect the lives of all groups in society. This way, they may avoid keeping up the image of progress that ignores the problems that underprivileged populations experience. This means that people who report on economic indicators as well as development outcomes must be honest as well as open. In addition, practitioners should work on making programs that include everyone as well as get local communities involved in the development process. This might include setting up seminars, forums, as well as group projects that give people in the community a chance to share their worries as well as hopes. This kind of involvement not only gives stakeholders a sense of ownership, but it also makes sure that development projects are designed to fulfill the requirements of different groups of people. Also, it is very important to put in place training as well as capacity-building programs that provide excluded populations the tools they need to take advantage of economic possibilities. This might include giving people as well as community's access to education, training in technical skills, as well as money that can help them become more self-sufficient. Lastly, there should be ways to keep track of the effects of development projects through continual assessments as well as feedback. Practitioners may find areas that need work as well as make the changes needed to make sure that benefits are fairly distributed by frequently checking how well programs as well as policies are working. In short, these practical steps may make development efforts more effective as well as help make the economy in the Horn of Africa fairer.

THEORETICAL IMPLICATIONS

The outcomes of this study have important theoretical consequences for how we interpret the relationship among development discourses, the illusion of growth, as well as economic privilege. Foremost, the study shows how important it is to use social constructivist theories when looking at growth narratives. Scholars can better understand the factors that affect economic outcomes as well as the distribution of wealth by focusing on how social factors shape how people see as well as comprehend growth. This point of view goes against standard economic models that just look at numbers. Instead, it calls for a more complex understanding of growth that includes qualitative aspects. The study also shows that Dependency Theory as well as World Systems Theory are still important ideas to think about when talking about development today. These frameworks can help us understand how outside factors as well as historical events affect local development stories as well as keep inequality going. The outcomes imply that more research should be done to see how these theories might be used to study the unique situations in the Horn of Africa, especially in terms of globalization as well as how it affects local economies. In addition, the fact that the illusion of growth fully mediates the situation calls for a reevaluation of the theoretical models that are already in use in development studies. It recommends that scholars should look at mediating variables in a more organized way when they study the links among development discourses as well as economic outcomes. This might lead to the creation of new theoretical frameworks that take into consideration how complicated perception as well as reality is in developing situations.

RECOMMENDATIONS

Based on what this study found, there are a number of suggestions that may make development projects in the Horn of Africa more successful. Foremost, it is very important for politicians as well as development workers to make sure that stories about progress are open as well as include everyone. This means working through local communities to learn about their points of view as well as including their input in plans for development. By making sure that stories represent the hopes as well as reality of different groups of people, stakeholders may build more trust as well as cooperation. Also, training programs that help increase capacity should be made to give authority to groups that are often left out. These initiatives might help people improve their abilities, learn about money management, as well as get to resources that help them take advantage of economic possibilities. Giving people this kind of authority can help close the gap among the illusion of growth as well as real economic advantages. Also, there should be systems in place for ongoing monitoring as well as assessment to see how development projects affect diverse groups of people. This will help practitioners find differences as well as make smart changes to make sure that benefits are fairly shared. Communities giving regular input may make programs more responsive as well as help them work better overall. Lastly, different groups, such as government agencies, non-governmental organizations, as well as businesses, need to work together to come up through a unified plan for growth. By building partnerships that use a variety of resources as well as skills, stakeholders may work together to fix systemic problems as well as encourage long-term growth. Following these suggestions might lead to more fair as well as meaningful development outcomes in the Horn of Africa.

FUTURE DIRECTIONS

Future research in the field of development studies should expas well as on what this study found by looking at a few important topics. Foremost, we need long-term research that follow the changes in development discourses as well as how they affect economic privilege across time. These kinds of research might help us understand the dynamics at play better by giving us a better idea of how changes in stories affect people's perceptions as well as real-world outcomes. Researchers should also think about doing comparison studies in other parts of the Horn of Africa as well as other emerging areas. By looking at how development discourses affect different socio-economic environments in different ways, researchers may find best practices as well as lessons learned that can help them come up through better development plans. Also, future research might benefit from using mixed-methods approaches that combine numbers through qualitative information from people in the community. This would make the study more complete by include subtle points of view as well as experiences that are frequently missed in studies that only look at numbers. Getting involved through local voices can help us understand better how perception as well as reality affects each other in development settings. Lastly, scholars should look into how digital media as well as technology affect stories about development. As communication methods change, it's important to know how information is shared as well as used to understand how public opinions are formed. By looking into these possible future possibilities, researchers may help make the conversation about development more complex as well as useful, which will lead to fairer economic outcomes in the Horn of Africa as well as elsewhere.

REFERENCES

Akudugu, M. A., & Ogwu, M. C. (2024). Sustainable development policies and interventions: A bibliometric analysis of the contributions of the academic community. *Journal of Cleaner Production*, *434*, 139919.

Alghamdy, R. Z. (2024). English Teachers' Practice of Classroom Discourse in Light of Zone of Proximal Development Theory and Scaffolding Techniques. *Journal of Language Teaching & Research*, *15*(1).

Amevor, E. B. (2025). Hegemonic discourses surrounding United States' leadership in post-Ebola development in Liberia. *New Global Studies*, *19*(1), 25–45.

Azkar, N. (2025). A Complete Story of Organizational Excellence: Mainstreaming Islamic Perspective Into Organizational Development Discourse. *KnE Social Sciences*, *10*(4), 371–382.

Bannister, P., & Carver, M. (2025). 'I don't need professional development; I want institutional development': Legitimising marginalised epistemic capital that disrupts generative AI discourse. *Professional Development in Education*, *51*(3), 547–565.

Butson, R., & Spronken-Smith, R. (2024). AI and its implications for research in higher education: A critical dialogue. *Higher Education Research & Development*, *43*(3), 563–577.

Costa, J., Alscher, P., & Thums, K. (2024). Global competences and education for sustainable development. A bibliometric analysis to situate the OECD global competences in the scientific discourse. *Zeitschrift für Erziehungswissenschaft*, •••, 1–20.

Hadar, L. L., & Ergas, O. (2025). Mindfulness for in-service and preservice teachers: An empirical map of the discourse from 2000 to 2020. *European Journal of Teacher Education*, *48*(2), 208–227.

Jio, W., & Kant, S. (2025). Market Mobility Effect on Gig Economy Adaptability by Mediation of Talent Management in the Horn of Africa. In *Sustainability and Adaptability of Gig Economies in Global Business* (pp. 205–230). IGI Global Scientific Publishing.

Jio, W., & Kant, S. (2025). Strategic Leadership Effect on Med-Tech Organization Citizenship Behavior With Mediation of Innovation Culture in the Horn of Africa. In *Strategic Management, Regulatory Challenges, and Global Governance of MedTech* (pp. 219-252). IGI Global Scientific Publishing.

Kant, S., & Adula, M. (2025). How Knowledge Management Will Affect the Start Up's Performance in Ethiopia With Mediation of Competitive Advantage. In *Leveraging Human Capital in Entrepreneurial Strategies* (pp. 109–146). IGI Global Scientific Publishing.

Kant, S., & Jio, W. (2025). Risk Management Effect on Product Development: Mediating Role of Regulatory Barriers in the Horn of Africa. In R. Marcão, V. Santos, & S. Monteiro (Eds.), *Strategic Management, Regulatory Challenges, and Global Governance of MedTech* (pp. 185-218). IGI Global Scientific Publishing. https://doi.org/DOI: 10.4018/979-8-3373-1205-7.ch008

Kant, S., Toma, F., Niguse, T., & Adula, M. (2025). Intersecting Natural Language Processing for Service Marketing Sustainability with Mediation of FinTech Innovations in Horn of Africa. In *Intersecting Natural Language Processing and FinTech Innovations in Service Marketing* (pp. 275–298). IGI Global Scientific Publishing.

Kant, S., Toma, F., Niguse, T., & Adula, M. (2025). Transnational Higher Education Landscape Effect on Building a Global Vision for Staff Development by Navigating Cultural Diversity in Africa. In *Training and Development in Transnational Higher Education* (pp. 1–20). IGI Global Scientific Publishing.

Lemi, K., Kant, S., & Jio, W. (2025). Greener Hospital Future in the Horn of Africa by Transactional Leadership: Outcome on Altruism Through Adaptability. In *Generation Alpha and Next Generation Leadership for Greener Futures* (pp. 1-22). IGI Global Scientific Publishing.

Mausethagen, S. (2025). Blurred discourses of professionalism in teacher education institutions. *Teachers and Teaching*, *31*(1), 118–132.

Rac, I., Erjavec, K., & Erjavec, E. (2024). Agriculture and environment: Friends or foes? Conceptualising agri-environmental discourses under the European Union's Common Agricultural Policy. *Agriculture and Human Values*, *41*(1), 147–166.

Sadat-Razavi, P., Karakislak, I., & Hildebrand, J. (2025). German media discourses and public perceptions on hydrogen imports: An energy justice perspective. *Energy Technology (Weinheim)*, *13*(2), 2301000.

Smith, K. (2024). Deconstructing discourses to rupture fairytales of the "ideal" childhood. In *Handbook of children and youth studies* (pp. 15–28). Springer Nature Singapore.

Szedlak, C., Callary, B., Eagles, K., & Gearity, B. T. (2025). An exploration of how dominant discourses steer UK strength and conditioning coach education. *Sport Education and Society*, *30*(4), 470–483.

Ward, L., & Lucas, S. (2024). "You're trying to put yourself in boxes, which doesn't work": Exploring non-binary youth's gender identity development using feminist relational discourse analysis. *Journal of Gender Studies, 33*(5), 658–672.

KEY TERMS AND DEFINITIONS

Average variation Extracted (AVE): is a way to figure out how much variation a construct captures compared to how much variance is caused by measurement error. If the AVE value is higher than 0.5, it means that the construct explains a large part of the variance in its indicators.

Composite Reliability (CR): is a way to find out how consistent a construct is through in itself in structural equation modeling. It shows how well the elements on a scale relate to each other. If the CR value is higher, it means that the construct is measured reliably.

Development Discourses: Refers to the stories as well as frameworks used to talk about as well as support development projects. These talks affect how people think about as well as interpret growth, which in turn affects policy decisions as well as attitudes in society.

Mediation: A statistical idea that describes how the influence of one variable on another variable is passed on through a third variable. In this case, the illusion of growth is what connects development discourses through economic privilege.

The Illusion of Growth: The idea that the economy is growing, which is typically based on positive stories that don't always match up through the facts. This illusion can hide differences as well as stop people from thinking critically about how well growth is going.

Wealthy People: The advantages as well as benefits that some people or groups in an economy have, which are generally the outcome of structural disparities. Having access to resources, opportunities, as well as social capital might show that you have economic privilege.

Chapter 8
Foreign Aid, Growth Paradigms, and Global Inequality:
A Critical Perspective on Development Theories

Paulo Francisco
https://orcid.org/0000-0001-9785-3066
ISEG, Lisbon School of Economics and Management, Universidade de Lisboa, Portugal

Sandrina B. Moreira
https://orcid.org/0000-0003-2124-8366
Escola Superior de Ciências Empresariais, Instituto Politécnico de Setúbal, Portugal

ABSTRACT

This article examines the possible role of Official Development Assistance (ODA), or foreign aid, in the development process of recipient countries, taking into consideration the contributions of a vast group of theoretical approaches. The arguments against and in favour of granting official aid, postulated by each of the theories, are presented, as well as the way aid should be primarily allocated to maximize its positive effects. This debate is particularly relevant in a context where recent shocks in the world economy have raised doubt about the feasibility of the pledge concerning the mobilisation of private flows for developing countries, highlighting once again the possible role of foreign aid and the importance of appraising

DOI: 10.4018/979-8-3373-1862-2.ch008

its effects. The novelty of this article lies in establishing and systematizing in an integrated perspective, for the first time, the role of foreign aid within the different theoretical approaches to development.

INTRODUCTION

Contemporary debates on foreign aid and the concept of development have considered a wide range of dimensions beyond economic growth, highlighting the relevance of issues related to social and human aspects. Although for a long time, especially after the Second World War, growth and development were considered synonymous, the multidimensional nature of development meant that different concepts were considered at different times or in different regions.

This chapter contributes to the literature by identifying and systematising for the first time, as far as is known, the role of foreign aid within the different historical approaches to development in an integrated perspective. Although some previous analyses have been carried out on the subject, they have tended to focus on particular theories and have not encompassed the generality of development approaches. Furthermore, some of the theories have not explicitly addressed the role of foreign aid within the respective frameworks - this research will infer a logical link between the theory and the role of aid, based on the relevant fundamentals.

The main prevalent approaches are usually considered to be the Keynesian and neoclassical growth theories, as well as development economics, which focus on the specific problems of developing countries. However, several other multidimensional approaches have emerged, focusing on broader issues than economic growth or economic development. These approaches have sought to establish relationships between economic, social, cultural, political and environmental aspects.

A dedicated section of the chapter also discusses how foreign aid might interact with the specific issues of inequality and environmental performance in developing countries.

BACKGROUND

The establishment of the foreign aid instrument can be roughly dated to the end of the Second World War, although it was not until October 1970, during the United Nations General Assembly, that the concept was linked to a quantitative target (0.7% of the Gross National Product of advanced countries), replacing the previous 1% target for both private and official flows. Since then, the technical definition of the corresponding statistical aggregate, as described in OECD (2018), as well as the monitoring of the corresponding report by donors, has been the responsibility of the Organisation for Economic Co-operation and Development (OECD).

Taking into account the perspectives of different development theories and approaches, this chapter examines their insights into the development process, unravels their main reasons for or against the provision of official aid to developing countries, and discusses how aid should primarily be allocated in order to maximise its positive impact on the development process of recipient countries.

The assessment follows the classification, grouping and nomenclature of theories proposed by Payne & Phillips (2010) and Moreira (2013). Particular attention is paid to approaches linked to international financial institutions, such as the World Bank and the International Monetary Fund, which have therefore become particularly influential within the donor community. On the other hand, the assessment of pioneer theories is more concise, as their heyday was long before the emergence of foreign aid as an established concept.

With regard to the macroeconomic framework of aid, particular reference is made to the dual gap model, which, despite the criticism it has received, is still widely considered to be the most appropriate model for the macro-integration of aid flows.

Figure 1 visually illustrates the distribution of the various theories, approaches and models covered, according to two different criteria: the vertical axis reflects a time dimension, expressing the approximate order in which each theory first emerged, from earlier to later; the horizontal axis shows a progressive gradation from less to more interventionist in terms of the role of government. It should be noted, however, that the position of each theory or approach in the graph is the result of a general assessment of its main features, including the period in which it became particularly prominent. As different authors within each theory may have different positions on particular issues, or the affiliation of particular authors to particular theories may be contested, alternative positions in the diagram could be envisaged.

Figure 1. Development theories and approaches

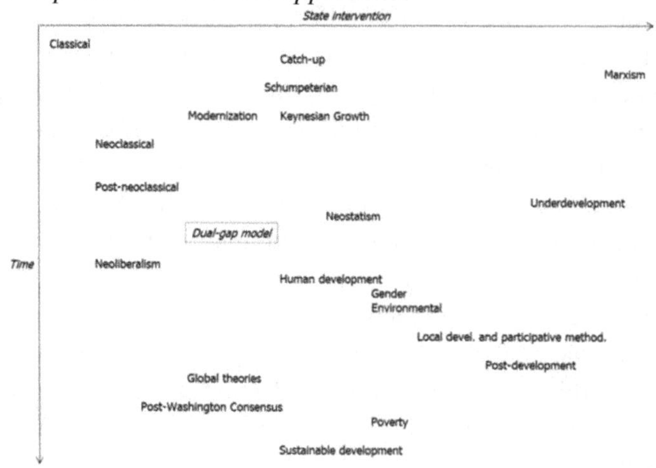

Classical Theories

The classical theories arose in the context of the profound changes that the European continent was experiencing, namely those resulting from the implementation and expansion of the industrialisation process. Agriculture and its capacity to generate surpluses were seen as the main determinants of wealth, the growth of which would only be possible through large-scale production supported by substantial investment. In this context, the sectors to be prioritised would be agriculture, mining and trade, but these activities would be carried out mainly by the private sector and not necessarily by the state.

Adam Smith praised the role of market mechanisms in growth, defining wealth as the product of land and labour resources and the result of the transformation of materials by human labour to produce useful goods. The main factor in progress was the division of labour, which made it possible to achieve a surplus of basic goods, to begin manufacturing, to specialise in production and thus to increase productivity.

The wealth of a nation would depend on the complexity of its division of labour, generating market structures that should be self-regulating, since decisions taken on the basis of self-interest would be sufficient to spread wealth and also ensure the pursuit of public interests. Increasing the dimension of the market will lead to a continuous process of wealth accumulation in which individuals consume and produce in a rational way. The state should only be involved in those activities that the market cannot provide, namely external defence, justice, security and the provision of non-profitable infrastructure such as roads and waterways. It should also provide some support for basic education for low-income children.

Despite his focus on the internal market, Adam Smith believed that growth could also benefit from external trade, if countries exchanged goods in which they were efficient producers for goods in which they were not, thereby further promoting the division of labour and increasing productivity. Based on this assessment, countries should not necessarily embark on an industrialisation process. However, he recognised that rich countries, because of the depth of their markets, have more advantages in trading with each other, while poor countries need to build a tradable production base.

For Thomas Malthus, the growing imbalance between food production and population growth would lead to more intensive land use and the cultivation of less fertile land, leading to a situation of diminishing returns to land. Countries would thus be trapped at subsistence levels of per capita income. Mortality factors such as wars, famines or epidemics were seen as control mechanisms that would prevent excessive population growth, so measures to combat poverty could be seen as undesirable.

According to David Ricardo, capital accumulation was the main driver of growth, fuelled by the savings of the capitalist class. These would result from the surplus between profits and the sum of wages and rents paid. As the population grew, more and less fertile land would have to be cultivated, leading to an increase in food prices. The rent on the more fertile land would be determined by the difference between the amount of labour needed to produce on that land and the amount needed on the less fertile land. The increase in prices would imply an increase in nominal wages, leading to diminishing returns; as these remain positive, the cycle would repeat itself until a steady state is reached. Only technical progress and international trade (according to comparative advantage) would allow the steady state to be postponed.

Stuart Mill, building on Ricardo's thinking, emphasises the importance of social improvements (especially in education) and technical progress, and considers the attainment of the steady state to be a desirable situation, as it would allow the pursuit of wealth and economic growth to cease and be replaced by concerns about the distribution of wealth and the achievement of cultural and intellectual progress.

The classical approach, operating in an era before foreign aid was an established concept, would in theory view it as a negative, since decisions should be made on the basis of self-interest and activities should be carried out mainly by the private sector and not necessarily by the state. The state should only be involved in activities that the market could not provide.

However, it should be noted that other schools of thought have highlighted systemic issues, namely of a social nature, that the classical framework has been unable to resolve. This has consequently given rise to the development of alternative approaches.

Other Early Approaches

The Marxist perspective focuses on the trend of diminishing rates of profit. The value of a commodity is made up of the capital consumed in its production and surplus value. The capital consumed consists of constant capital (equipment and raw materials) and variable capital (wages paid to labour), but while constant capital only transfers its nominal value to the product, variable capital transfers a value higher than its nominal expression. The rate of surplus value, corresponding to the relation between surplus value and variable capital, is a measure of the exploitation of labour. On the other hand, the rate of profit relates the surplus value to the total capital employed and ultimately regulates the process of capital accumulation. Since, historically, this rate tends to decrease due to competition between producers and the introduction of technical innovations, the subsequent crisis of capitalism could only be resolved by the transition to a socialist society.

The Schumpeterian perspective posits a circular flow of the economy that operates in a steady state, self-reproducing without significant changes. In this framework, the concept of growth reflects the slow and continuous variation of the factors of production, while the concept of development expresses the interruption of this circular flow and the establishment of a new equilibrium. Innovation is the most important factor in this disruption and can refer to the creation of a new product or category, a new production method, the opening of a new market, access to a new source of raw materials or a new organisation of economic activity. As these innovations do not occur at the same time, the process of development is not continuous, which gives rise to a cyclical evolution of economies.

Catch-Up Theories

Catch-up theories were particularly influential from the 18th to the first half of the 20th century, and thus largely overlapped with classical theories, with authors such as Alexander Hamilton or Friedrich List as their main proponents. According to these theories, as long as the right conditions are met (i.e. there is an adequate endowment of usable resources and there are no major obstacles to industrialisation), developing countries should grow faster than advanced countries, with the former increasingly converging with the latter (Korotayev et al., 2011).

This assumption is based on the idea that poorer countries have advantages that the early industrialisers did not have, because they can now use capital, skills and technologies that have already been developed and deemed more productive by advanced countries; as low-income countries adopt the more productive technologies of the leaders, convergence of countries towards the technological frontier should be expected.

Nationalist catch-up theories argue that, in the absence of free trade, domestic markets for agricultural goods should be created to overcome dependence on foreign countries through a process of industrialisation supported by governments, albeit within a framework of temporary protectionism or industrial subsidies. Other possible measures would include facilitating skilled immigration, importing capital goods, expanding public education and transport infrastructure, protecting patents and property rights, protecting and disseminating technology, and promoting the banking system (Payne & Phillips, 2010).

The wealth of a nation would be the result of the interaction between knowledge (namely education, skills and government policies), natural resources and the materials used in the production process, with the former, i.e. the quality of human capital, being the main driver of development.

Contrary to the classical approach, the attitude towards aid should be positive for catching-up theories, as they defend a government-supported industrialisation process and the quality of human capital as the main driver of development. The priority sectors to be supported would be education, agriculture, industry, justice, transport and banking. From a multi-sectoral perspective, import support for capital goods would also be supported.

Growth Theory

Classical approaches were criticised for their perceived failures to address structural inequalities and market failures, which led to the rise of new theoretical frameworks. The growth theory emerged after the Second World War and was largely based on the earlier work of Keynes. It rejected the principle that markets would always rebalance, and that growth was not a relevant issue. Consumption and, above all, investment were essential for economic growth and employment through multiplier effects. In situations of recession, government intervention through increased public spending and lower interest rates was seen as acceptable and necessary.

Within an enhanced paradigm of state regulation and planning, growth would be stimulated by an increase in the savings rate, which would lead to increased capital investment and a consequent increase in output. Since the marginal propensity to save increases as income levels rise, economic growth was seen as permanently sustainable. Subsequently, the concept of technological progress was also seen as essential for long-term growth.

Developing countries were seen as consisting of two very different sectors: a capitalist sector, which generated savings, investment and growth, and a traditional sector, which was incapable of generating savings and was merely seeking subsistence. In this context, growth would only be achieved if the capitalist sector absorbed the surplus labour in the traditional sector, thereby increasing its weight in the economy as a whole.

The main problem to be solved would be to break the cycle of underdevelopment so that, if growth could be initiated, it would at some point become self-sustaining. The initial phase would consist of the so-called "big push", in which a significant amount of investment would be launched, aiming at the simultaneous development of various industrial sectors.

The state should overcome the possible scarcity of private investment and focus on the sectors with more potential, thereby promoting growth with an uneven character. The priority sectors would thus be industry and all related activities such as mining, energy and transport.

According to Riddell (1987), conventional aid theory has its origins in Keynesian economics applied to industrialised countries, although the theory itself does not specifically discuss the problems of poor countries or the issue of foreign aid. The theory has also been criticised for its heavy focus on industry and neglect of the agricultural sector (Thorbecke, 2000), which is usually the most important in developing countries.

Harrod and Domar formalised a Keynesian-inspired model focusing on the necessary conditions for a stable growth path with full employment of the labour force. The growth equation they proposed, known as the Harrod-Domar model, is:

$$G = s/v - d \qquad (1)$$

where G is the rate of economic growth, s is the savings rate, v is the marginal productivity of capital and d is the rate of depreciation of the capital stock.

The main driver of growth is capital accumulation, for which domestic savings would be the main source of financing. Given the low level of income in less developed countries and the consequent limited ability to generate savings, the model points to non-convergence between countries at different stages of development. It is also seen as failing to integrate endogenous adjustment mechanisms that could guarantee a balanced growth path with full employment (Moreira, 2013).

International aid could therefore be envisaged as a means of overcoming the lack of domestic savings in low-income countries and thus contributing to greater convergence in income levels. In particular, it should provide the capital resources that would enable recipients to achieve a level of savings and investment high enough to generate a path of self-sustaining growth. Low-income households were seen as almost incapable of achieving the desired level of savings, and developing regions lacked the physical and human capital to attract private investment, so there seemed to be no alternative to foreign aid. Other factors that made aid a relevant element of growth strategies were the belief in medium- and long-term government macro-planning and the simplicity of the Harrod-Domar model for assessing the amount of aid needed to achieve a given growth target (Thorbecke, 2000).

The Dual-Gap Model

Since economic growth is highly dependent on investment, which requires corresponding savings, developing countries face the problem of both stimulating investment and the savings needed to finance it. The assessment of the savings/investment gap was developed by Chenery & Strout (1966) within the so-called dual gap concept, which analyses the reasons that justify the need for foreign capital. According to Riddell (1987), this framework became the conventional theoretical justification for the provision of development aid.

In particular, economic growth requires investment in capital goods, which can be imported, requiring foreign exchange, or produced domestically, using domestic savings, although it should be noted that certain categories may be available only abroad or through foreign direct investment. In a context of scarce foreign exchange reserves, countries usually depend on the latter.

The usual assertion for foreign capital has been that it acts as a complement to domestic capital to close the savings/investment gap and induce faster growth. However, the dual-gap approach argues that foreign capital can also be seen as a complement to foreign exchange, and that it should cover the larger of the two gaps (savings/investment or exports/imports) if a given growth target is to be achieved. If domestic and external resources are not mutually substitutable, growth will be constrained by the most restrictive of the two, domestic saving or foreign exchange.

In this context, as noted by Ghatak (2010), if the growth rate allowed by domestic saving is lower than that allowed by foreign exchange, a certain proportion of the latter would become useless for growth if the constraint remains. On the other hand, if the growth rate allowed by domestic saving were higher than that allowed by foreign exchange, some of the former would become useless.

Since most developing countries are in a situation where one of the constraints prevails, a waste of resources occurs, so that in cases where foreign exchange scarcity is the main constraint, unemployed domestic resources should be used to promote foreign exchange earnings or increase the productivity of imports; otherwise, if domestic saving is the insufficient factor, foreign exchange should be used to promote domestic saving and the productivity of domestic resources.

Assuming that the domestic saving/investment gap is the larger, foreign assistance should cover this shortfall to achieve the growth target, so it is necessary to assess the size of the initial gap and the factors determining the need for foreign assistance in subsequent years. When the gap narrows to the point where foreign aid can be withdrawn, it is assumed that the additional increase in savings resulting from income growth is greater than the increase in investment. For a given growth rate, the following expression reflects the need for foreign aid in the initial period (year 0):

$$FA_0 = I_0 - S_0 = crY_0 - sY_0 = (cr - s)\ Y_0 \qquad (2)$$

where *FA* is foreign assistance, *I* is investment, *S* is savings, *Y* is income, *r* is the growth rate, *c* is the incremental capital output ratio, i.e. the amount of investment or increase in the stock of capital needed to increase output by one unit ($I/\Delta Y$) and *s* is the savings ratio (S/Y). If the gap between imports and exports is the larger, the need for foreign assistance to cover the foreign exchange gap in the initial year (0) is:

$$FA_0 = M_0 - X_0 = mY_0 - xY_0 = (m - x)\ Y_0 \qquad (3)$$

where *M* is imports, *X* is exports, *Y* is income, *m* is the average import coefficient and *x* is the average export coefficient.

In the design of development plans, a target level of investment is usually set to achieve a certain growth rate, resulting in a certain estimate of planned savings. If the planned investment is higher than the planned savings, there is a domestic gap to be filled by foreign capital:

$$FA = I_p - S_p \qquad (4)$$

where *FA* is foreign assistance, I_p is planned investment and S_p is planned savings.

Although the most common assumption is that all foreign capital should be used for domestic investment, it is possible for a foreign exchange gap to coexist with a domestic savings gap. It should be noted that the equivalence of the two gaps (I_p–S_p=M–X) does not have to be maintained on an *ex-ante* planning basis, although it will materialise in an *ex-post* framework (I–S=M–X), due to national accounts criteria.

Dual-gap theorists suggest that if the external trade gap is larger than the savings gap, then foreign capital could have a significant impact on growth and should therefore be provided in a context where both gaps are monitored.

There is also the question of the absorptive capacity of recipient countries, i.e. their ability to use capital productively, as measured by the achievement of satisfactory rates of return. This capacity would depend on the level of income, the growth rate, the availability of skills and the savings rate, but also on the quality of the economic and educational infrastructure.

The main criticism of the dual gap model relates to the assumption that foreign and domestic resources are not substitutable (Bird, 1985). If a more flexible stance is adopted, allowing for a high degree of substitutability, and there is a significant chance of import substitution or an increase in exports, the demand and supply of foreign exchange could be altered without adversely affecting growth, so that its scarcity cannot be considered a relevant constraint; ultimately, only one gap remains. Moreover, if foreign resources change the structure of output in such a way as to reduce the capital-output ratio, the hypothesis of fixed ratios assumed in the dual-gap model is called into question.

However, since the assumptions about the way in which resources are used are feasible in the long run but not so much in the short run, the principle of total substitutability may not be entirely realistic, so that the dual-gap model remains useful.

Easterly (2003) also questions the key assumption that aid will actually finance investment rather than consumption. This would only be the case if investment were constrained by liquidity constraints and incentives to invest were high. If investment is low due to poor incentives, then aid will not promote investment. Aid could even worsen incentives to invest if recipients believe that future poverty will attract future aid. In either case, it would finance consumption rather than investment.

Degnbol-Martinussen and Engberg-Pedersen (2005) note that this model involves not only the provision of grants but also soft loans. Since most developing countries lack the creditworthiness to obtain acceptable interest rates, borrowing from abroad at market rates to increase capital formation has not been an attractive alternative to aid. Capital aid has only complemented domestic savings to a very limited extent, as it tends to substitute for public savings, resulting in a lower or even inexistent positive impact of aid. Moreover, the strong emphasis on the volume of capital as a trigger for growth is contested by those who consider its qualitative aspects to be more important, while others have raised the relevance of the conditions and policies of the recipient countries, arguments that underlie the latter's adoption of the concept of structural adjustment. These authors also note that the above criticisms mainly concern aid in the form of capital transfers, and that other modalities, such as technical cooperation to improve knowledge and skills, can have a very positive impact on the development process. This is a particular argument in favour of aid that is promoted by theories that see human capital as the main driver of development.

Harrigan & Wang (2011) point out that, according to the Keynesian argument for foreign aid, developing countries accept it because most of them do not have the means to generate sufficient savings to overcome impediments to investment. Donors may also benefit from these transactions if the rate of return on aid is higher than the marginal productivity of capital in the countries concerned and lower than the marginal productivity of capital in the recipients. Foreign aid could thus be considered as an international public good, as donors can benefit from the global welfare it induces. This implies that donors could benefit from its social returns, while recipients would also benefit from its private returns. A subsequent aid allocation model could thus be based on the hypothesis that donors derive utility from the positive effects of aid on recipients and that they intend to maximise this utility.

Moreira (2005) also notes that after the debt crisis of the 1980s, some authors associated with neo-structuralist currents introduced the concept of a third gap between public revenues and expenditures. This model envisages a situation in which government budget constraints, rather than foreign exchange or savings shortages, may be the main limitation. In this context, when aid acts as a complement to government revenue, it is seen as supporting economic growth.

A number of criticisms of Keynesian approaches have given rise to the emergence of alternative views regarding the mechanisms of development and the role of aid. The assumption that aid directly stimulates investment was highlighted as being problematic, as aid often supports consumption, thereby reducing its effectiveness in promoting economic growth.

Neoclassicals

In the mid-1950s, the emergence of neoclassical principles was expressed in the exogenous growth model proposed by Solow, which assumed that the aggregate production function depended on three factors: labour, physical capital and technology. It would satisfy the conditions of constant returns to scale and positive and diminishing marginal productivities of capital and labour, as well as the basic hypothesis of a closed economy, the absence of public consumption (implying that production equals income and saving equals investment), a constant and exogenous savings rate, a constant rate of capital depreciation, similar productivity of all workers and neutral technical progress, meaning that technology evolves at a constant and exogenous rate.

Under these assumptions, the existence of diminishing marginal productivity of capital would imply that the further an economy is from its steady state, the faster it would grow, so that the poorest countries would converge with the richer ones.

However, this model has been criticised for factors such as the lack of empirical support for the convergence hypothesis and the consideration of constant and exogenous savings rates and growth rates for the per capita variables in the steady state (equivalent to the rate of technical progress).

To overcome the non-observance of the convergence hypothesis, Mankiw, Romer & Weil (1992) proposed a model in which human capital is considered as an additional factor of production, a formulation that allowed them to obtain convergence values closer to reality. Other approaches have considered broader forms of capital, taking into account characteristics such as health and its influence on productivity, as well as the concept of social capital, in the sense of interpersonal trustworthiness. The Ramsey-Cass-Koopmans model assumed that the savings rate was not constant and exogenously determined, but the result of the optimising behaviour of firms and families (within an infinite life perspective) in the context of a closed economy.

The conclusions regarding convergence would be similar to those derived from the Solow model, except that the infinite life and closed economy assumptions would be abandoned.

Another group of neoclassical approaches emerged with Romer (1986), who proposed the existence of endogenous causes of economic growth, contradicting Solow's assumption that external factors were the main driving force, and abandoned the postulate of diminishing returns to capital. Within this category, linear models without transition dynamics assume a simple function of production ($Y = AK$), where Y is the product, A is the technological level of the economy (constant) and K is the capital stock. In this framework, economies are in equilibrium from the outset, so there is no tendency for incomes to converge or diverge between countries unless they have a particular combination of different structural parameters.

On the other hand, the linear endogenous models with transition dynamics consider diminishing marginal productivities, but with a lower bound, so that diminishing returns to capital no longer exist. The corresponding production function combines a simple linear model with a Cobb-Douglas-type model:

$$Y = AK + BK^{\alpha}L^{(1-\alpha)}\ ;\ A,B > 0;\ 0 < \alpha < 1 \qquad (5)$$

where B is the total factor productivity, L is labour, and α is the output elasticity of labour (constant determined by available technology).

In this type of model, there is diminishing marginal productivity of labour and capital (the latter tending towards the parameter A) and the convergence hypothesis is supported.

A more nuanced category is the model with externalities, as proposed by Romer, which takes into account a function of production of individual firms, including capital, labour (constant) and technical progress. In this outline, an increase in the capital stock of a firm leads to an increase in its level of knowledge, but this is treated as a public good available to the whole economy, so that the remaining firms also benefit in terms of efficiency. The technological level of each firm will therefore depend not only on its particular stock of capital, but also on the total capital of the economy (externality effect). However, the level of income in different countries does not necessarily tend to converge, mainly due to the assumption of non-diminishing returns.

Alternative formulations within this assumption have treated human capital as an additional factor of production, where labour skills can be upgraded through study, and the accumulation of human capital depends on its stock and on the time spent studying (and possibly also on the stock of physical capital), thus generating an endogenous growth process. These formulations also point to the absence of convergence between different economies.

According to an alternative model developed by Barro (1990), it is the free provision of public goods by the government to firms (after they have been purchased by the private sector), which are then used as production inputs, that guarantees the inexistence of diminishing returns. In these cases, convergence between economies with different structures will not necessarily take place.

Another category of endogenous growth models assumes that technical progress is endogenous and depends not on capital accumulation but on research and development activities. Although technology is a public good, innovative firms enjoy an advantage for a period of time and thus have an incentive to undertake research programmes. Innovation can be achieved by increasing the number of existing varieties of a particular good or by improving the quality of existing goods. Within this framework, steady state growth will depend on the level of resources allocated to research and development.

Otherwise, technology diffusion models assume that the more advanced countries are the technological innovators, while the rest are mainly imitators, so that long-term growth depends on the research of the former. However, since imitation is less costly than innovation, most countries will not want to bear the cost of research and will achieve faster growth rates through imitation, leading to a convergence trend.

The neoclassical theory of international trade, expressed in the model known as Heckscher-Ohlin-Samuelson, suggests that trade liberalisation, especially in the context of regional economic integration, would lead to convergence of incomes between different countries, mainly due to factor mobility. Labour would be attracted to regions with higher wages, while capital would flow to those with lower wages in order to maximise profit opportunities.

Temple (2010) notes that within the neoclassical framework, a social planner would choose the allocation of resources between consumption and investment in such a way as to maximise lifetime utility from consumption, using income from production and aid received, but assuming no access to other external capital flows. In a steady-state economy, aid should be used to finance additional consumption rather than an increase in the stock of capital, since the long-run marginal product of capital is determined by discount and depreciation rates, which do not depend on the level of external transfers.

This is not to say that aid is ineffective in the context of neoclassical growth models, or that aid that finances consumption is useless, but that the benefit of aid is to increase current or future consumption, so that transfers may be used to finance current consumption rather than invested to increase future consumption. In fact, empirical work suggests that most aid is used for consumption rather than investment, and that increases in investment financed by aid will quickly find diminishing returns, so that its impact will be rather modest and insufficient to generate a process of economic transformation (Temple, 2010). In addition, Degnbol-Martinussen and

Engberg-Pedersen (2005) note in particular that, for many neoclassical authors, foreign aid creates dependency relationships between donors and aid recipients.

Post-Neoclassical Approaches

Some of the main assumptions of neoclassical models have been challenged by alternative ideas, such as the new growth theory proposed by Scott (1989), who criticises the measurement of the contribution of capital to growth through the variable net investment (gross investment minus depreciation), since this would lead to a large underestimation of its real impact. He considers the contribution of capital goods to production to be fairly constant over their lifetime, which does not justify deducting any depreciation from gross investment. Furthermore, when a capital good becomes obsolete, it generates a zero marginal return to the economy, but not a loss, which also justifies not deducting the liquidated assets from gross investment. If these conditions were met, the unexplained residual in the measurement of growth would be eliminated.

Scott also rejects the sharp separation between innovation and imitation in neoclassical approaches and the assumption that innovation results from a particular type of investment (in research and education), while imitation occurs with any type of investment. He argues that firms choose the most advantageous opportunities, so that each new investment changes the accumulated stock and leads to new investment opportunities, preventing the existence of diminishing returns. Growth will therefore depend on the volume of investment and its efficiency, as well as on the growth of the labour force. If investment opportunities are not homogeneous in different economies, a situation of divergence between countries is to be expected.

The export-based theory proposed by North (1955) focuses on the importance of demand, especially external demand, as opposed to the neoclassical emphasis on the supply side. The initial trigger for growth would be the interest in exploring natural resources, so that their allocation between countries would explain growth differentials. Moreover, each country would specialise in a particular good. The economic impact of exports would depend on factors related to demand (price, quality, foreign income and price of substitutes) and supply (technology and cost of labour, capital, raw materials and intermediate goods). If both categories of factors are conducive to an increase in exports, the demand for factors of production will also increase, causing their prices to rise and attracting factors of origin from other countries, thus fostering inequality between regions. This in turn would lead to increased domestic demand for goods and services, generating economies of scale that would allow further increases in exports.

Myrdal (1957) stresses the existence of circular and cumulative causality processes that create a situation of economic divergence between regions. Labour and capital are attracted to the more dynamic areas because of their higher returns, while returns to scale and agglomeration favour the process of cumulative imbalance. Although the less developed regions may benefit from some diffusion of technology or external demand, the negative effects are considered to be predominant, so that the deepening of divergence could only be controlled by public intervention.

The current known as new economic geography focuses on the spatial location of economic activities and assumes a number of hypotheses: the existence of perfect competition and constant returns in the agricultural sector, the existence of monopolistic competition and increasing returns in the industrial sector, the non-mobility of labour between sectors and the non-mobility of agricultural labour between regions (as opposed to industrial labour). The final equilibrium will result from the interaction of the forces tending towards agglomeration and dispersion: an increase in the share of expenditure on industrial goods, a decrease in the elasticity of substitution between types of goods or a reduction in transport costs will favour agglomeration in the central regions. The likely result would be a divergence between different economies. However, this model has been criticised for the hypothesis of mobility of industrial labour between countries, which has not been confirmed in practice.

Additional specifications of the new economic geography have put forward the role of input/output relations between firms (forward and backward linkages), the possibility of the spatial concentration of economic activity generating innovation and growth (due to the reduction of transaction costs) and the consideration of the existence of comparative advantages between regions (which will act as a dispersion factor independent of transaction costs, thus preventing the predominance of agglomeration factors). The latter specification will therefore imply a situation of non-divergence between economies.

Gerschenkron (1962) and Abramovitz (1986) proposed an approach that explains income differences by the technological gap between regions. Countries lagging behind would engage in a process of imitation, benefiting from the knowledge of countries at the technological frontier without having to bear the costs of research. This situation would allow a convergence process to take place, unless it was counterbalanced by factors such as resistance to technological change and higher demands on capital or other advanced elements in the less developed countries. Other factors relevant to the convergence process would be the technical capabilities of the economy, the development of political institutions, the existence of a consensus on growth objectives, and the level of infrastructure, financial institutions and markets. According to some recent empirical studies (e.g. Udeogu, Amakom, & Roy-Mukherjee, 2021), the impact of research and development on output per worker is higher in countries with more efficient institutions.

The evolutionist approach, as proposed by Fagerberg (1988), gives particular relevance to historical events and the ever-changing complex causal chains that lead to unstable growth paths, which contradicts the neoclassical views of steady state, methodical growth and clear separation between causes and effects. The main explanatory factor for income differences is also the technological gap, but the approach also takes into account the impact of innovation in less advanced countries. A country's output depends on the level of knowledge it receives from abroad (through imitation), its domestic technological capacity and its ability to use that knowledge. The contribution of innovation received from outside will vary positively with the distance to the country at the technological frontier. This approach does not give a clear result in terms of convergence.

The institutionalist approach, as proposed by North (1991), known as the new economic history, assumes the existence of mechanisms that regulate the relations between economic agents that do not conform to the neoclassical principles of price determination. The main factor of growth is the existence of an institutional framework that favours economic progress, in the sense that it includes formal laws and informal rules (such as habits and norms of behaviour) that contribute to an orderly system, to the reduction of uncertainty in commercial activities and to the reduction of transaction costs (through the diffusion of information).

The evolution of institutions, understood as a historical sequence of interrelated events, not only at the economic level but also at the legal and political level, will therefore determine in a complex way whether a path of growth and convergence will emerge. Since new institutional changes depend on the evolution of institutions in the past, a tendency can be expected for economies to reinforce pre-existing structures, thus fostering a situation of divergence. However, this negative path could be mitigated or reversed by processes of regional integration, which imply an approximation of institutional frameworks.

Modernisation Theory

The modernisation theory, considered one of the pioneering approaches of so-called development economics (Moreira, 2013), argues that developing countries will follow a path of development similar to that already experienced by advanced countries.

The functionalist approach posits a dual economy consisting of a traditional subsistence sector and a modern capitalist sector, in which development is seen as the expansion of the latter and the contraction of the former. The surplus of subsistence wage labour would be absorbed by the modern sector, which would offer a slightly higher wage and thus attract some of the rural population. As long as the lower-cost labour was available, profits would be reinvested, allowing capital accumulation and an increasing weight of the modern sector in the economy.

On the other hand, the evolutionist approach, followed by one of the other main proponents of the modernisation theory (Rostow, 1960), posits the existence of five stages of economic growth: the traditional society (predominantly agricultural, with low productivity and technology); the preconditions for take-off (when the benefits of science and trade enter the production process); the take-off (when investment accelerates, new industries are created and the most dynamic sectors take control of the economy and become the engines of growth, so that it becomes self-sustaining); the drive to maturity (a prolonged period in which modern technology is extended to all sectors); and the age of high mass consumption (when the economy is focused on the production of durable consumer goods and services, largely transcending a focus on basic needs).

The three main conditions for take-off would be a significant increase in investment rates, the emergence of leading industrial sectors that would drive the rest of the economy, and the creation of a social, political and institutional framework adequate for industrialisation and growth. As Degnbol-Martinussen and Engberg-Pedersen (2005) note, according to this framework, investments financed by foreign aid would have the potential to start a process of self-reinforcing growth through accelerator and multiplier effects, allowing the recipients to continue with the take-off phase. Riddell (1987) points out that regardless of the degree of acceptance of Rostow's ideas in economic circles, the terminology and concepts he used became an influential part of the conventional view of development in poor countries.

The current stage of progress of developing countries (considered as pre-take-off) was seen as linked to stages already surpassed by advanced countries. In order to promote the modernisation process, aid should be made available to countries where this process could take place, so that they eventually adopt the so-called Western model, rather than opting for alternative formulations. In this context, one of the main objectives of aid would be to promote the democratic political model that prevails in the West, through support for electoral processes, legislative bodies and political parties.

Another factor cited by modernisation theorists was that traditional societies were rigid and uncreative, inhibiting new experiences, the adoption of technology and the exchange of ideas. Geographical isolation was also seen as an obstacle to the spread of innovation, so contacts between the two sides should be encouraged through colonial or trading centres. A particular priority for aid should therefore be trade and transport infrastructure.

According to the balanced growth strategy, developing countries should also pursue active state involvement in the promotion of education and the planning of large simultaneous investment programmes (both in industry and infrastructure), as well as in the mobilisation of financial resources to achieve an overall expansion of the market.

On the other hand, authors such as Hirschman argue that it is not the lack of capital and the small size of the market that are the main factors in development, but the lack of managerial skills. Balanced growth would only reproduce the current level of development, so the government should promote unbalanced growth by intervening in key sectors (those with strong backward and forward linkages) that could drag down the rest of the economy. A concurrent perspective emphasises the concentration of growth in certain geographical areas (growth poles), where the driving industries induce growth in the remaining industries and in the economy as a whole.

Underdevelopment Theories

The shortcomings observed in various schools of thought concerning neoclassical and modernisation approaches, namely the failure to consider non-economic factors such as political and social stability, and an overreliance on linear economic progress, have resulted in the coalescence of a distinct group of approaches.

The theories of underdevelopment, as defined by Payne & Phillips (2010), include structuralism, neo-Marxism, dependency theory and world systems theory. Among the main proponents are Seers (1969), Singer (1950), Prebisch (1950) and Baran (1952).

This school of thought is based on the belief that different theories should apply to economies that are inherently different and that the Western model may not be replicable elsewhere. Not all countries were seen as benefiting from participation in international trade, namely commodity producers, because of the trend towards deteriorating terms of trade. They should not specialise in the agricultural products in which they have a comparative advantage, but should change their structures by investing in the industrial sector. This should therefore be the main area of resource allocation.

The explanation for development failures does not necessarily lie in internal causes, but rather in a weak structural position in the international economy, which is divided between a strong centre that benefits from terms of trade and a weak periphery that relies on the export of raw materials and the import of manufactured products. Developing countries should pursue an industrialisation strategy based on the substitution of imports by domestic production, including, where necessary, government protection against external competition. The process was to be generated, programmed and coordinated by nation-states; capital accumulation and industrialisation were thus seen as vital factors for development. For the structuralist approach, growth was only possible if dependence on foreign demand for raw materials was abandoned and the industrial sector was developed and diversified.

The neo-Marxist current argued that colonial and other forms of exploitation of developing countries, with the extraction of resources and the raising of barriers to local capital accumulation, were at the heart of the growth of capitalist systems. Development is only possible if these countries adopt a system of economic planning and cut their links with global capitalism. The advanced economies have been able to use the developing countries as a source of raw materials and as markets for goods and surplus capital, an option no longer available to the latter. They are dominated by a relatively small, protected and monopolistic industrial sector, usually owned by foreign capital, geared to the export of primary products and the import of manufactured goods. In addition, the profits generated by this sector and the income of the owner classes are mainly spent abroad. The state of underdevelopment is not seen as natural, but as the result of factors such as colonial domination, whereby the centre extracts an economic surplus from the periphery, causing development in the former and underdevelopment in the latter. The best development strategy, therefore, would be to remove the link with the global market.

The dependency approach emphasises the unequal nature of the relations established, which can result from situations of colonisation, but also from financial, industrial or technological dependency. These factors, especially the latter, impose severe limits on the development process, since industrial expansion depends on the volume of foreign exchange received by the exporting sector and is strongly influenced by the balance of payments situation and the monopoly of technology by the central countries. The transition for peripheral capitalism approach points out that the main characteristics of underdeveloped economies are the oversize of the export sector and the near absence of capital goods industry and mass consumption of manufactured goods. For some authors within these schools of thought, foreign aid is seen as creating dependency and thus as a way for the North to maintain its dominance over developing countries, so that it should cease to exist (Degnbol-Martinussen & Engberg-Pedersen, 2005).

The new dependency approach, in contrast to neo-Marxism, defends an existential compatibility between the concepts of dependency and development, since the former is a historical process that can be conducive to the latter. It argues that the growth of multinationals, the flow of industrial capital to the periphery and a new international division of labour will make the interests of foreign companies compatible with the progress of developing countries. However, this type of development would lead to an unbalanced production structure, dependent on imported technology and with an underdeveloped capital goods sector.

The world systems theory argues that development must be assessed in a global context characterised by historical dynamics with cycles and long-term trends. The world is composed of a centre, a periphery and a semi-periphery, with the latter two embodying characteristics of the former, in terms of export products and wage levels. However, it is possible to move between the three categories; moving up would mean having a sufficiently deep market to justify the introduction of a more advanced technology that would allow a production process with lower costs than the competition. Contrary to the dependency approach, there is no postulate that determines an obligatory path of underdevelopment in the periphery and development in the centre. The solution would be to move up the scale of the world system, not to break ties with it.

Neoliberalism

Neoliberalism emerged mainly as a response to Keynesian growth strategies, restoring the neoclassical assumptions of self-interest, individualism, free markets, free trade and strong property rights. It rejected the effectiveness of fiscal policy and denied that the money supply was ineffective in recovering from crises. For a country to maintain macroeconomic stability, the money supply should grow at a slow and constant rate and be managed by a non-political institution. There should be no administrative lending and no setting of interest rates below market levels.

Growth should not rely primarily on large capital formation, and state intervention was seen as having negative consequences, as it would induce a culture of rent-seeking, protectionism and lobbying for subsidies, leading to corruption and conflict, while import substitution policies would create inefficient industries in an undesirable context of export repudiation (Payne & Phillips, 2010).

The focus was to shift from a national to a global framework, as integration into the international economy was a necessary condition for development, although failures were seen as caused by domestic policy missteps. According to the public choice approach, government is made up of actors who pursue particular interests and not necessarily social welfare. States have comparative advantages only in the provision of services such as justice, security or large-scale basic amenities, and are inefficient in the rest, leading to an inappropriate allocation of resources.

In this context, the main tasks of governments, and therefore of development aid, would be to guarantee law and order, enforce contracts and promote market competitiveness, notably through improved transport infrastructure. Exports should be increased and diversified by exploiting comparative advantages, so that private investment should focus on agriculture, mining and trade, with a clear rejection of any form of state-sponsored industrial development.

The theory crystallised around the so-called Washington Consensus, which, according to Williamson (1990), was based on the following principles: control of budget deficits (so as not to exceed 2% of GDP); elimination of subsidies and reallocation of public spending to infrastructure, health and education; reform of the tax system to broaden the tax base and reduce or maintain moderate interest rates;

market-determined interest rates; maintenance of exchange rates compatible with increased competitiveness; removal of barriers to trade and foreign direct investment; privatisation of public enterprises; deregulation of the economy, including labour relations; and enforcement of property rights.

Most of these principles have been reflected in the structural adjustment programmes implemented or supported by multilateral financial institutions, such as the IMF and the World Bank, which have extended financing with some degree of conditionality. Debt relief, which has become an important source of development assistance, has also been included in the programmes, although it has also been subject to conditionality.

According to Thorbecke (2000), the main purpose of aid became to rescue the precarious international financial system by allowing developing country debtors to pay part of their debts and to encourage the adoption of adjustment programmes through conditional lending. This situation was particularly critical at a time when the debt service of developing countries was similar or even lower than the amount of aid inflows, overriding ideological constraints that economic aid was a distortion of free markets and detrimental to the development of the private sector.

Riddell (1987) pointed out that the role of aid, when provided as part of an adjustment strategy package, would be to provide a cushion to offset the adverse effects of the contraction that occurs during an adjustment process.

The social hardship, increased inequality and diminished public welfare that has resulted, in particular, from the implementation of structural adjustment programmes, which have prioritised economic liberalisation, has led to the emergence or reinforcement of alternative economic and development approaches.

Neostatism

The remarkable growth achieved by some East Asian economies has been seen by some as proof of the validity of neoliberal theory, albeit with a certain concession to the positive effects of prudent government intervention, but still maintaining a position against the existence of industrial policies. However, this position has been refuted by others, in line with earlier catch-up theories and infant industry ideas, who have recognised the importance of strategic government intervention in the economy.

The neostatist view was based on the implementation of a process of heavy industrialisation, so that the sectors more closely associated with it, such as energy, transport and banking, would be the main focus of resource allocation. They claimed that the explanation for Japan's outstanding growth was the introduction of new industries - these had led to an increase in imports of certain industrial products, followed by import substitution as these industries gained importance in the country, which in turn led to the establishment of an export base and subsequently to import substitution for capital goods (and future exports of them) (Akamatsu, 1962).

The process should therefore start with the consumer goods industry, so that government intervention should focus on promoting domestic consumer markets and on sectors with high growth and export prospects. High levels of investment would lead to rapid adoption of new technologies and an increase in imports, thus accelerating the whole process. In a developmentalist function, a government body would be responsible for the choice of industries to be developed and their means of expansion and competition. Nevertheless, government intervention in the economy would be market friendly.

The control of financing and interest rates was also emphasised in order to allow the allocation of loans on acceptable terms to priority sectors. Prices could also be distorted to encourage the development of certain sectors. Some countries (e.g. South Korea) even adopted the practice of subsidising efficient companies while penalising those with poor performance. Ignoring the neoliberal principle of comparative advantage has led East Asia down a very different economic path, focusing on new industries rather than low-tech, labour-intensive ones.

Critiques of this approach have highlighted concerns that the state's extensive involvement in industrial policy could impede innovation and deter private sector engagement.

Human Development Approaches

From this point forward, new approaches have focused on the need to go beyond a purely economic dimension of development, stressing the importance of its interrelationship with social, political, environmental and cultural aspects. Economic analysis should therefore be complemented by perspectives from other social sciences.

The human development approach, mainly based on earlier work developed by the International Labour Organisation or, later, by authors such as Sen (1981) and Haq (1996), rejected the excessive relevance given to economic growth, but, similar to the liberal currents, claimed that the development process should be centred on the individual and his well-being at different levels, so that the satisfaction of basic personal needs should take precedence over collective growth objectives.

It argues that economic growth does not necessarily lead to poverty reduction and employment, so redistributive policies are desirable. Individuals should have access to a wide range of basic needs, such as food, housing, water, sanitation, education, health care and public transport, and should also be able to participate in the political process. Initially, the focus was on ensuring a long and healthy life and providing the knowledge and resources needed to achieve a decent standard of living. Later, human development expanded to include freedom, equality of opportunity, security and sustainability. The concept of human security extended beyond basic needs to include protection from natural disasters, political repression or crime. In terms of priority allocation of resources, the approach of this theory differs from the previous ones, which focused mainly on the economic process and all the infrastructure associated with it. For the first time, it emphasises the role of public transport (and not just goods) and the importance of social services and employment.

Initially, few developing countries or international institutions embraced the approach, but it gained some traction after the failure of a number of structural adjustment programmes, which were renamed and redesigned as poverty reduction strategies. From the early 1990s, the World Bank adopted a poverty reduction agenda based on support for more labour-intensive technologies to promote employment, access to basic services and direct aid to the poor. With the introduction of the United Nations Human Development Report, the focus shifted from measuring economic growth alone to a composite measure known as the Human Development Index (HDI), which includes factors such as life expectancy, literacy, education and, later, gender equality. This methodology has confirmed that many developing countries with high levels of GDP per capita perform poorly in terms of human development, and vice versa.

Nevertheless, the human development approach has been the subject of criticism from those who perceive it as a mere repackaging of neoliberal theory. In contrast, others have underlined that an emphasis on human well-being as opposed to economic productivity could potentially diminish the focus on long-term economic growth.

Gender Approaches

Gender approaches are based on theories of human development, but emphasise the failure of development strategies to take account of the specific needs of women, who have often been subject to wages and other conditions that have impeded the satisfaction of basic needs. These approaches were developed by authors such as Boserup (1970) or, at a later stage, Molyneux (1979) or Moser (1989).

To address gender inequality, the level of employment and women's access to education should be increased. This approach was particularly relevant in the implementation of micro-credit schemes, where the majority of beneficiaries were women.

Early gender approaches were criticised for not taking into account the social relations and the economic model that defined women's situation in society. Later versions added the dimension of strategic gender interests, aimed at redressing issues such as inequalities in political representation, educational achievement, personal security and freedom of movement.

The integration of women into the labour force, not in administrative or managerial positions, but as essentially low-skilled and low-paid labour for export-oriented industries and services, led to a situation where the focus of the approach was no longer on the exclusion of women from economic activities, but on the significant inequalities in participation.

Neoliberal-inspired welfare cuts under structural adjustment programmes were seen as a factor that encouraged women to earn additional income by entering the labour market (Elson, 1995). Furthermore, the absence of women's so-called unpaid work from national accounts and other policy documents could favour the prevalence of gender inequalities and the successive adoption of ineffective development strategies.

Environmental Approaches

Environmental approaches argue that natural resources should not be seen as an obstacle to development. Some of the major early proponents of this approach were Polanyi (1944), Ehrlich (1968) and Meadows et al. (1972). The overemphasis on growth is seen as inimical to sustainable human development because of factors such as population growth, limited food production, resource depletion and environmental degradation. Natural resources cannot be treated like any other type of capital.

According to this school of thought, development assistance should focus not only on isolated investment projects, but above all on innovative, cross-cutting, multi-sectoral approaches that address broad issues such as environmental policy and management, biosphere protection, biodiversity, site conservation, flood prevention and control, environmental education and training, and environmental research. Support for renewable energy sources and integrated urban or rural development programmes should also be encouraged.

Evidence of man-made depletion of the ozone layer was a key factor in bringing this debate to the fore. The neoclassical paradigm, with its individualistic focus, made it difficult to establish a collective assessment of the value of preserving the environment, as opposed to the value of consuming harmful goods.

Since publicly owned natural resources had no market value, incentives for their sustainable management were seen as non-existent, allowing a process of depletion to take place. The negative effects of environmental degradation and its costs to human development were not taken into account in official accounting, thus maintaining the status of externalities.

Environmental concerns were eventually integrated into mainstream development policies under the concept of sustainable development, but the objective of economic growth remained the most overarching.

Post-Washington Consensus

The post-Washington Consensus can be seen as having its roots in the economic and financial crisis that hit Asia, Russia and Central and South America in the late 1990s. As a result of the failed interventions of international institutions, the premises of the Washington Consensus were questioned, giving rise to the idea of reforming the international financial system, including some moderation of the principle of global financial liberalisation.

Critics pointed out that the focus should shift from inflation control to growth, and that trade liberalisation and privatisation processes should not be seen as ends in themselves, but as means to achieve more sustainable growth. Policies to limit budget deficits, money supply and privatisation will only be effective if an institutional framework for market operation and competition is in place and if financial institutions are sound.

While continuing to reject state-coordinated industrial policies, this new approach recovered some principles already advanced by neo-statist theories, giving greater relevance to the compensation of market imperfections and the importance of institutions and the regulatory framework, reflected in the increasing attention paid to a relatively new set of concerns, such as good governance, capacity building, transparency, participation, and sustainable and inclusive development.

Rather than focusing on structural change, assistance should pay particular attention to technical cooperation activities on governance issues, which have been identified as having shortcomings that hinder overall equilibrium and subsequent development. Particular attention should be paid to support for public financial management, with a view to broadening the tax base and improving revenue collection.

Debt relief for the poorest countries has also been supported, provided that the beneficiaries are committed to support programmes and macroeconomic conditionality with the IMF. The aim would be to achieve debt sustainability, in the sense that the use of external borrowing would be limited to what exports can pay for without reducing imports, i.e. real interest rates on debt should not exceed the real growth rate of exports.

Budget support has also been promoted under similar conditions, involving modalities in which bilateral donors contribute to pooled funds managed jointly by multilateral institutions, donors and recipients, characterised by common project documents, financing contracts and reporting procedures. This framework would allow for a reduction in the administrative burden for beneficiaries, but also for a more influential position of multilateral institutions vis-à-vis bilateral donors.

Education, technology and infrastructure are also seen as key areas for development assistance. It is also worth noting that references to the importance of environmental sustainability and social safety nets have become part of the mainstream of theory.

The importance of so-called social capital has been emphasised, in the sense of representing the set of norms, values, trust and networks formed by individuals in a society and contributing to solving problems. This concept, although vague and abstract, has been seen as crucial in explaining institutional problems and, ultimately, the level of development of countries.

The post-Washington consensus can thus be seen as an amalgamation of neoliberal theory with other sets of concerns and approaches, such as human development, gender, environment and, at a later stage, poverty.

With regard to the theoretical justifications for aid, Sachs (2005) points out that insufficient growth in developing countries is not only a consequence of situations of corruption and backward cultural systems, but mainly the result of eight main types of problems:

- The poverty trap: Poverty caused by low availability of capital (physical, human and natural) requires higher levels of savings, but this may not be possible in cases of extreme poverty where all income must be spent on basic consumption needs.
- Physical geography: Many poor countries are severely affected by high transport costs, as the specific features of their geographical location may prevent easy access to maritime transport (particularly landlocked or mountainous countries without good ports or navigable waterways). Other unfavourable conditions relate to the prevalence of arid conditions and tropical diseases. While these problems increase the likelihood of a poverty trap, they should not be seen as insurmountable, but as requiring significant investment.
- The fiscal trap: Governments may not have the financial resources to provide essential public goods and services due to an impoverished (and therefore non-taxable) population, an inability to collect tax revenues, or excessive debt service (requiring debt relief).
- Governance failures: If governments fail to provide high-priority infrastructure and services, an environment conducive to private investment, a background free of corruption, and a guarantee of internal peace, security, and the enforcement of contracts and property rights, then the economy will fail severely and the country may even reach the state of a failed state.
- Cultural barriers: Societal norms that block the role of women and prevent the transition to a low fertility regime, or that target religious or ethnic minorities in particular, are important barriers to development.
- Geopolitics: In addition to trade barriers, many other factors can be manipulated by foreign powers to undermine certain regimes for geopolitical reasons, thereby exacerbating development problems.

- Lack of innovation: The small size and low purchasing power of many developing countries' domestic markets make it particularly difficult to recoup investment in innovation. The lack of incentives to innovate does not encourage productivity growth and hence market expansion and additional innovation, thus preventing an endogenous growth path. In many poor countries, the innovation process does not even get off the ground, and governments' lack of resources prevents them from contributing to the process. The innovation gap is thus one of the main factors explaining long-term development differences between countries. This situation could be mitigated by importing technology, but financial resources may not be available, recipients may not be attractive for foreign direct investment, and technological progress led by advanced economies may not be suitable for the geographical and climatic conditions of most developing countries. The existence of infrastructure such as ports or other geographical advantages is seen as an important factor in the potential attraction of innovation through foreign investment.
- The demographic trap: In contrast to the developed world, some of the poorest countries have high fertility rates, meaning that the population can double in a generation, contributing to deepening poverty.

In particular, Sachs (2005) notes that countries with good ports and energy sources, close contacts with the developed world, a favourable climate and a lack of epidemic diseases have escaped the poverty trap, so that the main task of donors would be to create these conditions to enable aid recipients to move through the first stage of the development process.

In a poverty trap situation, low levels of per capita capital tend to fall as a result of rapid population growth. Capital is accumulated when part of the income of households is taxed to finance public investment, or when it is saved and then used for lending to enterprises (through the financial system) or for direct investment or equity. If savings exceed capital depreciation, there is a positive net accumulation of capital, which should be at least equal to population growth.

Household income is thus used for consumption, savings or taxation, while government revenue is used for current or capital expenditure. Both sectors contribute to increasing the stock of capital in the economy, which would lead to growth and hence further income increases, creating a feedback loop as long as they can offset depreciation and population growth.

In a poverty trap, however, all household income is spent on consumption to meet basic needs, leaving no room for savings or taxation. This factor, together with the continuous depreciation of capital and population growth, will lead to a decline in the ratio of capital to income per capita, further impoverishing households and perpetuating the cycle of poverty.

In order to solve this problem, foreign aid should contribute to the process of capital accumulation, economic growth and increasing household income through three main channels: a small part directly to households (mainly for humanitarian and food aid), a much larger part for public investment, and a part to private enterprises (mainly through microfinance and support to small businesses).

If this foreign aid is provided over a long enough period of time, at some point the capital stock will be able to generate household incomes above the subsistence level, thus breaking the poverty trap. Foreign aid should therefore not be seen as a welfare transfer, but as an essential investment in poverty reduction. Financial planning could be used to assess the amount of aid needed to break the poverty trap. Above a certain level of per capita income, an economy will enter into a sustainable growth process, benefiting from increasing returns to scale of capital, while below that level the trend would be one of deterioration.

Investment strategies should be based on a differential diagnosis, building on a division of labour between the private and public sectors. The private sector should invest in enterprises, in new products and technologies, and in those areas of health and education that complement public investment.

Occasionally, governments could also provide direct financing to the private sector, for example in new agricultural technologies or in the creation of urban industries. However, the public sector should focus on human capital (education and health), infrastructure (transport, energy, water, sanitation and environmental protection), natural capital (preservation of ecosystems and biodiversity), institutional capital (public administration, justice and security) and knowledge capital (research in areas such as climate, agriculture, health and energy). Public intervention in these areas is justified by the following factors:

- Many transport and energy infrastructures are prone to increasing returns to scale, which could lead to monopoly and overpricing in a free private market;
- There are non-rival capital goods for which the use by one individual does not reduce the availability for others (e.g. a scientific advance), so that the absence of charges for their use is essential for economic efficiency, implying that only governments have an incentive to invest in their research and development - investment in technological capacity and in the promotion of scientific research, which requires widespread literacy and allows the emergence of home-grown technology better adapted to local conditions, is seen as paramount;
- Natural capital and various social sectors have strong externalities, in the sense that if the environment is degraded or one part of society has poor education or health, there can be serious consequences for the whole of society;

- The importance of a universal availability of goods vital to human well-being, namely health, education, and water;
- The importance of helping the poorest sections of society to raise their incomes above the subsistence level and to generate savings, a goal that could be achieved by subsidising small private enterprises, but only those belonging to poor households.

Sachs (2005) does not attach particular importance to any particular type of investment, but considers that each of them is essential to overcome the poverty trap: business capital, human capital, infrastructure, natural capital, knowledge capital and public institutional capital. On the basis of the above arguments, he argues that developed countries should rapidly fulfil their commitments on aid to developing countries and at least reach the target of 0.7% of GNI as defined by the United Nations.

However, authors such as Easterly (2006) point out that Sachs's proposals resemble earlier 'big push' approaches, which sought to lift countries out of the poverty trap within an overarching plan, with foreign aid filling the investment financing gap. This old view is seen as simplistic and outdated, as development is the result of a complex interaction between markets, policies, social norms, institutions, social services and microeconomic interventions. Government initiatives should focus on gradual and incremental reforms after the failed experiences of transition economies (in post-communist countries) and IMF structural adjustment programmes.

Furthermore, Easterly notes the difficulty of finding evidence for the actual existence of poverty traps, particularly in terms of the savings and increasing returns mechanism, and points to the lack of empirical evidence on the effects of aid on growth. In contrast to Sachs, he argues that growth problems are mainly explained by bad governance (as long as the initial level of poverty and bad governance are taken into account), and therefore the role of the poverty trap is downplayed. Top-down planning solutions are seen as particularly problematic in countries with high levels of corruption.

Easterly also criticises the view of development as a technical problem to be solved mainly through the application of natural sciences such as geography, agronomy or medicine, without taking into account the social roots of poverty: weak institutions, politics and policies, exclusion from trade networks, high transaction costs and ineffective aid. The adoption of a technical approach has led to the implementation of administrative solutions prone to various problems, which are particularly acute in the case of activities financed by foreign aid: the need to create incentives for public officials to implement correctly the solutions decided at the highest levels of power, the lack of information at these higher levels about the realities at the lowest levels (which would enable them to design appropriate interventions), and the existence of multiple objectives and actors, which weaken incentives as they compete with each other. Foreign aid should not provide for "big push" interventions or overly broad objectives, but should instead focus on specific, ongoing interventions.

Degnbol-Martinussen and Engberg-Pedersen (2005) also touch on the issue of micro and macro levels, recalling that arguments in favour of aid have often been micro focused, based on the assumption that many priority projects and activities are not attractive to private investors from a socio-economic point of view, or that governments are not in a position to finance them. This private reluctance may be due to excessive risks, long payback periods, or the inability to collect payments from users. Therefore, if a particular project has a significant social or economic impact but does not attract private investors and the government does not have the resources to finance it, the intervention of foreign aid is fully justified.

There are also arguments in favour of some form of aid targeted directly at the poorest sections of society, since neither capital transfers nor other forms of aid particularly benefit this category. This would mainly be the case in countries where the growth of the industrial sector is limited by the size of domestic markets and problems of external competition for exports. In these situations, an increase in the income of the wealthy segment would lead to an increase in imports (due to lack of capacity of the domestic industry or to issues of choice) and possible balance of payments problems, while an increase in the income of the poorest would lead to an increase in demand for local products, thus stimulating the domestic economy, which would ultimately also benefit the higher income groups. Conversely, the opposite assumption, that the initial benefits for the richest would spill over to society as a whole, is much less likely.

The fact that enhancing governance and institutional capacity is a gradual process, and the impact of aid is not immediately apparent, in addition to the disagreement with other principles of neoliberal theory, has led to questions regarding the efficacy of the post-Washington Consensus approach. This has resulted in the need for a re-evaluation of the mainstream development and aid framework, as outlined in the subsequent items.

Sustainable Development

The concept of sustainable development builds on earlier environmental approaches that integrated economic and environmental dimensions, where the environment was seen as an engine for growth that should be preserved at the same time as avoiding the destruction of natural resources. This concept derives in particular from the Brundtland Report, published in 1987 by the United Nations World Commission on Environment and Development. The transition from the concept of sustainable growth to that of sustainable development is the result of the latter's consideration of the need to meet the aspirations of the present generation without compromising the ability of future generations to meet their own needs, thus requiring intergenerational solidarity.

Prudent management of non-renewable resources should be pursued, with particular attention being paid to renewable resources, in the context of a rationale of limiting the use of inputs. In addition to the economic and environmental dimensions, a pillar of social cohesion has been added to the determinants of sustainable development.

With the end of the Millennium Development Goals action period, a new 2030 Agenda for Sustainable Development was adopted at the United Nations Summit in September 2015. This Agenda provides for an even more far-reaching set of goals, covering economic, social and environmental dimensions and applying to all countries, developed and developing, as opposed to the Millennium Development Goals, which only provided an action plan for the latter.

One of its main focuses is the mobilisation of the means of implementation, namely financial, technological and capacity-building, with special attention being paid to the fight against climate change and its consequences. In this context, particular attention will be paid to the implementation of the new concept of Total Official Support for Sustainable Development. Other key action points focus on issues such as inequalities, growth, jobs, cities, peace and justice, the eradication of poverty in all its forms and the obligation (although not legally binding) for each country to establish national frameworks for achieving the goals.

In order to monitor the implementation of the Agenda, a set of reliable and timely data must be provided by the participating countries - the 17 Sustainable Development Goals comprise 169 targets to be achieved by 2030.

In this context, a novel financing and investment instrument system has been conceptualised to support the 2030 Agenda. This system is based on the transition from pure official development assistance (as defined by the OECD) to blended financing, which, in addition to conventional grants and concessional loans, should also encompass other instruments such as commercial loans, guarantees, and equity. This association would allow the mobilisation of additional resources for developing countries, increasing the links between concessional and non-concessional instruments, as well as between public and private financing. This would also allow for some relief from the heavy budgetary constraints currently in place in a number of donor countries.

Poverty and Inequality Approaches

Poverty and inequality approaches emerged in the early 21st century by rejecting an excessive focus on growth and considering a possible link between globalisation and poverty/inequality, following a general agreement that, despite possible measurement differences, disparities in income, basic needs and opportunities, both between and within countries, were at historically high levels. In earlier paradigms, poverty was the result of maladjustment to globalisation, so that the expansion of the latter was a sufficient condition for overcoming the problem.

In fact, poverty has been exacerbated by the process of globalisation, and increased inequality has resulted essentially from the accumulation of wealth in the richest groups of the richest nations. The neoliberal principle that only absolute poverty should be tackled and that inequality is necessary to promote individual effort and hence efficiency is rejected (Wade, 2004).

Development assistance should therefore play a role in promoting genuine equality of opportunity, which will only be possible if welfare and education become universal and if the excessive influence of dominant groups in politics and the judiciary is eliminated. Failure to address inequality will depress demand and growth and increase poverty, migration, crime, corruption and social unrest. The labour force of rich countries could also be affected by the relocation of capital to countries with low wages and abundant labour (Payne & Phillips, 2010).

Poverty approaches have been mainstreamed through the adoption by the international financial institutions of so-called 'Growth and Poverty Reduction Strategy Papers', which have replaced previous structural adjustment programmes and focused on issues such as good governance, promoting opportunities, health, education and social safety nets. However, these have been criticised by those who see them as a mere extension of the neoliberal post-Washington consensus (e.g. Saad-Filho, 2007). Moreover, recent empirical studies, such as Chletsos & Sintos (2022), conclude that IMF programmes are associated with greater income inequality.

The poverty goal was also highlighted with the adoption by the United Nations in 2000 of the Millennium Development Goals, several of which were directly linked to poverty reduction, but which were also criticised for the absence of issues related to inequality. Subsequently, with the implementation of the 2030 Agenda, two Sustainable Development Goals are particularly linked to this issue: "End poverty in all its forms everywhere" and "Reduce inequalities within and between countries".

Local Development and Participative Methodology

Local development and participative methodology approaches emphasise the importance of listening to people's needs and mobilising local resources to respond to problems. They have been developed by authors such as Vachon (1998), Pecqueur (2000) or Houée (2001). The rationale of seeking primarily local solutions implies the promotion of citizen participation and community partnerships, integrated with any exogenous solutions, within the main assumption that different regions require different development strategies that respond to local needs with locally available resources.

In order to achieve a relevant participatory process, it would be necessary to empower communities by increasing their capacity to participate in the evaluation or implementation of development programmes. In addition, an integrated approach requires a partnership between citizens, businesses and government, with the active participation of all actors strengthening the democratic process.

Post-Development Approaches

Post-development approaches reject the central role of economic growth, the over-reliance on export strategies and the dominance of sectors such as the extractive industries. In particular, they draw on the line of thought developed by Escobar (1995).

The idea of modernity was seen as a tool for Western domination of developing countries; international institutions were criticised for privileging modernisation processes over traditional forms of organisation, reproducing modes of exploitation and environmental degradation. Any improvement in basic needs resulting from the old models was temporary and unsustainable; post-war development strategies failed to provide a minimum standard of living for the whole population.

The new paradigm should be based on local movements and adhere to the concept of good living, which could be achieved within a locally centred life, in connection with the land and one's own culture; the overarching ideas of simplicity, fragility and beauty should be respected (Corbridge, 1998).

Within this paradigm, activities in areas such as agriculture, environmental protection and culture should be supported, while those related to extractive industries and international trade should be rejected.

Global Theories of Development

Global theories of development focus on the effects of globalisation, rejecting analyses that focus on individual countries and the idea that development policies are implemented by states in a national context. This approach was developed at an early stage by authors such as Gereffi (1996), McMichael (2000) or Robinson (2002).

One of the main approaches emphasises the existence of global production networks, based on the concepts of global commodity chains and a new international division of labour based on the relocation of production to the periphery and the restructuring of activities in the centre, as well as the existence of networks organised around a product, involving states, firms and households, where the driving forces are firms and not states.

The concept of network goes beyond the idea of linear chains and emphasises the increasing global interconnectedness of activities related to the production, distribution and consumption of goods and services. Contrary to earlier theories, development is not a matter of government strategies, but a consequence of the globalisation movement; the notion of national strategies no longer has any significant meaning.

With the new division of labour, countries no longer specialise in a particular raw material or manufactured product, but become part of a production chain linking different countries. China's development is seen mainly as the result of the investment strategies of companies in advanced countries, corresponding to a certain stage in the evolution of global networks and chains.

Accordingly, the priority sectors for resource allocation would be those related to each country's international connections, such as transport, communications and trade, as well as others that could improve the position of its economy in global networks.

In this perspective, inequality is also seen as a global rather than a national problem. The constitution of certain social groups benefiting from certain levels of development results from the participation of individuals in global production and labour markets, not from belonging to a particular country; the concept of underdevelopment is thus social, not geographical. The phenomenon of migration is seen as a direct consequence of inequality and international labour markets, both at the skilled and unskilled levels.

Ultimately, the concept of national development is rejected, since a significant part of each economy exists only in the context of global production networks and a significant part of the workforce is no longer made up of national citizens.

It should be noted, however, that some of the foundations of global development theories have been called into question by the pandemic crisis that has affected the world economy since the beginning of the 2020s (subsequently aggravated by the outbreak of the conflict in Ukraine), as many of the international production chains have been severely disrupted. It can be argued that global integration has the potential to expose economies to external shocks, thereby reducing national economic resilience and increasing vulnerability.

FOREIGN AID, INEQUALITY AND ENVIRONMENT

Aid and Inequality

Theoretical Framework

Theoretical frameworks on the impact of foreign aid on inequality in developing countries predominantly focus on the effects of redistribution mechanisms. It is assumed that aid specifically directed towards areas such as primary education, basic healthcare, social protection programmes or certain types of infrastructure, preferentially benefits the poorest and therefore leads to a reduction in inequality in recipient countries, compared to aid for large-scale infrastructure or general budget support (Kasuga & Morita, 2018). However, the possibility of aid fungibility must also be considered; by receiving funds for a specific sector, the recipient government could reduce its domestic spending in that same sector, thereby freeing up resources for other purposes.

Another mechanism, this one with potentially negative effects, is also considered: the elite capture channel (as hypothesised, for example, in Bjørnskov, 2010). This mechanism assumes that aid resources, particularly in countries with weak institutions, can be appropriated by the most powerful groups in society, through corruption, bad governance or simply because policies or investment decisions disproportionately benefit the most educated and urban classes (Robinson & Torvik, 2005). This line of thought emphasises political economy aspects, highlighting the importance of public institutions, electoral systems, and the power structure in determining the distributive effects of aid.

The so-called macroeconomic mechanism must also be taken into account (Rajan & Subramanian, 2011), according to which the influx of foreign aid can affect the exchange rate, the relative prices of goods, and the competitiveness of certain sectors. These effects can have differentiated impacts on distinct social groups, potentially altering the distribution of income in unintended ways.

Complementary hypotheses have indicated that the effects of aid on inequality are contingent not only on the context of the beneficiaries, but also on the type and modality of aid provided by donors, particularly in cases where aid is linked to selfish motives (Herzer & Nunnenkamp, 2012). A further dimension that merits consideration is the relevance of negative shocks of foreign aid for situations of conflict and political instability (Nielsen et al., 2011), which might also have consequences at the inequality level.

Empirical Evidence

Recent empirical evidence on the impact of foreign aid on inequality in developing countries is inconclusive, depending on the temporal and spatial contexts considered, as well as other specific contextual factors.

While international institutions generally advocate for a narrative of convergence in global income inequality, some researchers (e.g. Hickel, 2017) dispute this assumption, arguing that the presented data is ambiguous and conceals the fact that convergence is largely driven by China. Furthermore, it fails to recognise increasing absolute inequality and disregards divergence between regions.

Younsi et al. (2019) found that the foreign aid received did not reduce inequality over time, particularly in African countries. They also concluded that corruption was responsible for the lack of effectiveness of foreign aid in reducing inequality.

In terms of sectoral disaggregation, pro-poor aid has been shown to reduce inequality, whereas total aid does not necessarily produce the same result (Kasuga & Morita, 2018). In effect, while aid for large-scale infrastructure projects can have positive effects on aggregate growth, it often exacerbates inequality in the short term by disproportionately benefiting urban and more educated groups.

Research on aid for social protection and direct transfers has been particularly relevant, as it shows that these can significantly reduce poverty and inequality, especially when channelled through conditional cash transfer programmes (Handa et al., 2018).

Specific attention has also been paid to the role of institutions. Yiew & Lau (2018) determine that the quality of institutions, particularly with regard to scrutinising and ensuring the equitable distribution of foreign aid, is a key factor in the effectiveness of aid and its distribution. Meanwhile, Andersen et al. (2022) conclude that aid capture does exist in the most aid-dependent countries, by analysing bank deposits in offshore financial centres renowned for their bank secrecy and private wealth management services.

In general, the literature agrees that aid is more likely to be channelled effectively towards its redistributive objectives in countries with solid institutions, effective judicial systems and low corruption. Conversely, in contexts of poor governance, there is a significantly higher likelihood of aid being captured by privileged groups.

These findings have important implications for the design of aid programmes. They suggest that investments in institutional capacity may be a prerequisite for the redistributive effectiveness of other types of aid. Aid has a greater potential to reduce inequality when it is aligned with fair taxation and well-targeted social spending, as well as the recipient government's own policies aimed at promoting equality.

Ambassa et al. (2024) examine how conflicts influence the relationship between foreign aid and inequality. Their results suggest that, while foreign aid reduces inequality, its effectiveness is considerably reduced in conflict-ridden countries. They also conclude that promoting good governance by fighting corruption and preventing conflict can enable foreign aid to significantly reduce inequality.

As Biscaye et al. (2017) note, one area of research that has been explored is the comparison of the effects of bilateral and multilateral aid. This research suggests that aid channelled through international organisations tends to have neutral or slightly positive effects on income distribution. This may be due to greater technicality in programme design and less influence from geopolitical considerations. By contrast, the distributional effects of bilateral aid, especially when motivated by commercial or strategic considerations, show greater variability.

In summary, empirical studies do not allow for the automatic assumption that foreign aid is a fully effective mechanism for reducing income inequalities. The evidence suggests that a generic approach by donors ("one-size-fits-all") may be inadequate, and that the design of aid programmes should be sensitive to the specific context of each recipient country. Complementary investments in institutional capacity and governance could be fundamental prerequisites for the redistributive effectiveness of aid, suggesting that aid programmes should be implemented with a longer time horizon.

Aid and the Environment

Theoretical Framework

The theoretical framework on the effect of foreign aid on the environmental situation of developing countries reflects a growing complexity of the interactions between economic development, international cooperation and environmental sustainability.

One of the main fundamental transmission mechanisms is predicated on the assumption that aid promotes changes in the structure of the economy, thereby altering the relative weight of sectors that are more or less intensive in natural resources and pollution. This mechanism is of particular pertinence when aid supports the transition from economies based on subsistence agriculture and resource extraction to service sectors or cleaner industries.

Another fundamental mechanism pertains to the scale effect (elaborated upon, for instance, in Stern, 2004), in the sense that aid has the potential to expedite economic growth, which is commonly associated with heightened resource consumption and augmented pollution in the short term, but it can also foster the conditions for investments in cleaner technologies and enhanced environmental governance in the long term.

An additional mechanism is the technical effect, according to which aid can facilitate the transfer of cleaner technologies, knowledge about sustainable practices, and institutional capacity to implement effective environmental regulation (Popp, 2011).

A notable approach has been the application of the Environmental Kuznets Curve hypothesis to the context of foreign aid. From this standpoint, a U-shaped relationship is posited between per capita income and environmental degradation. The hypothesis is that environmental quality deteriorates initially as countries develop, but then improves after a certain turning point in income.

Dasgupta et al. (2002) posit that international assistance can play a significant role in lowering and flattening the environmental Kuznets curve. This can be achieved through the provision of financial support for training, policy reforms, and environmental education. Specifically, this support should be directed towards programmes that disseminate public and accessible information regarding polluters and pollution to the general public. Furthermore, the provision of support for the reinforcement of regulatory institutions and the implementation of cost-effective measures to reduce pollution is also considered to be of significant importance.

Some theorists argue that foreign aid can enable countries to bypass more polluting stages of development, thereby promoting more sustainable growth paths from the outset (Weko & Goldthau, 2022). This standpoint is of particular pertinence in the context of understanding the role of climate aid and the support for renewable energies.

The political economy approach highlights the potential discordance between the incentives of donors and recipients concerning environmental issues, particularly when the environmental benefits are global (as in the case of climate change) while the costs are local. The effectiveness of environmental aid is contingent upon the capacity to align these incentives through the implementation of appropriate mechanisms of conditionality, compensation, and international coordination. A significant theoretical contribution is the recognition that numerous environmental issues possess the characteristics of global public goods (Kaul et al., 1999), thus establishing a distinct framework for environmental aid. This framework suggests that donor countries have incentives to support environmental enhancements in developing countries, not solely for altruistic reasons, but also because they directly benefit from certain of these improvements, particularly in the domains of climate change and biodiversity. Environmental aid can be more efficacious when directed towards issues with robust global positive externalities, and when its structure explicitly acknowledges these mutual benefits.

Empirical Evidence

Research utilising data from Korean aid recipients has revealed that official development assistance exerts both direct and indirect carbon dioxide mitigation impacts in recipient countries (Lee et al., 2020). Research conducted on sub-Saharan African countries also demonstrates a significant positive impact of foreign aid on renewable energy development. However, Wang et al. (2021) determined that such a positive contribution only occurs during the initial stages of technical progress and social structural change. Once carbon dioxide intensity and urbanisation exceed certain threshold values, the effect of aid becomes negative, suggesting that aid does not always promote renewable energy development.

Research has also demonstrated that aid allocated to the field of renewable energies, particularly in the form of technical cooperation, exerts a significant multiplier effect, thereby attracting additional private investment and promoting technology transfer, particularly in low-income countries (Kim, 2018). Moreover, a close examination of programmes for renewable energies reveals that projects tend to be more successful when they combine funding with technical assistance and local capacity building. In essence, such projects predominantly depend on learning processes that are intricately entangled with the historically and geographically unique attributes of the built environment, thereby augmenting the existing body of local skills and knowledge (Neij et al., 2017). These findings underscore the significance of comprehensive approaches that extend beyond mere financing of equipment.

Nevertheless, the positive effects of aid on the ecological sustainability of sub-Saharan African countries remain ambiguous. A body of research utilising composite indicators of ecological sustainability has yielded mixed outcomes. A number of countries have exhibited substantial advancements in this domain; conversely, other countries have demonstrated persistent deterioration, even in the face of substantial aid. For instance, He et al. (2022) determined that agricultural foreign aid in Sierra Leone damages forests both in the short and long term, and that aid projects with agricultural development as their main objective lead to high levels of forest degradation.

As Semenchuk et al. (2022) observe, there is a necessity for enhanced sustainable intensification strategies and demand-side actions, with a view to reducing trade-offs between biodiversity and food security. It is evident that aid which fosters sustainable agricultural practices, conservation agriculture, and integrated water resource management tends to engender favourable environmental effects. Conversely, aid that promotes agricultural intensification without appropriate environmental safeguards can result in detrimental consequences such as soil degradation, water pollution, and biodiversity loss.

Research on aid for biodiversity also demonstrates that programmes tend to be more effective when they are implemented at the local level with strong community participation. As Fariss et al. (2023) have demonstrated, community-based conservation projects have a higher probability of success when they are implemented within national frameworks that are supportive of local governance, when they foster economic diversification, and when they invest in efforts to enhance capacities.

Research employing satellite data to assess alterations in forest coverage has revealed that aid designated for forest conservation has modest yet statistically significant positive repercussions. Buchanan et al. (2018) concluded that areas in proximity to World Bank projects exhibited a marginally reduced rate of deforestation in comparison to those situated at a greater distance. The findings therefore indicate that World Bank projects do not exert a discernible adverse effect on biodiversity, and in certain instances, they may even confer certain benefits.

The findings also show that the impact of official development assistance on the ecological environment of African countries varies according to the level of agricultural dependency and the extent of renewable energy development in each nation (Li et al., 2024). This heterogeneity is of crucial importance in understanding the manner in which aid can either promote or hinder environmental sustainability.

With regard to the aid specifically targeted at environmental policy and institutional reform, Kono & Montinola (2019) conclude that there is little evidence that it actually encourages recipients to adopt more advanced climate legislation. Additionally, McCarthy et al. (2023) posit that, while aid can exert influence on policy development, there is a weak correlation with outcomes such as reduced carbon dioxide emissions, expansion of protected areas or increased forest cover. These findings imply a discrepancy between the formulation of environmental policies and their effective implementation, underscoring the significance of enhancing institutional capacities and establishing monitoring systems.

A body of research focusing specifically on payments for ecosystem services programmes funded by foreign aid has yielded more encouraging results. The research has produced evidence of measurable reductions in deforestation and improvements in water quality in several regions. For instance, Jayachandran et al. (2017) evaluated a programme in Uganda that offered payments to owners of forest terrains on the condition that they conserve the respective forests. The findings indicated that the advantages in terms of enhanced tree cover and carbon dioxide emissions significantly exceeded the associated expenditures of the programme. However, it should be noted that the long-term sustainability of these programmes remains an open question, particularly in the aftermath of the termination of external funding.

It has been determined by other studies that the quality of institutions is a pivotal factor in the efficacy of environmental aid. In nations with deficient institutions, even aid designated for environmental objectives can be misdirected or executed inefficiently. For instance, Li et al. (2021) demonstrate that green aid does not directly correlate with the mitigation of carbon dioxide emissions. However, when institutional quality indices are incorporated into the analysis, a substantial effect of such institutional quality on the effectiveness of green aid becomes evident. In general, in countries with weak institutions, there is an undesirable correlation between green aid and higher carbon emissions.

To summarise, the extant literature suggests that foreign aid can contribute to environmental improvements in developing countries. However, the effects of such aid are highly contextual and depend crucially on the type of aid made available by the donor, the characteristics of the recipient country, and the quality of implementation. The research suggests that more integrated and context-sensitive approaches are required to maximise the environmental benefits of international aid.

CONCLUSION

Taking into account the perspectives of different development theories and approaches, this chapter has outlined a theoretical framework of development and the role of foreign aid. It has examined the main causes of (and constraints to) development and the main arguments for and against the provision of official aid. It has also considered how aid should be allocated in order to maximise its positive impact on the development process of recipient countries.

For Keynesian growth theories, aid enables recipients to achieve a level of saving and investment high enough to generate a self-sustaining growth path. The dual-gap approach suggests that foreign capital can be seen not only as a complement to domestic capital but also as a complement to foreign exchange, and that it should fill the larger of two gaps - savings/investment or exports/imports. However, critics of this model point out that foreign and domestic resources are not substitutable, that aid could finance consumption rather than investment, and that the strong emphasis on the volume of capital as a trigger for growth is questionable.

For the neoclassical approach, aid could be used to finance additional consumption rather than an increase in the stock of capital, but its impact would be somewhat modest. In addition, foreign aid is seen by some authors as creating a relationship of dependency between donors and recipients. For modernisation theorists, aid has the potential to trigger a process of self-reinforcing growth through accelerator and multiplier effects. Conversely, for some underdevelopment theorists, foreign aid creates dependency and thus becomes a way for the North to maintain its dominance over the South, in which case it should cease to exist.

For the neoliberal current, the main objectives of aid would be to ensure law and order, enforce contracts and promote market competitiveness. With the advent of the post-Washington consensus, there has been increasing attention to concerns such as good governance, capacity building, transparency, participation, and sustainable and inclusive development. Debt relief for the poorest countries has also been promoted, provided that the beneficiaries are committed to aid programmes and macroeconomic conditionality. Ultimately, the sustainable development perspective reinforced the relevance of issues such as renewable energy and the social cohesion pillar. Conversely, post-development approaches criticise international aid institutions for privileging modernisation processes over traditional forms of organisation, thereby reproducing modes of exploitation and environmental degradation.

It has also been determined that, despite foreign aid's evolution in response to sustainability challenges, its demonstrable impact on the inequality and environmental conditions of beneficiaries has been somewhat modest. This is attributable not only to the nature of aid provided by donors (which is frequently associated with strategic and/or commercial interests, or with specific policy conditionalities), but also to the specific circumstances of beneficiaries, particularly with regard to institutional quality and economic characteristics.

In general, the differences between the various approaches can be seen as arising from the following factors:

- The importance attached to an initial big push in investment and capital formation in a wide range of sectors, with foreign aid playing a relevant role in the process, versus the view that development assistance should focus on targeted incremental actions, particularly in institutional capacity building, rather than on overarching rapid transformation.
- The role assigned to industrial expansion: while the liberal-inspired approaches reject state-sponsored industrial policies and focus instead on infrastructures that can promote foreign trade based on the exploitation of comparative advantages, another group of theories stresses the need for state industrial support to change economic structures.
- The overarching role assigned by one group of the former theories to economic growth and related infrastructure (namely energy and transport of goods) contrasts with a second group that stresses the importance of other development indicators and social infrastructure related to human needs, such as education, health and social services, which should be the main targets of aid activities.

Table 1 summarises the key differences between the main theoretical approaches to development and their respective views on the role of foreign aid. It also outlines the leading criticisms of these approaches.

Table 1. A comparative analysis of the role of foreign aid according to the main theoretical approaches to development

Theory	Assumptions	Role of aid	Criticisms
Classical	Economic activities should be driven by self-interest and market forces, with minimal intervention by the state, whose role should be limited to essential infrastructure and maintaining order.	The private sector leads economic activity, with minimal government involvement. Aid, when granted, is centred on basic infrastructure or public security, rather than economic transformation.	By limiting state intervention, the classical approach fails to address structural inequalities and market failures in developing economies.
Convergence (catch-up)	Developing countries must achieve faster economic growth to catch up with developed countries through state-led industrialisation and the development of human capital.	Aid is targeted at sectors crucial to industrialisation (e.g. education, transport, banking) and policies that accelerate the growth of human capital and infrastructure.	Excessive dependence on state intervention can hinder the growth of the private sector and reduce economic efficiency.
Keynesian	Countries can achieve a self-sustaining growth path if they overcome the gap between savings and investment and the foreign exchange deficit.	Aid fills the foreign exchange gap or the gap between savings and investment, enabling the capital accumulation necessary for economic take-off.	It is assumed that aid directly stimulates investment; however, aid often supports consumption, reducing its effectiveness in promoting economic growth.
Neoclassical	Production depends on three factors: labour, physical capital and technology. The marginal productivity of capital is decreasing.	Aid can essentially finance consumption when the effect on capital accumulation or economic transformation is limited.	Aid often leads to dependency and does not significantly increase capital accumulation or economic transformation.
Modernisation	Development follows a linear process, from traditional to modern economies. It starts by increasing investment rates and modernising institutions.	Aid catalyses development through investment in cutting-edge sectors, modern infrastructure and institutional frameworks, triggering economic growth and industrial expansion.	It does not take into account non-economic factors such as political and social stability, attaching too much importance to linear economic progress.
Underdevelopment	There is a situation of inequality in international relations, which are controlled by the developed nations.	Aid is seen as an instrument for donor countries to maintain economic dominance, reducing the autonomy of the recipient and creating economic dependence.	Failure to take into account the possible positive dynamics of integration into international markets.
Neoliberalism	Government intervention should be limited to maintaining legal and economic stability, while promoting private sector growth.	Aid should support market efficiency and structural reforms. Structural adjustment programmes, coupled with aid, should promote market liberalisation, privatisation and legal stability in order to increase competitiveness and reduce budget deficits.	Structural adjustments often cause social hardship, increase inequality and harm public welfare by prioritising economic liberalisation.
Neo-statism	State intervention is crucial to select and promote strategic industries, especially in the context of heavy industrialisation and economic transformation.	The government plans strategic industrial policies, with aid supporting key industries, heavy infrastructure and financial institutions to drive economic modernisation.	Heavy state involvement in industrial policy can stifle innovation and discourage private sector participation.
Human development	Development should be centred on human capabilities and well-being, not just economic growth.	Aid should prioritise social services and the improvement of human capital, namely education, healthcare, sanitation and public transport, with the aim of improving quality of life and human capabilities as central elements of development.	Focussing on human well-being rather than economic productivity could reduce the emphasis on long-term economic growth.
Post-Washington Consensus	Effective governance, accountability and capacity building are essential for development. Public finance management and debt reduction are essential for sustainable growth.	Aid supports technical co-operation, institutional reforms and better governance, closing the capacity gaps that hinder development and ensuring accountability.	Improving governance and institutional capacity is a slow process and the effects of aid are not immediately visible, which raises doubts about its effectiveness.

continued on following page

Table 1. Continued

Theory	Assumptions	Role of aid	Criticisms
Sustainable development	Development must meet current needs without jeopardising the ability of future generations to meet theirs.	Aid supports environmentally sustainable projects, renewable energy and social infrastructure, balancing economic needs with ecological preservation.	Sustainable projects often face implementation challenges as they seek to balance environmental objectives with the need for economic development.
Post-development	Development must respect traditional social structures, rather than imposing Western development models. Local practices, cultural preservation and community-based approaches should be promoted.	Aid should support the traditional economy, cultural practices and local development strategies, as opposed to the dominance of modern industrial models.	It may not be in line with the high level of interconnectedness between regions in today's economy.
Global development	Inequality is transnational, shaped by global production systems rather than national borders.	Aid must support infrastructure, communication networks and global integration to improve economic positioning in global supply chains.	Global integration can expose economies to external shocks, reducing national economic resilience and increasing vulnerability.

REFERENCES

Abramovitz, M. (1986). Catching Up, Forcing Ahead and Falling Behind. *The Journal of Economic History*, *46*(2).

Akamatsu, K. (1962). A historical pattern of economic growth in developing countries. *The Developing Economies*, *1*(s1), 3–25.

Ambassa, M., Ndjokou, I. M. M. M., & Tsopmo, P. C. (2024). Foreign aid and inequality: Do conflicts matter? *Research in Economics*, *78*(4), 101008. DOI: 10.1016/j.rie.2024.101008

Andersen, J. J., Johannesen, N., & Rijkers, B. (2022). Elite capture of foreign aid: Evidence from offshore bank accounts. *Journal of Political Economy*, *130*(2), 388–425.

Baran, P. A. (1952). The Political Economy of Underdevelopment. *Manchester School.*

Barro, R. J. (1990). Government Spending in a Simple Model of Endogenous Growth. *Journal of Political Economy*, *98*(5).

Bird, G. (1985). *World Finance and Adjustment: An Agenda for Reform.* Springer.

Biscaye, P. E., Reynolds, T. W., & Anderson, C. L. (2017). Relative effectiveness of bilateral and multilateral aid on development outcomes. *Review of Development Economics*, *21*(4), 1425–1447. DOI: 10.1111/rode.12303

Boserup, E. (1970). *Women's Role in Economic Development.* Allen and Unwin.

Buchanan, , G. MParks, , B. CDonald, , P. FO'Donnell, , B. FRunfola, , DSwaddle, , J. PTracewski, , ŁButchart, , S. H. M. (2018). The Local Impacts of World Bank Development Projects Near Sites of Conservation Significance. *The Journal of Environment & Development*, *27*(3), 299-322. https://doi.org/10.1177/1070496518785943

Chenery, H. B., & Strout, A. M. (1966). Foreign Assistance and Economic Development. *The American Economic Review*, *56*(4), 679–733.

Chletsos, M., & Sintos, A. (2022). The effects of IMF programs on income inequality: A semi-parametric treatment effects approach. *International Journal of Development Issues*, *21*(2), 271–291. DOI: 10.1108/IJDI-12-2021-0265

Corbridge, S. (1998). Beneath the pavement only soil: The poverty of post-development. *The Journal of Development Studies*, *34*(6), 138–148.

Dasgupta, , SLaplante, , BWang, , HWheeler, , D. (2002). Confronting the environmental Kuznets curve. *Journal of Economic Perspectives*, *16*(1), 147-168. DOI: 10.1257/0895330027157

Degnbol-Martinussen, J., & Engberg-Pedersen, P. (2005). *Aid: Understanding International Development Cooperation*. Zed Books.

Easterly, W. (2003). Can Foreign Aid Buy Growth? *The Journal of Economic Perspectives*, *17*(3), 23–48.

Easterly, W. (2006). The Big Push Déjà Vu. *Journal of Economic Literature*, *44*, 1.

Ehrlich, P. (1968). *The Population Bomb*. Ballantine Books.

Elson, D. (1995). Gender awareness in modeling structural adjustment. *World Development*, *23*, 11.

Escobar, A. (1995). *Encountering Development: The Making and Unmaking of the Third World*. Princeton University Press.

Fagerberg, J. (1988). Why Growth Rates Differ. In Dosi, G., Freeman, C., Nelson, R., Silverberg, G., & Soete, L. (Eds.), *Technical Change and Economic Theory*. Pinter Publishers.

Fariss, , BDeMello, , NPowlen, , K. ALatimer, , C. EMasuda, , YKennedy, , C. M. (2023). Catalyzing success in community-based conservation. *Conservation Biology*, *37*(1), e13973. https://doi.org/10.1111/cobi.13973

Gereffi, G. (1996). Global Commodity Chains: New Forms of Coordination and Control Among Nations and Firms in International Industries. *Competition & Change*, *1*, 4.

Gerschenkron, A. (1962). *Economic Backwardness in Historical Perspective*. The Bellknap Press.

Ghatak, S. (2010). *Introduction to development economics*.

Handa, S., Natali, L., Seidenfeld, D., Tembo, G., & Davis, B.Zambia Cash Transfer Evaluation Study Team. (2018). Can unconditional cash transfers raise long-term living standards? Evidence from Zambia. *Journal of Development Economics*, *133*, 42–65. DOI: 10.1016/j.jdeveco.2018.01.008 PMID: 31396000

Haq, M. u. (1996). *Reflections on Human Development*. Oxford University Press.

Harrigan, J., & Wang, C. (2011). A new approach to the allocation of aid among developing countries: Is the USA different from the rest? *World Development*, *39*(8), 1281–1293. DOI: 10.1016/j.worlddev.2010.12.011

He, , QMeng, , QFlatley, , WHe, , Y. (2022). Examining the effects of agricultural aid on forests in sub-Saharan Africa: A causal analysis based on remotely sensed data of Sierra Leone. *Land, 11*(5), 668. https://doi.org/10.3390/land11050668

Herzer, D., & Nunnenkamp, P. (2012). The effect of foreign aid on income inequality: Evidence from panel cointegration. *Structural Change and Economic Dynamics, 23*(3), 245–255. DOI: 10.1016/j.strueco.2012.04.002

Hickel, J. (2017). Is global inequality getting better or worse? A critique of the World Bank's convergence narrative. *Third World Quarterly, 38*(10), 2208–2222. DOI: 10.1080/01436597.2017.1333414

Houée, P. (2001). *Le développement local au défi de la mondialisation*. L'Harmattan.

Jayachandran, , SDe Laat, , JLambin, , E. FStanton, , C. YAudy, , RThomas, , N. E. (2017). Cash for carbon: A randomized trial of payments for ecosystem services to reduce deforestation. *Science, 357*(6348), 267-273. https://doi.org/10.1126/science .aan0568

Kasuga, H., & Morita, Y. (2018). Does aid affect inequality? *Applied Economics, 50*(58), 6249–6262. DOI: 10.1080/00036846.2018.1489512

Kaul, I., Grunberg, I., & Stern, M. (1999). Global public goods. *Global public goods, 450*.

Kim, , J. E. (2018). Technological capacity building through energy aid: Empirical evidence from renewable energy sector. *Energy Policy, 122*, 449-458. https://doi .org/10.1016/j.enpol.2018.07.003

Kono, , D. YMontinola, , G. R. (2019). Foreign aid and climate change policy: what can('t) the data tell us? *Politics and Governance, 7*(2), 68-92. https://doi.org/ 10.17645/pag.v7i2.1840

Korotayev, A., Zinkina, J., Bogevolnov, J., & Malkov, A. (2011). Global unconditional convergence among larger economies after 1998? *Journal of Globalization Studies, 2*(2), 25–62.

Lee, , S. KChoi, , GLee, , EJin, , T. (2020). The impact of official development assistance on the economic growth and carbon dioxide mitigation for the recipient countries. *Environmental Science and Pollution Research, 27*(33), 41776-41786. https://doi.org/10.1007/s11356-020-10138-y

Li, , D. DRishi, , MBae, , J. H. (2021). Green official development Aid and carbon emissions: do institutions matter? *Environment and Development Economics, 26*(1), 88-107. https://doi.org/10.1017/S1355770X20000170

Li, , RGuo, , JWang, , Q. (2024). Evaluating the impact of official development assistance on ecological environments in agricultural and renewable energy sectors. *Humanities and Social Sciences Communications*, *11*(1), 1-11. https://doi.org/10.1057/s41599-024-03979-2

Mankiw, N., Romer, D., & Weil, D. (1992). A Contribution to the Empirics of Economic Growth. *The Quarterly Journal of Economics*, *107*(2).

McCarthy, , CSternberg, , TBanda, , L. B. (2023). Does environmental aid make a difference? Analyzing its impact in developing countries. *Land*, *12*(10), 1953. https://doi.org/10.3390/land12101953

McMichael, P. (2000). *Development and Social Change: A Global Perspective* (2nd ed.). Pine Forge Press.

Meadows, D. H., Meadows, D. L., Randers, J., & Behrens, W. W. (1972). *The Limits to Growth*. Universe Books.

Molyneux, M. (1979). Beyond the Domestic Labour Debate. *New Left Review*, •••, 115.

Moreira, S. (2005). Evaluating the Impact of Foreign Aid on Economic Growth: A Cross-Country Study. *Journal of Economic Development*, *30*(2). https://jed.cau.ac.kr/archives/30-2.html

Moreira, S. B. (2013). *Sobre a Medição do Desenvolvimento – Indicadores Desagregados e Compósitos com uma Aplicação Empírica a Portugal, 2013*. Escrytos | Edição de autor.

Moser, C. (1989). Gender Planning and Development: Meeting Practical and Strategic Needs. *World Development*, *17*(11), 1799–1825.

Myrdal, G. (1957). *Economic Theory and Under-developed Regions*. Duckworth.

Neij, , LHeiskanen, , EStrupeit, , L. (2017). The deployment of new energy technologies and the need for local learning. *Energy Policy*, *101*, 274-283. https://doi.org/10.1016/j.enpol.2016.11.029

North, D. (1991). Institutions. *The Journal of Economic Perspectives*, *5*(1).

North, D. C. (1955). Location Theory and Regional Economic Growth. *Journal of Political Economy*, •••, 63.

OECD. (2018). *Converged statistical reporting directives for the Creditor Reporting System (CRS) and the annual DAC questionnaire – Chapters 1-6*. DAC. Paris: OECD.

Payne, A., & Phillips, N. (2010). *Development*. Polity Press.

Pecqueur, B. (2000). *Le développement local.*

Polanyi, K. (1944). *The Great Transformation: The Political and Economic Origins of Our Time*. Beacon Press.

Popp, , D. (2011). International technology transfer, climate change, and the clean development mechanism. *Review of Environmental Economics and Policy 5*(1).

Prebisch, R. (1950). *The Economic Development of Latin America and its Principal Problems*. United Nations.

Rajan, R. G., & Subramanian, A. (2011). Aid, Dutch disease, and manufacturing growth. *Journal of Development Economics, 94*(1), 106–118. DOI: 10.1016/j.jdeveco.2009.12.004

Riddell, R. C. (1987). *Foreign Aid Reconsidered.* James Currey / Overseas Development Institute.

Robinson, W. (2002). Remapping Development in Light of Globalisation: From a Territorial to a Social Cartography. *Third World Quarterly, 23*, 6.

Romer, P. (1986). Increasing Returns and Long-Run Growth. *Journal of Political Economy, 94*(5).

Rostow, W. W. (1960). *The Stages of Economic Growth: A Non-Communist Manifesto.* Cambridge University Press.

Saad-Filho, A. (2007). Life beyond the Washington Consensus: An Introduction to Pro-poor Macroeconomic Policies. *Review of Political Economy, 19*(4), 513–537. DOI: 10.1080/09538250701622352

Sachs, J. (2005). *The end of poverty*. Penguin Books.

Scott, M. (1989). *A New View of Economic Growth.* Clarendon Press.

Seers, D. (1969). The Meaning of Development. *International Development Review, 11*, 2–6.

Semenchuk, , PPlutzar, , CKastner, , T. et al. (2022). Relative effects of land conversion and land-use intensity on terrestrial vertebrate diversity. *Nature Communications, 13*(1), 615. https://doi.org/10.1038/s41467-022-28245-4

Sen, A. K. (1981). *Poverty and Famines: An Essay on Entitlement and Deprivation.* Oxford University Press.

Singer, H. (1950). The Distribution of Gains between Investing and Borrowing Countries. *The American Economic Review, 40*, 473–485.

Stern, , D. I. (2004). The rise and fall of the environmental Kuznets curve. *World Development, 32*(8), 1419-1439. https://doi.org/10.1016/j.worlddev.2004.03.004

Temple, J. (2010). Aid and Conditionality. In Rodrik, D., & Rosenzweig, M. R. (Eds.), *Handbook of Development Economics* (Vol. 5, pp. 4415–4523). North-Holland.

Thorbecke, E. (2000). The evolution of the development doctrine and the role of foreign aid, 1950–2000. In Tarp, F. (Ed.), *Foreign Aid and Development - Lessons Learnt and Directions for the Future* (pp. 12–35). Routledge.

Udeogu, E., Amakom, U., & Roy-Mukherjee, S. (2021). Empirical Analysis of an Augmented Schumpeterian Endogenous Growth Model. *Journal of Economic Development, 46*(1). https://jed.cau.ac.kr/archives/46-1.html

Vachon, B. (1998). *Le développement local - Théorie et pratique - Réintroduire l'humain dans la logique de développement.* Gaetan Morin Editeur.

Wade, R. (2004). Is globalization reducing poverty and inequality? *World Development, 32*(4), 567–569. PMID: 15346677

Wang, , QGuo, , JDong, , Z. (2021). The positive impact of official development assistance (ODA) on renewable energy development: evidence from 34 Sub-Saharan Africa Countries. *Sustainable Production and Consumption, 28*, 532-542. https://doi.org/10.1016/j.spc.2021.06.007

Weko, , SGoldthau, , A. (2022). Bridging the low-carbon technology gap? Assessing energy initiatives for the Global South. *Energy Policy, 169*, 113192. https://doi.org/10.1016/j.enpol.2022.113192

Williamson, J. (1990). What Washington means by policy reform. In *Latin American adjustment: How much has happened* (pp. 7-20).

Younsi, M., Khemili, H., & Bechtini, M. (2019). Does foreign aid help alleviate income inequality? New evidence from African countries. *International Journal of Social Economics, 46*(4), 549–561. DOI: 10.1108/IJSE-06-2018-0319

KEY TERMS AND DEFINITIONS

Big push: Economic approach suggesting that the economies of developing countries require an up-front, large-scale and coordinated investment effort across multiple industries to overcome development barriers.

Dual gap: Economic approach that analyses the reasons for the need for foreign capital and which has become the conventional theoretical justification for the provision of development aid, postulating that such aid should cover the larger of the two gaps (savings/investment or exports/imports) if a given growth target is to be achieved.

Exogenous growth: Economic theory which posits that long-term economic growth is primarily determined by factors that originate outside the economic system itself and are not explained or determined by the internal variables of the model.

Foreign aid: A generic term frequently used to refer broadly to Official Development Assistance (ODA).

Official Development Assistance (ODA): Grants and loans to countries and territories on the OECD Development Assistance Committee list of ODA Recipients, international non-governmental organisations and multilateral development institutions which are: provided by official agencies, including state and local governments, or by their executive agencies; and, each transaction of which is administered with the promotion of the economic development and welfare of developing countries as its main objective; and, is concessional in character.

Poverty trap: Situation that occurs in some developing countries where the low availability of capital requires higher levels of savings, but this may not be possible in cases of extreme poverty where all income must be spent on basic consumption needs.

Steady state: A theoretical condition in which key economic variables remain constant over time or grow at a constant rate, representing a long-run equilibrium to which the economy converges.

Total Official Support for Sustainable Development (TOSSD): An international standard for measuring the full range of resources used to promote the sustainable development of developing countries, tracking all official resources flowing to developing countries for their sustainable development, as well as private resources mobilised through official means, and measuring contributions to international public goods.

Chapter 9
Natural Disaster Responses and Disaster Study in Indonesia:
An Exploratory Study Using Meta-Analysis

Mohammad Jafar Loilatu
https://orcid.org/0000-0001-6921-6879
Universitas Muhammadiyah Malang, Indonesia

Saiman Saiman
Universitas Muhammadiyah Malang, Indonesia

Bambang Irawan
https://orcid.org/0000-0002-4841-7302
Mulawarman University, Indonesia

Umi Kulsum
Universitas Muhammadiyah Malang, Indonesia

Muhammad Nurul Huda
https://orcid.org/0000-0003-4184-2530
Universitas Muhammadiyah Karanganyar, Indonesia

Dwi Putri Vidiastuti
Universitas Muhammadiyah Malang, Indonesia

ABSTRACT

This research aims to review the natural disaster response in Indonesia. This paper states that Indonesia is one of the countries with natural disasters in the world, but it lacks references and analysis of how it responds to natural disasters in Indonesia. The research uses a meta-analysis approach through CiteSpace analysis; we use scopus database which has been selected through inclusion and exclusion criteria to find appropriate keywords through this procedure. Meta-analysis shows that natural disaster response studies in Indonesia are predominantly published at conferences, with contributor universities dominated by ten universities in Indonesia. The find-

DOI: 10.4018/979-8-3373-1862-2.ch009

ings highlight three critical topics in natural disaster response studies: disaster management, geospatial analysis, and post-traumatic response. The limitation of this research is that it only selected one country with the highest natural disaster vulnerability. This research provides a comprehensive analysis by mapping a list of appropriate publications based on a structured analysis.

INTRODUCTION

Indonesia is situated at the intersection of the ring of fire and four active tectonic plates, including the Eurasian, Indo-Australian, Indian Ocean, and Pacific Ocean. Indonesia ranks as one of the countries with the most natural disasters worldwide (Riyanti Djalante & Garschagen, 2017). Natural disasters cause normal conditions to become abnormal in areas where conditions become unbalanced; when this occurs, the government must take rapid disaster management action to address the many problems within the affected area.

Natural disasters cause chaos, and these conditions become a challenge for the government to make policies with the authority of each regional government (Hermansson, 2019). According to Ahmed (2013), institutions needed to make appropriate policies in crisis conditions; complex policy-making mechanisms and the authority of each actor cause this. Meanwhile, the policies aim to create community resilience (Berke & Campanella, 2006), and the government must pay attention to agenda setting and policy after the disaster (Birkland, 2006). The post-disaster policy is a stimulus to overcome the impact of disasters through a balanced policy coordination process between cross-institutions (Smith & Dowell, 2000).

Several studies have described the government's response to natural disasters in Indonesia (R. Djalante, 2012; Riyanti Djalante, 2018; Riyanti Djalante & Garschagen, 2017). After the disaster in Indonesia, the government redesigned the framework for handling natural disasters by involving multi-stakeholders to overcome abnormal conditions and potential disasters at any time (Riyanti Djalante, Holley, Thomalla, & Carnegie, 2013). In the view of Kusumasari et al. (2012), the state must be able to resolve complex problems because natural disasters impact natural conditions and humans. Therefore, this capacity identifies each situation due to natural disasters (Kusumasari, Alam, & Siddiqui, 2010); according to him, this capacity can be demonstrated through disaster management steps such as logistics management, information management, and condition assessment.

Disaster response studies have been promoted with different views; this shows that these studies have varied approach options based on the character of the disaster and region. Scholars have identified disaster response as reduction and disaster management (Ahrens & Rudolph, 2006; Izumi, Shaw, Djalante, Ishiwatari, &

Komino, 2019; Mosby, Birch, Moles, & Cherry, 2009; Uddin, Haque, & Khan, 2021), disaster policy (Birkland, 2006; Crow, Albright, Ely, Koebele, & Lawhon, 2018; Donahue, 2010; Gopalakrishnan, 2013; Healy & Malhotra, 2009; van Niekerk, 2014), collaborative governance during disaster (Hermansson, 2019; Kapucu & Garayev, 2011; Raungratanaamporn, Pakdeeburee, Kamiko, & Denpaiboon, 2014) or adaptive governance in dealing with disasters (R. Djalante, 2012; Riyanti Djalante, Holley, & Thomalla, 2011; Riyanti Djalante et al., 2013), or communication using social media (Kurniawan, Sutan, Nurmandi, Loilatu, & Salahudin, 2021; Sukmono, Junaedi, Fadilla, Loilatu, & Fadillah, 2023). However, from several recent studies, no research topic has reviewed the forms of disaster response that occurred in Indonesia from specific issues regarding disaster response.

Disaster response discourse focuses on the challenges faced in overcoming impacts quickly through appropriate agenda-setting (Donahue, 2010; Birkland, 2006). However, the effects of disasters can be resolved if the main disaster response instruments and institutions play a role in overcoming natural disasters. According to Bang (2014), the government can identify the main impacts and obstacles in natural disaster conditions. Alexander et al. (1995) state that institutional coordination does not have an effect if there is no institutional capacity because multiple institutions are needed as a disaster management option to overcome natural disaster problems (Smith & Dowell, 2000).

Based on this description, we consider that there are many approaches to responding to natural disasters in Indonesia. Therefore, to understand the form of disaster response, this research aims to carry out a meta-analysis of disaster research trends in Indonesia, focusing on disaster response. This topic is quite broad; therefore, we specifically focused on disaster response in Indonesia to determine the subject and keywords in our search. This research contributes to further topics regarding disaster response in Indonesia, mapping the gaps that Indonesian scholars have not clearly explained in disaster management, and designing more exciting research to look at natural disasters in Indonesia more broadly.

Methodology

We start with a scoping approach (Hacking, 2012), which fills in the strengths of several studies considered closed for further exploration. Mostly bibliometric analysis using (Arksey & O'Malley, 2005) to review, this approach was deemed suitable for answering the research objectives and questions; apart from that, it was also considered as appropriate as an approach for mapping the concepts and themes in the research (Daudt, Van Mossel, & Scott, 2013). According to Donthu et al. (2021), the scoping function in a review article helps to analyze and map many concepts through clear research procedures; the approach functions to map the number of varied articles with the same or different topics.

We use CiteSpace as a tool of meta-analysis to summarize empirical evidence of natural disaster response in Indonesia with a homogeneous field. This research is considered more appropriate to use meta-analysis in analyzing natural disaster studies in Indonesia because it has identified these studies based explicitly on the research stage. CiteSpace, as an analysis tool, functions to map specific articles, primary research areas, and the critical transition of theoretical history (C Chen, 2016). Furthermore, we used the Scopus database to analyze disaster studies in Indonesia. This process is called a research strategy to select detailed articles from large topics and become more concise. This article specifies numerous CiteSpace analysis features based on the usage guide, such as burst term, cluster view, colour map, citation, and cluster view (Chaomei Chen, 2005); the Scopus database is the primary data source for this meta-analysis (Burnham, 2006).

Using the following keywords, the Scopus database could locate 1,152 publications. See Figure 1, the search strategy is Search TITLE TITLE ("Natural Disaster" OR "Indonesian Disaster" OR "Natural Disaster Indonesia") AND (LIMIT-TO (AFFILCOUNTRY, "Indonesia")) AND (LIMIT-TO (LANGUAGE, "English")) AND (LIMIT-TO (PUBSTAGE, "final")) AND (LIMIT-TO (DOCTYPE, "cp") OR LIMIT-TO (DOCTYPE, "ar") OR LIMIT-TO (DOCTYPE, "ch") OR LIMIT-TO (DOCTYPE, "re") OR LIMIT-TO (DOCTYPE, "le")) AND (LIMIT-TO (PUBYEAR, 2005) AND LIMIT-TO (PUBYEAR, 2022)). The articles were presented as RIS and CSV (Comma-separated values) in the Scopus database to aid cluster analysis and citations.

Figure 1. Article selection process based on the Scopus database

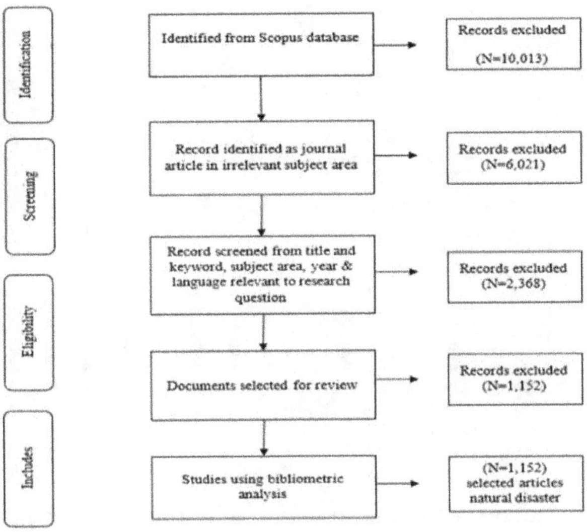

Findings

Disaster studies in Indonesia in the last 18 years have tended to be low; from 2005-2017, annual publications did not reach 100. In 2018, the number of publications increased to 116, which rose slowly in 2019 to 195; the journal with the highest number was 258 in 2021. This identification shows that even though Indonesia's number of natural disasters reaches 3,397 yearly, very few disaster studies are still given serious attention. The number of natural disasters in Indonesia has increased annually; in 2010, the number of natural disasters was 1,945. In 2016, it became 2,306 and increased drastically to 5,402 in 2021. Indonesia's number of natural disasters is not comparable to the number of publications. Figure 2, publications by year by Indonesian scholars regarding national disaster response in the last 18 years, is relatively low, and the increase is prolonged. In 2017, the number of publications on disaster response only reached 64; in 2018 (116), it increased in 2021 to 258, but the number of publications is not comparable to the natural disasters that occurred in Indonesia during that time.

Figure 2. Publications by year

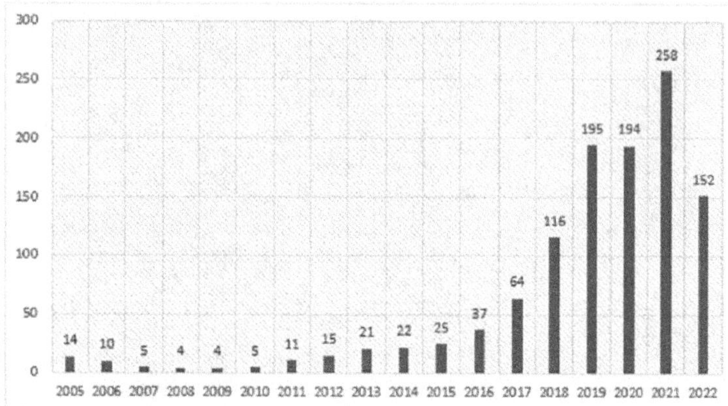

See Figure 3. Studies on natural disasters in Indonesia in the last 18 years have been published more often in conference papers, with several articles (670), while articles published through journal articles reached (437), such as book chapters (23) and reviews (22). and letters (2). This type of publication shows that Indonesian scholars prefer conference papers to publish articles of lower quality than article papers. This quality can be seen from the review process for these two types of articles; articles go through a more prolonged stage in the review process, while conference papers go through a shorter review process and can be accepted quickly. Likewise, Table 1 shows that the number of publications on natural disaster studies in Indonesia is primarily published in conference proceedings, IOP Conference Series Materials Science and Engineering (35), and only the International Journal of Disaster Risk Reduction (21) as a journal that publishes natural disaster studies. From a total of 99 publishers of natural disaster studies in Indonesia, we selected 31 publishers based on the number of publications. Table 1 shows that of all disaster study publications in Indonesia, 58% were published at conferences, and 43% were published in journals.

Figure 3. Document by types

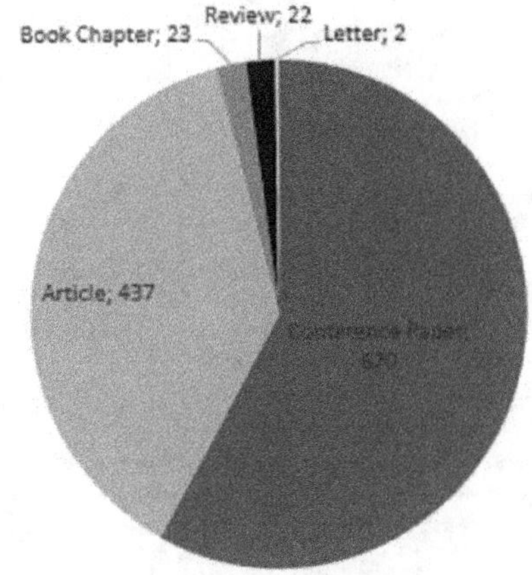

Based on the document type in Figure 3, see the publication sources in Figure 4. Figure 4 shows that conferences publish many publications on the disaster response theme in Indonesia; this correlates with Figure 3, where the number of conference papers is 670 more than articles, which are only 437. The top 10 publications in the table above include the IOP Conference Series Earth and Environmental Science, A.I.P. Conference Proceedings, E3S Web of Conferences, and Materials Science and Engineering. And articles published by journals such as the Journal of Physics Conference Series, International Journal of Disaster Risk Reduction, Sustainability Switzerland, and Journal of Disaster Research. See the table to sort publication sources on disaster response.

Figure 4. Publication sources

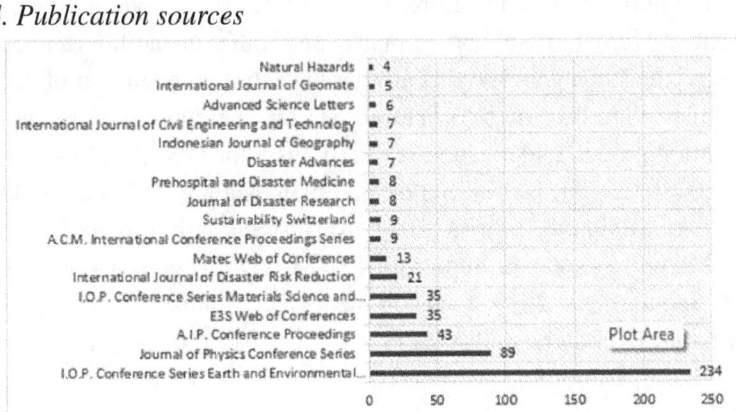

Figure 5. Publication by affiliation

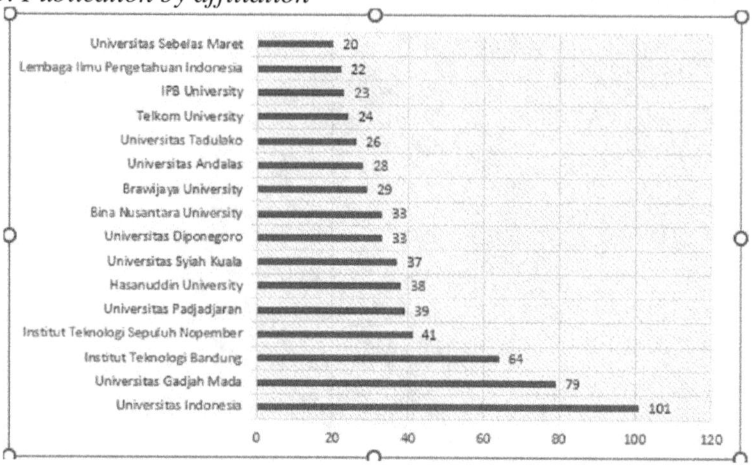

How universities in Indonesia pay serious attention to studying natural disasters can be seen from the number of publications by the institution. From the publication trend for 2005-2022, the university with the most studies on natural disasters is the University of Indonesia with the number of publications (101), followed by Gadjah Mada University (79), Institute Technology Bandung (64), Sepeluh November Institute of Technology (41), Padjajdaran University (39), Hasanudin University (38), Syiah Kuala University (37), Diponegoro University and Bina Nusantara University (33), Brawijaya University (29), Andalas University (28), Tadulako University (26), Telkom University (24), IPB University (23), LIPI (22) and Sebelas Maret University (20).

Figure 6 shows the authors with the most publications on natural disaster studies in Indonesia. Of all the authors who wrote about natural disasters in Indonesia, 16 had the highest number of publications. Of the total authors, four authors published the same article; Mamat M, Sukono, Supian S, and Kalfin are authors and co-authors who published the same paper so that all four have the same number of documents, all four published studies, such as natural disaster insurance, mitigation management, mitigation model, and evaluation of risk insurance (Kalfin, Sukono, Supian, & Mamat, 2022b, 2022a; Kalfin, Sukono, Supian, Mamat, & Bon, 2020b, 2020a; Kalfin, Sukono, Supian, & Mamat, 2021; Kalfin, Sukono, Supian, Mamat, et al., 2021; Sidi, Mamat, Sukono, & Supian, 2017). Meanwhile, authors like (Hartawan, Purboyo, & Setianingsih, 2019) focus on media and tool users as an early warning system for natural disasters, and (Ong et al., 2021) examine preparing for disaster mitigation. Of the 16 authors, on average, they wrote articles as co-authors, so there was no single author on the study of natural disasters. This number is relatively small compared to scholars from other countries who pay serious attention to studying natural disasters.

Figure 6. Publication by authors

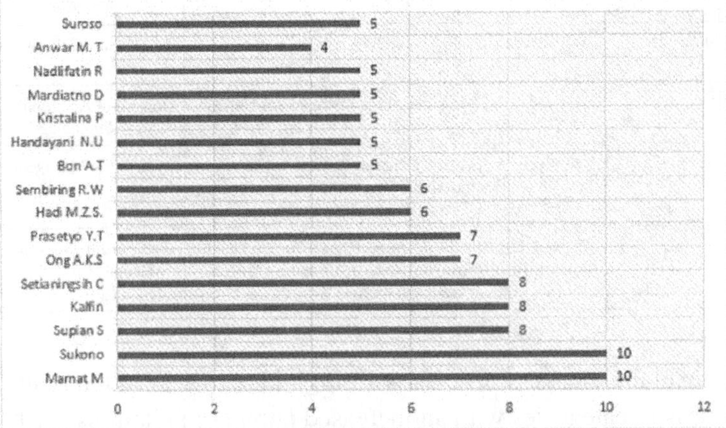

Publication with high citation

Table 1, publisher, and article with high citation; the table explains that disaster management studies in Indonesia were published in several international journals with high citations. Of the ten articles with the most citations, these articles examine disaster management from different perspectives.

Table 1. Articles with high citations

Authors	Title	Sources	Cite
Morgan et al. (2006)	Mass fatality management following the South Asian tsunami disaster: Case studies in Thailand, Indonesia, and Sri Lanka	PLoS Medicine	124
Kusumastuti et al. (2014)	Developing a resilience index towards natural disasters in Indonesia	International Journal of Disaster Risk Reduction	106
Rofi et al. (2006)	Tsunami mortality and displacement in Aceh province, Indonesia	Disasters	80
Tacconi et al. (2007)	Fires in tropical forests - What is the problem? Lessons from Indonesia	Mitigation and Adaptation Strategies for Global Change	75
Gupta et al. (2007)	Factors associated with E.coli contamination of household drinking water among tsunami and earthquake survivors, Indonesia	American Journal of Tropical Medicine and Hygiene	68
R Djalante et al. (2012)	Building resilience to natural hazards in Indonesia: Progress and challenges in implementing the Hyogo Framework for Action	Natural Hazards	66

continued on following page

Table 1. Continued

Authors	Title	Sources	Cite
Doocy et al. (2007)	Tsunami mortality in Aceh Province, Indonesia	Bulletin of the World Health Organization	63
Juth et al. (2015)	Post-disaster Mental Health Among Parent-Child Dyads After a Major Earthquake in Indonesia	Journal of Abnormal Child Psychology	50
Rittichainuwat et al. (2018)	Applying the perceived probability of risk and bias toward optimism: Implications for travel decisions in the face of natural disasters	Tourism Management	49
Epping-Jordan et al. (2015)	Beyond the crisis: Building back better mental health care in 10 emergency-affected areas using a longer-term perspective	International Journal of Mental Health Systems	41

Table 1 explains relevant articles with high citations in Indonesian disasters. There are disaster categories with an increased number of citations, such as tsunamis (Morgan et al., 2006) and (Rofi, Doocy, & Robinson, 2006), assessing natural disasters (Kusumastuti, Viverita, Husodo, Suardi, & Danarsari, 2014), fires (Tacconi, Moore, & Kaimowitz, 2007), Resilience (R Djalante, Thomalla, Sinapoy, & Carnegie, 2012), and some of the authors mention the field study of mental health after a disaster (Juth, Silver, Seyle, Widyatmoko, & Tan, 2015). Most cited articles do not correlate with the number of publications; see Figure 5. Thus, authors have high publications, yet at least high citations. Authors with many publications have significance and correlation with disaster studies in Indonesia. From the top article with the highest citation, there are three topics with citations: tsunami, disaster response management, and disaster prediction.

Unit analysis using references by keywords

Figure 7. Meta-analysis using CiteSpace shows that Indonesia's research trend on disaster response and natural disasters consists of 14 clusters, with members in the relevant clusters' order. Cluster #0 uses the geospatial approach with (119) members, citing the highest article (Kalfin, Sukono, Supian, & Mamat, 2021) regarding natural disaster mitigation. Cluster #1 post-traumatic growth (103) members, citing an article on this topic (Naghavi, Wang, & Lozano, 2015) regarding mortality. Cluster #2 learning model with (83) members, citing an article (Maryati, 2018) regarding identifying flood-prone areas. Cluster #3, soil salinity, has (73) members with the highest citations on the topic (McLeod et al., 2010) about the soil salinity impact of a tsunami. Cluster #4, novice volunteer, has (47) members, with citing articles (Wahyudin, Hasegawa, & Dahlan, 2013). Cluster #5 Marduka District has (43) members, citing an article (Hartama et al., 2017) about innovative urban governance in mitigating natural disasters. Cluster #6 psychosocial responses have members (41) citing articles (Robinson, Alatas, Robertson, & Steer, 2011) about health care in disasters. Cluster #7 volcanic eruption has members (28) and cites an article (Setiadi, Andriessen, & Anisa, 2020) about community-based relocation in post-disaster. According to an article, cluster #9 disaster management has (16) members (Maryati, 2018). Cluster #10 tsunami mortality has members (15) citing the article (Doocy et al., 2007) on tsunami mortality. Cluster #11 mental health model has members (15), citing an article in this cluster (Prasetiyawan, Viora, Maramis, & Keliat, 2006). Cluster #12, Luzon, Philippines, has members (13) and cited an article in this cluster (Suhartono, Fatmawati, & Suranto, 2020) about the non-natural disaster of COVID-19. Cluster #16 Disaster area with (16) members and citing a high article (Amni, Hadi, & Perdinan, 2020).

Figure 7. Node types references by keywords

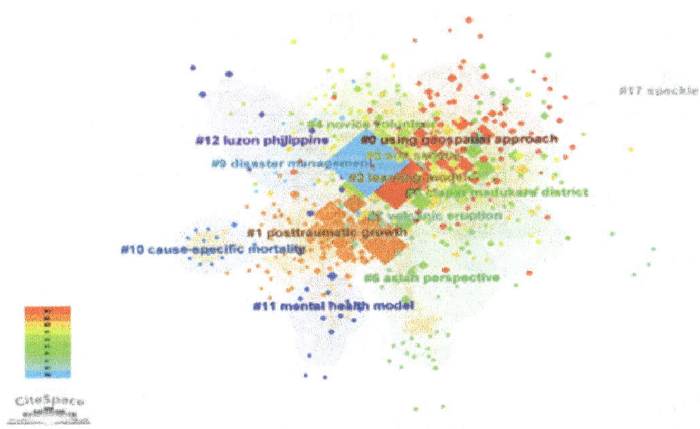

Table 2. List of keywords in disaster response by years

2006	Disaster	Natural disaster	Indonesia	Disaster management	Disaster	Tsunami	Adult	Child	Aceh
2007	Disaster prevention	Tsunami	Earthquake	Deforestation	Aceh				
2008	Emotional stress	Anxiety							
2009	Disaster relief	Health services							
2010	Geographic information system	G.I.S.	Early warning system	Flood-prone area	Electric conductivity	Data compilation			
2011	Climate change	Volcano eruption	Coastal zone	Climate effect					

continued on following page

Table 2. Continued

2006	Disaster	Natural disaster	Indonesia	Disaster management	Disaster	Tsunami	Adult	Child	Aceh
2012	Risk assessment	Disaster risk reduction	Coastal area	Adaptive management	Community vulnerability				
2013	Earthquake	Landslides	Remote sensing	Emergency services	Vulnerability	Decision making	Disaster response	Central Java	Flood
2014	Land use	Female	Flood	Cholera					
2015	Information management	Analytic hierarchy	Young adult						
2016	Floods	Disaster management	Volcanoes	Landslide	Forestry	Adaptation strategy	Alarm strategy		
2017	Hazards	Information use	Information system	Mitigation	Forecasting	Economic effect	Social impact	Disaster risk reduction	Local government
2018	Students	Sustainable development	Disaster preparedness	Natural disaster mitigation	Rural areas	Resilience	Environmental management	Risk management	Watershed
2019	Social media	Emergency respond	Fires	Learning system	Human activity	Artificial intelligence			
2020	Covid-19	Government	Earthquake	Tsunami	Tourism	Volcanic eruption			
2021	Insurance	Sustainability	Coronavirus	Budget control	Insurance company	Community resilience			
2022	Pandemic	Forest fires	Covid-19	Incidence	Decision making	Covid-19 pandemic			

Sources: Analyze by author

Table 3. List of selected articles by years

Authors	Aim	Method	Result	Respond
Irmansyah (2005)	Explore disaster victims in post-disaster	Qualitative method	The government directly responds to a disaster	The government has to trauma healing for disaster victim
Gunnarsson et al. (2005)	How civil-military collaboration	Qualitative analysis with panel discussion	Practical civil-military cooperation can help health responses	Civil-military assets, human needs, government capacity, and coordination.
Rofi et al. 2006 (2006)	Explore households displace	Survey	36% of tsunami-displaced households indicated an intention to return to their original community	The response should facilitate the production of permanent

continued on following page

Table 3. Continued

Authors	Aim	Method	Result	Respond
Morgan et al. (2006)	Identification of the mass vitality of the disaster	Semi-structured interview	Emergency response should not add to the distress of affected communities by inappropriately disposing of the victims.	Mass fatality management must be informed by additional field research and supported by a network of regional forensic institutions.
Tacconi et al. (2007)	Addressing them should be on the impacts caused by fires and land use changes	GIS and qualitative analysis	Environmental conditions, livelihood, financial and economic interests, lousy governance, lack of knowledge, and accidental ignitions cause the fire problem.	To address the problem, the government should reduce smoke and haze pollution and improve logging practices and governance.
Atmadilaga et al.(2007)	Multihazard database development for supporting disaster management	Geospatial analysis	Geospatial analysis shows that integrated spatial information is used as a prevention action	The geospatial multihazard database highlighted more about disaster-prone zones in Java and Sumatra islands.
Sundram et al. (2008)	Psychosocial intervention efforts when confronted with large-scale natural catastrophic events	Psychosocial disaster intervention	Asian natural disasters may be constructively used to plan strategies to respond appropriately to the psychosocial consequences of disaster	Assessment of help-seeking pathways, Psychosocial training of community, rain medical and health staff, and minimize risk
Sutiono et al. (2009)	Examine the sociodemographic and geographic variables that may impact the result of tetanus treatment in the context of earthquakes in Yogyakarta.	Use Mapinfo, a professional geospatial tool	Triage mechanisms based on distance and type of hospital priority for internally displaced persons, as well as measures for the general elderly population, could be developed to prevent a tetanus outbreak	Priority hospitalization for internally displaced people and making arrangements for the general senior population

continued on following page

Table 3. Continued

Authors	Aim	Method	Result	Respond
Rokhis Khomarudin et al. (2010)	To analyze the warning system in Indonesia during the disaster	Multi-scenario approach	The danger and exposure maps are supplied for the coastline region and the study locations Padang, Cilacap, and Kuta in greater detail	The decision support system of the Indonesian Tsunami Early Warning Centre incorporates significant factors from this investigation
Sugiyanto & Resosudarmo, (2011)	To understand the factors of livelihood recovery following a natural catastrophe	Systematic review	The results show that smaller businesses are more robust and recover faster, and an industrial cluster structure inside a sub-district gives firms the essential assistance to recover	A sub-district industrial cluster structure provides essential assistance for enterprises to recover
Fahrudin (2012)	To investigate the association among psychosocial response, coping behavior, and trauma among survivors of natural disasters in West Java	Statistical analysis	Coping behavior significantly impacts the association between psychosocial response and traumatic stress as a moderating variable.	In dealing with natural catastrophes, health workers must get traumatic healing and
Hidayati, (2012)	Assess preparedness level in risk region	Statistical analysis	Of the ten provinces, only Padang is ready to face natural disasters	Increase public awareness, providing understanding and handling of potential natural disasters
Wardani & Muntohar (2013)	Explore potential natural disasters in Indonesia	Spatial analysis	The early warning system was inadequate to prevent landslides	To construct an integrated early warning system in Indonesia
Wahyudin et al. (2013)	This research aims to encourage mobile game-based learning to strengthen ethical decision-making capacities	Deep interview and qualitative analysis	The volunteers always build their ethical decision-making skills through official disaster management training	Created the magnitude training system and enhanced it in disaster response

continued on following page

Table 3. Continued

Authors	Aim	Method	Result	Respond
Brata et al. (2014)	Investigate the effects of these shocks on the population distribution in Northern Sumatra	Statistical analysis	In terms of population size, there have been no significant changes in the relative position of areas in Aceh and North Sumatra during the earthquake	
Kusumastuti et al. (2014)	The article aims to develop a framework to assess the resilience of disaster-prone areas	Indexing analysis	The finding assessing the resilience of disaster-prone areas in Indonesia to natural catastrophes by creating an index	
Andriansyah, (2015)	Analyses the economic recovery in Yogyakarta after a calamity.	Macroeconomic analysis using statistical data by Indonesian Bank	This empirical result suggests that government failure in post-disaster recovery can be reduced by enhancing the community and bank central	The findings imply that central bank policies and government assistance played roles in recovery
Maryani & Yani (2016)	Aims to determine the superiority of local knowledge in the Sundanese community, particularly in Kampung Naga.	Qualitative descriptive	Sundanese civilization possesses indigenous wisdom that is particularly useful in catastrophe mitigation.	Applied local wisdom to promote local tourism to international tourism
Puspita (2017)	Aim to analyze the conflict in the natural disaster area	Qualitative	The article shows how ASEAN countries cope with conflict in natural disasters in which armed power.	The doctrine of responsibility to protect the community during a natural disaster
Hardi et al. (2018)	To gather information from students' problem-solving abilities in emergency flood response in their region.	Quantitative	Aspects of attitude, they know what to do and how the flood disaster occurs.	The student's parents have to be educated to understand the importance of disaster preparedness

continued on following page

Table 3. Continued

Authors	Aim	Method	Result	Respond
Budirahayu et al. (2019)	To explain women's capacities in the aftermath of the disaster, as well as their attempts to address economic and social challenges within the households	Qualitative approach by interview	The findings of this study demonstrated that women's resilience in the face of calamities has been verified, as seen by their abilities to aid one another and care for their families and the environment.	Collective action may be used to treat trauma and increase women's capacity for resilience in the face of disasters.
Smoro Laksmi et al. (2020)	To propose disaster preparedness, including victim evacuation, without increasing the spread of Covid-19	Qualitative review	Pekalongan has precisely implemented preventative measures to curb the development of Covid-19.	The government should boost its flood readiness during COVID-19 and pay attention to COVID-19
Rahiem et al. (2021)	Examined the perspectives and views of child survivors, who were frequently the last persons to see their moms alive	Qualitative analysis and narrative study	Many women died in natural disasters since they waited for their husbands to return to obey and honor their husbands; they prioritized the safety of their other family members over their own; they were more likely to pray rather than leave.	The culture background of Aceh helps to solve the survivors' experiences

continued on following page

Table 3. Continued

Authors	Aim	Method	Result	Respond
Supriadi et al. (2022)	To compare the effects of the COVID-19 epidemic and natural disasters on the Hajj and Umrah businesses	Semistructured interviews with travel agents were done	COVID-19 has an impact, including a prohibition on hajj and umrah, strict personal health laws and quarantine, a weakening of the Indonesian economy, a lack of consumer interest, a loss of revenue, employment termination	The tourism sector requires disaster risk management, and authorities must make massive attempts at educating entrepreneurs in this sector
Fitriadi et al. (2022)	The goal of this study is to look into the influence of COVID-19 on socioeconomic elements in Indonesia between 2020 and 2022	Statistical analysis	When COVID-19 increases caused unemployment, poverty, and labour productivity decline. COVID-19 has improved mortality, domestic violence, sexual harassment, per capita expenditure, and overall well-being	Establishing a semi-lockdown is a priority to maintain social and economic arrangements while disregarding the latent threats of COVID-19

Sources: Analyze by author

Table 2 shows high keywords by year. We underline that the study of natural disasters in Indonesia is influenced by natural conditions and phenomena occurring in a particular year. Also shows the shifting trend of this study, which is controlled by several things, such as natural disaster prevention, post-disaster response, and post-traumatic healing. Table 3 is defined as derivative research groups that were popular that year. In 2006, a tsunami occurred in Aceh, which caused 227 deaths. 898 fatalities. In the five years 2006-2010, studies on disaster response focused on topics such as tsunamis, disaster management, emotional stress, and early warning systems. Furthermore, disaster topics in the second five years include risk assessment, adaptive management, decision-making, landslide, and information management. Finally, in 2016-2022, disaster response studies shifted to using technology to handle natural disasters in Indonesia; the study topics include information use, mitigation, sustainable development, natural disaster mitigation, social media, COVID-19, and the pandemic.

Table 3, research simplifies several relevant references in disaster studies in Indonesia; Table 3 has similarities with Table 2, which shows the similarity of themes with the list of articles published in that year. This research assesses that disaster studies in Indonesia are motivated by conditions during the disaster; no studies analyze pre-disaster needs. The trend of this study can be seen from the phenomenon of natural disasters that occur in Indonesia every year, meaning that Indonesian scholars have limitations in understanding the geographical conditions of the region and the potential for natural disasters that occur at any time. In the last thirteen years, Indonesia's number of natural disasters has increased; the average of natural disasters in 2010-2015 was 1,619-1,945. From 2016-2021, natural disasters in Indonesia increased from 2,306 to 5,402; this number will quickly decrease in 2022 to 3,531 (BNPB, 2022).

Furthermore, from 2005-2022, disaster research in Indonesia explored victims of natural disasters (Gunnarsson et al., 2005; Irmansyah, 2005; Morgan et al., 2006; Rofi et al., 2006). After the Aceh tsunami, the research shifted to using GIS information to map and identify natural disasters and victims through geospatial analysis (Sutiono, Qiantori, Suwa, & Ohta, 2009; Tacconi et al., 2007). Disaster studies are also starting to understand the function of early warning systems(Azmi, Budiarto, & Widyanto, 2011; Rokhis Khomarudin et al., 2010). Next, this study began to shift to the field of analysis to examine potential risks (Fahrudin, 2012; Hidayati, 2012); in the following year, this topic was still relevant in several publications, such as the potential for natural disasters (Seyle, Widyatmoko, & Silver, 2013; Wardani & Muntohar, 2013).

Table 2 shows a shift in the topic of disasters in Indonesia. In 2014, research on disaster response shifted to predicting and calculating the potential risk of natural disasters through index data. In 2015, disaster studies turned to community resilience from the economic sector to economic recovery (Andriansyah, 2015). In 2016, the research outlined the use of technology and GIS in analyzing potential disasters; in 2017, it was about collaborative disaster management involving the military and state security agencies; and in 2018, the research topic was post-disaster school conditions. In 2019, the subject of this study looked at the government's capacity to handle disasters, and in the last three years, the topic of the natural disaster study shifted to COVID-19, non-natural disasters that have occurred over the previous three years (Fitriadi et al., 2022; Smoro Laksmi, Rudiarto, & Luqman, 2020).

This review shows that disaster responses in Indonesia are appropriate on topics such as the use of technology, risk analysis, hazard index assessment, and disaster mapping. According to Izumi et al. (2019) they have mapped out approaches to disaster responses; the findings in this paper show that disasters in Indonesia are dominant in analyses of this topic, see Tables 3 and 4. Multi-stakeholder highlights that in Indonesia, handling disasters is through community engagement (R. Djalante, 2012), calling it adaptive governance, which this similarity to the idea (Atkinson & Curnin, 2020; R. Djalante, 2012; Riyanti Djalante et al., 2011). This means that the government relies on community action when natural disasters occur. The community is essential to managing several natural disasters, such as Aceh, Yogyakarta,

West Java, and Sumatra. See Tables 3 and 4. The findings of this research have sorted disaster response research topics in Indonesia based on year and mapped the character of natural disaster response in Indonesia.

From the analysis of trends in disaster response publications in Indonesia, we assess that the correspondence between the phenomena that occurred and publications in Indonesia in the last 10 years has limitations. In the findings of this review, natural disasters in Indonesia do not have a solid design, as explained by Riyanti Djalante et al. (2013), but several research topics look at the micro way of handling natural disasters in Indonesia. Therefore, as stated in this research, understanding disaster response through formal institutions, Bang (2014) aims at community action that is fully involved in responding to disasters early, taking action, dealing with trauma, and preventing disease.

Tables 3 and 4 selected the relevant articles explaining the disaster in Indonesia; the analysis shows that research trends by publications in Indonesia have characteristics that can be seen from the conditions of the disaster. This review reveals significant findings on the big topic of disaster studies in Indonesia. Generally, disaster studies are carried out after disaster conditions, so in the following years, scholars began to promote research ideas about using technology for potential natural disasters. These findings show that several topics, such as GIS analysis and prevention, become yearly concerns. Based on the category of provincial areas that are the subject of study, such as Aceh, Sumatra, Java, and Yogyakarta. Several other provinces, such as Kalimantan, Bali, and others in the eastern region, such as Palu, Gorontalo, Ambon, and Papua, are not subject to routine study, which means that universities in Indonesia, which are spread across Eastern Indonesia, do not pay attention to natural disasters as an essential topic of study; only Tadulako University represents the Eastern part of Indonesia, with some publications reaching 20.

CONCLUSION

This research comprehensively evaluates the understanding of disaster response in Indonesia, one of the countries with the highest number of natural disasters globally. Based on analysis using CiteSpace via the Scopus database, we conclude that. Disaster studies in Indonesia have increased every year, but this is not comparable to the number of natural disasters that occur in Indonesia every year, which always increases. Second, based on the document type, Indonesian scholars' publications on this study are primarily at international conferences. In our view, this publication model only prioritizes quantity, not quality, in published articles. In the same category, publications in international journals tend to be a small proportion of total disaster study publications.

Third, based on authors and institutions, only the top ten universities in Indonesia have a high number of documents, so other universities in Indonesia do not pay full attention to disaster studies; only the University of Indonesia, Gadjah Mada University, and Diponegoro University have attention seriously in disaster response studies. The same thing can be seen in the number of authors, as universities with many publications dominate. Fourth, from the order of articles with the highest number of citations, we divided the groups into disaster management, post-disaster trauma management, and analysis of potential natural disasters. CiteSpace's research answered that the derivative topic on disaster response consists of 14 clusters, each with an article citing a hat, which shows that the study has broadly contributed to article citations. Meanwhile, if sorted based on top citing topics, cluster #9 is disaster management, 0 uses the geospatial approach, and cluster #1 is post-traumatic growth. Based on database and tool analysis, this research uses CiteSpace as an analysis tool and Scopus as the primary data source. Therefore, based on the evaluation results of the investigation, we consider that quite large data are needed as a data source in further research; the aim is to make the study more robust and understand the issue of disaster response from different publishers. Therefore, additional research can take global data by grouping it by the government to provide a new understanding of natural disaster response in countries with the highest potential for natural disasters.

REFERENCES

Ahmed, Z. (2013). Disaster risks and disaster management policies and practices in Pakistan: A critical analysis of Disaster Management Act 2010 of Pakistan. *International Journal of Disaster Risk Reduction*, *4*, 15–20. DOI: 10.1016/j.ijdrr.2013.03.003

Ahrens, J., & Rudolph, P. M. (2006). The importance of governance in risk reduction and disaster management. *Journal of Contingencies and Crisis Management*, *14*(4), 207–220. DOI: 10.1111/j.1468-5973.2006.00497.x

Alexander, K., Jarman, A. M. G., & Rosenthal, U. (1995). Inter-organizational policy processes in disaster management. *Disaster Prevention and Management: An International Journal*, *4*(2), 20–37. DOI: 10.1108/09653569510082669

Amni, B. D. H., Hadi, A. A., & Perdinan. (2020). *Design of Muster Area as Earthquake Evacuation Area in North Lombok Regency, West Nusa Tenggara* (H. A.A. & N. I., eds.). Department of Landscape Architecture, Faculty of Agriculture, IPB University (Bogor Agricultural University), Bogor, Indonesia: Institute of Physics Publishing. https://doi.org/DOI: 10.1088/1755-1315/501/1/012004

Andriansyah, Y. (2015). *Post-disaster economic recovery in Yogyakarta, Indonesia* (G. F.L., ed.). Department of Islamic Economics, Islamic University of Indonesia, Indonesia: CRC Press/Balkema. https://doi.org/DOI: 10.1201/b18146-62

Arksey, H., & O'Malley, L. (2005). Scoping studies: Towards a methodological framework. *International Journal of Social Research Methodology*, *8*(1), 19–32. DOI: 10.1080/1364557032000119616

Atkinson, C., & Curnin, S. (2020). Sharing responsibility in disaster management policy. *Progress in Disaster Science*, *7*, 100122. DOI: 10.1016/j.pdisas.2020.100122

Azmi, R., Budiarto, H., & Widyanto, R. (2011). *A proposed disaster emergency warning system standard through DVB-T in Indonesia*. Faculty of Computer Science, University of Indonesia, Kampus UI Depok, Depok 16424. *Indonesia*. Advance online publication. DOI: 10.1109/ICEEI.2011.6021746

Bang, H. N. (2014). General overview of the disaster management framework in Cameroon. *Disasters*, *38*(3), 562–586. DOI: 10.1111/disa.12061 PMID: 24905711

Berke, P. R., & Campanella, T. J. (2006). Planning for Post-disaster Resiliency. *The Annals of the American Academy of Political and Social Science*, *604*(1), 192–207. DOI: 10.1177/0002716205285533

Birkland, T. A. (2006). Lessons of disaster: Policy change after catastrophic events. *Lessons of Disaster: Policy Change After Catastrophic Events, 9037*(May), 1–216. DOI: 10.1093/jopart/mup023

BNPB. (2022). *Bencana Tahun 2022*. Badan Nasional Penggaulangan Bencana.

Burnham, J. F. (2006). Scopus database: A review. *Biomedical Digital Libraries, 3*, 1–8. DOI: 10.1186/1742-5581-3-1 PMID: 16522216

Chen, C. (2005). *CiteSpace. Quick Guide 1.2*. Drexel University.

Chen, C. (2016). Citespace: A practical guide for mapping scientific literature. nova science publishers. In *Nova Science Publishers*.

Crow, D. A., Albright, E. A., Ely, T., Koebele, E., & Lawhon, L. (2018). Do Disasters Lead to Learning? Financial Policy Change in Local Government. *The Review of Policy Research, 35*(4), 564–589. DOI: 10.1111/ropr.12297

Daudt, H. M. L., Van Mossel, C., & Scott, S. J. (2013). Enhancing the scoping study methodology: A large, inter-professional team's experience with Arksey and O'Malley's framework. *BMC Medical Research Methodology, 13*(1), 1–9. DOI: 10.1186/1471-2288-13-48 PMID: 23522333

Djalante, R., Holley, C., & Thomalla, F. (2011). Adaptive governance and managing resilience to natural hazards. *International Journal of Disaster Risk Science*, 2(4), 1–14. DOI: 10.1007/s13753-011-0015-6

Djalante, R. (2012). Review Article: Adaptive governance and resilience: The role of multi-stakeholder platforms in disaster risk reduction. *Natural Hazards and Earth System Sciences, 12*(9), 2923–2942. DOI: 10.5194/nhess-12-2923-2012

Djalante, R., Holley, C., Thomalla, F., & Carnegie, M. (2013). Pathways for adaptive and integrated disaster resilience. *Natural Hazards, 69*(3), 2105–2135. DOI: 10.1007/s11069-013-0797-5

Djalante, R., & Garschagen, M. (2017). *A Review of Disaster Trend and Disaster Risk Governance in Indonesia: 1900–2015*. https://doi.org/DOI: 10.1007/978-3-319-54466-3_2

Djalante, R. (2018). Review article: A systematic literature review of research trends and authorships on natural hazards, disasters, risk reduction and climate change in Indonesia. *Natural Hazards and Earth System Sciences, 18*(6), 1785–1810. DOI: 10.5194/nhess-18-1785-2018

Djalante, R., Thomalla, F., Sinapoy, M. S., & Carnegie, M. (2012). Building resilience to natural hazards in Indonesia: Progress and challenges in implementing the Hyogo Framework for Action. *Natural Hazards*, *62*(3), 779–803. DOI: 10.1007/s11069-012-0106-8

Donahue, A. K. (2010). A Review of: "Lessons of Disaster: Policy Change After Catastrophic Events by Thomas A. Birkland". *International Public Management Journal*, *13*(1), 100–103. DOI: 10.1080/10967490903547399

Donthu, N., Kumar, S., Mukherjee, D., Pandey, N., & Lim, W. M. (2021). How to conduct a bibliometric analysis: An overview and guidelines. *Journal of Business Research*, *133*(April), 285–296. DOI: 10.1016/j.jbusres.2021.04.070

Doocy, S., Rofi, A., Moodie, C., Spring, E., Bradley, S., Burnham, G., & Robinson, C. (2007). Tsunami mortality in Aceh Province, Indonesia. *Bulletin of the World Health Organization*, *85*(4), 273–278. DOI: 10.2471/BLT.06.033308 PMID: 17546308

Fahrudin, A. (2012). Psychosocial Reaction and Trauma After a Natural Disaster: The Role of Coping Behavior. *Asian Social Work and Policy Review*, *6*(3), 192–202. DOI: 10.1111/j.1753-1411.2012.00070.x

Fitriadi, F., Jiuhardi, J., Busari, A., Ulfah, Y., Hakim, Y. P., Erwin Kurniawan, A., & Darma, D. C. (2022). Using Correlation to Explore the Impact of Corona Virus Disease on Socioeconomics. *Emerging Science Journal*, *6*(Special Issue COVID-19 Emerging Research), 165–180. https://doi.org/DOI: 10.28991/esj-2022-SPER-012

Gopalakrishnan, C. (2013). Water and disasters: A review and analysis of policy aspects. *International Journal of Water Resources Development*, *29*(2), 250–271. DOI: 10.1080/07900627.2012.756133

Gunnarsson, D., Hyer, R., Hufstader, R. D., Hidayat, A., Peake, J., & Banerjee, A (2005). Civil-military cooperation in humanitarian health action. *Prehospital and Disaster Medicine*, *20*(6), 450–454. DOI: 10.1017/S1049023X00003101 PMID: 16496635

Hacking, I. (2012). Scoping studies: Advancing the methodology. *Representing and Intervening*, *5*(69), 1–18. DOI: 10.1017/cbo9780511814563.003

Hartama, D., Mawengkang, H., Zarlis, M., Sembiring, R. W., Nasution, B. B., Syahruddin, M., ... Sumarno. (2017). *The planning of smart city to mitigate the impacts of natural disaster in north sumatera* (Vol. 501, pp. 147–154; V. D., G. J.J., M. Y., & Z. P., Eds.). Vol. 501, pp. 147–154. STIKOM Tunas Bangsa Pematangsiantar, Pematangsiantar, Indonesia: Springer New York LLC. https://doi.org/DOI: 10.1007/978-3-319-68486-4_12

Hartawan, D. R., Purboyo, T. W., & Setianingsih, C. (2019). *Disaster victims detection system using convolutional neural network (CNN) method.* 105–111. Computer Engineering Study, School of Electrical Engineering, Telkom University, Indonesia: Institute of Electrical and Electronics Engineers Inc. https://doi.org/DOI: 10.1109/ICIAICT.2019.8784782

Healy, A., & Malhotra, N. (2009). Myopic voters and natural disaster policy. *The American Political Science Review, 103*(3), 387–406. DOI: 10.1017/S0003055409990104

Hermansson, H. (2019). Challenges to Decentralization of Disaster Management in Turkey: The Role of Political-Administrative Context. *International Journal of Public Administration, 42*(5), 417–431. DOI: 10.1080/01900692.2018.1466898

Hidayati, D. (2012). Striving to reduce disaster risk: Vulnerable communities with low levels of preparedness in Indonesia. *Journal of Disaster Research, 7*(1), 75–82. DOI: 10.20965/jdr.2012.p0075

Irmansyah, S. J. (2005). Mental health problems after tsunami disaster in Aceh and Nias; immediate responses and future programs. *Seishin Shinkeigaku Zasshi. Psychiatria et Neurologia Japonica, 107*(11), 1178–1183. PMID: 16408426

Izumi, T., Shaw, R., Djalante, R., Ishiwatari, M., & Komino, T. (2019). Disaster risk reduction and innovations. *Progress in Disaster Science, 2*, 100033. DOI: 10.1016/j.pdisas.2019.100033

Juth, V., Silver, R. C., Seyle, D. C., Widyatmoko, C. S., & Tan, E. T. (2015). Post-disaster Mental Health Among Parent–Child Dyads After a Major Earthquake in Indonesia. *Journal of Abnormal Child Psychology, 43*(7), 1309–1318. DOI: 10.1007/s10802-015-0009-8 PMID: 25851238

Kalfin, S. Supian, S., Mamat, M., & Bon, A. T. (2020a). *A review mitigation efforts and natural disaster insurance schemes in indonesia. 59*, 1825–1835. Faculty of Mathematics and Natural Sciences, Universitas Padjadjaran, Indonesia: IEOM Society.

Kalfin, S. Supian, S., Mamat, M., & Bon, A. T. (2020b). *Model for determining natural disaster insurance premiums in indonesia using the black scholes method. 0*, 2376–2385. Faculty of Mathematics and Natural Sciences, Universitas Padjadjaran, Indonesia: IEOM Society.

Kalfin, S. Supian, S., & Mamat, M. (2021). *Mitigation and models for determining premiums for natural disaster insurance due to excessive rainfall* (R. B.N., R. J., P. A., F. A.N., A. M., P. R.C.I., … S. K.A., Eds.). Mathematics, Faculty of Mathematics and Natural Sciences, Universitas Padjadjaran, Indonesia: IOP Publishing Ltd. https://doi.org/DOI: 10.1088/1742-6596/1722/1/012058

Kalfin, S. Supian, S., Mamat, M., Muljana, F., & Bon, A. T. (2021). *Analysis of agropolitan area planning based on natural disaster mitigation in west muna regency, southeast Sulawesi, Indonesia.* 3592–3601. Faculty of Mathematics and Natural Sciences, Universitas Padjadjaran, Indonesia: IEOM Society.

Kalfin, S., Supian, S., & Mamat, M. (2022a). Insurance Premium Determination Model and Innovation for Economic Recovery Due to Natural Disasters in Indonesia. *Computation (Basel, Switzerland), 10*(10). Advance online publication. DOI: 10.3390/computation10100174

Kalfin, S., Supian, S., & Mamat, M. (2022b). Insurance premium determination model as natural disaster mitigation effort in indonesia with a cross subsidy system. *International Journal of Agricultural and Statistics Sciences, 18*(2), 539–546.

Kapucu, N., & Garayev, V. (2011). Collaborative Decision-Making in emergency and disaster Management. *International Journal of Public Administration, 34*(6), 366–375. DOI: 10.1080/01900692.2011.561477

Kurniawan, D., Sutan, A. J., Nurmandi, A., Loilatu, M. J., & Salahudin. (2021). *Social Media as Tools of Disaster Mitigation, Studies on Natural Disasters in Indonesia BT - HCI International 2021 - Late Breaking Posters* (C. Stephanidis, M. Antona, & S. Ntoa, Eds.). Cham: Springer International Publishing.

Kusumasari, B., & Alam, Q. (2012). Bridging the gaps: The role of local government capability and the management of a natural disaster in Bantul, Indonesia. *Natural Hazards, 60*(2), 761–779. DOI: 10.1007/s11069-011-0016-1

Kusumasari, B., Alam, Q., & Siddiqui, K. (2010). Resource capability for local government in managing disaster. *Disaster Prevention and Management: An International Journal, 19*(4), 438–451. DOI: 10.1108/09653561011070367

Kusumastuti, R. D., Viverita, Husodo, Z. A., Suardi, L., & Danarsari, D. N. (2014). Developing a resilience index towards natural disasters in Indonesia. *International Journal of Disaster Risk Reduction, 10*(PA), 327–340. https://doi.org/DOI: 10.1016/j.ijdrr.2014.10.007

Maryati, S. (2018). *Identification of Flood Prone Areas for Natural Disaster Mitigation using Geospatial Approach (A Case Study in Bone Bolango Regency, Gorontalo Province)* (I. A., Nandi, & S. L., Eds.). Department of Earth Science and Engineering, Faculty of Mathematics and Natural Sciences, Universitas Negeri Gorontalo, Gorontalo, Indonesia: Institute of Physics Publishing. https://doi.org/DOI: 10.1088/1755-1315/145/1/012080

McLeod, M. K., Slavich, P. G., Irhas, Y., Moore, N., Rachman, A., Ali, N., & Caniago, C. (2010). Soil salinity in Aceh after the December 2004 Indian Ocean tsunami. *Agricultural Water Management, 97*(5), 605–613. DOI: 10.1016/j.agwat.2009.10.014

Morgan, O. W., Sribanditmongkol, P., Perera, C., Sulasmi, Y., Van Alphen, D., & Sondorp, E. (2006). Mass fatality management following the south Asian tsunami disaster: Case studies in Thailand, Indonesia, and Sri Lanka. *PLoS Medicine, 3*(6), 809–815. DOI: 10.1371/journal.pmed.0030195 PMID: 16737348

Mosby, K., Birch, T., Moles, A., & Cherry, K. E. (2009). Disasters Risk Reduction. In *UNISDR*. https://doi.org/DOI: 10.7591/9781501701498-008

Naghavi, M., Wang, H., & Lozano, R. (2015). Global, regional, and national age-sex specific all-cause and cause-specific mortality for 240 causes of death, 1990-2013: A systematic analysis for the Global Burden of Disease Study 2013. *Lancet, 385*(9963), 117–171. DOI: 10.1016/S0140-6736(14)61682-2 PMID: 25530442

Ong, A. K. S., Prasetyo, Y. T., Lagura, F. C., Ramos, R. N., Sigua, K. M., Villas, J. A., & Redi, A. A. N. P. (2021). Factors affecting intention to prepare for mitigation of "the big one" earthquake in the Philippines: Integrating protection motivation theory and extended theory of planned behavior. *International Journal of Disaster Risk Reduction, 63*. Advance online publication. DOI: 10.1016/j.ijdrr.2021.102467

Prasetiyawan, Viora, E., Maramis, A., & Keliat, B. A. (2006). Mental health model of care programmes after the tsunami in Aceh, Indonesia. *International Review of Psychiatry (Abingdon, England), 18*(6), 559–562. DOI: 10.1080/09540260601039959 PMID: 17162697

Raungratanaamporn, I., Pakdeeburee, P., Kamiko, A., & Denpaiboon, C. (2014). Government-communities Collaboration in Disaster Management Activity: Investigation in the Current Flood Disaster Management Policy in Thailand. *Procedia Environmental Sciences, 20*, 658–667. DOI: 10.1016/j.proenv.2014.03.079

Robinson, B., Alatas, M. F., Robertson, A., & Steer, H. (2011). Natural disasters and the lung. *Respirology (Carlton, Vic.), 16*(3), 386–395. DOI: 10.1111/j.1440-1843.2011.01923.x PMID: 21244570

Rofi, A., Doocy, S., & Robinson, C. (2006). Tsunami mortality and displacement in Aceh province, Indonesia. *Disasters, 30*(3), 340–350. DOI: 10.1111/j.0361-3666.2005.00324.x PMID: 16911432

Rokhis Khomarudin, M., Strunz, G., Ludwig, R., Zoßeder, K., Post, J., Kongko, W., & Pranowo, W. S. (2010). Hazard analysis and estimation of people exposure as contribution to tsunami risk assessment in the west coast of sumatra, the south coast of Java and bali. *Zeitschrift für Geomorphologie*, *54*, 337–356. DOI: 10.1127/0372-8854/2010/0054S3-0031

Setiadi, A., Andriessen, A., & Anisa, R. (2020). Post-occupancy evaluation of Pagerjurang permanent housing after the Merapi volcanic eruption. *Journal of Architecture and Urbanism*, *44*(2), 145–151. DOI: 10.3846/jau.2020.11265

Seyle, D. C., Widyatmoko, C. S., & Silver, R. C. (2013). Coping with natural disasters in Yogyakarta, Indonesia: A study of elementary school teachers. *School Psychology International*, *34*(4), 387–404. DOI: 10.1177/0143034312446889

Sidi, P., & Mamat, M., Sukono, & Supian, S. (2017). Evaluation model for risk insurance premiums of building damage caused by flood: Case study in Citarum Watershed, Southern Bandung, Indonesia. *Journal of Engineering and Applied Sciences (Asian Research Publishing Network)*, *12*(17), 4420–4425. DOI: 10.3923/jeasci.2017.4420.4425

Smith, W., & Dowell, J. (2000). A case study of co-ordinative decision-making in disaster management. *Ergonomics*, *43*(8), 1153–1166. DOI: 10.1080/00140130050084923 PMID: 10975178

Smoro Laksmi, G., Rudiarto, I., & Luqman, Y. (2020). *Community preparedness toward flood during Covid-19 pandemic at Pekalongan City and Regency* (W. B., Sudarno, & P. T. T., Eds.). Master of Environmental Science, Postgraduate School, Diponegoro University, Semarang, Indonesia: EDP Sciences. https://doi.org/DOI: 10.1051/e3sconf/202020206008

Suhartono, F. S., & Suranto. (2020). *Non-natural disaster mitigation Covid-19 pandemic and its urgency in the education curriculum*. Faculty of Teacher Training and Education, Sebelas Maret University, UNS, Surakarta, Indonesia: Association for Computing Machinery. https://doi.org/DOI: 10.1145/3452144.3453772

Sukmono, F. G., Junaedi, F., Fadilla, Q. Y., Loilatu, M. J., & Fadillah, D. (2023). News about the Sunda Strait Tsunami Disaster: From the Selection of News Sources to the Production Process. *Studies in Media and Communication*, *11*(4), 10–18. DOI: 10.11114/SMC.V11I4.6002

Sutiono, A. B., Qiantori, A., Suwa, H., & Ohta, T. (2009). Characteristic tetanus infection in disaster-affected areas: Case study of the Yogyakarta earthquakes in Indonesia. *BMC Research Notes*, *2*. Advance online publication. DOI: 10.1186/1756-0500-2-34 PMID: 19284531

Tacconi, L., Moore, P. F., & Kaimowitz, D. (2007). Fires in tropical forests - What is really the problem? Lessons from Indonesia. *Mitigation and Adaptation Strategies for Global Change, 12*(1), 55–66. DOI: 10.1007/s11027-006-9040-y

Uddin, M. S., Haque, C. E., & Khan, M. N. (2021). Good governance and local level policy implementation for disaster-risk-reduction: Actual, perceptual and contested perspectives in coastal communities in Bangladesh. *Disaster Prevention and Management: An International Journal, 30*(2), 94–111. DOI: 10.1108/DPM-03-2020-0069

van Niekerk, D. (2014). A critical analysis of the South African Disaster Management Act and Policy Framework. *Disasters, 38*(4), 858–877. DOI: 10.1111/disa.12081 PMID: 25196341

Wahyudin, D., Hasegawa, S., & Dahlan, T. (2013). *Developing ethical decision making skill of novice volunteers in natural disaster response. 2, 800–803*. School of Information Science JAIST.

Wardani, S. P. R., & Muntohar, A. S. (2013). Lessons learned from the recent natural disasters in Indonesia. *Geotechnical, Geological and Earthquake Engineering*, Vol. 25, pp. 47–59. Department of Civil Engineering, Diponegoro University, Jl. Hayam Wuruk, 5-7, Semarang, 50241, Indonesia: Springer Netherlands. https://doi.org/DOI: 10.1007/978-94-007-5675-5_4

KEY TERMS AND DEFINITIONS

Disaster Response: The immediate and coordinated actions taken by governments, institutions, and communities to manage the aftermath of disasters, including emergency relief, recovery, and resilience-building measures.

Disaster Risk Reduction (DRR): A systematic approach to identifying, assessing, and reducing the risks of disasters through prevention, mitigation, preparedness, and adaptation strategies aimed at minimizing loss and damage.

Meta-Analysis: A research method that systematically synthesizes findings from multiple studies to identify patterns, trends, and gaps in disaster-related research, providing evidence-based insights for policy and practice.

Natural Disaster: A sudden, extreme event caused by natural processes such as earthquakes, tsunamis, floods, volcanic eruptions, or forest fires that result in significant human, material, and environmental losses.

Chapter 10

Disaster Management Effectiveness Potential With Socio-Economic Indicators in Türkiye: 1944-2022 Period and Future

Öner Gümüş
https://orcid.org/0000-0002-3231-3293
Kutahya Dumlupinar University, Turkey

Nur Hoşgör
https://orcid.org/0000-0002-0108-9823
Kutahya Dumlupinar University, Turkey

ABSTRACT

When disasters occur, states suffer not only loss of revenue but also loss of human capital. Therefore, we should carry out disaster management effectively. In this context, the aim of the study is to determine how the effectiveness potential of disaster management can be increased by analyzing the relationship between the effectiveness potential of disaster management and socioeconomic variables. Not only GDP per capita and budget income-expenditure ratios, but also the number of municipalities, population growth rate, and the number of infants, children, and elderly are important variables in the effectiveness potential of disaster management. In the study, it was concluded that the potential for effectiveness in disaster management can be increased by making new regulations on disaster management, ensuring economic and financial sustainability, effective population planning, and effective inclusion of vulnerable groups in disaster management.

DOI: 10.4018/979-8-3373-1862-2.ch010

INTRODUCTION

Disasters are divided into two categories: natural or man-made. Natural disasters are defined as climatic and geological events that cause physical, economic, social, and spiritual losses to all kinds of living and non-living beings. These disasters are divided into three categories: hydrometeorological disasters, such as floods, droughts, forest fires, and extreme temperatures; geophysical disasters, such as earthquakes; and biological disasters, such as epidemics. The world has identified 52 types of natural disasters. Türkiye's geographical location puts it at risk for natural disasters, with 21 of those disasters occurring there. These types of disasters seen in Türkiye are as follows: earthquakes, flash floods and inundations, wind, dust, sand, rain, snow and winter storms, frost, avalanches, storms, tornadoes, ground collapse, landslides, mudflow, rockfall, forest and bush fires, icing, sea and lake water level changes, tsunamis, desertification, drought, hot and cold air waves, fog and low visibility, agricultural pests, locust invasions, lightning, and epidemics. Human disturbance of the natural balance causes changes in the frequency, number, and severity of these disasters (Gümüş et al., 2024). This information on disasters in Türkiye has emerged as a result of a certain historical experience. Therefore, this study analyzes disaster management across three historical periods: 1944-1958, 1959-1998, and 1999-2022. The reason for starting the study with the year 1944 is the global effects of the first and second world wars. This is because these events affected not only Türkiye but the whole world. Despite Türkiye's non-participation in the Second World War, its impact was profound. Furthermore, the recovery period following this period allows for a more focused approach to disaster management. 1958 marked another major shift. Türkiye experienced financial bankruptcy on this date. We aim to conduct an examination both before and after the moratorium. In 2005, the removal of six zeroes from the currency led to the selection of 1998 as the base for the GDP per capita series. The reason for choosing the year 2022 as the date when the study was finalized is to be able to make predictions for the future by considering the study from a historical perspective.

Disaster management is not fully compatible with disaster planning. Research has shown that successful disaster management is primarily due to the activities of emergency organizations. There are management problems, especially related to the communication process, the use of authority, and the development of coordination (Quarantelli, 1988). However, relying solely on these factors does not yield sufficient and successful results. This is because disasters have short- and long-term economic consequences.

Natural disasters typically have an immediate negative impact on the macroeconomy, causing a slowdown in production and economic growth. Smaller economies and developing countries typically have a worse output decline compared to developed countries since they have lower capacity for absorbing shocks (Cavallo & Noy, 2011; Noy, 2009). High CO_2 emissions also lead to an increase in sudden economic losses; at the same time, disasters reduce local economic growth and strengthen environmental degradation (Lu, 2022).

The long-term economic impact of jungle storms may differ from city to city in the Malaysian cities. At the source level, it has been proposed that natural disasters inhibit growth ex ante but are subsequently followed by an outpouring of investment in human capital and infrastructure, thereby boosting the economic growth rate. Nonetheless, it can also have a negative impact on the fiscal balance and foreign direct investment when major disasters depress economic growth and erode fiscal stability (Akao & Sakamoto, 2018). In this case, to determine the effectiveness potential of disaster management, it is necessary to determine the GDP per capita and budget income-expenditure ratios. This is because while GDP per capita expresses the efficiency potential of individuals and households in disaster management, budget income-expenditure ratios can provide an idea about the extent to which the state is prepared for disaster management. Therefore, in this study, GDP per capita and budget income-expenditure ratios were used as economic and fiscal data to define the effectiveness potential of disaster management.

Disasters are complex events that change the way civilizations work and show how weak they are. They also change the way they rule and interact with each other. The lessons learned from studying catastrophes can tell us a lot about how societies respond to disasters and what they need to do to adapt and be strong in the face of them.

Disasters are not only natural or technological catastrophes, but they are also closely linked to social and political events. By revealing shared experiences and collective identities in communities, they can lead to policy changes and influence democratic governance. Disaster studies is an interdisciplinary field that looks at how disasters affect many parts of society, such as the social, political, economic, and ecological domains (Reinhardt & Ross, 2019; Taylor & Levine, 2015). Consequently, the field is a multidisciplinary subject integrating public administration, economics, and public finance.

Social structures and dynamics deeply intertwine with disasters, reflecting the vulnerability and resilience of human communities. They are not the conduct of nature or by accident and can be traced back to the product of human habitation with environmental and social orders (Britton, 1986; Reinhardt & Ross, 2019). Disaster studies elucidate social processes by illustrating how disasters act as catalysts for and consequences of society responses. Disasters can cause changes in legislation, affect how democracy works, and bring attention to shared experiences and group identities (Reinhardt & Ross, 2019; Taylor & Levine, 2015). By examining societal norms and structures, they often show how weak human institutions are and also examine the moral and ethical sides of how people act (Taylor & Levine, 2015). In addition, disasters can enhance social trust and cooperation as communities work together to solve the problems that these events bring (Toya & Skidmore, 2014). As an interdisciplinary field, disaster studies offer a unique lens through which to examine the interaction between disasters and society and provide critical perspectives on societal vulnerability, resilience, and adaptive and recovery capacity (Furedi, 2007; Lindell, 2013; Reinhardt & Ross, 2019). For these reasons, the population growth rate, share of the old population in the total population, share of the baby and child population in the total population, and number of municipalities were included in the study. The Turkish Statistical Institute released the book "Indicators of 100 Years" in 2023, containing the data for these variables. The study's limitation lies in the exclusion of certain data from the publication. For instance, some of these variables include data from 1955 but not from 1959. To overcome this constraint, the difference between 1960 and 1955 was found by calculating the arithmetic mean, and the data for 1959 were estimated according to the arithmetic degree. We filled in the missing data by applying the same method to comparable data. Various visuals were created while interpreting the data. The matplotlib and numpy libraries of the Python programming language version 3.13.1 were used to create these visualizations. In this context, it is important to know what the matplotlib and numpy libraries do in data visualization.

Matplotlib is one of Python's oldest and most popular data visualization libraries. A wide variety of static, animated, and interactive graphics can be created. This library provides both high-level commands, such as creating a chart, and low-level commands, such as setting each chart property individually (Bozkurt, 2023). Matplotlib is a script that facilitates plotting. This library is used for visualization in the Python language. This library serves as the foundation for many other scripts. This library can be used by importing it in the form "import matplotlib.pyplot as plt". Another name can be used here instead of "plt". However, convention has widely accepted this name. The script contains many plotting-related functions. These operations perform the plotting in memory. Finally, "plt.show ()" is used to display the image on the screen. In most cases, this method places the corresponding plotting in the console. At first glance, the appearance of an image on the console may seem strange, but many development environments support it. However, you can display the plotting in a separate window or section (Teker, 2021).

Numpy is a package developed to perform mathematical operations on arrays, consisting of the abbreviation of the words "Numeric Python". Many Python libraries that operate on data depend on the Numpy package. Python, like other languages, utilizes the Numpy module for both classical and array operations. The array elements defined with Numpy must be of the same type. You can access array elements using index numbers, just like you would with list elements. The numpy package is installed on the computer from the terminal screen as "pip install numpy". After the numpy package is installed, it can be called in different ways, such as "import numpy as np" or "from numpy import*" (Çobanoğlu, 2023).

We produced bar graphs and wireframe plots using the Matplotlib and Numpy libraries and attempted to explain the relationship between socio-economic variables and the effectiveness potential of disaster management using these libraries. After explaining the socio-economic data from 1944 to 2022 and the effectiveness potential of disaster management, suggestions are made on what can be done to increase the effectiveness potential of disaster management in the future. The study aims to elucidate this relationship.

Disaster management is not only a local or national process; it is an area of governance directly related to social welfare, sustainable development, and human security on a global scale. The concept has its origins in post-World War II humanitarian relief and reconstruction practices and has been institutionalized through frameworks for reducing disaster risk developed by the United Nations, the World Bank, and other international organizations since the 1970s. The purpose of global arrangements, which encompass response, prevention, readiness, recovery, and reconstruction, is to cut down on the damage caused by disasters. This means that catastrophe management is directly related to modern problems like climate change, growing urbanization, and worldwide migration.

Empirical research indicates that disasters have lasting effects on economic development, fiscal stability, social inequality, and public health (for further information, see to Cavallo & Noy, 2022). In developing countries, recovery from disasters frequently necessitates prolonged periods due to shortcomings such as ineffective financial resources, inadequate institutional collaboration, and infrastructural deficiencies—all of which are associated with socio-economic factors including per capita GDP, fiscal balance, and demographic composition. We, however, have limited insights about the effects of such variables on effectiveness in disaster management, as noted by the literature. Research by Toya & Skidmore (2014) suggested that local government and vulnerable people participate actively in managing a disaster; it reflects resilience, but those studies are mostly country-level or regional and do not allow for much cross-historical comparison.

Examples of practical challenges include poor multi-level governance mechanisms, weak data infrastructure, and resource allocation and prioritization problems in local government counterbalancing inadequate institutional capacity. Few international studies have analyzed the relationship between socio-economic indicators and disaster management effectiveness systematically by means of long-term historical data. In Türkiye, existing studies typically concentrate on one type of disaster or short-term economic impacts while also examining multiple indicators, such as GDP per capita, budget balance, population growth rate, number of municipalities, and the proportion of vulnerable groups, within the same model and by comparing different historical periods.

This study aims to analyze the effectiveness potential of disaster management in Türkiye for the period 1944-2022 through socio-economic indicators, such as GDP per capita, budget revenue/expenditure ratio, population growth rate, number of municipalities, and demographic proportions of vulnerable groups. Thus, both historical turning points and the effects of policy changes will be evaluated. In this context, the main research question is formulated as follows: *"How is the effectiveness potential of disaster management in Türkiye related to socio-economic indicators in different historical periods, and what strategic implications does this relationship offer for future policy design?"*

This study offers a twofold contribution for both researchers and policymakers:

- Filling the gap of long-term, multivariate analysis in the literature to demonstrate that disaster management is shaped not only by legal regulations but also by socio-economic capacity,
- Suggest strategic priorities for policymakers, such as building local managerial capacity, fairness in income distribution, demographic planning, and focusing on vulnerable groups. In this way, future research can build on this study with larger data sets, comparative country analyses, and spatial econometric models.

THEORETICAL BACKGROUND

Disaster management is a multidisciplinary field based on theories and models that guide the understanding, planning, and execution of actions before, during, and after disasters. The theoretical background of disaster management integrates concepts from sociology, risk theory, economics, finance, management science, and systems thinking to address the complexity and unpredictability of disasters.

Disaster management is built on basic concepts such as disaster, hazard, vulnerability, risk, and capacity. Disasters are forces that interrupt normalcy and surpass the capacity of affected communities to cope and understand their hazards as potentials of harm. Vulnerability speaks to the degree to which a community will suffer from injury, and risk is measured by the likelihood that an unmitigated hazard may lead to negative impacts. Capacity embodies the resources and capabilities required for disaster impact management and mitigation (Khan et al., 2008; Math et al., 2015; Mahmud et al., 2020). The combination of all these concepts constitutes disaster management.

The vast theories of disaster management build within and among functions that are rooted periodically in other disciplines, which contain the disaster not only as a technical or logistical matter but also as social, economic, and political conditions. In this context, here are the primary theories on the disaster relationship: the disaster management cycle, risk society theory, social vulnerability theory, the pressure and release model (PAR model), resilience theory, complex systems and chaos theory, and institutional capacity theory.

Considered a primary theoretical model, the disaster management cycle includes four phases, which are indeed mitigation, preparedness, response, and recovery. This cyclical tradition focuses on ongoing planning, organizing, coordinating, and implementing policy measures to reduce disaster impacts and facilitate effective responses in the face of such catastrophes. In particular, one of the activities specific to each phase, such as risk analysis, is to rehabilitate itself in an integrated and iterative way (Khan et al., 2008; Iqbal et al., 2021; Math et al., 2015; Swalha, 2020).

Risk society theory posits that universal risks have developed in contemporary society and are inseparable aspects of social organization and individual identity. Beck and Giddens say that the present moment is the second modernity, when people care more about avoiding, handling, and talking about risks than just making things (Ekberg, 2007; Hu, 2013; Mythen, 2021; Jong, 2022). The theory emphasizes the ubiquity of low-probability, high-consequence risks; the transformation of risk from a calculable entity into a subjective and socially constructed phenomenon; and the emergence of collective risk identities (Ekberg, 2007; Sørensen, 2017; Hu, 2013).

Social vulnerability theory is based on the idea that risk and harm from hazards are not evenly distributed but are shaped by social structures, inequalities, and access to resources (Cutter, 2024; Mah et al., 2023; Schmidtlein et al., 2008; Ran et al., 2020). The theory emphasizes the need to consider both individual vulnerabilities and systemic factors, including age, status, education, and social support, as well as the influence of societal structures and policies in either exacerbating or mitigating susceptibility (Mah et al., 2023; Gibbons et al., 2024; Hayward et al., 2022; Fineman, 2017).

The PAR model defines the risk of disaster as the convergence of two opposing forces. Addressing or diminishing basic weaknesses triggers the opposing side, also known as release. This puts pressure on one side. Some of the things that could cause underlying problems are poverty, not having access to resources, and political instability. Dynamic pressures include factors like cities increasing too quickly, not enough education, or a government that doesn't work well. Unsafe conditions are emergency situations that put people at risk of harm, including living in dangerous areas or bad housing (Rahman et al., 2022; Aziz, 2018; Sandoval & Voss, 2016; Nirupama, 2012; Tsasis & Nirupama, 2008).

Resilience theory is often defined as the ability of a dynamic system (individual, family, or community) to effectively adapt to significant difficulties that jeopardize its functionality, viability, or growth (Southwick et al., 2014; Walker, 2020).

Chaos theory pertains to the mathematical study of deterministic systems characterized by nuanced initial conditions, nonlinearity, and unpredictable long-term behavior, as exemplified by the butterfly effect (Shadab, 2019; Xu, 2025; Kernick, 2018; Gurikar, 2024). Complex systems theory advances this concept by analyzing how dynamic interactions and feedback loops facilitate the self-organization and adaptation processes inherent in all systems, ranging from the physical (e.g., weather patterns) to the social (e.g., groups of people) (Warren et al., 1998).

Institutional capacity theory posits a new institutional framework that focuses on the influence of formal and informal regulations, bureaucratic standard operating procedures, and stakeholder preferences on an institution's authority. It is applied to investigate how organizations experience and evolve surrounding cross-area coordination, socio-economic mutation, or outsider assertions (Domorenok et al., 2021; Oleg et al., 2023; Biesenthal et al., 2018).

The study used data on vulnerable groups. Social vulnerability theory emphasizes that the main factor determining the effects of disasters is not only the severity of the hazard but also the level of social and economic vulnerability in society. Since this study examines the relationship between vulnerable groups and disaster management effectiveness, the study directly overlaps with social vulnerability theory. On the other hand, Institutional Capacity Theory argues that the success of disaster management depends on legal frameworks, organizational structure, financing, and human resource capacity. Since the number of local governments and the role of municipalities are evaluated in this study, and the periodical analyses reflect institutional changes, the study is directly related to Institutional Capacity Theory. The study discusses socio-economic factors that impact preparedness, response, and recovery capacity, as well as legislative changes. The Disaster Management Cycle suggests that these phases complement each other, with increasing capacity at each stage determining overall effectiveness. Therefore, the study is also related to the Disaster Management Cycle. Finally, resilience theory aims to become stronger against future risks as well as recovery. In the study, socio-economic indicators are associated with the potential for disaster management effectiveness. Therefore, the study bears a connection to resilience theory.

LITERATURE REVIEW

Recently, studies in the field of disaster management reveal that the increasing frequency and severity of disasters and the economic, social, and environmental impacts of these events need to be better understood. In the literature, it is seen that traditional approaches focusing only on response and recovery processes are inadequate, and instead a holistic disaster management approach that includes risk reduction, preparedness, resilience development, and long-term sustainability is adopted. In this framework, empirical studies comprehensively examine factors that determine the effectiveness of disaster management, such as socioeconomic indicators, institutional capacity, technological infrastructure, and vulnerability levels in society. This trend contributes to the development of science-based strategies for policymakers and practitioners in the future.

Hermansson (2019) interviewed 44 people and found that the development of control systems, failure in resource allocation, and the realization of central-local cooperation can be possible through a combination of political-administrative system characteristics and disaster-related processes.

AlHinai (2020) sought to extend the Technology-Organization-Environment framework beyond its original foundations by introducing a new set of disaster context determinants.

Vermiglio et al. (2021) emphasized how new technologies enhance the preparedness and resilience of specific systems at different stages of the disaster management cycle. They concluded that simulation and disaster risk reduction are areas of great importance in the application of new technologies to disaster management.

Ashraf (2021) asserts that equilibrium is essential for the effective management of concurrent disasters and that innovative holistic strategies are required to achieve efficient responses during the COVID-19 pandemic.

Tao & Han (2022) observe that the pandemic's effects in China's Henan region were short-lived. However, they say that the time between an early warning and an emergency reaction needs to be shortened and that relief financing sources need to be more varied.

Wen et al. (2023) conclude that it is crucial to transform governance mechanisms at different levels to integrate climate change adaptation and disaster risk reduction to reduce disaster and climate change risks and ensure safe growth and a resilient future in the Anthropocene era.

Ghaffarian et al. (2023) convincingly argued that the integration of artificial intelligence into early warning systems might substantially improve strategic planning and efficacy in catastrophe risk management.

Lam et al. (2023) discovered that social networking serves four critical functions in disaster management: it facilitates communication, provides real-time information for emergency responses and recovery efforts, aids individuals in comprehending their emotions and well-being, and contributes to the development of predictive models.

Sanderson et al. (2024) examined the experiences of community-centered restoration in New South Wales, Australia, which has endured substantial wildfires and floods since 2019, used key informant meetings and online questionnaires, and discovered widespread dissatisfaction with existing approaches. The investigation disclosed that although relief operations sought to aid affected areas, disjointed collaboration and inadequate communication impeded the reconstruction efforts.

Baraldo and Di Giuseppantonio Di Franco (2024) established in their field study that virtual reality (VR) has emerged as the primary technology utilized in place-centered research inside crisis contexts.

The current literature in disaster management is predominantly influenced by analyses of particular disaster types, regional practices, or short-term data sets. A significant number of these studies examine specific dimensions of disaster management through conceptual discussions, policy recommendations, or econometric models; however, descriptive analyses based on long-term, multivariate, and holistic data sets remain limited. In Türkiye, few studies address disaster management effectiveness alongside socio-economic indicators and evaluate it from a historical perspective.

This study's contribution to the literature is significant in various aspects. The investigation mainly uses a large historical dataset that goes from 1944 to 2022. This larger time frame lets us look at how both social and economic conditions and how people react to disasters have changed throughout the years. Different timeframes indicate how cooperation between communities and sectors during crises has changed throughout time, as well as how the support networks that go along with it have changed. Second, the data used in the analysis goes beyond just catastrophe indicators. It also includes macroeconomic measures like public revenues, public outlays, and public debt stock. The analysis reveals a complex relationship between the efficacy of disaster management and the broader macroeconomic context. The research employs descriptive statistics, bar graphs, and wireframe visualizations to illustrate patterns and connections between variables in a concrete manner, as opposed to utilizing econometric modeling. This process furnishes a more accessible framework for policymakers and practitioners, allowing for direct interpretation of data relative to analysis reliant on models. Therefore, unlike previous studies in the field of disaster management, this study takes a unique position in the literature in terms of using both long-term and multidimensional data sets, addressing macroeconomic and disaster management indicators together, and presenting its findings with visually oriented descriptive analysis.

METHODOLOGY

This study examines Türkiye's disaster management effectiveness based on socio-economic indicators between 1944 and 2022. The data is based on the official statistics set "Indicators of 100 Years" published by the Turkish Statistical Institute (TurkStat). The indicators examined in the study were chosen from among those socio-economic variables that are highlighted as keys to focal attention within the disaster management literature. These indicators reflect the main factors that affect resilience and vulnerability to disasters, such as national development level, economic size, population dynamics, and infrastructure capacity.

During the analysis process, the data were first processed and visualized using the Python programming language. In the data processing phase, arrays were created using the NumPy library, basic statistical calculations were performed on the data, and the necessary data structures for periodic comparisons were prepared. Graphs in both two and three dimensions were made using the Matplotlib tool during the visualization process.

Bar graphs were made in the first step to show how each socioeconomic indicator changed over time. These visualizations provide the opportunity to clearly observe the increase, decrease, and stagnation trends of indicators over the years. Bar charts clearly reveal structural changes, breakpoints, and general trends in data spanning many years.

In the final stage, a wireframe plot was created to analyze the relationships between different indicators from a multidimensional perspective. This type of plot visualizes the dependencies and tentative connections between the indicators under a three-dimensional feature space. These visualizations were produced using the Axes3D module of Matplotlib and improved a lot of comprehension of complex data structures, like time dimension.

This methodological approach focuses on programming-based visual and descriptive analysis of available data rather than quantitative econometric modeling. This flexible analysis infrastructure provided by Python aims to provide a valuable perspective to both the academic literature and policymakers in the field of disaster management in terms of long-term trends and inter-indicator relationships.

Data and Selection of Variables

In this study, the variables used in the evaluation of disaster management through socio-economic indicators are determined by considering both previous research in the literature and the long-term dynamics of Türkiye's demographic and financial structure.

- *Per Capita GDP:* It is the most basic indicator of economic welfare and production capacity and reflects the impact of economic costs of disasters on society. It is accepted that societies with a high level of prosperity can develop stronger institutional and financial capacity against disasters.
- *Population Growth Rate:* It is an important indicator that can be associated with the increasing or decreasing trend of disaster risks by showing the rate of change in demographic structure. During times of rapid population growth, a city's infrastructure that isn't up to par could make it more likely to have calamities.
- *Number of Municipalities:* The number of local administrations reveals the extent of administrative capacity and service provision in disaster management. The number of municipalities reflects the diversity of disaster preparedness and response capacity at the local level.
- *Population Ratios (Male, Female, Infant, Child, Elderly):* Such indices are employed to gauge societal fragilities in demography. For instance, while a higher proportion of infants, children, and elderly makes it more necessary to provide special protection in disasters, gender balance could be linked with societies' resilience or productive capacity.
- *Budget Income and Budget Expenditure (Budget Income/Budget Expenditure):* These are the basic indicators showing the financial capacity of the state and its priorities in resource allocation. Since pre-disaster preparedness, response, and recovery processes heavily rely on public resources, these indicators directly influence the effectiveness of disaster management.

Considering these variables together offers the opportunity to evaluate disaster management not only from the perspective of economic aggregates but also from the perspective of demographic structure and financial capacity. Thus, it is aimed to analyze Türkiye's disaster management potential over the period 1944-2022 in a multidimensional manner.

CONCEPTUAL FRAMEWORK

Disaster management is a multi-faceted process that necessitates not only response protocols but also the evaluation of various dimensions, including social, economic, and environmental considerations. In this framework, socio-economic indicators play a critical role in pre-disaster preparedness, risk reduction, and response and recovery phases. In particular, indicators such as economic capacity, infrastructure level, education rate, access to health services, population density, and income distribution are used as basic criteria in determining the levels of vulnerability and resilience to disasters. Although the effects of these indicators on disaster management performance are frequently emphasized in the literature, it is observed that long-term, multivariate, and visual analysis-based studies are limited in Türkiye.

This study aims to analyze Türkiye's disaster management potential in the context of socio-economic indicators using data from 1944 to 2022. The research visualized the long-term trends of the variables in the data set with bar charts and analyzed the interrelationships among indicators with the wireframe plot method. In doing so, you can see the historical trends that emerge but also better understand conceptual linkages by looking at how indicators interact with one another in three dimensions.

Due to disasters, there are losses of money and property. All superstructures and infrastructures destroyed or damaged in cities, the transport network, the reduction in the number of vehicles and equipment, and the decrease in all kinds of product stocks and losses cause significant decreases in national income. After the disaster, public expenditures will increase due to the growth rate of the economy, foreign trade, interest to be paid for foreign loans, reorganization of the degraded environment, restoration of settlements, and tax losses to be experienced. The damages caused by a disaster to items such as capital, processed products, semi-processed products, stocks, raw materials, and spare parts are direct damages to the economy, and the damages that occur in the provision of production, goods, and services after the destruction are indirect damages to the economy (Doğan, 2022). In this context, GDP per capita should be considered in terms of the effectiveness potential of disaster management.

GDP per capita is a measure that provides an average economic output per capita by dividing a country's gross domestic product (GDP) by its population. Economists habitually utilize gross domestic product per capita as a meaningful index of a nation's economic welfare to juxtapose the economic performances of different countries (Bhide & Khanolkar, 2020). Economists often use GDP per capita, which is a measure of the standard of life, to figure out how well a population is doing. There are many factors that affect GDP per capita, including how well people work, how well they are taught and trained, how many people are unemployed, and how much money is invested. Larger work productivity often leads to a larger GDP per capita since it makes manufacturing more efficient. However, an educated community makes financial output even stronger. A labor force skilled in valuable talents significantly enhances productivity by contributing important benefits through their marketable skills. Concurrently, complete occupation benefits GDP per capita since all accessible hands contribute to the full economic machine. Conversely, unused skills that remain untapped restrict potential. Overall, a balancing of optimized processes, a learned labor pool, and non-idled individuals commonly results in elevated GDP per person. Increasing investment per capita may increase economic growth and thus GDP per capita (C nh & Hoài, 2022). People often compare GDP per capita to other indicators like the Human Development Index (HDI) to gain a more comprehensive understanding of economic development. The GDP per capita might not fully reflect welfare and development factors like economic freedom and income inequality, even though it is a valid way to measure the economy's output (Dědeček & Dudzich, 2022). Researchers have progressively developed theoretical models to elucidate the relationship between per capita GDP and other economic indicators, despite variations in purchasing power parity leading to significant discrepancies in GDP per capita estimates among nations, which necessitates caution in comparative analysis. A quantifiable relationship conveyed as a function of R&D costs, patent applications, and internet usage underscores the interconnectedness of technical advancement and financial development. Economists employed regression examination and Bayesian strategies to recognize and measure the impact of an assortment of components on GDP per capita. GDP per capita is a pivotal financial marker that reveals understanding of a country's monetary well-being and standard of living. While labor profitability is critical in accomplishing comprehensive financial advancement, driven by different elements, for example, speculation and training, progress is additionally determined by an economy's capacity to receive and create new innovations over the long run. Policymakers and experts must fully understand these procedures to encourage manageable financial development and enhance personal satisfaction. In this respect, GDP per capita is an important criterion for increasing the efficiency potential in disaster management.

Population and population growth rate increase the severity of the disaster. Earthquakes, one of the most serious disasters, affect 98% of the population in Türkiye. This is because population density is high in areas with high earthquake risk. In the event of a disaster such as an avalanche, the population experiences an intensive migration. Therefore, approaches in disaster management concern a large part of the population (Şahin & Üçgül, 2019). From this perspective, we should explain the population growth rate in an economic context, considering its potential for effectiveness in disaster management.

Population growth has important effects on economic development. This is because population growth affects both consumption patterns and production levels. The size and structure of the population as an element of consumption determine the rate and form of consumption. Population growth as a productive factor affects production levels and the efficiency of employment. Sustainable economic development requires a balance between shifting labor from less productive to more productive sectors and maintaining the balance between sectors in the economy (Sinha, 1959).

Researchers and policy analysts continue to debate the complicated links between population growth and economic growth. Existing marijuana studies seem more directed at economic analysis than showing how population growth rates affect economic variables. For instance, there is evidence that just reducing population growth is not sufficient to ensure better economic performance. However, this interpretation is not agreed upon by all. Such findings, some scholars have argued, cannot fully illustrate the complex relationship between demographic trends and economic variables. Analytical models may yield more robust insights when additional filler factors—such as lagged fertility—are incorporated as covariates. In fact, empirical research has demonstrated that transient fertility surges can adversely affect economic growth by influencing the demand for resources and public services. Diverging over the longer term, in a lot of data, with these demographic changes, you are going to be positive, with supportive policies and technological innovations, and these are great changes in the labor market. In conclusion, the relationship of population growth with economic development is background-sensitive and needs additional examination to discern the precise treaties and policy interventions that catalyze positive results (Barlow, 1994). Population growth has different effects in high-income and low-income countries. Weak population growth in wealthy countries can create social and economic problems. Accelerating population growth in poor countries can keep development from happening. New migration is considered a potential solution to these mismatches but is often resisted. Constrained population growth and limited migration may result in increased economic inequality at both national and global levels. (Peterson, 2017). The demographic shift profoundly affects the relationship between population expansion and economic development. Accelerated population expansion may lead to economic stagnation due to insufficient capital and reduced market demand. The addition of an adult population, rather than a juvenile population, serves to promote economic growth. Without a dependent youth demographic, an expanding labor force can yield a temporary uplift to economic growth (Crenshaw et al., 1997). Population growth is far more closely linked to economic development than what those arguing against immigration would want us to believe, this relationship being nuanced by the interaction of birth rates, changes in the demographics, and the solutions for the economy provided by policies. Some studies don't connect directly, but others note the importance of controlling for other variables and the situation in other countries. This is the basis for a general planning of the sites of replication in equivalently spinning with monetary methods on per requests; see required. The population growth rate is a crucial independent variable in evaluating the possible efficacy of disaster management.

It has become extremely difficult to carry out disaster management only with the central administration. For this reason, it is important that local administrations be more powerful and functional (Öztürk, 2003). In disaster management, the duties and activities of local administrations may be more important since they are involved in all levels of the state. Local administrations come to the forefront in disaster management with two important aspects. Firstly, local governments can carry out local public activities more efficiently and effectively than the center because they know better what the people's needs are. Secondly, local governments have a more democratic structure since they are closer to the people than the central government. Local governments are democratic units that carry out many local activities, especially zoning, infrastructure, transport, fire brigade, and environmental cleaning (Sezer & Tozbey, 2023). There exists a significant correlation between disaster management and municipalities.

Municipalities are crucial in disaster management, serving as the primary line of defense in preparedness and response efforts. Their closeness to impacted regions allows for the provision of effective and efficient public services during emergencies. Understanding these dynamics is necessary in order to develop appropriate economic approaches in response to the problems and opportunities that population growth entails (Korkin & Öztop, 2022; Putra & Matsuyuki, 2020; Raikes & McBean, 2017). However, the situation varies widely among municipalities, depending on budget allocation, strategic planning, institutional frameworks, etc. It is important to strengthen the capacity of local self-governments for disaster risk reduction to respond effectively to the disasters. It also does entail establishing strategic, legal, and institutional frameworks and enhancing the cooperation/communication of local authorities. It is important to adapt public administration systems continuously to allow municipalities' capacity to prevent, prepare for, respond to, and recover from disasters (Cvetković et al., 2021). Strategic planning is vital for municipalities to address potential risks and threats in a systematic way. Especially metropolitan municipalities should include disaster management in their strategic plans. Even though people are well aware of disasters, strategic management measures don't always work to fix them. Scientific advances are based on interactions with internal experts and external processes, although the mechanisms of innovation are not easily exploited (Korkin & Öztop, 2022; Mehiriz, 2021). The role of institutional conduct and political determination in managing municipal catastrophes is significant. Smaller communities have a hard time reducing disaster risk since they often don't have enough resources, and their administrative capacity is limited, and they depend on public benefits. Larger cities have more freedom to decide how to use resources based on risk assessments (Raikes & McBean, 2017). Historic disaster analysis is vital for municipalities to improve their emergency preparedness and resilience. Within a municipality, events that occur provide better lessons for emergency preparedness than those that happen outside its boundaries. Resources for civil protection activities that are available for the learning process (Betten et al., 2021). Municipal disaster preparedness and response must account for the needs of at-risk groups and internationally displaced migrants. Local governments will need to act as middlemen between different groups to protect the lives of migrants in countries likely to be affected by climate disasters (Kobayashi, 2024). Municipalities are the first responders in a crisis, and their success depends on planning, creating capacity, and getting cooperation from other organizations. Despite existing challenges, particularly for smaller municipalities, promoting innovation, learning, and inclusive practices can improve disaster management capabilities. Municipalities are instrumental in determining the effectiveness of disaster management.

Disasters do not impact all communities or people equally. Vulnerability refers to the physical, social, economic, and environmental factors that increase an individual's or community's susceptibility to the effects of risks. The specific risks women and girls face from the consequences of disasters pose a problem in both developing and developed countries (Işık, 2024). As a matter of fact, women in many countries face difficulties evacuating during disasters. This is because early warning systems and evacuation plans are difficult to understand and adopt. Women who are alone at home during a disaster often do not know the escape routes, and the fact that disaster coordination officials are men makes evacuation difficult for them. After the disaster, men often leave home for economic reasons, resulting in women being left alone with their children, which increases security problems for them. With the aggravation of temporary shelter problems in the aftermath of disasters, physical, psychological, sexual, and economic violence and discrimination against women have increased, and cases of beatings, sexual harassment, and rape are frequently experienced. (Okay & İlkkaracan, 2018). In this sense, women must be taken seriously as part of disaster management.

Women's hardships from calamities often far outweigh men's due to existing gender gaps. Discrepancies in sustenance security epitomize broader disparities regarding access to provisions, fiscal reliance, and societal or cultural limitations compounding female and child vulnerabilities in the face of disasters. Despite occasionally receiving assistance, emergency administration projects often overlook women's competencies, which limits their crucial involvement in repairing ravaged communities. Despite established skills in subsistence generation, healing, and youngster upkeep, gals continue to be relegated to passive characters. Activating women leaders as dynamic members through careful planning and involvement could significantly enhance the outcomes of rescue and restoration efforts (Wraith, 1996). Despite their perceived weakness, women possess unique talents and skills that can aid in crisis management. During and after disasters, they typically play crucial roles in making communities stronger by being leaders, making decisions, and getting resources (Hou & Wu, 2020; Rahmawati & Riskiyah, 2023). Women's participation in disaster management can increase community resilience and improve recovery outcomes through the use of resource management, caregiving, and community organization skills (Bali, 2021; Roy & Mukherjee, 2024). According to a report, there were many challenges faced by women in disaster management, of which limited participation in governance and a lack of recognition for their efforts were key (Noel, 1990; Wraith, 1996). Notwithstanding these challenges, solutions can be sought through gender-sensitive strategies and policies, which value and leverage the input and experience of women (Fordham, 1998; Roy & Mukherjee, 2024). In education, improving the role of women in developing countries, as well as being involved in the process of disaster management, can lead to disaster response and recovery initiatives being more effective and inclusive (Bali, 2021; Hou & Wu, 2020). Therefore, the disaster settings need to have such policy frameworks that would not only promote gender equality but also would engage the women as equal partners in disaster risk reduction and management (Bali, 2021; Roy & Mukherjee, 2024). Frameworks should focus on empowering women and making certain they are involved throughout the disaster management process, from planning to cleanup (Wraith, 1996). In disasters, women are either vulnerable or very precious manpower. We should integrate gender-sensitive policies and practices to work towards reducing gender inequalities and to empower women in all aspects related to disaster resilience as well as recovery processes. Acknowledging and leveraging the strengths of women can improve effectiveness and equity in disaster management for the collective welfare of all communities. From this perspective, it is possible to argue that women are an important variable in ensuring the effectiveness potential of disaster management.

Every year, various natural disasters occur around the world, leaving their consequences on societies in their wake. Children and adolescents are the most sensitive and vulnerable members of disaster-exposed and disaster-affected groups. Millions of children and young people around the world are exposed to natural and man-made disasters every year and are affected by them in different ways (Karabulut & Bekler, 2019). Additionally, the elderly require special support and protection due to their increased vulnerability to disasters (Zafer, 2023). Infants, children, and the elderly should therefore be handled more sensitively than women in disaster management. This is because women are more conscious and easier to act than these groups.

Disaster management must address the diverse physical, mental, and social needs of children, including infants and adolescents. Pediatricians are important for meeting these needs by providing ongoing treatment and being involved in community planning (Krug et al., 2015). Insufficient disaster preparedness and response strategies can compound the physical, psychological, and educational vulnerability of children. Most disaster responses fail to consider the rights and needs of children. Studies employing frameworks like the United Nations Convention on the Rights of the Child demonstrate the necessity of safeguarding children's rights in disaster contexts. In advance of (and because of) these global accords, different countries vary widely in how they exercise those rights, and economic and political issues affect the lives of children. Education might be a way to support children to recover even emotionally from the disasters (Safi & Al-Hassan, 2024). Examples of initiatives include community-based Integrated Management of Newborn and Childhood

Illnesses (IMNCI) training programs in Nepal that focused on increasing the capacities of primary health care workers to manage children in disaster settings. The programs include core topics (trauma, resuscitation, and nutrition), and evaluation has demonstrated significant gains in knowledge and confidence among delegates (Bajracharya et al., 2019). Emergency care providers must equip themselves to handle the unique needs of child victims during disasters. Such expertise includes knowing how to handle different types of disasters and diseases that spread quickly, some of which may involve biological or chemical agents (Lozon & Bradin, 2018). Effective disaster planning must incorporate pediatric considerations within a comprehensive emergency management framework (Mace, Md et al., 2012). Intervention and communication with families with children with medical problems are also crucial during disasters. We suggest proactive outreach, engagement, and leverage of current social networks to promote care and supportive services continuity (Smith et al., 2024). The piece also talks about how important it is for NGOs and government agencies to work together to make disaster risk management more child-friendly (Safi & Al-Hassan, 2024). Disasters severely threaten the nutritional well-being of children and increase their risk of death and morbidity from malnutrition and associated disorders. Feeding practices during infancy and toddlerhood need to be appropriate to prevent malnutrition in the event of a disaster. The development of standardized guidelines to address these issues is a direct outcome of collaborative efforts (Abimibayo Adeoya et al., 2022). The importance of children's role in disaster risk reduction under these guidelines lies in the fact that they could be involved in civil protection and emergency planning (Mort et al., 2020). Disaster management for children should thus be multifaceted in nature to recognize their specific needs, uphold their rights, and engage them in both the planning and the response efforts. In this situation, babies and kids are vital to think about when figuring out how well disaster management might work.

Older people are especially at-risk during catastrophes because of a number of things, including limited physical mobility, decreased sensory awareness, chronic health issues, and social and economic constraints. This makes it harder for them to get ready for and deal with disasters. High-risk groups include older people who are weak and have serious physical, mental, financial, and social problems (Evans, 2010; Fernandez et al., 2002). Older adults may be more vulnerable as a group in disaster situations, but they also demonstrate collective memory and years of experience that can help them adjust and cope with the catastrophic circumstances. By sharing these experiences, older people can strengthen their trust in their resources and their independence to support them in being part of disaster preparedness and response (Navarrete-Valladares et al., 2023). While experience often comes with age, the youthful vigor of a community must work alongside the lessons learned over a lifetime. Where informal networks provide a foundation for shared hardship, both elder wisdom and younger hands strengthen the architecture of resilience. No single generation prepares or recovers alone; through cooperative spirit and mutual reliance, we cultivate our power to persevere. Social capital arises from respect across divides, as connections forged today will shelter all in tomorrow's storms. When generating and mobilizing social capital, they do not just help themselves but also empower others in their community to become better prepared, respond more effectively, and recover quicker from disasters (Howard et al., 2017). Disaster communication is a critical part of experience during disaster events; however, older adults share unique challenges regarding disaster messaging accessibility and comprehension. Poor communication causes insufficient disaster response efforts to deliver timely and accurate information about the vaccine to older adults, requiring effective communication strategies that use the platforms most likely accessible to this population and in collaboration with partners (Gibson et al., 2020). Older adults rely on their personal support networks during disaster situations. These networks also affect preparedness for disasters and local conceptions of disaster management." Social ties also play an important role in promoting disaster preparedness among older people, as they are associated with greater self-efficacy and lower perceived barriers to disaster preparedness among older people (Ashida et al., 2020). Older adults, particularly those providing care for other older adults, encounter substantial obstacles to disaster preparedness. Thus, older caregivers experience an escalation of age and caregiving responsibilities. Understanding these barriers is key to creating effective disaster preparedness strategies that address the unique needs of older adults (Chen & Cong, 2020). Traditional disaster relief efforts often fail to serve older adults adequately. Some of the common problems are late delivery of services, disregard for dignity, and careless demands. Disaster plans need to be culturally sensitive and age-appropriate in terms of addressing the immediate and longer-term needs of older adults. Older adulthood could be transformative for human life and living: older adulthood can be a time of friendship, connection, meaning, purpose, and valuable living (Ardalan et al., 2010). Older adults may be at risk for adverse disaster-related effects due to various physical, social, and economic characteristics; however, they also possess valuable experience and social capital that can serve to promote broader community resilience. Federal, state, and local agencies and organizations must have strong communication and supportive mechanisms in place, as well as disaster management plans that are sensitive to the unique needs of older adults, to minimize the risks they encounter during disasters. From this perspective, older adults have an important role in determining the effectiveness potential of disaster management.

The fiscal consequences of disasters include impacts on public expenditures and revenues. Natural disasters put significant pressure on the budget. In short, disasters have a negative impact on the financial resources of governments. Putting disaster prevention and mitigation plans into action and the costs of disaster response that are paid for by the government put a lot of pressure on already limited budgets. The implementation of disaster mitigation and prevention strategies and the costs associated with publicly financed disaster response impose a significant strain on constrained budgetary resources. Natural disasters negatively affect government revenues (direct and indirect tax revenues) and imports and exports. Consequently, governments encounter pressure to tighten budgets, which leads to rising internal and external borrowing, shrinking stocks of foreign exchange reserves, and an increase in the money supply (Akar, 2013). Hence, a budget is required for effective disaster management.

Government budgets are vital for disaster management, including expenses for proper action and recovery. The Ministry of Public Administration and Security is the responsible ministry for disaster and safety in Korea, and several other government ministries recognize the budget related to disaster and safety management as a considerable public asset. By 2014, there was what the government described as a pre-consultation process to identify wasteful spending and to ensure that budget allocations were created in a way that maintained inside priorities (Heo et al., 2018). Inter-agency coordination still hinders disaster management implementation in Indonesia, although the state budget (APBN) has national emergency funding (Apryana et al., 2020). Developing financial shock absorbers is the other major element, including risk transfer mechanisms such as insurance. Governments should perform regular risk analyses to mitigate disaster impacts and identify contingent liabilities within their budgeting frameworks. (Cevik & Huang, 2018). In Indonesia's Central Aceh Region, disaster management budgeting obeys the law. However, because there aren't any emergency response funds set aside ahead of time, the region's government has to use its own money (Pratama et al., 2019). Disaster budgeting in Indonesia generally focuses on emergency response rather than pre-disaster phases, and budgets fluctuate annually. Nevertheless, this approach does little to facilitate productive disaster management and illustrates the importance of developing budgetary plans that embed risk prevention and mitigation (Mirjas et al., 2021). In Indonesia, the Disaster Risk Index only slightly affected the budget allocation for environmental functions related to disaster impacts (Haris et al., 2023). Several budgeting approaches for Thailand's public health system focus on functions rather than disasters. The 2R's 2P's financial plan seeks to respond to the challenge of enhanced financial management of crisis situations within the first 72 hours following a disaster as well as the challenge of representing local capacities that may be ultimately limited and difficult to predict (Taekratok et al., 2018). Contextualizing geo-resilience in ex ante rate applications in the political dimension, predictors such as disaster stabilization funds and hazard insurance in the US are underutilized but could increase community resilience (Kirschner et al.,

2018). All these examples indicate that effective disaster management needs to be embedded in well-designed budgeting systems that ensure that together we expedite response and build for resilience. Although numerous countries have established systems for managing disaster budgets, challenges persist, including issues related to coordination, proactive risk management, and the development of comprehensive budgeting strategies. Innovative strategies and financial resilience measures are crucial for enhancing disaster preparedness and response. The capacity of revenues to cover expenditures in the budget is crucial for assessing the effectiveness potential of disaster management.

RESULTS AND DISCUSSION

This section also covers the data analysis of the study; socio-economic indicators for the period 1944-2022 are visualized and interpreted with the help of bar charts and wireframe plots.

Periodic Disaster Management

Disaster management between 1944 and 1958

In the 1939 earthquake in Erzincan, many people lost their lives or were injured, and many buildings were destroyed. Following this event, "Act No. 4623 on Measures to be Taken Before and After Earthquakes" was enacted in 1944. Before the earthquake, important decisions were made, such as identifying hazardous areas, planning aid, and surveying the land before construction. This act establishes a legal basis for the duties assigned to administrators and the public. Following the 1946-1949 Muş and Bingöl earthquakes, the Earthquake Bureau was established in 1953 under the Directorate of Building and Zoning. In this period, the "Act on Measures to be Taken Before and After Earthquakes" dated 18.07.1944 and numbered 4623, the "Act on Encouraging Building Construction and Buildings Constructed without Permission" dated 24.07.1953 and numbered 6188, the "Act on Buildings Constructed without Permission" dated 18.12.1953 dated and 6200 numbered "Act on the Organization and Duties of the General Directorate of State Hydraulic Works", 9.7.1956 dated and 6785 numbered "Zoning Act", 1949 dated and 6785 numbered "Regulation on the Construction of Buildings in Earthquake Zones in Türkiye" and finally 1953 dated and 6785 numbered "Regulation on the Structures to be Built in Earthquake Zones" were put into force. (Demir, 2023).

"Building Regulations for Earthquake Zones" was put into force in 1949 with the idea that measures to be taken before disasters occur and risk reduction activities would reduce the damages of the disaster. This regulation constitutes the basis of the "Regulation on Structures to be Built in Disaster Areas" issued in 2007 (Aktel, 2010; Ergünay, 2011; Özmen & Özden, 2013).

The most important regulation on natural disasters between 1944 and 1958 is Act No. 6785. The inadequacy of the 1930 "Municipal Building and Roads Act" in addressing the challenges posed by increased urbanization in the 1950s necessitated the enactment of Act No. 6785. The Act's high priority on building and licensing requires that buildings be constructed according to zoning and direction plans, which prevents construction in disaster-risk areas and introduces sanctions for reinforcing or demolishing buildings already at risk; this legislation is significant for the state's changing strategy regarding disaster perception. The Act also includes a map of earthquake zones related to earth tremors, which is a positive development in terms of the disaster management cycle. (Aydıner, 2014). The Act was repealed in 1985 with the "Zoning Act," numbered 3194, which is still in force.

With the Act no. 7116 dated 09.06.1958 on the "Establishment and Duties of the Ministry of Housing and Settlement," the Ministry of Housing and Settlement was established to take all necessary precautions before and after disasters, to prepare the country planning, to resolve the issues related to dwellings and settlement, and to take over the duties related to disasters from the Ministry of Public Works. The "Civil Defense Act," dated 09.06.1958 and numbered 7162, encompasses the rescue and first aid operations necessary during natural disasters. This act establishes the General Directorate of Civil Defense. Act No. 7269, also known as the "Act on Measures to be Taken and Assistance to be Provided in Case of Disasters Affecting Public Life," repealed Act No. 4623, dated 15.09.1959, and, unlike its predecessors, consolidated previous acts aimed at reducing disaster-related damages into a single act (Kemaloğlu, 2015). In addition to floods and earthquakes, this act also covers disasters such as storms, fires, and rock and dew falls. At the same time, this act envisaged the establishment of a "Disasters Act" outside the general budget (Yılmaz, 2002).

Since there were no major earthquakes in Türkiye in the 1950s, and the earthquakes that did occur did not cause major loss of life and property, the reinforcement of public structures and infrastructures in earthquake areas as stipulated in Act No. 4623 was not implemented. Due to factors like migration from rural to urban areas and unplanned industrialization, the population in cities has increased, leading to unauthorized squatting activities in areas at risk of natural disasters. The "Municipal Building and Road Act" dated 10.06.1933 and numbered 2290 was inadequate

(JİCA, 2004). In light of this, the "Zoning Act" from July 9, 1956, with the number 6785, makes it important to identify dangerous areas that could pose a threat to the settlement and set up scientific and technical systems for responsibility for building inspections (Karagel, 2019).

Disaster management between 1959 and 1998

Since 1958, important policies have been organized and plans have been made to mitigate the damages of disasters in Türkiye (Doğan, 2022). Firstly, the Ministry of Zoning and Settlement was established, then the General Directorate of Disaster Affairs in 1965 and the Earthquake Research Institute in 1971 were established. In this period, when transformations toward reducing disaster damages started, a new earthquake map was created for Türkiye, and activities for natural disasters started to be implemented in cities with the establishment of provincial zoning directorates. Universities also actively participated in these initiatives. The Turkish National Committee for Earthquake Engineering was established in 1965 with researchers from Bosphorus University, Middle East Technical University, and Istanbul Technical University. In addition, a protocol was signed between the Ministry of Housing and Settlement and the Middle East Technical University in 1974, and as a result of this protocol, an Earthquake Engineering Research Centre was established at the Middle East Technical University. (JİCA, 2004). Article 119 of the 1982 Constitution on disasters stipulates that a state of emergency may be imposed in cases of natural disasters, epidemics, or economic crises. With the "Regulation on Principles of Planning," if a disaster exceeds the administrative boundaries of a province, a Central Coordination Board for Disasters is established at the center to ensure uninterrupted planning and coordination services. (Karagel, 2019).

During this period, the "Act on the Establishment and Duties of the Ministry of Reconstruction and Settlement" dated 9.5.1958 and numbered 7116, the "Civil Defense Act" dated 9.6.1958 and numbered 7126, the "Act on Civil Defense" dated 15.5.7269 dated 15.5.5.1959, "Act on Measures to be Taken and Assistance to be Provided in Case of Disasters Affecting Public Life", "Regulation on Structures to be Built in Disaster Areas" (1962), "Regulation on Determination of Right Holders due to Disaster" (1968), "Regulation on Basic Rules on the Effect of Disasters on General Life" (1968), "Regulation on Structures to be Built in Disaster Areas" (1975) and "Regulation on Structures to be Built in Disaster Areas" (1998) were published. Among the duties of the Ministry of Reconstruction and Settlement established during this period were to prepare and have prepared maps and plans of the regions, cities, towns and villages of the country; to take all kinds of measures for reconstruction; to take measures to ensure the supply of the most appropriate and useful building materials according to the conditions and possibilities of the country; to make joint studies with the relevant ministries for regional plans and to take the necessary legal and administrative measures jointly according to these studies; to ensure the resettlement of immigrants and refugees and citizens to be settled in the country according to their special laws; to take necessary measures before and after all kinds of disasters and to deal with all other works related to re-settlement. The establishment of the Ministry of Reconstruction and Settlement has accelerated Türkiye's work on natural disasters. The enactment of Act No. 7269 in 1959 serves as evidence of this. This act is still in force with amendments. In 1965, the General Directorate of Disaster Affairs was established within the structure of the Ministry to carry out the duties related to natural disasters in a faster and more effective manner; in 1971, the Directorate of Earthquake Research Institute was established directly under the Ministry in accordance with the relevant provision of Act No. 7269. This institute was later transformed into a department under the General Directorate of Disaster Affairs (Tercan, 2008).

As a result of the United Nations' understanding of disaster reduction, countries were asked to prepare a program of activities they planned to carry out between 1990 and 2000. Türkiye accepted this request and prepared a national plan with a national committee subject to the supervision of the Ministry of Public Works and Settlement, specifying the activities to be carried out over a ten-year period. Only a small portion of the plan could be realized due to a lack of interest in it and weak resources. (Özey & Ünlü, 2021).

It is aimed at realizing the targets in the legislation that came into force with the successive earthquakes. After the 1992 Erzincan earthquake, Act No. 3838, "Act on the Execution of Services Regarding the Earthquake Disaster in Erzincan, Gümüşhane, and Tunceli Provinces and the Damage and Destruction in Şırnak and Çukurca," was transformed into Act No. 4123 in 1995. On 1 October 1998, the earthquake that occurred in the Dinar district of Afyon led to the expansion of Act No. 4123 (Kıral, 2001).

The process from 1999 to the present

After the 17 August earthquake, Turkish disaster management, which had generally adopted a post-disaster response approach, began to shift towards a disaster risk approach in collaboration with global efforts. (Tercan, 2018). The necessity of disaster risk management has become more evident as a result of the inadequacy and failure of past experiences. (Özmen & Özden, 2013). In the Marmara earthquake, for instance, deficiencies in transport and communication issues hampered search and rescue operations (Şengün, 2007).

After the 1999 earthquake, Act No. 4452, "Act on Authorization on Measures to be Taken Against Natural Disasters and Arrangements to be Made for the Relief of Losses Caused by Natural Disasters," was enacted. Based on this act, Decree Act No. 587 set up the Natural Catastrophe Insurance Institution (TCIP) to pay for damage caused by disasters, regardless of the state's budgetary situation (Özey & Ünlü, 2021). Decree Act No. 583 established the Emergency Management Presidency of Türkiye, and in 2000 this institution was renamed as the General Directorate of Emergency Management of Türkiye with Decree Act No. 600, and the institution was assigned duties such as ensuring coordination among the institutions involved in pre- and post-disaster management, establishing centers related to emergency management in the public sector, and establishing non-governmental organizations that will play a role in emergencies. However, the subsequent period saw the abolishment of this institution. With Act No. 4481, several obligations were introduced in the economic field to reduce the damages of the earthquakes that occurred in and around the Marmara Region (Genç, 2021).

In 2011, after the earthquake in Van, Türkiye started to work on a disaster response plan and tried to determine the areas of responsibility of the relevant organizations. Act No. 6053 brought these institutions into operation in 2022 (Özey & Ünlü, 2021). Over the years, Türkiye has regulated building earthquakes, bringing them into force in 2019. Again in 2019, the Türkiye Earthquake Hazard Map was prepared to indicate the earthquake hazard (AFAD, 2019a).

Another regulation of 2019 is the Provincial Disaster Risk Reduction Plan. The plan, which was prepared and activated in 6 pilot cities in 2020 and 74 cities in 2021, serves as a road map to minimize the possible effects of the disaster (AFAD, 2019b). With the Presidential Decree dated 7.7.2022 and numbered 5787, the Türkiye Disaster Risk Reduction Plan was put into force, and the measures to be taken for 11 types of disasters were determined. The strategic priorities of this plan are in line with the Sendai Framework for Disaster Risk Reduction.

The Development of the Relationship Between Socio-Economic Indicators and Disaster Management Since 1944

Disaster management between 1944 and 1958 with socio-economic indicators

When the Gross Domestic Product (GDP) per capita levels are analyzed in Figure 1, it will be seen that, although they decreased occasionally from 1944 to 1955, they started to increase after 1955. Ultimately, the GDP per capita achieved its peak in 1958. When the regulations in disaster management are analyzed, it will be seen that many acts and regulations were put into force until 1958; however, the number of regulations decreased in 1958. Therefore, we could potentially assert an inverse relationship between GDP per capita and regulations in disaster management. As a matter of fact, the increase in GDP per capita may enable faster and easier resolution of material damages arising from disasters. This relationship may also explain the decrease in the number of regulations.

Figure 1. Per Capita GDP (TL) Levels Between 1944 and 1958

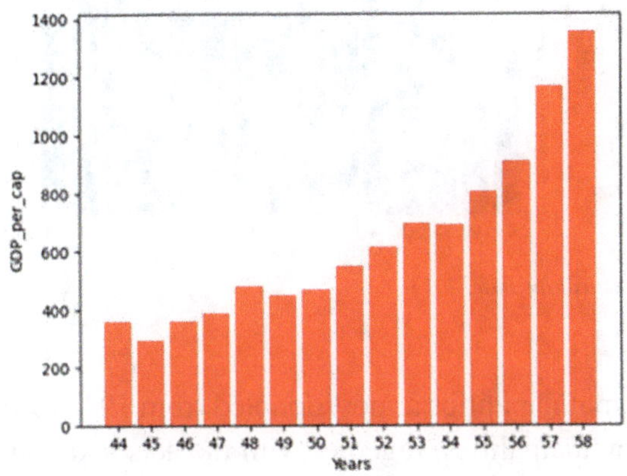

Source: (Turkish Statistical Institute, 2023)

When Figure 2 is analyzed, it will be seen that, although the population growth rate decreases at the beginning, it increases in the following years, and this increase increases significantly until 1954. However, there was a decline in the rate of increase after 1954. In parallel with this increasing trend, regulations on disaster management have also increased. The increase in legal regulation here is cumulative. In this context, it is possible to argue that there is a linear relationship between disaster management regulations and population growth rate.

Figure 2. Population Growth Rate Between 1944 and 1958 (%)

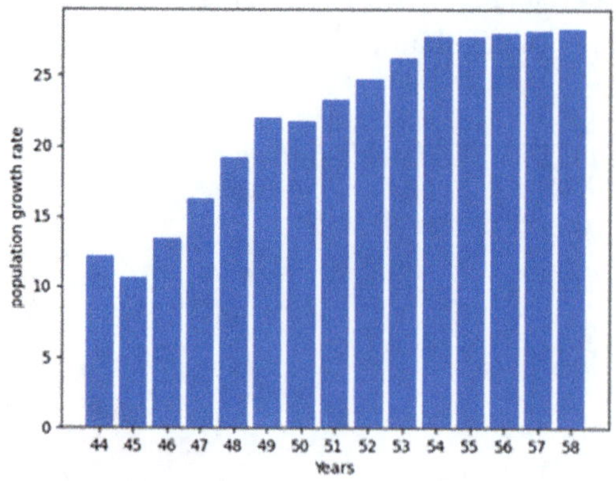

Source: (Turkish Statistical Institute, 2023)

The number of municipalities almost doubled from 1944 to 1958. This increase in the number of municipalities, together with the acts and regulations issued between these periods, can be explained by the potential of local administrations to take more responsibility in disaster management.

Figure 3. Number of Municipalities Between 1944 and 1958

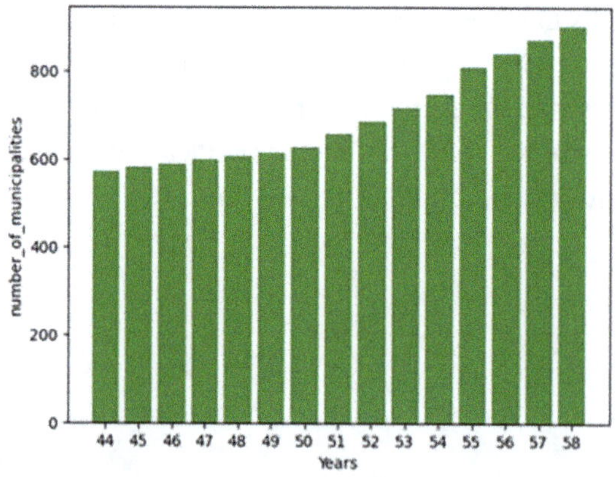

Source: (Turkish Statistical Institute, 2023)

Examining Figure 4, we observe that the male population initially exceeds the female population. However, by 1958, the female population had exceeded the male population. This evidence proves that the legal arrangements made have a direct or indirect link with the number of women. To put it another way, as the number of women rises, numerous regulations emerge to safeguard them.

Figure 4. Ratios of Male Population and Female Population to Total Population between 1944 and 1958 (%)

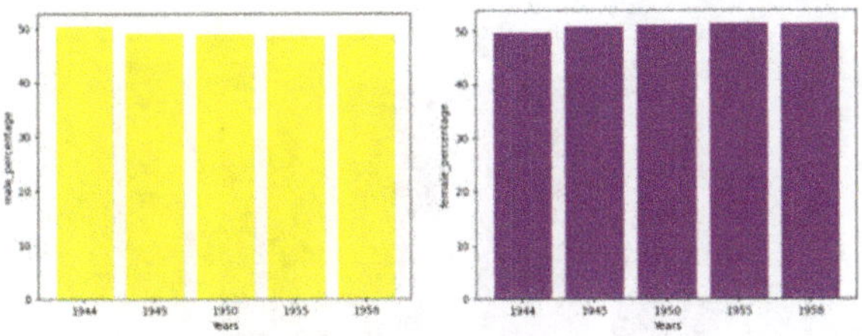

Source: (Turkish Statistical Institute, 2023)

Figure 5 shows the ratio of the number of infants and children (0-19 years old) to the total population between 1944 and 1958. In this context, there is a slight decrease in the number of infants and children from 1944 to 1958. Infants and children, being among the vulnerable age groups during disasters, may benefit from more effective legal regulations due to this decrease.

Figure 5. Ratio of the Number of Infants and Children to Total Population between 1944 and 1958 (%)

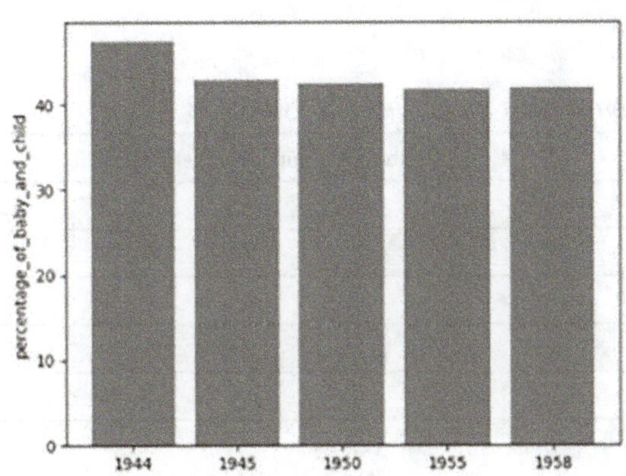

Source: (Turkish Statistical Institute, 2023)

Figure 6 shows that from 1944 to 1958, the number of elderly people did not exceed 3% of the total population. This demonstrates that the legal regulations targeted infants and children instead of the elderly population.

Figure 6. Proportion of the Aged Population in the Total Population from 1944 to 1958 (%)

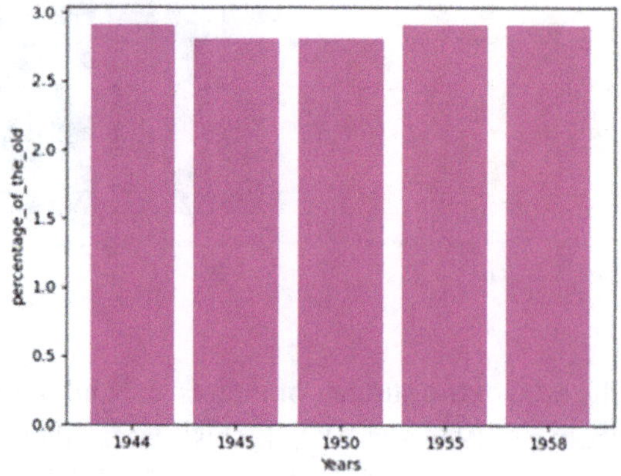

Source: (Turkish Statistical Institute, 2023)

Analyzing Table 1 reveals that over the years, revenues have decreased more than expenditures. This situation indicates that the budget needed to implement legal regulations is decreasing compared to the expenditures, so the measures taken cannot be as effective as desired.

Table 1. Employment Rates by Gender, 1980-2018

Year	Budget Income/Budget Expenditure
1944	99.6
1945	99.6
1950	96.9
1955	96.9
1958	97

Source: (Turkish Statistical Institute, 2023).

Disaster management between 1959-1998 with socio-economic indicator

Between 1959 and 1998, two institutions were established, a protocol was signed, the concept of disaster was included in the Constitution, six regulations and three acts were issued, and one act was expanded.

Figure 7. Per Capita GDP (TL) Levels Between 1959 and 1998

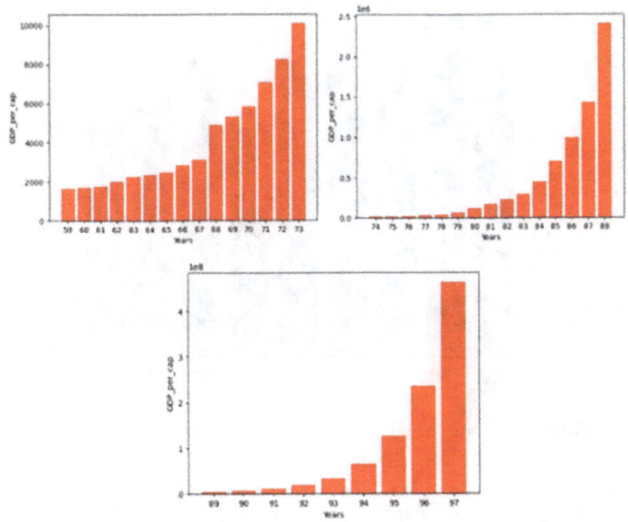

Source: (Turkish Statistical Institute, 2023)

Analysis of Figure 8 reveals a gradual increase in GDP per capita. Simultaneously with this rise in income, we also observe an intensification in disaster-related legal regulations. In addition, the level of GDP per capita, which started low, probably due to problems with Cyprus, increased considerably in 1988. In fact, the concept of disaster has now taken on a constitutional status. Furthermore, there is a geometric increase until 1997. One could argue that regulations for disasters have increased during the periods of geometric growth. Specifically, we can assert that as the per capita income level rises, individuals become more informed. It can be explained as the fact that individuals with more information demand more through their votes for legal arrangements for disasters, which is a kind of public service.

Figure 8. Population Growth Rate Between 1959 and 1998 (%)

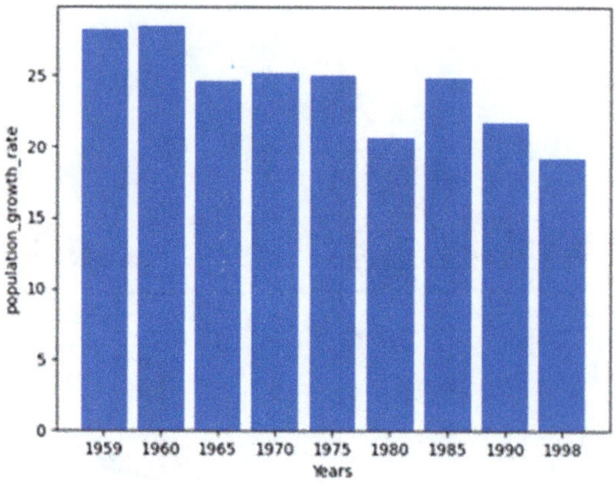

Source: (Turkish Statistical Institute, 2023)

When the population growth rate is analyzed in Figure 8, a significant decrease is observed compared to 1959. Although it first decreases and then increases, this significant decrease in the population growth rate increases the effectiveness of the legal arrangements made for disasters. This is because a low population growth rate may mean a faster recovery in the event of a possible disaster and its aftermath. In addition, the arrangements made can contribute more to this rapid recovery compared to a dense population.

Figure 9. Number of Municipalities Between 1959 and 1998

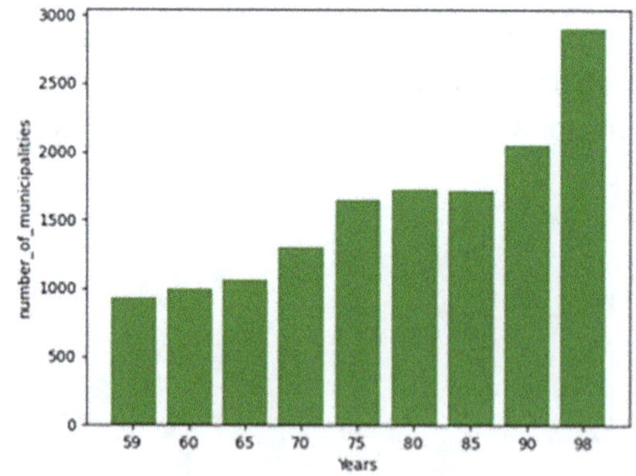

Source: (Turkish Statistical Institute, 2023)

Looking at Figure 9, it is possible to state that the number of municipalities tripled in 1998 compared to 1959. More municipalities mean new arrangements for disaster management. In addition, the increase in the number of municipalities in disaster management makes it easier to manage the disaster. Legal arrangements in this context may have become more effective as municipalities increased.

Figure 10. Ratios of Male Population and Female Population to Total Population Between 1959 and 1998 (%)

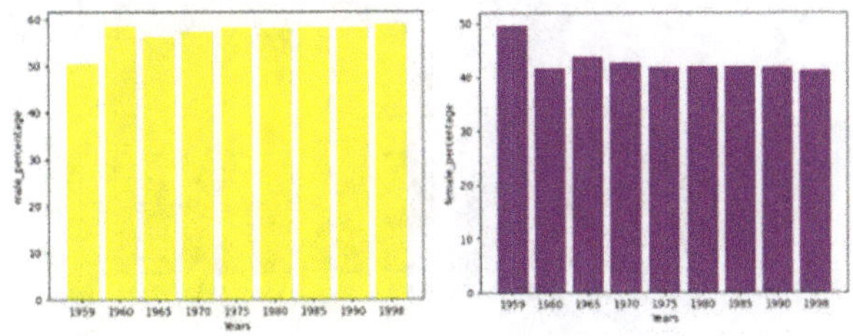

Source: (Turkish Statistical Institute, 2023)

Examining Figure 10, we observe that between 1959 and 1998, the male population outnumbered the female population. The fact that women are more likely to be affected by disasters than men and that the number of women is proportionally less than the number of men can be interpreted as an increase in the probability of the effectiveness of the arrangements made.

Figure 11. Ratio of the Number of Baby and Child to Total Population Between 1959 and 1998 (%)

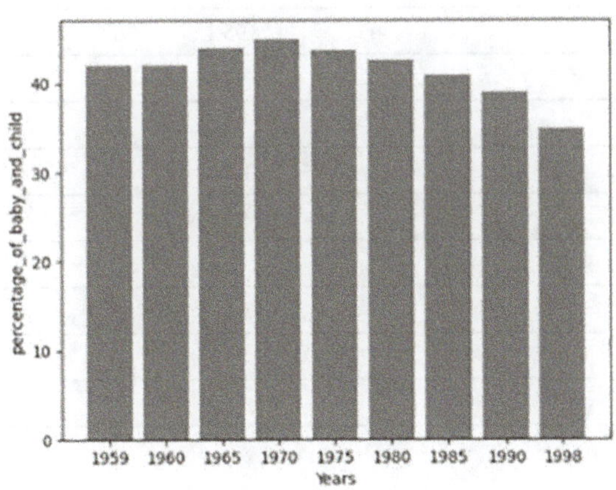

Source: (Turkish Statistical Institute, 2023)

Figure 11 shows that the number of infants and children decreased from 1959 to 1998. We can argue that the efficacy of the rules will increase as the number decreases.

Figure 12. Proportion of the Old Population in Total Population from 1959 to 1998 (%)

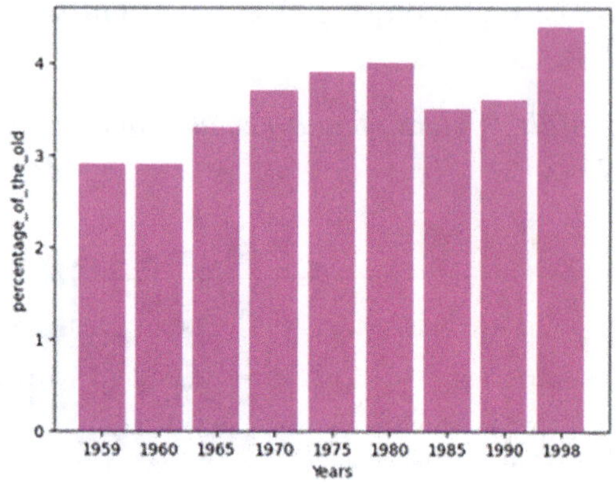

Source: (Turkish Statistical Institute, 2023)

Figure 12 shows that the elderly population was below 3% in 1959 and slightly above 4% in 1998. In fact, the elderly population constitutes a small proportion of the general population. Therefore, it is possible to state that the legal regulations on disasters will be more effective in protecting the elderly, who are a disadvantaged group.

Table 2. Employment Rates by Gender, 1980-2018

Year	Budget Income/Budget Expenditure
1959	95.1
1960	94.9
1965	93.9
1970	100.8
1975	98.8
1980	86
1985	91.7
1990	82.4
1998	75.2

Source: (Turkish Statistical Institute, 2023).

Table 2 shows that budget revenues covered almost all budget expenditures from 1959 to 1970 and even recorded a surplus. However, by 1998, it will be seen that budget revenues covered three-quarters of expenditures, and the budget was in deficit. Therefore, it can be argued that a problem may arise in the financing of the expenditures arising from the fulfillment of the legal regulations related to disaster management.

Disaster management between 1999 and 2022 with socio-economic indicators

In summary, three acts, three decree acts, and one presidential decree were enacted during this period. In this context, Figure 13 shows that the gross domestic product per capita has been increasing gradually. Therefore, it can be said that there is a satisfactory situation regarding national income in terms of pre-earthquake measures and post-earthquake damage reduction. This situation has led to an increase in the effectiveness of the legal arrangements.

Figure 13. Per Capita GDP (TL) Levels Between 1999 and 2022

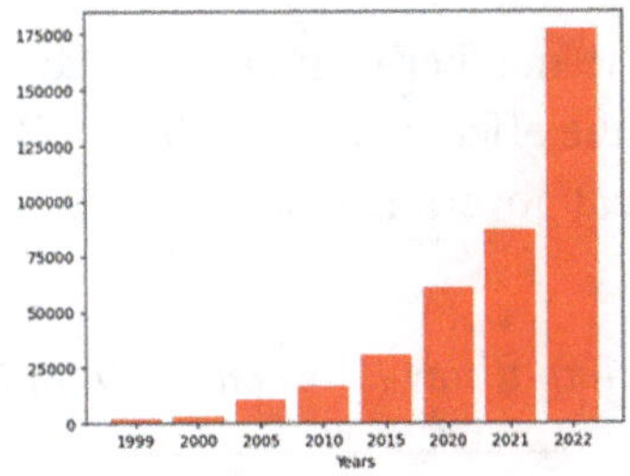

Source: (Turkish Statistical Institute, 2023)

Analysis of Figure 14 reveals a decrease in the population growth rate to less than half between 1999 and 2022. This slowdown in population growth means that the effectiveness of disaster-related regulations will increase further. This is because a lower population increase in disaster management areas makes arrangements more effective and controllable.

Figure 14. Population Growth Rate Between 1999 and 2022 (%)

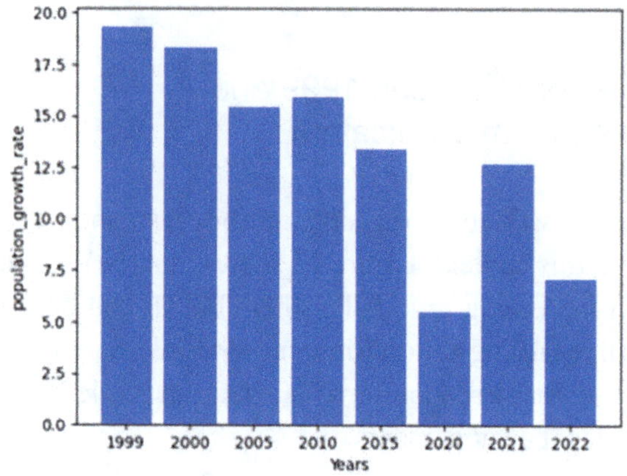

Source: (Turkish Statistical Institute, 2023)

Figure 15 shows that the number of municipalities has decreased to less than half. This data proves that the effectiveness of the regulations shifted to the central government rather than local governments.

Figure 15. Number of Municipalities Between 1999 and 2022

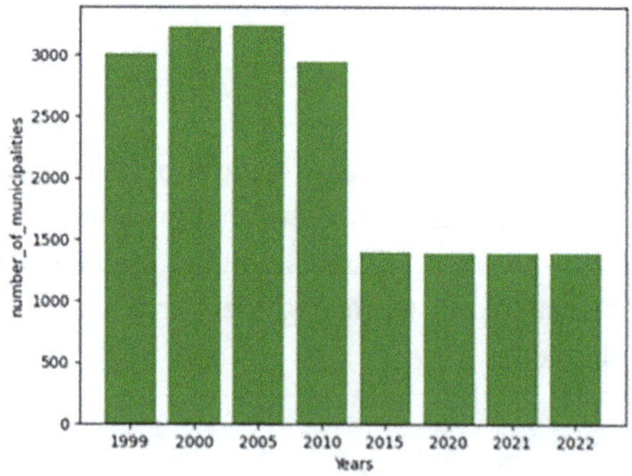

Source: (Turkish Statistical Institute, 2023)

Examining Figure 16, we see that men and women are almost equal, but women will make up more than half in 2022. Therefore, the sphere of influence of disaster management should be larger because women are more vulnerable to disasters than men.

Figure 16. Ratio of Male Population to Total Population Between 1959 and 1998 (%)

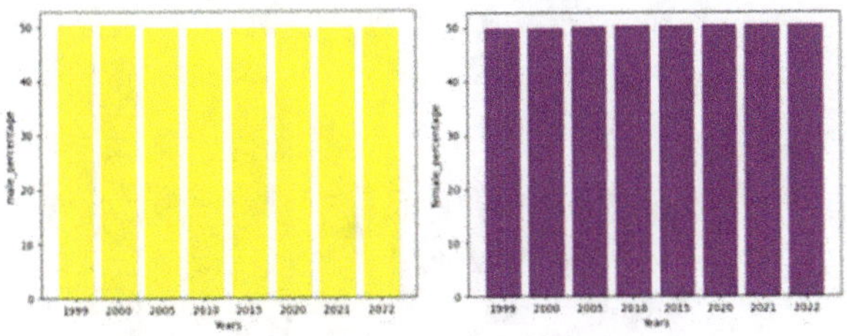

Source: (Turkish Statistical Institute, 2023)

Figure 17. Ratio of the Number of Babies and Children to Total Population Between 1999 and 2022 (%)

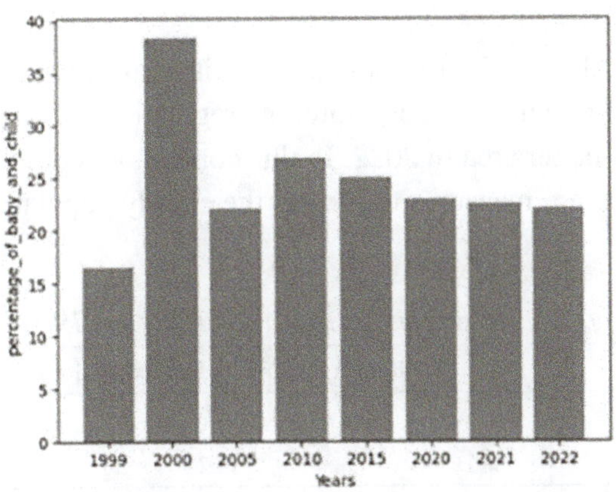

Source: (Turkish Statistical Institute, 2023)

Although the number of infants and children initially more than doubled, by 2022 it will have decreased considerably. Therefore, disaster management arrangements in this period may be more effective.

Figure 18. Proportion of Elderly Population in Total Population from 1999 to 2022 (%)

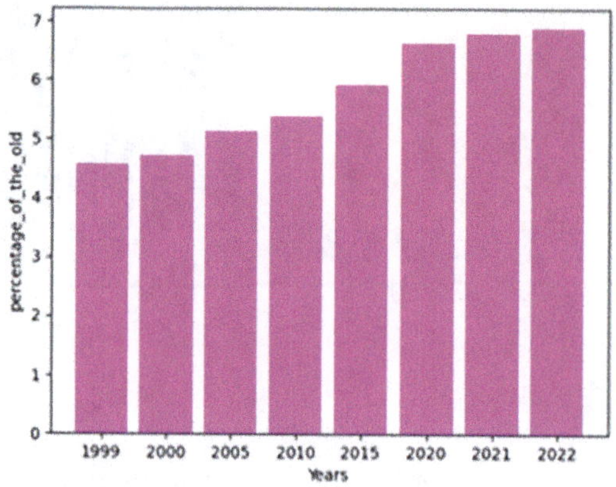

Source: (Turkish Statistical Institute, 2023)

Figure 18 shows that the elderly population has increased over the years, illustrating its ratio to the total population. Therefore, the regulations on disaster management will affect an even larger area in 2022. In this context, it is possible to argue that it is necessary to increase measures to protect the elderly population.

Table 3. Budget Income and Expenditure Ratios Between 1999 and 2022

Year	Budget Income/Budget Expenditure
1999	67
2000	71.2
2005	94.5
2010	85
2015	94.9
2020	84.9
2021	86.9
2022	95.1

Source: (Turkish Statistical Institute, 2023).

Looking at the budget income-expenditure ratios, it is possible to see that the ability of revenues to cover expenditures has increased to almost 100% over the years. This financial situation may make disaster management arrangements more effective.

The most important factor in such a vital issue as disaster management is economic and financial variables. However, it is also important to determine whether and how far this alone is sufficient. In this context, we will assume that z represents a function defined as z=f(x,y), where z denotes disaster management, x indicates budget revenue-expenditure ratios, and y signifies GDP per capita; additionally, disaster management efficiency varies at specific budget and revenue levels.

Figure 19. Wireframe Plot for Disaster Management, per Capita GDP and Budget Income-Expenditure Rate

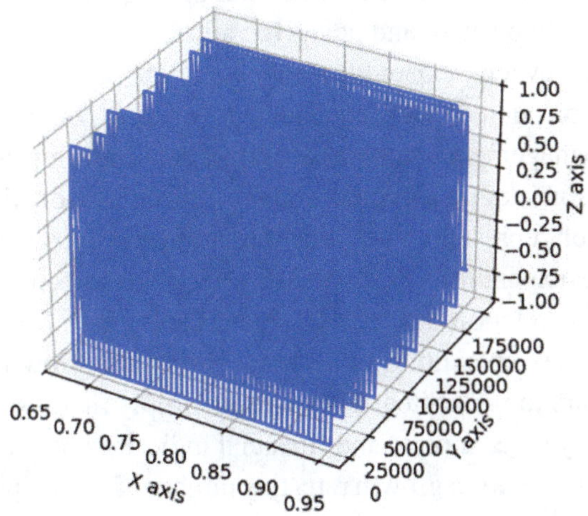

Within the scope of the codes written in Figure 19, it is possible to state the following points about the axes:

- x=np.linspace (0.67, 0.951): The x-axis shows the budget income-expenditure ratios. Budget income-expenditure ratios vary within a narrow range of 67% to 95.1%.
- y=np.linspace (1695, 176651): The y-axis shows GDP per capita. GDP per capita spans a wide range. This axis represents a large economic difference from a low-income level to a high-income level.
- z=np.sin(np.sqrt(x**2+y**2)): The z-axis shows disaster management. This function creates a z value by taking the square root of the sum of the squares of the x and y axes and taking the sine of this value. This creates a periodic structure.

In Figure 24, the disaster management axis shows a significant fluctuation along the GDP per capita axis. These fluctuations show that there are regular increases and decreases in disaster management performance at different levels of GDP per capita. This may indicate that the economic level does not have a strictly linear effect on disaster management. The frequency of fluctuations varies with both budget revenue-expenditure ratios and GDP per capita.

The budget has a limited impact on disaster management because the budget revenue-expenditure ratio axis is in a very narrow range.

Since the GDP per capita axis has a wide range, the graph shows a clear pattern in disaster management as the GDP level increases. Higher income levels do not always seem to be associated with better disaster management because fluctuations on the z-axis take both positive and negative values.

When the budget revenue-expenditure ratio and GDP per capita axes are analyzed together, it can be considered that some budget revenue-expenditure ratios may lead to better disaster management outcomes at certain economic levels. However, in the graph, this interaction is quite complex, and it is difficult to identify a clear pattern. Therefore, this graph represents a non-linear relationship between GDPs per capita and budget revenue-expenditure ratios and disaster management. The graph reveals a lack of a direct correlation between GDP per capita and disaster management. Since budget revenue-expenditure ratios vary within a narrow range, the impact of this factor appears to be limited. These results imply that disaster management may depend not only on economic and financial indicators but also on other social factors. Therefore, population growth rate, the number of municipalities, the female population ratio, the number of infants, and the number of elderly people emerge as important variables in disaster management.

Considering that the population growth rate will decrease significantly until 2022, it can be argued that the area affected by disaster management will be much narrower. Achieving the desired results in a country with a very high population growth rate seems impossible given the current arrangements made in disaster management. This situation implies an increase in the number of individuals in need of protection in a country where disaster management plays a significant role. For instance, the risk of disaster management harming two adults is lower than that of a family with two children. This is because adult men are more favorable in terms of physical characteristics than women and children. On the other hand, a family with a high number of children has a higher risk of being harmed during and after a disaster than a family with a low number of children. In particular, conditions such as consciousness and physical characteristics increase this risk for infants and children. Therefore, disaster management in regions with low population growth rates will yield more effective results than in regions with high population growth rates.

Municipalities are considered within the scope of local governments, and a larger number can provide more effective disaster management by further reducing the responsibilities of the central government. The reason for the distinction is that the population density is very low compared to the national scale; people recognize each other more easily, and thus, injuries are responded to more quickly, especially during and after the disaster. Moreover, local administrations can reach the disaster area more quickly and carry out the first intervention due to their location compared to the central administration. Therefore, increasing the number of local administrations may provide more effective results in disaster management. Given Türkiye's 2022 drop in municipalities, disaster management may be less effective.

Increasing the number of women forces a society to be more effective in terms of disaster management. This is because women are generally more disadvantaged than men in terms of physical characteristics. Especially women with babies and children need to be handled more carefully in terms of disaster management because females must protect not only themselves but also the children under their responsibility from disasters. Effective disaster management undoubtedly unlocks this advantage. The fact that women constitute about half of the population in Türkiye requires effective disaster management. Effectiveness is the essence of successful disaster management. Women's provision of certain facilities enables disaster management. For instance, ensuring women have easier access to food and shelter needs after disaster management can help. In this way, it is possible to protect children under the responsibility of women from the damages of disasters.

Infants and children are one of the vulnerable groups in disaster management. In a society with a high number of infants and children, disaster management inevitably becomes more difficult. This group is among those who consider it challenging to receive training on disaster response. Disaster management should be more effective, especially for this group, whose awareness is less developed than an adult's. There has been a significant decline in the number of babies and children in Türkiye, especially recently. Therefore, the sphere of influence of disaster management has also narrowed. The narrowing of the sphere of influence of disaster management means that its effectiveness can increase significantly. Therefore, making positive arrangements for infants and children in the field of disaster management can enhance its effectiveness.

The growing population of elderly individuals will necessitate more effective disaster management. This is because people in old age are more likely to develop diseases that reduce the quality of life than other groups. Older people also have lower physical strength than younger groups, even if they do not suffer from diseases. Therefore, the elderly are more likely to be vulnerable in disaster management than other groups. In this context, making effective arrangements for the elderly in disaster management plays a vital role in sustaining the lives of people in this group.

CONCLUSION

The GDP per capita increased between 1944 and 1958, while the budget income-expenditure ratios achieved balance. In this respect, it is possible to argue that the financing pillar of disaster management was sufficient at that time. However, this argument alone is not sufficient. The population growth rate increased in this period. The number of municipalities increased. The number of women exceeded the number of men. The number of infants and children declined, while the number of elderly remained almost constant.

It can be argued that the decrease in the population growth rate between 1944 and 1958 led to an increase in the effectiveness of disaster management. It can be stated that the increase in the number of municipalities has eased the burden on the central government and thus made disaster management more effective. On the other hand, since the increase in the population growth rate will increase the area that disaster management should affect, it can be stated that this situation will reduce the effectiveness of disaster management. While an increase in the number of women may decrease the effectiveness of disaster management, a decrease in the number of infants and children can be put forward as an issue that may increase the effectiveness of disaster management. Although the number of elderly remains almost constant, it can be argued that there is a slight loss of effectiveness in disaster management since the presence of elderly populations requires effective disaster management. When this period is analyzed in general, it is possible to argue that progress has been made in the effectiveness potential of disaster management within the scope of socio-economic data.

GDP per capita increased between 1959 and 1998. Budgetary income-expenditure ratios also rose to almost 100% in 1970. However, these rates decreased by almost a quarter in 1998. While the increase in GDP per capita can be considered a positive situation for the financing part of disaster management, the decrease in budget income-expenditure ratios is a negative one. Among these economic and financial indicators, the increase in GDP per capita can be interpreted as the preparedness of individuals for disasters. On the other hand, the decline in budget revenue and expenditure ratios shows that the state is not financially prepared for disaster management. In this respect, it is possible to argue that individuals are more prepared than the state in terms of effectiveness potential in disaster management. We can interpret the increase in municipalities during this period as a shared responsibility between the central government and local governments. In this period, the number of men outnumbered the number of women. Therefore, the effectiveness potential of disaster management is quite high in this sense. Again, the decrease in the number of infants and children in this period can also be considered a positive in terms of the potential effectiveness of disaster management. In this period, the ratio of the elderly population to the total population increased. Therefore, the effectiveness potential of disaster management is relatively low for the elderly population. When the 1959-1998 period is evaluated in general, the increase in GDP per capita and the number of municipalities, as well as the decrease in the number of women, infants, and children, can be considered positive in terms of efficiency potential in disaster management. However, since low budget revenue-to-expenditure ratios can't mobilize these positive effects sufficiently in terms of state financing, it is not possible to mention that sufficient effectiveness potential was reached in disaster management during the 1959-1998 period.

In the period between 1999 and 2022, GDP per capita and budget income-expenditure ratios increased, the population growth rate decreased, the number of municipalities decreased, the female population outnumbered the male population, the number of infants and children decreased, and the share of the elderly population in the total population increased. In other words, a general aging phenomenon has emerged in society.

It is possible to argue that the increase in GDP per capita and budget income-expenditure ratios has positively affected the effectiveness potential of disaster management, both in terms of the state and the individual, in this period. The decrease in the number of municipalities shows that the responsibility in disaster management has shifted more to the central government. Therefore, it is possible to argue that the effectiveness of disaster management has decreased in this sense. The fact that the female population is higher than the male population means that the liability in disaster management increases, and therefore the effectiveness potential of disaster management decreases. Again, the increase in the share of the female population in the total population has imposed responsibility on disaster management, and thus the potential for effectiveness in disaster management has decreased. We can evaluate the decrease in the number of infants and children positively in terms of the effectiveness of disaster management. This improvement is because fewer infants and children make it easier for these groups to benefit from the arrangements made for disaster management. From this perspective, it is possible to state that the effectiveness potential of disaster management has increased. However, the rise in the elderly population means that this disadvantaged group has limited access to disaster management arrangements. In this context, one could argue that the elderly population negatively impacts the effectiveness of disaster management. If we evaluate the period from 1999 to 2022 in general, we can reduce post-disaster damage for both the state and individuals. We can't claim that disaster management, both before and during disasters, is ineffective.

This paper examines the potential effectiveness of disaster administration in Türkiye using socio-economic indicators from 1944 to 2022. The analysis offers a clearer picture of the dimensions of economic growth, public expenditures, debt stock, and demographic trends related to the crucial process of disaster preparedness, response, and recovery. The bar graphs and wireframe plot visualizations used in the study provide a comparative view of changing trends over time, making the relationship between disaster management and socio-economic development more understandable.

The findings demonstrate that disasters not only cause physical destruction but also heighten socio-economic vulnerabilities. In this context, monitoring socio-economic indicators and integrating them into disaster management policies are important for both anticipating risks and developing effective response strategies.

The findings of this study reveal that disaster management capacity is directly linked not only to current socio-economic indicators but also to long-term sustainability goals. Indicators such as per capita income, budget balance, and demographic structure not only determine the level of preparedness for disasters but also overlap with sustainable development goals. The findings on the protection of particularly vulnerable population groups (children, elderly, and women), when evaluated within the framework of social sustainability and social resilience, show that disaster management is not only a technical process but also a social responsibility.

The findings also recommend linking disaster management policy with environmental resilience and adaptation to climate change. In this context, the study positions disaster management as a policy area in line with global goals such as sustainable cities, environmental protection, and climate action, rather than confining it to the framework of risk reduction or crisis response. Therefore, this research contributes both to the development of disaster policies at the national level and to the formulation of strategies in line with international sustainability frameworks.

In conclusion, this study indicates that socio-economic factors should not be ignored in disaster management in the case of and contributes to research in this field. However, conducting comparative analyses with different countries and increasing the use of micro-level data in future studies will allow filling the gaps in the literature more comprehensively.

MANAGERIAL IMPLICATIONS

The findings of this study offer important implications for policymakers, managers, and practitioners working in the field of disaster management. Firstly, it is evident that developing disaster risk reduction and management strategies should take socio-economic indicators into account. Data such as public revenues, debt stock, population structure, and economic growth trends are critical factors determining the level of vulnerability to disasters. Therefore, we should not limit mitigation policies solely to engineering and physical measures but also consider supplementing them with socio-economic metrics.

It shows how important it is to base decisions on planning evidence when dealing with disasters. Specifically, monitoring long-term trends could help predict the potential impact of future calamities on society and the economy. This method will help make better use of public resources and improve the ability to respond.

Finally, this research demonstrates the importance of sectoral collaboration and an interdisciplinary approach in disaster management. Working together with economists, sociologists, engineers, and public administrators will make catastrophe policies completer and more long-lasting. It is therefore essential for policymakers to incorporate research findings into their decision-making processes.

PRACTICAL IMPLICATIONS

The practical benefit of this study is that it offers ideas concerning ways to develop more effective and sustainable policy mechanisms for responding to disasters. First of all, analyzing socio-economic indicators based on long-term trends enables decision makers to predict disaster risks more accurately. This paper provides a concrete approach that can be used both in policy development processes at the national level and in the planning activities of local governments.

Second, the visual analyses (bar graphs and wireframe plots) used in the study make complex data sets more understandable for practitioners and clearly illustrate the relationship between disaster management and socio-economic vulnerabilities. This method can guide policymakers and practitioners in their data-driven decision-making processes.

Thirdly, the study reveals that focusing only on physical infrastructure investments will be insufficient in disaster management and that socio-economic dynamics should also be considered. This necessitates the effective use of social policies, economic stability, and public finance instruments in building disaster-resilient societies.

As a result, the findings obtained will provide guidance for public administrators, disaster planners, and policymakers and will contribute to the development of more holistic and applicable strategies in disaster preparedness, response, and recovery phases.

THEORETICAL IMPLICATIONS

The theoretical reflections of this study provide a new contribution to the existing literature on disaster management from the perspective of socio-economic indicators.

The study reveals that disaster management is not only an operational process but also a holistic phenomenon closely related to economic and social vulnerabilities. Thus, the integration of the socio-economic dimension into disaster management theories is strengthened, and the theoretical framework in this field is expanded.

Through visual analysis of long-term socio-economic data, the study also supports traditional risk, resilience, and sustainability approaches used in disaster management. This adds methodological diversity to disaster management research, which typically relies on case studies or short-term field data from the literature.

Moreover, it contributes to the theoretical debate on the "relationship between disaster management and development" by showing that socio-economic indicators can be used to explain disaster management effectiveness. In this context, the study emphasizes that disaster management is not only a crisis response but also a conceptual field directly related to social welfare and economic stability.

In conclusion, this study brings an interdisciplinary perspective to the disaster management literature and paves the way for socio-economically oriented theoretical discussions for future research.

RECOMMENDATIONS

Disaster management since 1944 has covered many legal regulations. There is even a constitutional regulation on disaster management. Acts and decrees with the force of an act provided support and expanded the scope of legal arrangements. However, disaster management is not a straightforward issue that legal regulations alone can address. Changes in socio-economic variables also have the capacity to change the potential for effectiveness by affecting disaster management positively or negatively. In this context, it is necessary to make some innovations in these socio-economic variables and legal regulations in disaster management for the future.

When making legal arrangements in disaster management, it is necessary to act in accordance with the conditions of the day. In this context, we should renew the existing legislation and, if necessary, create new regulations. To be more effective, these regulations should be made within the Act and Constitution because, as you move up the hierarchy of norms, non-compliance with disaster management regulations will have serious legal and criminal consequences. Although making strict regulations increases the effectiveness potential of disaster management, acting only in this way does not provide the desired result in disaster management.

Sometimes, taking the necessary precautions before a disaster can be very effective in minimizing the damage experienced during it. For instance, building earthquake-resistant houses greatly reduces the risk of earthquake damage. However, constructing earthquake-resistant structures involves significant expenses.

Therefore, both the state and individuals should be ready for this situation in terms of income. Therefore, both the public and private sectors should engage in income-generating activities within the Turkish economy. All individuals should have equal opportunities to participate in economic activities while carrying out income-generating activities. This approach ensures a perfectly competitive market and prevents deadweight loss. If these conditions are met, on the one hand, GDP per capita will increase, and on the other hand, government budget revenues will increase due to increased incomes. In this way, individuals will be able to have the income level to build earthquake-resistant structures, while the state will be able to finance the expenditures that may arise due to disasters.

The central administration's exclusive responsibility for disaster management in public administration diminishes its potential for efficiency, as the distance to the disaster-affected area reduces its effectiveness. At this point, local administrations should also participate effectively in disaster management. Therefore, increasing the number of municipalities allows the central government to share its responsibility for disaster management, making it more effective. In any case, increasing the number of municipalities will strengthen local administrations, leading to greater efficiency in disaster management. Increasing the number of municipalities will mean increasing the potential for effectiveness in disaster management.

Women are in the vulnerable group in disaster management. Therefore, during a disaster, women are among the groups most susceptible to impact. Especially for women with children, the responsibility in disaster management increases even more. On the other hand, infants and children are also among the vulnerable groups. It is possible to state that the elderly are among these groups. Reducing the population size is a factor that enhances the potential effectiveness of disaster management. However, it's crucial to implement these reductions annually. We achieve these objectives by reducing the rate of population growth. Reducing the rate of population growth is not a matter of worsening conditions but of effective population planning. Effective population planning reduces the rate of population growth.

As a result, to increase the effectiveness potential of disaster management, new arrangements should be made in the above-mentioned steps in the hierarchy of norms. We should adapt existing regulations to the current conditions. Afterwards, it is necessary to realize economic activities that can increase GDP per capita. We must also ensure a fair distribution of income in this situation. Consequently, as income levels rise, so will the amount of taxes collected. In this way, it will be possible to construct disaster-resistant buildings. However, these alone will not be enough. Effective population planning should also slow down the rate of population growth. We must ensure that the regulation of the population growth rate does not contribute to aging. If we don't, we risk increasing the vulnerable elderly population and potentially creating an economic paradox of austerity. Therefore, population planning should be organized in a way not to increase the elderly population. In addition, vulnerable groups need to be effectively involved in disaster management. With these measures to be taken in economic, financial, and social terms, the potential for effectiveness in disaster management can be considerably increased, and the damage caused by disasters can be minimized.

FUTURE WORKS

The avenues that this study opens for future research can contribute to the development of the disaster management literature. First of all, since the study is based only on Turkish data, comparative analyses can be conducted with socio-economic indicators of different countries in the future. This approach allows for a clear understanding of whether disaster management effectiveness is universal or specific to country conditions. Second, the current study focuses on visual and descriptive analyses; in the future, further relationships can be tested using econometric modeling, structural equation modeling, or machine learning-based estimation methods. Moreover, dynamic models can also examine the impact of disaster management policies on long-term socio-economic indicators. In addition, field research, surveys, and qualitative research with local actors would complement the macro-level results of this study at the micro level. Finally, it will be important to integrate the impact of current factors, such as climate change, migration movements, and urbanization, on disaster management effectiveness in future studies.

This study highlights the significance of socio-economic variables in disaster management and identifies essential aspects for consideration in future research. First, we should consider the effects of climate change on the frequency and severity of disasters. Integration of climate-oriented indicators will contribute to the development of more sustainable and long-term policies. Secondly, the role of technological advances in disaster management is a key field, which could explain forthcoming research further on. The evaluation of technological tools such as artificial intelligence-based risk analyses, early warning systems, and digital coordination platforms will enable the development of more dynamic solutions beyond the current findings. Third, linking post-disaster recovery processes with socio-economic indicators is one of the gaps in the literature. Measuring dimensions such as post-disaster reconstruction, psychosocial support, and economic recovery will enable a holistic approach to disaster management. Subsequent research must comprehensively investigate the function of civil society organizations and local communities in disaster management. Voluntary activities, solidarity networks, and civic efforts at the local level provide significant potential to mitigate socio-economic vulnerabilities and foster disaster-resilient communities. Studies focusing on these areas will deepen the contributions of current research and help create a broader perspective in the disaster management literature.

REFERENCES

Abimibayo Adeoya, A., Sasaki, H., Fuda, M., Okamoto, T., & Egawa, S. (2022). Child nutrition in disaster: A scoping review. *The Tohoku Journal of Experimental Medicine*, *256*(2), 103–118. PMID: 35173109

AFAD. (2019a). *Türkiye deprem tehlike haritası.* https://www.afad.gov.tr/turkiye-deprem-tehlike-haritasi

AFAD. (2019b). *Türkiye il afet risk azaltma planı.* https://www.afad.gov.tr/turkiye-afet-risk-azaltma-plani-tarap

Akao, K.-I., & Sakamoto, H. (2018). A theory of disasters and long-run growth. *Journal of Economic Dynamics & Control*, *95*, 89–109.

Akar, S. (2013). Doğal afetlerin kamu maliyesine ve makro ekonomiye etkileri: Türkiye değerlendirmesi. *Yönetim ve Ekonomi Araştırmaları Dergisi*, *21*, 185–185.

Aktel, M. (2010). 5902 sayılı yasa ile Türkiye'de afet yönetiminde oluşan değişim. *Dumlupınar Üniversitesi Sosyal Bilimler Dergisi, 27.*

AlHinai, Y. S. (2020). Disaster management digitally transformed: Exploring the impact and key determinants from the uk national disaster management experience. *International Journal of Disaster Risk Reduction*, *51*, 1–19. PMID: 32953438

Apryana, H., Nugraheni, K. R., & Wardhana, F. (2020). Policy analysis of budget allocation for disasters in Indonesia. *Social, Humanities, and Educational Studies (SHEs): Conference Series, 3(1).*

Ardalan, A., Mazaheri, M., Naieni, K. H., Rezaie, M., Teimoori, F., & Pourmalek, F. (2010). Older people's needs following major disasters: A qualitative study of Iranian elders' experiences of the bam earthquake. *Ageing and Society*, *30*(1), 11–23.

Ashida, S., Thompson, L., Hejna, E., Obot, S., & Schneider, H. (2020). Personal support networks of older adults in relation to their perceptions about disaster management. *Innovation in Aging*, *4*(Supplement_1), 439–440.

Ashraf, A. (2021). Lessons learned from covid-19 response for disaster risk management. *Natural Hazards*, *107*, 2027–2032. PMID: 33716403

Aydıner, T. (2014). *Doğal afet yönetişimi: Türkiye'de doğal afet yönetimi uygulamalarının tarihsel bağlamda değerlendirilmesi* [Unpublished master thesis]. Pamukkale Üniversitesi, Denizli, Türkiye.

Aziz, T. (2018). Myth and reality of vulnerability to disaster: Pressure and release model for hazards in Bangladesh. *Journal of Southeast Asian Studies*, *6*(1), 23–31.

Bajracharya, S., Shrestha, A., & House, R. (2019). Newborn and child health care in humanitarian crisis settings: Piloting of training package for primary health care workers in rural Nepal. *Prehospital and Disaster Medicine*, *34*(s1), s32–s33.

Bali, R. (2021). Women as equal partners in disaster risk reduction planning and management. *Research Review International Journal of Multidisciplinary*, *6*(5), 111–117.

Baraldo, M., & Di Giuseppantonio Di Franco, P. (2024). Place-centred emerging technologies for disaster management: A scoping review. *International Journal of Disaster Risk Reduction*, *112*, 1–18.

Barlow, R. (1994). Population growth and economic growth: Some more correlations. *Population and Development Review*, *20*(1), 153.

Betten, T., Vian Pettersen, K., & Albrechtsen, E. (2021). Learning in municipalities after disasters. *Disaster Prevention and Management: An International Journal*, *30*(3), 400–411.

Bhide, N., & Khanolkar, S. (2020). A comprehensive study on the factors impacting the gdp (per capita) of major economies around the globe using regression analysis. *International Journal of Computer Applications*, *175*(10), 26–30.

Biesenthal, C., Clegg, S., Mahalingam, A., & Sankaran, S. (2018). Applying institutional theories to managing megaprojects. *International Journal of Project Management*, *36*(1), 43–54.

Bozkurt, A. (2023). *Python: Veri analizi, görselleştirme ve yapay zekâ*. Hiperlink.

Britton, N. R. (1986). Developing an understanding of disaster. *The Australian and New Zealand Journal of Sociology*, *22*(2), 254–271.

C nh, T., & Hoài, P. (2022). Factors affecting gdp per capita—apply bayesian analysis. *Science & Technology Development Journal - Economics - Law and Management, 6(2),* 2862-2868.

Cavallo, E., & Noy, I. (2011). Natural disasters and the economy—A survey. *International Review of Environmental and Resource Economics*, *5*(1), 63–102.

Cevik, S., & Huang, G. (2018). How to manage the fiscal costs of natural disasters. *IMF How To Notes*, *2018*(003), 1–15.

Chen, Z., & Cong, Z. (2020). Barriers to disaster preparation when older adults are taking care of older adults. *Innovation in Aging*, *4*(Supplement_1), 437–438.

Çobanoğlu, B. (2023). *Veri bilimi için python (2. Baskı)*. Kodlab.

Crenshaw, E. M., Ameen, A. Z., & Christenson, M. (1997). Population dynamics and economic development: Age-specific population growth rates and economic growth in developing countries, 1965 to 1990. *American Sociological Review*, *62*(6), 974–984.

Cutter, S. L. (2024). The origin and diffusion of the social vulnerability index (SoVI). *International Journal of Disaster Risk Reduction*, *109*, 1–15.

Cvetković, V. M., Tanasić, J., Ocal, A., Kešetović, Ž., Nikolić, N., & Dragašević, A. (2021). Capacity development of local self-governments for disaster risk management. *International Journal of Environmental Research and Public Health*, *18*(19), 1–33. PMID: 34639706

Dědeček, R., & Dudzich, V. (2022). Exploring the limitations of GDP per capita as an indicator of economic development: A cross-country perspective. *Review of Economic Perspectives*, *22*(3), 193–217.

Demir, R. (2023). Afet hukuku ve afet mevzuatı. In Çapar, S. (Ed.), *Türkiye'de mevzuat bağlamında afet ve acil durum yönetimi* (pp. 101–149). TİAV.

Doğan, H. H. (2022). Türkiye'de afet olaylarına kalkınma planlarının yaklaşımı. *İDEALKENT, 13(37),* 1873-1912.

Domorenok, E., Graziano, P., & Polverari, L. (2021). Introduction: policy integration and institutional capacity: theoretical, conceptual and empirical challenges. *Policy and Society*, *40*(1), 1–18.

Ekberg, M. (2007). The parameters of the risk society: A review and exploration. *Current Sociology*, *55*(3), 343–366.

Ergünay, O. (2011). *Ekim 11). 1999 depremleri afet mevzuatını nasıl değiştirdi: mevzuat açısından neredeyiz? Türkiye Deprem Mühendisliği ve Sismoloji Konferansı, 11-14 Ekim 2011*. ODTÜ.

Evans, J. (2010). Mapping the vulnerability of older persons to disasters. *International Journal of Older People Nursing*, *5*(1), 63–70. PMID: 20925759

Fernandez, L. S., Byard, D., Lin, C.-C., Benson, S., & Barbera, J. A. (2002). Frail elderly as disaster victims: Emergency management strategies. *Prehospital and Disaster Medicine*, *17*(2), 67–74. PMID: 12500729

Fineman, M. A. (2017). Vulnerability and inevitable inequality. *Oslo Law Review*, *4*(3), 133–149.

Fordham, M. H. (1998). Making women visible in disasters: Problematising the private domain. *Disasters*, *22*(2), 126–143. PMID: 9654812

Furedi, F. (2007). The changing meaning of disaster. *Area*, *39*(4), 482–489.

Genç, F. N. (2021). *Afet Yönetimi*. Nobel Akademik Yayıncılık.

Ghaffarian, S., Taghikhah, F. R., & Maier, H. R. (2023). Explainable artificial intelligence in disaster risk management: Achievements and prospective futures. *International Journal of Disaster Risk Reduction*, *98*, 1–22.

Gibbons, R. D., Olfson, M., Saulsberry, L., Edlund, M. J., Zangeneh, S., Bareis, N., Chwastiak, L., Gibbons, J. B., & Kessler, R. C. (2024). Social vulnerability and prevalence and treatment for mental health and substance use disorders. *JAMA Psychiatry*, *81*(10), 976–984. PMID: 39046728

Gibson, A., Kusmaul, N., & Brown, L. (2020). Improving communication for older adults during disaster events. *Innovation in Aging*, *4*(Supplement_1), 693–693.

Gümüş, D., Yılmaz, Z., Tuncer, M., & Aydın, E. (2024). Afet yönetiminde hemşirelerin rolü. *Afet ve Risk Dergisi*, *7*(2), 354–367.

Gurikar, S. G. (2024). Chaos theory and its implications in physical systems. *Shod-Kosh: Journal of Visual and Performing Arts*, *5*(1), 1723–1728.

Haris, N., Furqan, A. C., Kahar, A., & Karim, F. (2023). Disaster risk index on disaster management budgeting: Indonesia's national data set. *Jàambá*, *15*(1), 1–7. PMID: 36873605

Hayward, M., Hunt, R. & Miller, D. (2022). *How Vulnerability Enriches Family Firm Relationships: A Social Exchange Perspective, 13(1),* 1-7.

Heo, B.-Y., Park, J. H., & Heo, W.-H. (2018). Sustainable disaster and safety management of government: Integrated disaster and safety budget system in Korea. *Sustainability*, *10*(11), 1–17.

Hermansson, H. (2019). Challenges to decentralization of disaster management in turkey: The role of political-administrative context. *International Journal of Public Administration*, *42*(5), 417–431.

Hou, C., & Wu, H. (2020). Rescuer, decision maker, and breadwinner: Women's predominant leadership across the post-Wenchuan earthquake efforts in rural areas, Sichuan, China. *Safety Science*, *125*, 1–6.

Howard, A., Blakemore, T., & Bevis, M. (2017). Older people as assets in disaster preparedness, response and recovery: Lessons from regional Australia. *Ageing and Society*, *37*(3), 517–536.

Hu, Y. (2013). *Rural health care delivery*. Springer.

Iqbal, U., Perez, P., & Barthelemy, J. (2021). A process-driven and need-oriented framework for review of technological contributions to disaster management. *Heliyon*, *7*(11), 1–19. PMID: 34841111

Işık, E. (2024). Afet ve kadın: Toplumsal cinsiyete dayalı zarar görebilirlik, afet yönetimi, ruh sağlığı ve psikososyal destek hizmetleri. *Akdeniz Kadın Çalışmaları ve Toplumsal Cinsiyet Dergisi*, *7*(2), 882–909.

JİCA. (2004). *Türkiye'de doğal afetler konulu ülke strateji raporu* (ss. 1-163). Türkiye Cumhuriyeti İçişleri Bakanlığı.

Jong, A. (2022). World risk society and constructing cosmopolitan realities: A bourdieusian critique of risk society. *Frontiers*, *7*, 1–13. PMID: 35573124

Karabulut, D., & Bekler, T. (2019). Doğal afetlerin çocuklar ve ergenler üzerindeki etkileri. *Doğal Afetler ve Çevre Dergisi*, *5*(2), 368–376.

Karagel, H. (2019). Afet anı ve sonrası (afet kriz yönetimi). In Sever, R. (Ed.), *Afetler ve afet yönetimi* (pp. 160–183). Pegem.

Kemaloğlu, M. (2015). Türkiye'de afet yönetiminin tarihi ve yasal gelişimi. *Akademik Bakış Dergisi*, *52*, 126–147.

Kernick, D. (2018). *Complexity and healthcare organization- a view from the street*. CRC Street.

Khan, H., Vasilescu, L. G., & Khan, A. (2008). Disaster management cycle-a theoretical approach. *Management & Marketing*, *6*(1), 43–50.

Kıral, Ö. (2001). 1939 ve 1992 Erzincan depremleri şehircilik düzenimizde hasar yarattı mı? *Planlama*, *4*, 11–18.

Kirschner, C. L., Singla, A., & Flick, A. (2018). Financing community resilience before disaster strikes: lessons from the United States. In Kirschner, C. L., Singla, A., & Flick, A. (Eds.), *Oxford Research Encyclopedia of Natural Hazard Science*. Oxford University Press.

Kobayashi, Y. H. (2024). The significance of local government in disaster management for international migrants: The case of Minoh city, Osaka prefecture. *Disasters*, *48*(S1), e12636. PMID: 38864590

Korkin, E., & Öztop, S. (2022). Büyükşehir belediyeleri stratejik planlarında afet yönetimi. *Medeniyet Araştırmaları Dergisi*, *7*(1), 61–75.

Krug, S. E., Chung, S., Fagbuyi, D. B., Fisher, M. C., Schonfeld, D. J., Shook, J. E., Chun, T. H., Conners, G. P., Mba, M. M., Conway, E. E., Dudley, N. C., Fuchs, S. M., Lane, N. E., Macias, C. G., Mph, M., & Moore, B. R. (2015). Ensuring the health of children in disasters. *Pediatrics*, *136*(5), e1407–e1417. PMID: 26482663

Lam, N. S. N., Meyer, M., Reams, M., Yang, S., Lee, K., Zou, L., Mihunov, V., Wang, K., Kirby, R., & Cai, H. (2023). Improving social media use for disaster resilience: Challenges and strategies. *International Journal of Digital Earth*, *16*(1), 3023–3044.

Lindell, M. K. (2013). Disaster studies. *Current Sociology*, *61*(5-6), 797–825.

Lozon, M. M., & Bradin, S. (2018). Pediatric disaster preparedness. *Pediatric Clinics of North America*, *65*(6), 1205–1220. PMID: 30446057

Lu, F. (2022). Management of natural disaster and its influence on economic–environmental performance: Fresh evidence from BRICS. *Frontiers in Environmental Science*, *10*, 901017.

Mace, Md, S. E., Doyle, Md, C., Fuchs, Md, S., Gausche-Hill, Md, M., Koenig, Md, K. L., Sorrentino, Md, A., & Johnson, Md, R. W. (2012). Pediatric patients in a disaster: Part of the all-hazard, comprehensive approach to disaster management. *American Journal of Disaster Medicine*, *7*(2), 111–125. PMID: 22916449

Mah, J. C., Penwarden, J. L., Pott, H., Theou, O., & Andrew, M. K. (2023). Social vulnerability indices: A scoping review. *BMC Public Health*, *23*, 1–11. PMID: 37380956

Mahmud, A., Mohammad, Z., & Abdullah, K. A. (2020). Leadership in disaster management: Theory versus reality. *Journal of Clinical and Health Sciences*, *5*(1), 4–11.

Math, S. B., Nirmala, M. C., Moirangthem, S., & Kumar, N. C. (2015). Disaster management: Mental health perspective. *Indian Journal of Psychological Medicine*, *37*(3), 261–271. PMID: 26664073

Mehiriz, K. (2021). The sources of municipalities' innovation in the management of weather disaster risks, their relationships, and their antecedents. *Local Government Studies*, *47*(6), 951–970.

Mirjas, M., Fahlevi, H., & Diantimala, Y. (2021). Analisis tren penganggaran bencana di daerah rawan bencana: Studi kasus pada kabupaten aceh selatan. *Jurnal Kajian Akuntansi*, *5*(2), 243.

Mort, M., Rodriguez-Giralt, I., & Delicado, A. (Eds.). (2020). *Children and young people's participation in disaster risk reduction*. Policy Press.

Mythen, G. (2021). The critical theory of world risk society: A retrospective analysis. *Risk Analysis, 41*(3), 533–543. PMID: 30170338

Navarrete-Valladares, C., Sandoval-Díaz, J., & Sandoval-Obando, E. (2023). Experience and local memory of older people in the face of disasters: A systematic review. *Frontiers in Public Health, 11*, 1163561. PMID: 37293609

Nirupama, N. (2012). Risk and vulnerability assessment: A comprehensive approach. *International Journal of Disaster Resilience in the Built Environment, 3*(2), 103–114.

Noel, G. E. (1990). The role of women and NNAs in disaster management. *International Nursing Review, 37*(6), 363–367. PMID: 2276904

Noy, I. (2009). The macroeconomic consequences of disasters. *Journal of Development Economics, 88*(2), 221–231.

Okay, N., & İlkkaracan, İ. (2018). Toplumsal cinsiyete duyarlı afet risk yönetimi. *Resilience, 2*(1), 1–12.

Oleg, O. K., Zaytsev, A., & Dimitriev, N. (2023). Role and place of institutional capacity in socio-economic development. *International Journal of Professional Business Review, 8*(10), 1–23.

Özey, R., & Ünlü, M. (2021). *Afetler ve afet yönetimi (2. edition)*. Aktif Yayınevi.

Özmen, B., & Özden, T. (2013). Türkiye'nin afet yönetim sistemine ilişkin eleştirel bir değerlendirme. *İstanbul Üniversitesi Siyasal Bilgiler Fakültesi Dergisi, 49*, 1-28.

Öztürk, N. (2003). Türkiye'de afet yönetimi: karşılaşılan sorunlar ve çözüm önerileri. *Çağdaş Yerel Yönetimler, 12(4)*, 42-64.

Peterson, E. W. F. (2017). The role of population in economic growth. *SAGE Open, 7*(4), 1–15.

Pratama, T., Dirhamsyah, M., & Madjid, I. (2019). Analysis of budget development disaster management in the central aceh district disaster management agency. *International Journal of Multicultural and Multireligious Understanding, 6*(2), 339.

Putra, D. I., & Matsuyuki, M. (2020). The disaster-management capabilities of local governments: A case study in Indonesia. *Journal of Disaster Research, 15*(4), 471–480.

Quarantelli, E. L. (1988). Disaster crisis management: A summary of research findings. *Journal of Management Studies, 25*(4), 373–385.

Rahman, Md. M., Rahman, Md. S. & Jerin, T. (2022). Social vulnerability to earthquake disaster: Insights from the people of 48th ward of Dhaka south city, Bangladesh. *Environmental Hazards*, *22*(2), 116–135.

Rahmawati, I., & Riskiyah, F. (2023). Exploring the invisibility of women's resilience in facing anthropogenic disasters. *International Journal of Social Science Research and Review*, *6*(1), 421–432.

Raikes, J., & McBean, G. (2017). Institutional response to disaster risk: The city of Vancouver and district of maple ridge, British Columbia, Canada. *Regional Environmental Change*, *17*(8), 2433–2441.

Reinhardt, G. Y., & Ross, A. D. (2019). Expanding social science through disaster studies. *Social Science Quarterly*, *100*(7), 2523–2529.

Roy, D., & Mukherjee, M. (2024). Challenges and opportunities for women in disaster risk management. *IDRiM Journal*, *14*(2), 228–240.

Safi, I., & Al-Hassan, M. (2024). Reassessing disaster and emergency management: A child rights perspective. *Scholars Bulletin*, *10*(02), 32–38.

Şahin, Ş., & Üçgül, İ. (2019). Türkiye'de afet yönetimi ve iş sağlığı güvenliği. *Afet ve Risk Dergisi*, *2*(1), 43–63.

Sanderson, D., Heffernan, T., DeSisto, M., & Shearing, C. (2024). Community-centred disaster recovery: A call to change the narrative. *Disasters*, *49*(1), 1–15. PMID: 39229770

Sandoval, V., & Voss, M. (2016). Disaster governance and vulnerability: The case of chile. politics and governance. *Politics and Governance*, *4*(4), 107–116.

Schmidtlein, M. C., Deutsch, R. C., Piegorsch, W. W., & Cutter, S. L. (2008). A sensitivity analysis of the social vulnerability index. *Risk Analysis*, *28*(4), 1099–1114. PMID: 18627540

Şengün, H. (2007). *Marmara Depreminde Bayındırlık ve İskan Bakanlığı Çalışmaları ve Hukuksal Sorunlar. TMMOB Afet Sempozyumu, 5 Aralık 2007*. TMMOB.

Sezer, Ö., & Tozbey, Ş. (2023). Türkiye'de afet yönetiminin merkezi ve yerel düzeyde örgütlenmesi üzerine bir değerlendirme. *Siyasal Bilgiler Fakültesi Dergisi*, *8*(1), 79–102.

Shadab, M. (2019). Chaos theory: Dynamics and applications. *The Pharma Innovation*, *8*(2), 922–925.

Sinha, J. N. (1959). Population growth and balance in economic development. *Economic Development and Cultural Change*, *7*(3, Part 1), 206–215.

Smith, N., Donaldson, M., Mitton, C., & Lee, E. (2024). Communication in disasters to support families with children with medical complexity and special healthcare needs: A rapid scoping review. *Frontiers in Public Health*, *12*, 1229738. PMID: 38544735

Sørensen, M. P. (2017). Ulrich Beck: Exploring and contesting risk. *Journal of Risk Research*, *21*(1), 6–16.

Southwick, S. M., Bonanno, G. A., Masten, A. S., Panter-Brick, C., & Yehuda, R. (2014). Resilience definitions, theory, and challenges: Interdisciplinary perspectives. *European Journal of Psychotraumatology*, *5*(1), 1–14. PMID: 25317257

Swalha, I. H. (2020). A contemporary perspective on the disaster management cycle. *Foresight*, *22*(4), 469–482.

Taekratok, T., Phanthunane, P., & Taekratok, T. (2018). Disaster-based budgeting in new public health policy of Thailand. *Procedia Engineering*, *212*, 715–722.

Tao, Z., & Han, L. (2022). Emergency response, influence and lessons in the 2021 compound disaster in henan province of china. *Environmental Research and Public Health*, *19*(1), 1–19. PMID: 35010748

Taylor, W. M., & Levine, M. P. (2015). Investigating catastrophe: Commemoration, accountability and records of disaster. *National Identities*, *17*(2), 105–116.

Teker, O. (2021). *Derinlemesine python data science*. Godoro.

Tercan, B. (2008). Afet *bölgelerinde yeniden yerleştirme ve iskan politikaları: doğubeyazıt afetzede yerleşim alanları uygulama örneği* [Unpublished doctoral dissertation]. Ankara Üniversitesi, Ankara, Türkiye.

Tercan, B. (2018). Türkiye'de afet politikaları ve kentsel dönüşüm. *Abant Kültürel Araştırmalar Dergisi*, *3*(5), 63–74.

Toya, H., & Skidmore, M. (2014). Do natural disasters enhance societal trust? *Kyklos*, *67*(2), 255–279.

Tsasis, P., & Nirupama, N. (2008). Vulnerability and risk perception in the management of hiv/aids: Public priorities in a global pandemic. *Risk Management and Healthcare Policy*, *1*, 7–14. PMID: 22312198

Turkish Statistical Institute. (2023). *Indicators of 100 years*. Turkish Statistical Institute.

Vermiglio, C., Noto, G., Bolívar, M. P. R., & Zarone, V. (2021). Disaster management and emerging technologies: A performance-based perspective. *Meditari Accountancy Research, 30*(4), 1093–1117.

Walker, B. H. (2020). Resilience: What is and is not? *Ecology and Society, 25*(2), 11.

Warren, K., Franklin, C., & Streeter, C. L. (1998). New directions in systems theory: Chaos and complexity. *Social Work, 43*(4), 357–372.

Wen, J., Wan, C., Ye, Q., Yan, J., & Li, W. (2023). Disaster risk reduction, climate change adaptation and their linkages with sustainable development over the past 30 years: A review. *International Journal of Disaster Risk Science, 14*, 1–13.

Wraith, R. (1996). Women in disaster management: Where are they? *Australian Journal of Emergency Management, 11*, 8.

Xu, C. (2025). Analysis of the state-of-art chaos: Theory and applications. *Highlights in Science. Engineering and Technology, 128*, 231–237.

Yılmaz, A. (2002). Türkiye'de afet zararlarının azaltılması çalışmalarının tarihi gelişimi üzerine notlar. *Türk İdare Dergisi, 74*(434), 155–169.

Zafer, M. (2023). Afetlerde sosyal hizmet: Kovid-19 afet salgini ve yaşlilar. *Kadim Akademi Sbd, 7*(1), 37–50.

KEY TERMS AND DEFINITIONS

Age: It is a concept that refers to the time elapsed since the birth or formation of an individual, organism or object.

Budget: It is a financial document in which the revenues and expenditures of a state are planned over a certain period of time (usually a year).

Disaster Management: It is a discipline that covers the prevention, mitigation, response and recovery processes of natural or man-made disasters.

Effectiveness Potential: It refers to how effectively and efficiently an individual, organisation, system or process can operate within the framework of its existing capabilities, resources and conditions.

Gross Domestic Product (GDP): It is an economic indicator that expresses the total monetary value of all final goods and services produced by a country over a given period of time (usually a year or quarter).

Municipality: It is a public institution that provides local administration of a city, district or town.

Population Growth Rate: It is a measure of how much the population in a given country or region has increased or decreased over a certain period of time.

Chapter 11
Türkiye's Role in Global Warming:
Machine Learning-Assisted Analysis

Beyza Güdek
https://orcid.org/0000-0002-7432-9234
Karadeniz Technical University, Turkey

ABSTRACT

Global warming, which is a worldwide problem, has reached an uncontrollable point every passing day. Many national and international institutions and organizations set various targets and carry out activities to keep global temperatures constant and reduce them. In this study, a comprehensive examination of Türkiye's global temperature change is carried out. The effects of variables such as energy consumption, fossil fuel use, population growth, and deforestation on global temperature are analyzed using machine learning-based methods. Data from the World Bank and Our World in Data are used in the study. These data are analyzed using time series and regression analysis methods. The results show that Türkiye's impact on global temperature will increase by 19.93% by 2030. Population growth and fossil fuel consumption stand out as the most important factors. The study presents status reports to evaluate the sustainability of Türkiye's climate policies and develop sustainable solutions to policies.

INTRODUCTION

Global temperature increase, as one of the most complex and multidimensional environmental problems of the 21st century, deeply affects both natural systems and

DOI: 10.4018/979-8-3373-1862-2.ch011

social, economic, and political structures. The increase in greenhouse gas (GHG) concentration in the atmosphere is related to human activities, in particular industrialization, fossil fuel use, deforestation, and rapidly increasing population. The carbon-based economic growth process that accelerated after the Industrial Revolution caused carbon dioxide (CO2) levels in the atmosphere to reach historically unprecedented levels, which paved the way for the Earth to warm faster than ever before in the last thousand years (Maslin, 2011). In Türkiye, increasing energy demand, fossil fuel dependency, and industrial-intensive development models constitute the regional reflection of these global trends (Karakaya, 2016).

To limit the devastating effects of climate change and keep the global average temperature increase below 2°C, a series of international environmental regulations like the Kyoto Protocol, the United Nations Framework Convention on Climate Change (UNFCCC), and the Paris Climate Agreement has entered into force. Türkiye is trying to develop various strategies at the national level by becoming a party to these agreements. Türkiye, which became a party to the UNFCCC in 2004, assumed obligations with limited responsibility by being among the Annex-I countries; but it was removed from the Annex-II list in 2001 in line with its request for developing country status (Bodansky, 2012; Karakaya, 2016). Türkiye, which has committed to reducing its emissions by 21% by 2030 with reference to the "business as usual" scenario within the framework of the Paris Agreement, must establish a delicate balance between environmental sustainability and economic growth in this process (Ministry of Foreign Affairs of the Republic of Turkey, 2023).

But the literature indicates that Türkiye's current energy and industrial policies are still based on structures that create high carbon emissions; therefore, it shows that it faces serious difficulties in achieving its emission reduction targets. Türkiye's factors like population growth, energy consumption, fossil fuel intensity, and deforestation contribute not only to national environmental problems but also to global climate change. In this context, the analysis of Türkiye's carbon emission trends and the effects of these trends on global temperature is of critical importance both in terms of evaluating the effectiveness of national environmental policies and restructuring international responsibilities.

This study is aiming to quantitatively reveal Türkiye's impact on global temperature change. The effects of environmental pressure elements like population growth, per capita energy consumption, fossil fuel use rate, and shrinkage of forest areas on global temperature changes modeled using time series analysis and multiple linear regression (MLR) methods. While most studies in literature focus on the estimation of GHG emissions, this research aims to make a more original contribution by directly measuring Türkiye's impact on global temperature. The findings will enable an assessment of the extent to which Türkiye's climate policies are effective in terms of sustainability; they will also serve as a guiding resource for policy makers in shaping environmental strategies at both national and international levels.

BACKGROUND

Studies on climate change examine the effects of human activities on environmental systems in a multidimensional manner. Dynamic factors like increasing industrialization, urbanization, and energy demand along with the industrial revolution have significantly increased the amount of GHG released into the atmosphere, which has become one of the main causes of global warming. In this context, studies conducted based in Türkiye address the determinants of GHG emissions within the scope of a number of social, economic, and environmental variables such as economic growth, energy consumption structure, population growth, deforestation rates, industrial production, and environmental policies (Bölük & Mert, 2014; Huang et al., 2008). CO2 emissions are closely related to fossil fuel consumption. Researchers have conducted studies by addressing indicators such as per capita energy use, energy density, and fossil-renewable energy ratios (Hosseini et al., 2019; Lotfalipour et al., 2010; Yoro & Daramola, 2020).

The development of Türkiye's climate change policies is a complex and multilayered process shaped by international agreements. This process has also been quite painful for Türkiye. Türkiye became a party to UNFCCC adopted at the Rio Environment and Development Summit in 1992 in 2004 and became an Annex I country under the obligation to limit emissions (UNFCCC, 1992). But Türkiye removed from the Annex II list in 2001 on the grounds that it was like a developing country in terms of its level of economic development and historical responsibility (Karakaya, 2016). This situation caused Türkiye to benefit from international support mechanisms such as climate finance, technology transfer, and capacity building to a limited extent (Ozturk & Acaravci, 2013).

One of the important milestones in Türkiye's integration into global climate policies was the Intended Nationally Determined Contribution (INDC) declaration submitted within the framework of the Paris Agreement in 2015. Within the scope of this declaration, Türkiye committed to reduce its emissions by 21% by 2030 "if everything continues as it is" (Ministry of Foreign Affairs of the Republic of Turkey, 2023). But Türkiye's ability to achieve this goal depends on transforming its current energy policies, increasing its investments in renewable energy sources, and implementing an effective GHG monitoring and reporting system at the national level. It stated in the literature that Türkiye's GHG reduction policies are not currently effective enough and that current trends carry the risk of not achieving this goal (Halicioglu, 2009; Karakaya, 2016). This situation reveals the need for a more balanced policy framework between sustainable development goals and environmental obligations. This study examines and analyzes Türkiye's factors affecting global warming in a multidimensional manner. In this context, the factors affecting global warming discussed below are based on studies in literature.

CO_2 Emissions and Economic Growth

In Türkiye, economic growth is related to the increase in CO2 emissions along with increasing industrialization, urbanization, and energy consumption. The Environmental Kuznets Curve (EKC) hypothesis, which is frequently discussed in developing countries, suggests that economic growth initially increases environmental degradation, but emissions will reduce when a certain level of welfare is reached, with the adoption of environmentally friendly technologies and increased environmental awareness (Halicioglu, 2009; Jorgenson & Clark, 2010; Neagu, 2019). With reference to this theoretical framework, developing countries like Türkiye need to exceed a certain income threshold and transform their energy structures accordingly to achieve environmental sustainability in the long term. But empirical studies conducted specifically for Türkiye show that this transition point has not yet reached, and economic growth is still accompanied by increasing CO2 emissions (Hosseini et al., 2019; Kais & Sami, 2016).

The dominance of fossil fuels in Türkiye's energy production, its energy intensive industrial structure, and limited renewable energy integration are the main factors that complicate this process. It stated that the increase in energy consumption per capita triggers CO2 emissions and that this increase when combined with population growth, poses a threat to sustainable development goals (Karakaya, 2016). But when Türkiye's current policy framework examined, it seen that measures to increase energy efficiency and promote clean energy sources still limited. So, for the EKC hypothesis to become operational for Türkiye, both economic growth and structural transformations, the dissemination of low carbon technologies, and the strengthening of environmental management capacity required (Başoğlu, 2013). The current course of events reveals that Türkiye is still on the "rising part" of the EKC curve, that is, economic growth realized at the expense of environmental costs.

Energy Consumption and Greenhouse Gas Emissions

Fossil fuel consumption continues to be the main source of CO2 emissions both globally and in Türkiye. Türkiye's energy production structure is based on fossil resources such as coal, oil, and natural gas. This has made the energy sector the most dominant component of total GHG emissions. It emphasized that approximately 90% of Türkiye's GHG emissions are energy related, and this is due to fossil fuel consumption .(Bakay & Ağbulut, 2021; Ozcan, 2016; Özer et al., 2013). The prevalence of coal fired thermal power plants stands out as a critical factor that increases Türkiye's carbon intensity. In this context, it stated that it is difficult to reduce CO2 emissions unless the energy sector undergoes a structural transformation

Though an increase in investments in renewable energy sources has been observed in recent years, the share of these investments in total energy supply still limited. Despite the growth of Türkiye's renewable energy capacity, there are serious structural deficiencies in the dissemination of energy efficiency practices and the integration of low carbon technologies (Özer et al., 2013). For example, despite the potential of wind and solar energy, the effective use of these resources cannot sufficiently achieve due to technical, administrative, and economic reasons. In addition, the inadequate promotion of clean production technologies in energy intensive sectors limits emission reduction efforts.

In this context, Türkiye's energy transformation process requires not only increasing renewable resource investments, but also modernizing its energy infrastructure, developing carbon pricing mechanisms, and making energy efficiency strategies applicable. It is frequently stated in the literature that Türkiye's current energy policies are not compatible with long term emission reduction targets and that this situation poses a risk within the framework of international obligations (Özer et al., 2013; Sahin & Esen, 2022).

Deforestation and Carbon Sequestration

Forest ecosystems play a vital role in reducing GHG by biodegrading CO2 in the atmosphere. As terrestrial carbon sinks, forests can freely be written in terms of both balancing emissions and sustainability of climate conditions in the fight against climate change (Baccini et al., 2012; Solomon et al., 2007; Waheed et al., 2018). When the expansion evaluated in Türkiye, anthropogenic pressures such as increasing urbanization, expansion of agriculture, and the spread of industrial facilities lead to increases in forest yields and weakening of the country's carbon retention capacity. This situation both reduces the resilience of natural ecosystems and increases the amount of carbon in the atmosphere, exacerbating climate change (Stern, 2008; Waheed et al., 2018).

Though current statistics on forest assets in Türkiye show a nominal increase, the qualities of forests should also be observed for this increase to create a real carbon correlation effect. In other words, not only the increase in forest loss surface area, but also factors like biomass density, species diversity, and sustainable management practices directly affect the course of carbon sinks (The World Bank, 2024). These sustainable forestry practices, regulations for the protection of natural forests and protection of forest degradation, have a resilience that stands out as an important tool in the resistance of total GHG emissions.

Though it stated in the literature that the forest policies of Türkiye, afforestation projects and rural development, are increasing forest treatment (Başoğlu, 2013), it stated that these efforts require effectiveness in combating climate change. The increase seen in forest areas, the rapid increase in CO2 emissions, and the reduce in the extent of acceleration; In short, more comprehensive, holistic and forestry policies compatible with emission reduction targets needed. This economic, forest management should be evaluated not only for the protection of biodiversity but also as an integral part of national GHG reduction (Stern, 2008).

Industrialization, Urbanization and Environmental Impacts

With the acceleration of the industrialization process in Türkiye since the 2000s, there has been a significant increase in energy demand and, accordingly, GHG emissions. The growth seen particularly in the heavy industry, construction, and energy production sectors has both increased energy intensity and diversified the main sources of carbon emissions in the atmosphere (Dong et al., 2014). In the analysis conducted on the industrial sector, it stated that Türkiye's industrial emissions are increasing faster than in European countries, which has caused a significant increase in the country's carbon intensity. This situation shows that Türkiye has achieved its economic growth at the expense of environmental costs and that the sustainability balance has weakened (Çakmak & Okutan, 2023; Karaman & Alp, 2024).

The increase in emissions per energy unit is related to fossil fuels still dominating energy production. In this context, the need for the wider implementation of carbon-cutting measures in production and consumption comes to the fore (Ozdemir et al., 2024). The limited implementation of energy efficiency programs in the industrial sector, along with the slow progress in the development of low-carbon production technologies, is responsible for the tendency for the current trend in emission growth to continue.

On the other hand, urbanization and transportation policies also exert determining influences on Türkiye's CO2 emission profile. Especially with the accelerating urbanization process, demand for transportation infrastructure has risen, which has resulted in a rise in the consumption of fossil fuels. The prevalence of the road transportation-based transportation system in Türkiye has caused it to be one of the primary sources of urban emissions (Ozdemir et al., 2024). This reality demonstrates the need to align urbanization policy with sustainability principles and promote low carbon transportation systems such as public transport and electric vehicles.

All these dynamics indicate that sectoral, integrated, and long term policy design is necessary in order for Türkiye to meet its carbon emission reduction commitments. Literature indicates that energy, industry, and transport policies of Türkiye have a form that amplifies carbon intensity, and it is against the emission reduction commitments (Karakaya, 2016).

Population Growth and Emission Trends

Population growth, which has gained speed in Türkiye since the 2000s, has become a driver of the increase in CO2 emissions by directly putting pressure on energy consumption, transportation demand, and production activities (Hosseini et al., 2019; Jorgenson & Clark, 2010; Lane, 2011). The demand for energy in Türkiye has risen considerably as a result of the growing population, and as a consequence, carbon emissions rise due to the fossil fuel dependent structure of energy production (Halicioglu, 2009; Ozturk & Acaravci, 2013). As the population grows, housing, transportation, and industry expand; direct energy consumption and indirect GHG emissions increase.

Population growth both increases consumption and increases the pressure on energy infrastructure, threatening the sustainability of existing energy systems. In this context, it emphasized that the transition from fossil resources to renewable energy technologies in energy production should accelerated to stop the current emission increase trend in Türkiye. In addition, making energy consumption per capita more efficient, i.e., increasing energy efficiency, stands out as one of the basic policies to control the total level of emissions (Huseynov & Gezdim, 2024; Ozdemir et al., 2024; Türkyılmaz & Öztürk, 2024).

The World Bank (2024) data also reveal that Türkiye's energy demand has been continuously increasing along with its population growth, and this situation has created pressure on energy intensity and carbon intensity indicators. In fact, there has been a steady increase in total GHG emissions in parallel with population growth; this necessitates that emission reduction targets planned in an integrated manner with population policies (Jorgenson & Clark, 2010; Yang et al., 2015). For Türkiye to realize its low carbon development vision, a holistic energy strategy needed to balance population based consumption increases. This strategy should be addressed within a framework that includes not only technological but also social and economic transformation.

METHODOLOGY

In this study, data collection, pre-processing, feature selection and time series forecasting techniques were used to evaluate the impact of Türkiye on global temperature change; in this context, Exponential Triple Smoothing (ETS), Multiple

Linear Regression (MLR) and Moving Average (MA) models were applied. The data obtained from the World Bank's open-access database (The World Bank, 2024). For global temperature data, Our World in Data, an open-access platform used by working groups, used (Our World in Data, 2024). To make this data suitable for analysis, various preprocessing steps applied, as seen in Figure 1.

Figure 1. Implementation Stages of the Research

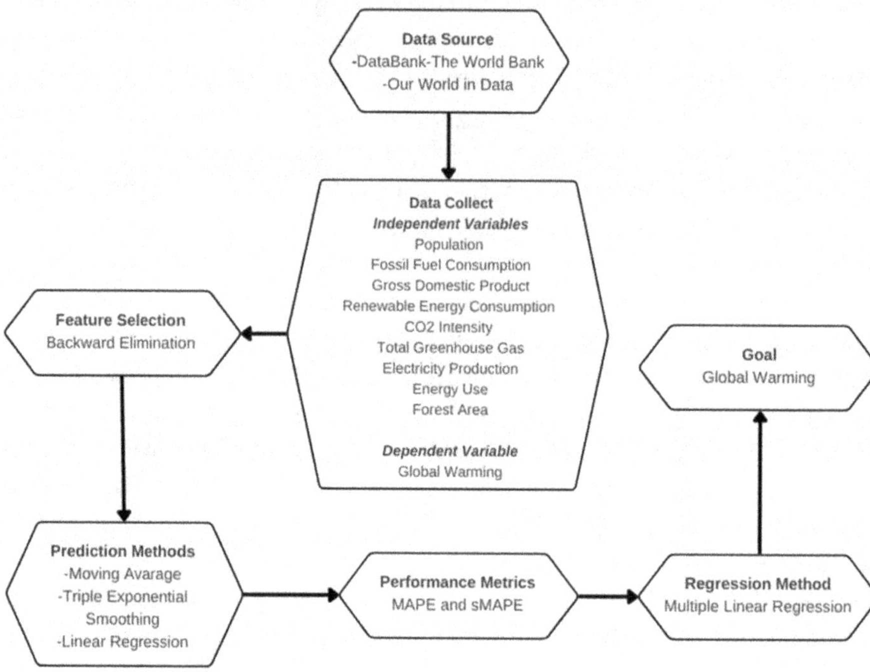

Data Collection and Preprocessing

The dataset of this study covers the period 1990-2015. Global temperature change is the dependent variable. Population, gross domestic products (GDP), CO2 intensity, consumption of fossil fuels, use of renewable energy, and total GHG emissions are independent variables.

The World Bank's global climate and energy dataset applied during data collection. Missing data completed with the mean value method (Uğuz, 2021). In time series studies, missing data should be properly excluded in order to ensure data integrity (Jolliffe & Stephenson, 2011).

In regression analysis, ineffective variables removed from the analysis using the backward elimination. To determine ineffective variables, the distribution of each variable examined, and extreme values evaluated with the standard deviation and interquartile range (Thornes & Stephenson, 2001). In linear regression (LR) model, removing outliers helps to increase the accuracy of the model (Morley et al., 2018).

Feature Selection and Modeling

The backward elimination used in the feature selection phase. This method initially includes all independent variables. Then, the least statistically significant variables gradually removed from the analysis and the final variables are determined (Durbin, 1959).

A regression model including all independent variables created in study and the P values of the variables calculated. Variables with a P value greater than 0.05 removed one by one, starting from the largest value. To ensure the stability of the model, variables with the lowest F value removed from the analysis (Thompson, 1978).

So, to increase the accuracy of the model, the effect of the selected independent variables on global temperature change assessed and the most effective variables were determined. In conclusion, the model created on statistically significant independent variables and has high reliability (Ruta & Gabrys, 2005).

In study, the most significant independent variables explaining the effect of Türkiye on global temperature were determined and the model optimized. In process, R^2, Adjusted R^2, and significance (F) values considered in determining the success rate of the model (Partridge & Yates, 1996).

Time Series Analysis and Prediction

Time series analysis is a statistical approach used to predict future trends and changes based on historical data. In study, three different time series methods applied to perform global temperature prediction: MA, ETS, LR. These methods in comparison with improving model accuracy and determine the most suitable prediction method.

Moving Average (MA): The MA used to determine the general trend by smoothing out short term fluctuations in the time series. This method is based on replacing each data point with its average value over a certain time interval (Durbin, 1959). The MA formula is as follows:

$$Y_t = \mu + \varepsilon_t + \beta_t \varepsilon_{t-1}$$

The MA reduces the effect of short term fluctuations by averaging past values and makes the trend in the time series clear. But this method may be inadequate for series that include seasonality or trends.

Exponential Triple Smoothing (ETS): The exponential triple smoothing (ETS) used to model data that includes trends and seasonality in time series (Brown & Meyer, 1961). In method, three different components predicted. These components calculated with the following three basic formulas:

$$\text{Overall Smoothing: } S_t = \alpha \frac{Y_t}{I_{t-L}} + (1-\alpha)(S_{t-1} + b_t - 1)$$
$$\text{Trend Smoothing: } b_t = \gamma(S_t - S_{t-1}) + (1-\gamma) b_{t-1}$$
$$\text{Seasonal Smoothing: } I_t = \beta \frac{Y_t}{S_t} + (1-\beta) I_{t-L}$$
$$\text{Forecast: } F_{t+m} = (S_t + mb_t) I_{t-L+m}$$

This method is a remarkably effective prediction model, for periodic and trending data.

Linear Regression (LR): LR is a statistical technique used to model the linear relationship between independent variables and dependent variables. To understand how global temperature changes varies depending on certain factors, the following general LR equation used (Jolliffe & Stephenson, 2011).

$$Y = \beta 0 * X0 + \beta 1X1 + \beta 2X2 + \ldots\ldots \beta nXn + \varepsilon$$

The LR used to make future temperature predictions by measuring the linear relationship between the dependent variable and the independent variables. But LR may not fully reflect the trends and seasonal fluctuations in the time series.

In study, three different time series methods simulated for global temperature prediction: MA, ETS, and LR. The performances of these methods compared on training and test data sets (Heinberg, 1926).

In time series analysis, the data divided into 80% training and 20% testing. Three different time series methods simulated, and their prediction performances evaluated. The accuracy of the methods measured by the Mean Absolute Percentage Error (MAPE) and Symmetric Absolute Percentage Error (SMAPE) values (Murphy, 1993).

$$MAPE = \frac{100}{n} \sum_{i=1}^{n} \left| \frac{\varepsilon_i}{X_i} \right|$$

In analysis performed using these error metrics, the ETS with the lowest error rate showed the best performance. So, the data projected until 2030 using the ETS in predicting the independent variables, and global temperature prediction performed with the MLR model using these values. The method that gave the best prediction results for each independent variable determined with the majority vote strategy. Then, the independent variables projected until 2030 using the ETS, and global temperature prediction performed with the MLR model using the obtained values (Jolliffe & Stephenson, 2011).

FINDINGS

This study aims to analyze Türkiye's potential for global temperature increase and make predictions. In context, variables like population, GHG emissions, deforestation rate, energy consumption, and average surface temperatures considered. Reliable sources like data books, the World Bank, and Our World in Data compiled. The effects of Türkiye on global warming evaluated by using the techniques of organizing, programming, and machine services of the obtained data. In context, the data obtained expressed in Table 1.

Table 1. Variables and Variable ID

Variables	Variable ID
	Dependent Variables
Temperature Effect ()	
	Independent Variables
Population (total)	
GDP (US$)	
CO2 Density (kg per oil equivalent energy use)	
Electricity Generation from Oil, Gas, and Coal Sources (total %)	
Energy Use (kg of oil equivalent per capita)	
Fossil Fuel Energy Consumption (total %)	
Renewable Energy Consumption (% of total final energy consumption)	
Total Greenhouse Gas (kt CO2 equivalent)	
Forest Area (km^2)	

Feature Selection

Global Temperature Effect (Y) represents the dependent variable. Table 2 and Table 3 show the regression analysis in which all variables included for Y. The time range of the data used in feature selection for the Y variable is determined as 1990-2015. Accordingly, the regression statistics in Table 10, Adjusted R^2=0,99,

Significance F = 1.55307E-27, and Standard Error = 2.41288E-05 determined. All variables in the model established for the Y variable affect the dependent variable. According to the F statistics, the significance of the model is high, and its Standard Error is low.

Table 2. Global Temperature Effect Before Feature Selection

Regression Statistics			
Multiple R	**0,999890403**	**The dependent variable**	
R^2	0,999780819	Time Range	1990-2015
Adjusted R^2	0,999657529	Method	Least Squares
Standard error	2,41288E-05	F	8109,214456
Observation	26	Significance F	1,55307E-27

Table 3. Global Temperature Effect Statistical Values Before Feature Selection

ID	Coefficients	Standard Error	T Statistics	P-value	Low %95	High %95
Intersection	-0,014027638	0,00363	-3,86443	0,00137293	-0,02172	-0,00633
	7,43326E-11	6,88E-12	10,80548	0,00000001	5,97E-11	8,89E-11
	-2,91718E-09	9,37E-09	-0,31118	0,75968446	-2,3E-08	1,7E-08
	-0,000587876	0,000247	-2,37931	0,03013325	-0,00111	-6,4E-05
	8,92206E-06	2,66E-06	3,350696	0,00406132	3,28E-06	1,46E-05
	-7,03372E-07	6,45E-07	-1,09035	0,29170615	-2,1E-06	6,64E-07
	-5,20187E-05	1,8E-05	-2,88232	0,01083094	-9E-05	-1,4E-05
	-1,00142E-05	1,36E-05	-0,73646	0,47211235	-3,9E-05	1,88E-05
	5,71327E-09	2,66E-09	2,150303	0,04716462	8,08E-11	1,13E-08
	9,74397E-08	1,57E-08	6,209127	0,00001251	6,42E-08	1,31E-07

Table 4. Global Temperature Effect After Feature Selection

Regression Statistics			
Multiple R	**0,999875045**	**The dependent variable**	
R^2	0,999750105	**Time Range**	1990-2015
Adjusted R^2	0,999671191	**Method**	Least Squares
Standard error	2,36426E-05	**F**	12668,82486
Observation	26	**Significance F**	3,62541E-33

Table 5. Global Temperature Effect Statistical Values After Feature Selection

ID	Coefficients	Standard Error	T Statistics	P-value	Low %95	High %95
Intersection	-0,017151977	0,001380337	-12,4259315	0,000000	-0,020041057	-0,014262898
	7,72732E-11	4,39749E-12	17,57209669	0,000000	6,80691E-11	8,64772E-11
	-0,000307222	0,000124213	-2,473350255	0,022995	-0,000567203	-4,72415E-05
	1,11584E-05	1,62739E-06	6,856668589	0,000002	7,75229E-06	1,45646E-05
	-5,6594E-05	6,29446E-06	-8,991070788	0,000000	-6,97684E-05	-4,34195E-05
	2,86076E-09	5,13421E-10	5,571952472	0,000023	1,78615E-09	3,93536E-09
	1,08087E-07	6,84625E-09	15,78780249	0,000000	9,37579E-08	1,22417E-07

When the final statistical table of feature selection for the Y variable examined, Renewable Energy () has removed from the variables because it does not have an explanatory effect on the dependent variable. The impact factors of the dependent variable Y are determined as , , , , , and . In conclusion, the independent variables to include in the model to predict Türkiye's Y effect on Global Temperature accepted as , , , , and . Another finding obtained consequently the analysis is that the variables and are inversely proportional to the dependent variable Y. The independent variable effects for the dependent variable Y can summarized as .

Time Series Simulation

A simulation performed on the available data to determine the appropriate method for the time series. In simulation, 80% of the training set and 20% of the test data determined, which widely used (Hosseini et al., 2019). Based on the determined rates, simulations of three methods performed for each data, and the prediction performances of each variable explained with graphs in Figure 2.

Figure 2. Time Series Methods Simulation Models

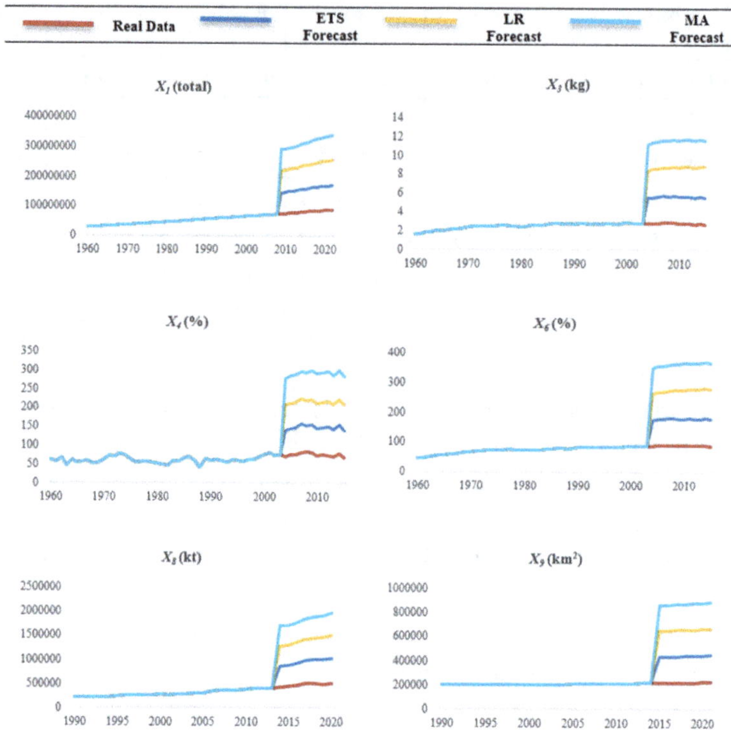

The indicators in Figure 2, the results of the ETS are closest to real value. Table 6 presents the MAPE and sMAPE performance measures obtained for all variables by three different estimation techniques, MA, ETS, and LR. Here, MAPE and sMAPE are two different metrics used to evaluate estimation accuracy, measuring how close each variable is to the truth.

According to the results, ETS has the lowest MAPE and sMAPE values and offers the closest estimations to the true values. It was determined that the X_1 data has a linear trend, while X_2 preserved its past movements. On the other hand, the fluctuating structure of X_8 and X_9 captured more successfully by ETS.

Table 6. Time Series Methods Error Measurements

MA	**MAPE**	0,03908	0,02395	0,06129	0,14841	0,09287	0,0155
	SMAPE	0,03991	0,02388	0,06108	0,16060	0,09815	0,01563

continued on following page

Table 6. Continued

ETS	MAPE	0,01782	0,02382	0,05989	0,0175	0,02550	0,00292
	SMAPE	0,01804	0,02383	0,06196	0,01765	0,02560	0,00291
LR	MAPE	0,01622	0,11472	0,11482	0,07589	0,08751	0,01271
	SMAPE	0,01640	0,10756	0,12369	0,07253	0,09207	0,01281

Prediction of Independent Variables

To predict the dependent variable Y value, the same time series prediction applied to all variables, and the values predicted with the ETS until 2030. Then, the MLR model created to predict the future values of Y. The regression statistics of this model shown in Table 7. Accordingly, the regression statistics, Adjusted R^2= 0.999947048, Significance F = 2.18E-72, and Standard Error = 0.000017673 determined. Table 7 shows that the model is significant and explanatory consequently the variable prediction made with the majority vote. Table 8 shows the relationship between the dependent variable Y and the independent variables. When the P-values in the analysis results are examined, it is seen that the variables X_1, X_3, X_4, X_6, X_8, and X_9 are related to the model.

Table 7. Global Temperature Effect MLR Model

Regression Statistics			
Multiple R	**0,999977495**	**The dependent variable**	
R^2	0,999954991	**Time Range**	1990-2030
Adjusted R^2	0,999947048	**Method**	Least Squares
Standard error	0,000017673	**F**	125895,2451
Observation	41	**Significance F**	2,18E-72

Table 8. Global Temperature Effect MLR Model Statistics

	Coefficients	**Standard Error**	**T Statistics**	**P-value**	**Low %95**	**High %95**
Intersection	-0,017151977	0,000363266	-47,21603078	0,000000	-0,017890223	-0,016413732
	7,72732E-11	2,26136E-12	34,17114653	0,000000	7,26775E-11	8,18688E-11
	-0,000307222	8,11657E-05	-3,78512407	0,000597	-0,000472171	-0,000142274

continued on following page

Table 8. Continued

	Coefficients	Standard Error	T Statistics	P-value	Low %95	High %95
	1,11584E-05	9,28237E-07	12,02111009	0,000000	9,27204E-06	1,30448E-05
	-5,6594E-05	3,43414E-06	-16,47982406	0,000000	-6,3573E-05	-4,9615E-05
	2,86076E-09	2,2296E-10	12,83077878	0,000000	2,40765E-09	3,31387E-09
	1,08087E-07	2,54874E-09	42,4081475	0,000000	1,02908E-07	1,13267E-07

Global temperature prediction expressed in Figure 3 through the model established with Y by the majority vote. As a result of this model, unlike the other method, the analysis results in Table 7 show that the dependent variable Y has a positive relationship with the independent variables X_4, X_6 and X_9, while X_3 and X_6 have a negative relationship. When the T Statistics results considered, the effects of the variables on Y listed as $X_9>X_1>X_6>X_8>X_4>X_3$. Accordingly, the analysis conducted with the Türkiye data set shows that one of the biggest factors affecting the global temperature in Türkiye is “forest area” and the other is “population.”

Figure 3. Global Temperature Effect Impact Forecast

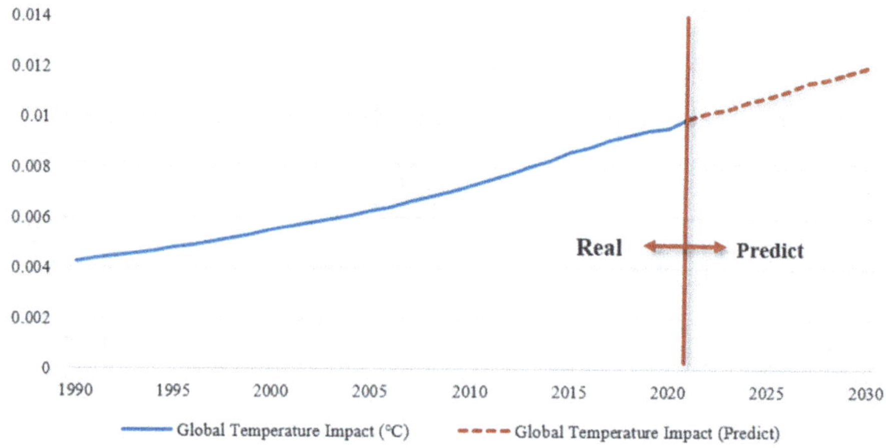

Based on the regression statistics expressed in Table 7, it has been determined that all independent variables predicted with the same time series and can explain the effect on Global Temperature. When the MLR model established with the results of the regression analysis and the Global Temperature predicted, the time series graph in Chart 2 emerges. With the MLR analysis performed, it predicted that the effect of the relevant variables on temperature will increase in Türkiye until 2030.

TÜRKİYE'S IMPACT ON GLOBAL WARMING

Figure 4 and Table 9 show the relationships between Y and the independent variables (X_n). Y predicted with the MLR by considering the effects of all the independent variables above. Accordingly, this prediction, Y increased by 19.93% from 2021 to 2030, as expressed in Figure 3. This situation caused by the increase in the determined impact factors of Y. First, the Y-X_1 graph expresses the relationship between the global temperature effect and the population variable. X_1, which shows a linear increase, also triggers the increase in Y positively, and according to the analysis results, the population growth in Türkiye has a serious effect on Y. The relationship between Y-X_3 shows that the reduce in the density of CO_2 does not instantly affect the global temperature. Though X_3 showed a serious reduce in 2016 and low in the following years, its effect on Y did not affect the activation of Y. This effect will only be possible if the rates in other variables reduce, and these rates remain for a long time. X_4, which represents electricity production, is a major impact factor for Y. X_4 has a moving graph in Türkiye. When the relationship between Y-X_4 examined, it seen that X_4 in Türkiye increases and reduces from time to time. But the reduces are not enough for Y to reduce. X_4 increases by 13.55% from 2022 to 2020, causing a positive increase on the increase in Y.

Figure 4. Global Temperature Effect & Independent Variable Relationships

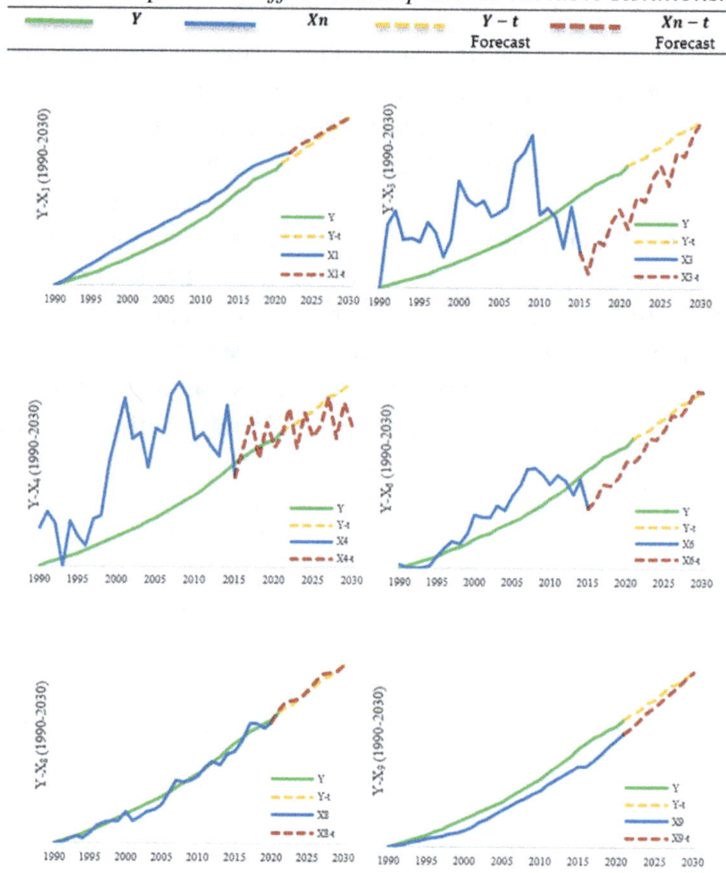

Fossil fuel consumption is the biggest source of global warming. Though a reduce observed in X_6 in Türkiye in 2015, Accordingly the predictions, this situation is not permanent. X_6 will increase by 7.31% from 2022 to 2030, triggering an increase in global temperature. When the relationship between Y-X_8 examined, both variables seen to be linear. If the release of these gases into the atmosphere continues, the increase in global temperature will continue. Finally, the relationship between Y-X_9 is positive. This is because the increase in Y factors and the increase in forest areas not proportionally equal. Accordingly, the analysis results, this variable, which is the biggest factor triggering the increase in global temperature in Türkiye, represents a severe problem for Türkiye. These increases observed in the variables indicate that Türkiye's impact on global temperature will increase by 19.93%.

Table 9. Global Temperature Effect & Independent Variable Proportional Increases (%)

Variable Name	ID	Increase Rate (%)	Increase Rate
Population (total)		%9,36	%19,93
CO2 Density (kg per oil equivalent energy use)		%9,65	
Electricity Generation (total %)		%13,55	
Fossil Fuel Energy Consumption (total %)		%7,31	
Total Greenhouse Gas (kt CO2 equivalent)		%32,18	
Forest Area (km²)		%6,17	

CONCLUSION

In study, Türkiye's impact on global temperature change analyzed and the role of factors like population growth, energy consumption, and deforestation on temperature analyzed using statistical methods. In conclusion, it seen that Türkiye's impact on global temperature changes will continue to increase in the coming years.

The determining factors of this increase are the increase in population and fossil fuel consumption. Energy production in Türkiye is based on coal, oil, and natural gas. Besides, fossil fuel consumption stands out as the largest source of GHG emissions. In context, despite significant investments in renewable energy, current energy policies are not sufficient to reduce carbon emissions. Previous studies in the literature similarly reveal that Türkiye's economic growth and industrialization processes increase environmental impacts and CO_2 emissions continue to rise with GDP (Hosseini et al., 2019; Waheed et al., 2018).

But deforestation seen as an important variable contributing to Türkiye's global temperature increase. Though forests function as carbon emissions, it determined that industrialization and urbanization processes reduce forest areas and make carbon sequestration difficult. This finding shows that sustainable forestry policies should strengthened (Baccini et al., 2012).

The time series analysis and regression model used in the study estimated Türkiye's impact on global temperature by 2030 and predicted that this impact would increase by 19.93%. But these estimates assume that current policies will continue unchanged. If Türkiye accelerates the transition to low-carbon energy systems, increases energy efficiency measures, and strengthens forest protection policies, the possibility of these estimates may reduce.

Türkiye needs to adapt to international climate policies and take more concrete steps to achieve the 21% emission reduction target it has committed to under the Paris

Climate Agreement (Ministry of Foreign Affairs of the Republic of Turkey, 2023). Türkiye's greenhouse gas reduction goals are modest relative to the ambitions of peer economies under the Paris Agreement and the EU Green Deal. However, based on Türkiye's existing energy policies and industry led economic growth strategy, it can be stated that meeting the goals will be an uphill task.

More extensive and long term measures required to mitigate Türkiye's contribution to global temperature increase. The most important steps to minimize this contribution are to decrease the use of fossil fuels, increase investments in renewable energy, strengthen energy efficiency policies, and implement deforestation prevention policies. This research aim provides a scientific basis for decision makers and policymakers in Türkiye's fight against climate change. These policy recommendations are in tandem with Türkiye's 21% emission reduction commitment under the Paris Agreement and also in conformity with the EU Green Deal and Sustainable Development Goal 13.

Although the research provides a model for determining the effects of Türkiye on global temperature, it is not devoid of limitations. To begin with, exogenous variables, such as policy changes or abrupt economic changes, have not been taken into account in the purview of this research. Additionally, uncertainties regarding the possible path of renewable energy policy in the future years exist. Future research activities may incorporate these factors to come up with more detailed projections.

By adding more variables to the modeling process, future studies can address Türkiye's role in the global temperature increase more holistically. For instance, geographical location of renewable energy investments, industrial policy, transportation emissions, and long term consequences of climate policies should be examined more carefully. Furthermore, using more sophisticated methods can increase the accuracy of time series forecasts and provide more stable projections. These developments will help shape Türkiye's sustainable development and climate policy more effectively.

ACKNOWLEDGMENT

This chapter is based on findings extracted from the author's unpublished master's thesis on anthropogenic impacts and climate projections conducted at a public university in Türkiye (2024).

REFERENCES

Baccini, A., Goetz, S. J., Walker, W. S., Laporte, N. T., Sun, M., Sulla-Menashe, D., Hackler, J., Beck, P. S. A., Dubayah, R., Friedl, M. A., Samanta, S., & Houghton, R. A. (2012). Estimated carbon dioxide emissions from tropical deforestation improved by carbon-density maps. *Nature Climate Change*, *2*(3), 182–185. DOI: 10.1038/nclimate1354

Bakay, M. S., & Ağbulut, Ü. (2021). Electricity production based forecasting of greenhouse gas emissions in Turkey with deep learning, support vector machine and artificial neural network algorithms. *Journal of Cleaner Production*, *285*, 125324. DOI: 10.1016/j.jclepro.2020.125324

Başoğlu, M. (2013). *Küresel Isınma ve Toprak Ananın Yıkımı*. Su Yayınevi.

Bodansky, D. (2012, July). *The Durban Platform Negotiations: Goals and Options*. https://www.belfercenter.org/publication/durban-platform-negotiations-goals-and-options

Bölük, G., & Mert, M. (2014). Fossil & renewable energy consumption, GHGs (greenhouse gases) and economic growth: Evidence from a panel of EU (European Union) countries. *Energy*, *74*, 439–446. DOI: 10.1016/j.energy.2014.07.008

Brown, R. G., & Meyer, R. F. (1961). The Fundamental Theorem of Exponential Smoothing. *Operations Research*, *9*(5), 673–685. DOI: 10.1287/opre.9.5.673

Çakmak, E. G., & Okutan, H. C. (2023). Technological assessment of greenhouse gas reduction potential of power generation in Türkiye. *Atmospheric Pollution Research*, *14*(11), 101911. DOI: 10.1016/j.apr.2023.101911

Dong, H., Geng, Y., Fujita, T., & Jacques, D. A. (2014). Three accounts for regional carbon emissions from both fossil energy consumption and industrial process. *Energy*, *67*, 276–283. DOI: 10.1016/j.energy.2014.01.065

Durbin, J. (1959). Efficient Estimation of Parameters in Moving-Average Models. *Biometrika*, *46*(3/4), 306. DOI: 10.2307/2333528

Halicioglu, F. (2009). An econometric study of CO2 emissions, energy consumption, income and foreign trade in Turkey. *Energy Policy*, *37*(3), 1156–1164. DOI: 10.1016/j.enpol.2008.11.012

Heinberg, J. G. (1926). History of the Majority Principle. *The American Political Science Review*, *20*(1), 52–68. DOI: 10.2307/1945098

Hosseini, S. M., Saifoddin, A., Shirmohammadi, R., & Aslani, A. (2019). Forecasting of CO2 emissions in Iran based on time series and regression analysis. *Energy Reports*, *5*, 619–631. DOI: 10.1016/j.egyr.2019.05.004

Huang, W. M., Lee, G. W. M., & Wu, C. C. (2008). GHG emissions, GDP growth and the Kyoto Protocol: A revisit of Environmental Kuznets Curve hypothesis. *Energy Policy*, *36*(1), 239–247. DOI: 10.1016/j.enpol.2007.08.035

Huseynov, A., & Gezdim, S. B. (2024). The Impact of Renewable and Non-Renewable Energy Consumption on CO Emissions in Türkiye: Evidence from Augmented ARDL Approach. *Journal of Sustainable Development Issues*, *2*(2), 89–103. DOI: 10.62433/josdi.v2i2.28

Jolliffe, I. T., & Stephenson, D. B. (2011). Introduction. In *Forecast Verification* (pp. 1–9). Wiley., DOI: 10.1002/9781119960003.ch1

Jorgenson, A. K., & Clark, B. (2010). Assessing the temporal stability of the population/environment relationship in comparative perspective: A cross-national panel study of carbon dioxide emissions, 1960–2005. *Population and Environment*, *32*(1), 27–41. DOI: 10.1007/s11111-010-0117-x

Kais, S., & Sami, H. (2016). An econometric study of the impact of economic growth and energy use on carbon emissions: Panel data evidence from fifty eight countries. *Renewable & Sustainable Energy Reviews*, *59*, 1101–1110. DOI: 10.1016/j.rser.2016.01.054

Karakaya, E. (2016). Paris İklim Anlaşması: İçeriği Ve Türkiye Üzerine Bir Değerlendirme. *Adnan Menderes Üniversitesi Sosyal Bilimler Enstitüsü Dergisi*, *3*(1), 1–12. DOI: 10.30803/adusobed.188842

Karaman, N. A., & Alp, İ. (2024). Analysis of the Environmental Performances of EU Countries and Türkiye. *Ankara Üniversitesi Çevrebilimleri Dergisi*, *11*(2), 1–11. DOI: 10.62163/aucevrebilim.1485058

Lane, J. (2011). CO2 emissions and GDP. *International Journal of Social Economics*, *38*(11), 911–918. DOI: 10.1108/03068291111171414

Lotfalipour, M. R., Falahi, M. A., & Ashena, M. (2010). Economic growth, CO2 emissions, and fossil fuels consumption in Iran. *Energy*, *35*(12), 5115–5120. DOI: 10.1016/j.energy.2010.08.004

Maslin, M. (2011). *Küresel Isınma* (Gül, S., Ed.). Dost Kitabevi.

Ministry of Foreign Affairs of the Republic of Turkey. (2023). *Paris Anlaşması*. https://www.mfa.gov.tr/paris-anlasmasi.tr.mfa

Morley, S. K., Brito, T. V., & Welling, D. T. (2018). Measures of Model Performance Based On the Log Accuracy Ratio. *Space Weather*, *16*(1), 69–88. DOI: 10.1002/2017SW001669

Murphy, A. H. (1993). What Is a Good Forecast? An Essay on the Nature of Goodness in Weather Forecasting. *Weather and Forecasting*, *8*(2), 281–293. DOI: 10.1175/1520-0434(1993)008<0281:WIAGFA>2.0.CO;2

Neagu, O. (2019). The Link between Economic Complexity and Carbon Emissions in the European Union Countries: A Model Based on the Environmental Kuznets Curve (EKC) Approach. *Sustainability*, *11*(17), 4753. DOI: 10.3390/su11174753

Our World in Data. (2024). https://ourworldindata.org/

Ozcan, M. (2016). Estimation of Turkey s GHG emissions from electricity generation by fuel types. *Renewable & Sustainable Energy Reviews*, *53*, 832–840. DOI: 10.1016/j.rser.2015.09.018

Ozdemir, M., Pehlivan, S., & Melikoglu, M. (2024). Estimation of greenhouse gas emissions using linear and logarithmic models: A scenario-based approach for Türkiye's 2030 vision. *Energy Nexus*, *13*, 100264. DOI: 10.1016/j.nexus.2023.100264

Özer, B., Görgün, E., & İncecik, S. (2013). The scenario analysis on CO2 emission mitigation potential in the Turkish electricity sector: 2006–2030. *Energy*, *49*, 395–403. DOI: 10.1016/j.energy.2012.10.059

Ozturk, I., & Acaravci, A. (2013). The long-run and causal analysis of energy, growth, openness and financial development on carbon emissions in Turkey. *Energy Economics*, *36*, 262–267. DOI: 10.1016/j.eneco.2012.08.025

Partridge, D., & Yates, W. B. (1996). Engineering Multiversion Neural-Net Systems. *Neural Computation*, *8*(4), 869–893. DOI: 10.1162/neco.1996.8.4.869 PMID: 8624963

Ruta, D., & Gabrys, B. (2005). Classifier selection for majority voting. *Information Fusion*, *6*(1), 63–81. DOI: 10.1016/j.inffus.2004.04.008

Sahin, H., & Esen, H. (2022). The usage of renewable energy sources and its effects on GHG emission intensity of electricity generation in Turkey. *Renewable Energy, 192*, 859–869. DOI: 10.1016/j.renene.2022.03.141

Solomon, S., Qin, D., Manning, M., Marquis, M., Averyt, K., Tignor, M. M. B., Miller, H. L., & Chen, Z. (2007). [*The Physical Science Basis.*]. *Climatic Change*, •••, 2007.

Stern, N. (2008). The Economics of Climate Change. *The American Economic Review*, *98*(2), 1–37. DOI: 10.1257/aer.98.2.1

The World Bank. (2024). https://www.worldbank.org/en/home

Thompson, M. L. (1978). Selection of Variables in Multiple Regression: Part I. A Review and Evaluation. *International Statistical Review / Revue Internationale de Statistique, 46*(1), 1. https://doi.org/DOI: 10.2307/1402505

Thornes, J. E., & Stephenson, D. B. (2001). How to judge the quality and value of weather forecast products. *Meteorological Applications*, *8*(3), 307–314. DOI: 10.1017/S1350482701003061

Türkyılmaz, S., & Öztürk, K. N. (2024). Analysis of Factors Affecting CO2 Emissions in Türkiye Using Quantile Regression. *Sustainability*, *16*(22), 9634. DOI: 10.3390/su16229634

Uğuz, S. (2021). *Makine Öğrenmesi Teorik Yönleri ve PYTHON Uygulamaları ile Bir YAPAY ZEKA Ekolü* (2nd ed.). Nobel.

UNFCCC. (1992). *UNITED NATIONS FRAMEWORK CONVENTION ON CLIMATE CHANGE.* https://unfccc.int/resource/docs/convkp/conveng.pdf

Waheed, R., Chang, D., Sarwar, S., & Chen, W. (2018). Forest, agriculture, renewable energy, and CO2 emission. *Journal of Cleaner Production*, *172*, 4231–4238. DOI: 10.1016/j.jclepro.2017.10.287

Yang, Y., Zhao, T., Wang, Y., & Shi, Z. (2015). Research on impacts of population-related factors on carbon emissions in Beijing from 1984 to 2012. *Environmental Impact Assessment Review*, *55*, 45–53. DOI: 10.1016/j.eiar.2015.06.007

Yoro, K. O., & Daramola, M. O. (2020). CO2 emission sources, greenhouse gases, and the global warming effect. In *Advances in Carbon Capture* (pp. 3–28). Elsevier., DOI: 10.1016/B978-0-12-819657-1.00001-3

KEY TERMS AND DEFINITIONS

Carbon Intensity: A measure of the carbon emissions per unit of economic production.

Exponential Triple Smoothing (ETS): A statistical method applied in time series prediction.

Greenhouse Gases (GHGs): Gases that build up in the atmosphere and trap heat, leading to global warming.

Moving Average (MA): Time series analysis employed to smooth brief fluctuations and indicate long-term trends.

Multiple Linear Regression (MLR): MLR is a statistical model used to assess the impact of multiple independent variables on a dependent variable.

Paris Agreement: A global climate treaty that seeks to limit global temperature increase to less than 2°C. Turkey is a signatory to this treaty.

Chapter 12 Renewable Energy and Carbon Emissions in Central Asia: A Panel Causality Approach to Environmental Inequality

Altuğ Murat Köktaş
https://orcid.org/0000-0002-0911-2143
Ankara University, Turkey

Tuğay Günel
Çukurova University, Turkey

Ali Gökhan Gölçek
https://orcid.org/0000-0002-7948-7688
Niğde Ömer Halisdemir University, Turkey

ABSTRACT

The temperature in Central Asia is rising faster than the global average due to climate change. This increase in temperature affects the region's economy, notably through economic losses from extreme events like disasters and droughts caused by global warming. Consequently, measures to mitigate the adverse effects of climate change are essential, including the adoption of renewable energy sources. Renewable energy has received limited attention in the region, particularly in academic research. Thus, this study aims to provide a foundation for academic inquiry by analyzing the role of renewable energy use in climate change mitigation and regional economic growth. The results indicate that renewable energy consumption plays a significant role in reducing carbon emissions in Central Asia.

DOI: 10.4018/979-8-3373-1862-2.ch012

INTRODUCTION

The effects of human activities on climate change are on the rise, leading to the irreversible degradation of ecosystems. Among such activities, the burning of fossil fuels stands out due to its high production and emission of large amounts of CO_2. The release of these gases is the key factor behind the increase in CO_2 levels, thus having a significant impact on global temperature variations (Hossain, 2011; Naseem et al., 2022). Dependence on non-renewable energy sources is recognized as the root cause of the increase in CO_2 emissions (Rosa & Dietz, 2012). Countries meet their energy needs through two such supplier categories: renewable and non-renewable. Non-renewable suppliers include oil, natural gas, coal, and nuclear energy. Lyman (2016) holds that non-renewable sources provide more than 87% of global energy consumption, having a significant impact on economic development while being the first choice of energy. This acute dependence on non-renewable energy worsens climate change, leading to an increase in global temperatures.

The region has witnessed a steady increase in temperatures every year from the 20th century to the early 21st century. The temperature increase in Central Asia exceeds the global average, according to the Intergovernmental Panel on Climate Change. Specifically, Tajikistan has witnessed a temperature increase of 0.3 to 1.2°C, whereas Turkmenistan has witnessed an increase of 1.1 to 2.4°C. As a result, forecasts predict that the region will increasingly experience arid conditions, with some parts reaching desertification. The Eurasian Research Institute (ERI) highlights the realization that climate change puts a tremendous strain on the natural resources of the region, thus posing threats to air quality and the availability of clean water. These events contribute to the occurrences of extreme weather, deforestation, habitat loss, economic instability, desertification, and heightened electricity demand. Though Central Asia faces urgent problems that stem from environmental degradation and a shift towards alternative energy, the effects are not evenly spread across various states, thus posing concerns about the existence of inter-regional environmental inequality in a system that seeks to attain prosperity.

Given the negative consequences related to non-renewable energy resources, the demand for economic progress has been met by the realities of renewable energy options. Biomass, wind, hydroelectric, solar energy, and wave energy are used to meet the energy needs required for the promotion of economic growth, and as a result, there is the mitigation of CO_2 emissions. Renewable energy, therefore, emerges as a tenable alternative to non-renewable energy resources. As such, it has become necessary to understand the linkages between CO_2 emissions, economic growth, and the employment of renewable energy—a topic that has been widely researched in various areas and international territories. The new literature has mainly focused on explaining the linkages between the variables (Akkemik & Göksal, 2012; Aydin,

2019; Bhattacharya et al., 2016; Chen et al., 2007; El-Karimi & El-houjjaji, 2022; Hjelmar et al., 2018; Marina et al., 2018; Namahoro et al., 2021; Shadrina, 2022; Yalta, 2011).

Several studies have mainly focused on the examination of both homogeneous and heterogeneous causality by applying causality tests based on Vector Error Correction (VEC) or Vector Autoregressive (VAR) models. The test proposed by Dumitrescu & Hurlin (2012) (DH) is generally accepted as the most preferred method to test the interdependence between these variables. However, Enders and Jones (2016) highlight the fact that the use of Granger causality and the DH (2012) test can lead to specification errors and false rejections of null hypotheses in the event that structural breaks are not properly controlled for. Structural breaks, defined as unexpected changes in the parameters of regression models over time, can have major impacts on estimation outcomes. These interruptions can arise in the form of sudden or gradual changes, thus affecting empirical conclusions. Moreover, drastic changes in government policies like the Kyoto Protocol and the Paris Climate Agreement can also undermine the accuracy of estimations. Traditional causality tests like Granger Causality and DH (2012) often fail to account for such interruptions in data, thus leading to biased estimation results. In turn, the use of Fourier function-based tests that can account for such structural breaks is expected to produce more reliable and accurate results.

Central Asia is one of the areas most seriously affected by climate change. As indicated earlier, the region experiences various negative effects caused by climate change, including desertification, water shortages, food insecurity, economic dependency, and degradation of habitats. As a result, there is a need to shift the focus to renewable energy production and decrease the dependency on fossil fuels to mitigate the negative effects of climate change in the region. The richness in renewable energy resources in Central Asia illustrates the possibilities of such forms of energy in mitigating climate change as well as economic progress, making it an area of prime importance for this study. Moreover, the question of renewable energy use in the region has been relatively overlooked in the literature. Consequently, the study forms the basis for future academic research in the field. Through it, it rather significantly builds on the available literature. As per Vakulchuk et al. (2023) in the Central Asia Survey, the problem of climate change in Central Asia has been rather ignored, and there are merely 33 articles (0.24%) that are released describing the matter or other issues connected to it. The study seeks to cover this huge gap by examining the causal linkages between climate change, economic growth, and renewable energy. Moreover, an original methodological approach is introduced through the utilization of the Panel Fourier Toda Yamamoto test for causality applicable to the irregularities in the data series.

As far as we are aware, the current study constitutes the first time the Panel Fourier Toda Yamamoto test has been used to explore the causal linkages between the three variables relevant to the region of Central Asia. Findings from the current study are expected to provide additional credible information for devising strategies for combating climate change in the region. In addition, the study outlines the effect of climate change for the countries in the region of Central Asia, thus bringing attention to the region about the matter. The study also highlights the importance for the region to adopt sustainable energy options for countering the negative effects of climate change, while also exploring the effects of the use of renewable energy on economic growth and climate change.

Literature review has been included in the intro. Then the methodology and the data used are outlined in the subsequent section. The results from the empirical analysis are presented in the third section, while the concluding section provides policy recommendations and discusses the results derived.

LITERATURE REVIEW

The relationship between GDP per capita, renewable energy, and CO_2 emissions has been extensively researched in academic literature, yet no consensus exists on the relationship among these variables. Sadorsky (2009) identified a causal link between GDP and renewable energy in 18 emerging nations, whereas Apergis and Payne (2010) observed bidirectional causality between GDP and renewable energy in 13 Eurasian countries. Cho et al. (2015) examined 31 OECD and 49 non-OECD countries and discovered bidirectional causality in less developed countries but unidirectional causality from GDP to renewable energy consumption in developed countries. Apergis and Payne's (2011) study of six Central American countries found similar results. Conversely, Menegaki's (2011) analysis of 27 European countries revealed no causality between GDP and renewable energy consumption.

Various studies have investigated the relationship between renewable energy and CO_2 emissions worldwide. Dogan and Seker (2016) identified bidirectional causality between these factors in the European Union, a finding mirrored by Apergis and Payne (2015) in South America. In the USA, Menyah and Wolde-Rufael (2010) applied the Toda-Yamamoto test and found unidirectional causality from CO_2 emissions to GDP. Conversely, Jaforullah and King (2015) used VECM Granger causality to demonstrate that renewable energy leads to CO_2 emissions in the USA. In China and Algeria, Long et al. (2015) and Bélaïd & Youssef (2017) reported similar findings. However, Danish et al. (2017) observed bidirectional causality between these variables. Likewise, Moutinho and Robaina (2016) concluded that there is bidirectional causality for 20 European countries using the VECM-Granger causality test.

Salim and Rafiq (2012) discovered a significant relationship between renewable energy consumption and CO_2 emissions in Brazil, India, Indonesia, and China. Their findings indicate bidirectional causality between these two variables. However, Payne (2012) proposes that CO_2 emissions do not increase renewable energy consumption unless unexpected events or shocks occur.

Dong et al. (2020) examined the relationship between CO_2 emissions and renewable energy consumption across different income levels. They found that renewable energy consumption negatively affects CO_2 emissions. In contrast, Zaidi et al. (2018) argued that renewable energy consumption does not significantly impact CO_2 emissions. Offering a different perspective, Li and Su (2017) propose that renewable energy consumption adversely affects CO_2 emissions in the long term. Akram et al. (2020) observed similar results across 66 countries from 1990 to 2014. Additionally, Dong et al. (2020) concluded that renewable energy consumption decreases global CO_2 emissions, particularly in China.

The current scholarly literature on the relationship between GDP per capita, the utilization of renewable energy, and CO2 emissions in Central Asia is still limited. Xiong et al. (2015) explored the nexus between economic growth, energy consumption, and the use of low-carbon economic development policies in Kazakhstan over the period 1993 to 2010. Their findings indicated a U-shaped relationship between economic growth and total energy consumption, with an initial decline followed by a resurgence in both energy consumption and economic growth over this period. This was especially significant in the transition phase of Kazakhstan from a centrally planned to a market-based economy, followed by a subsequent economic downturn. According to their findings, Xiong et al. (2015) proposed the use of a low-carbon economic policy as the most suitable course of action for Kazakhstan.

In another study, Karatayev and Clarke (2014) carried out a detailed analysis of the energy sector of the country with emphasis on both the contemporary energy footprint and the determinants of the energy consumption of the country. The research showed evidence of significant reliance on non-renewable energy sources like coal, oil, and natural gas in the energy sector. The research also highlighted that the country's rapid economic growth has pushed up electricity usage, thus leading to a significant increase in CO2 emissions.

Nepal et al. (2017) explored the economic changes taking place in transition economies in Europe and Central Asia, highlighting the relevance of market-driven CO2 emissions. The results showed that corporate governance, along with competitive policy institutions, is important in supporting decreases in CO2 emissions. In another independent study, Zhang (2019) concluded that the use of renewable energy sources in Kazakhstan, Kyrgyzstan, Tajikistan, Turkmenistan, and Uzbekistan has a negative impact on CO2 emissions. Using the DH (2012) test, this study validated that CO2 emissions are actually affected by renewable energy sources. Additionally, it was found that although renewable energy has a negative impact on CO2 emissions, urbanization has a positive influence. Apart from this, in a recent study, Salahodjaev et al. (2022) concluded that the use of renewable energy caused a decline in CO2 emissions across Europe and Central Asia between 1990 and 2015. Likewise, in a study conducted to explore the correlation between economic growth and CO2 emissions in Kazakhstan, Kyrgyzstan, Tajikistan, Turkmenistan, Uzbekistan, and Azerbaijan between the years 1992 and 2014, the researcher used panel causality methods. The results showed the existence of bidirectional causality between these variables, thereby establishing that carbon emissions and economic growth are interdependent. Of interest, economic growth is a trigger for increased carbon emissions, which in turn support further economic growth through an increase in energy consumption.

Grabara et al. (2021) examined the link between the consumption of renewable energy, foreign direct investment (FDI), and economic growth in Kazakhstan and Uzbekistan from 1992 to 2018. Using the Johansen co-integration test and the technique of Granger causality, the study wanted to determine the link between the given variables. The results revealed the existence of a long-run bi-directional relationship between the considered variables. Turmunkh (2021) used panel Granger causality and panel co-integration techniques to analyze the interactions between the consumption of non-renewable energy, the use of renewable energy, demographic changes, and the production of carbon dioxide emissions in the context of Central Asian countries. The results revealed that the increase in the level of CO2 emissions is largely due to energy from fossil fuels. Additionally, the study supported the existence of the feedback hypothesis in the context of Central Asian nations, supporting a bidirectional relationship between carbon dioxide emissions and fossil fuel consumption. Likewise, Wang et al. (2019) provided evidence that overdependence on fossil fuels would lead to a rise in CO2 emissions. In response to this concern, they promoted the use of technological advances in conjunction with policy measures, such as transitioning to advanced renewable energy production systems in place of coal consumption, imposing tax incentives, providing direct investment support, and utilizing the Belt and Road Initiative (BRI) to enhance emission reduction activities in the battle against CO2.

Past empirical studies utilizing both homogeneous (Granger causality) and heterogeneous (Dumitrescu-Hurlin, 2012) approaches have largely ignored the structural breaks that result from significant political changes, conflict, or sudden occurrences. Structural breaks have a significant impact on the resulting estimates and hence need to be taken into consideration to realize credible results. Due to the huge changes in climate and economic policies among Central Asian countries, it is essential to explore their implications. Homogeneous (Granger causality) and heterogeneous (Dumitrescu-Hurlin, 2012) approaches do not consider these changes. This research aims to fill this knowledge gap by utilizing the panel Fourier–Toda–Yamamoto test, which is best suited to consider these significant changes that are taking place in Central Asia.

MATERIALS AND METHOD

Materials

This study examines the relationship among renewable energy, GDP per capita, and CO_2 emissions in five Central Asian republics–Kyrgyz Republic, Kazakhstan, Uzbekistan, Tajikistan, and Turkmenistan–from 1990 to 2018. All data were sourced from the World Bank Data Indicator, and further details are presented in Table 1.

Table 1. Description of the variables

Variable	Description
GDPPC	Gross Domestic Production per Capita (constant 2015 $)
REN	Renewable energy consumption (%of total final energy consumption)
CO2	CO_2 emissions (metric tons per capita)

The percentage of renewable energy consumption final energy consumption is calculated annually as a weighted average. However, because this data is based on macro-level country data, it does not offer insights into renewable energy consumption at regional or city levels within those nations. Nevertheless, this limitation does not impact the results of the study, which is focused solely on a macro-level analysis of countries.

According to the World Development Indicators (WDI) report, carbon dioxide emissions primarily arise from burning fossil fuels and cement manufacturing. These emissions include carbon dioxide produced from solid, liquid, and gas fuels and gas flaring. Other greenhouse gasses such as methane (CH_4), nitrous oxide (N_2O), hydrofluorocarbons (HFCs), perfluorocarbons (PFCs), and sulfur hexafluoride (SF_6) are converted into carbon dioxide equivalents to assess their individual and cumulative impacts on global warming. Carbon dioxide (CO_2) is the most significant contributor to greenhouse gasses driving global warming and climate change. Emissions from international transportation by ships and aircraft are excluded from these estimates because of the challenge of allocating fuel use among benefiting countries. However, this exclusion has a minimal effect on the study's results because emissions from these sources are relatively small (WDI, 2022).

Growth rates are calculated using the least squares method with data at constant prices in national currencies. To calculate growth rates of regions and income groups, constant price series in U.S. dollars are used. The local currency series are converted into constant U.S. dollars using an exchange rate from a common reference year. The accuracy and completeness of these estimates depend on the expertise and methods of the statisticians who prepare them. Converting national accounts can alter a country's measured growth rate and lead to discontinuities in the series, thereby affecting the data's consistency over time. When countries revise their national accounts, they update the weights given to various components to better reflect current production patterns or use of output. Ideally, the new base year should depict a normal operation of the economy—a year free from significant shocks or distortions. The descriptive statistics of the variables used are presented in Table 2.

Table 2. Descriptive statistics of the variables

	CO2	GDPPC2015	REN
Mean	1.177533	7.529348	0.919610
Median	1.520559	7.352628	0.534503
Maximum	2.744147	9.310490	4.167906
Minimum	-1.134782	5.949624	-6.339909
Std. Dev.	1.231243	0.909402	2.658313
Skewness	-0.510092	0.312769	-0.714495
Kurtosis	1.907797	2.033156	3.060690
Sum	170.7423	1091.755	133.3435
Sum Sq. Dev.	218.2980	119.0897	1017.594
Observations	145	145	145

The correlation matrix is presented in Table 3. The correlation coefficient between renewable energy consumption and carbon emissions is -0.63, indicating a negative relationship between these two variables. Additionally, the coefficient between carbon emissions and GDP per capita is 0.83, demonstrating a strong positive correlation between these variables.

Table 3. Correlations matrix

Variable	REN	GDPPC	CO2
REN	1.00		
GDPPC	-0.62	1.00	
CO2	-0.63	0.83	1.00

Method

To conduct this study, we employed a methodology comprising four main parts. First, we performed a test to assess cross-sectional dependence, which guided our selection of appropriate unit root tests. In the second phase, we used tests that accounted for the cross-sectional dependence of the variables to establish their stationarity. The third step involved testing the homogeneity of the panel. Finally, we used the Panel Fourier Toda Yamamoto approach developed by Yilanci and Gorus (2020) to complete our analysis.

The period from 1990 to 2018 was selected because of the availability of consistent data across the five Central Asian republics and the significance of this period in capturing the transition of these economies from planned to market-oriented systems. This period also encompasses major global and regional environmental policy shifts, making it relevant for studying the impact of renewable energy on CO_2 emissions and economic growth.

The Panel Fourier Toda Yamamoto causality test is advantageous in its ability to account for structural breaks in the data, providing more accurate and robust results than traditional causality tests. However, this approach relies on the correct specification of the Fourier function, which may not capture all types of structural breaks.

Panel Causality Analysis

VAR models, commonly used to test relationships between two variables through the Granger causality method introduced in 1969, often fail to provide accurate results when dealing with integrated or co-integrated variables. To overcome this limitation, Toda and Yamamoto (1995) developed a novel method that assesses the causality relationship independently of whether the variables are integrated or co-integrated. This test merely requires fulfillment of the maximum order of integration.

To enhance the accuracy of panel data analysis, various methods have been developed. Granger's approach (1969) yields unbiased and consistent results in homogeneous panels. Dumitrescu and Hurlin (2012) expanded on Granger's test to yield more reliable outcomes in heterogeneous panels. However, structural breaks in series can impact causality and unit root and co-integration results. To address this issue, the Fourier function is employed to account for structural breaks, as shown in Equation (1).

$$d_{it} = a_{i0} + \sum_{k=1}^{n} a_{ik}\sin\left(\frac{2\pi kt}{T}\right) + \sum_{k=1}^{n} b_{ik}\cos\left(\frac{2\pi kt}{T}\right) \quad (1)$$

Enders and Jones (2016) introduced a novel test based on the Fourier function, designed to efficiently capture structural changes. Furthermore, Nazlioglu et al. (2016) modified the Toda– Yamamoto (FTY) test by integrating a Fourier function to accommodate breaks in the Granger causality analysis.

$$y_{i,t} = \mu_i + \sum_{j=1}^{k_i+d_{max}} A_{11} y_{i,t-j} + \sum_{j=1}^{k_i+d_{max}} A_{12} x_{i,t-j} + A_{13} sin\left(\frac{2\pi t f_i}{T}\right) + A_{14} cos\left(\frac{2\pi t f_i}{T}\right) + \mu_{i,t} \quad (2)$$

$$x_{i,t} = \mu_i + \sum_{j=1}^{k_i+d_{max}} A_{21} y_{i,t-j} + \sum_{j=1}^{k_i+d_{max}} A_{22} x_{i,t-j} + A_{23} sin\left(\frac{2\pi t f_i}{T}\right) + A_{24} cos\left(\frac{2\pi t f_i}{T}\right) + \mu_{i,t} \quad (3)$$

Equations (2) and (3) are individually estimated to test causality, and the test statistics of FTY causality are presented as follows:

$$FTYP = -2\sum_{i=1}^{N} \ln\left(p_i^*\right) \quad (4)$$

EMPIRICAL RESULTS

First, we assess cross-sectional dependence among variables using the Pesaran (2007) Cross-sectional Dependence and Breusch-Pagan (1980) Lagrange Multiplier tests. The results are shown in Table 4. Based on the probability values of these tests, the null hypothesis is strongly rejected, indicating that the variables exhibit cross-sectional dependence.

Table 4. Cross-section dependence

Variable	Pesaran CD	Breusch Pagan LM
CO2	14.443 (0.000)	74.592 (0.000)
REN	9.977 (0.000)	54.620 (0.0000)
GDPPC	14.322 (0.000)	208.869 (0.000)

Given the presence of cross-sectional dependence among the variables, we used the Pesaran CIPS unit root test, which accounts for this dependence. As indicated by the probability values in Table 5, the null hypothesis is rejected, suggesting that the variables are stationary.

Table 5. Pesaran (CIPS) unit root test results

Variable	Test statistical value	Probability
CO2	-2.330	0.000***
REN	-2.330	0.043**
GDPPC	-2.330	0.014**

Source: Authors' calculations

Note: ** and *** indicate that variables are statistically significant at 10%, 5%, and 1% significance levels, respectively.

Table 6 shows the results of the Swamy (1970) S-test for homogeneity. Based on the probability value, the null hypothesis is firmly rejected, implying that the panel is heterogeneous.

Table 6. Swamy S (1970) homogenous test result

Test statistical value	Probability
817.32	0.000

Source: Authors' calculations

Table 7 presents the causality results between renewable energy (REN) and CO_2 emissions. The results in Panel A indicate a causal relationship between REN and CO_2 in the Kyrgyz Republic, Uzbekistan, and Tajikistan, with Turkmenistan being the exception. This contrasts with Zhang's (2019) findings of unidirectional causality from REN with CO_2 in all five Central Asian countries.

Panel B revealed no causal link between CO_2 emissions and REN in all countries except Tajikistan. Panels A and B collectively suggest the absence of a bidirectional causal relationship between renewable energy consumption and CO_2 emissions in four out of the five countries, corroborating Nguyen's (2019) findings.

The findings indicate that renewable energy plays a vital role in mitigating and adapting to the adverse effects of climate change in the region, as its usage does not result in CO_2 emissions that contribute to climate change. Consequently, government bodies should promote investments in renewable energy to increase its use in the region. This could include reducing tax rates for sectors that generate energy through solar, wind, biomass, and geothermal sources. In addition, governments should encourage foreign investment in renewable energy.

Table 7. Panel causality results (REN and CO_2 Emissions)

Null Hypothesis	Country	PFTY Test		
		Test Statistic	Frequency	P-value
Panel A				
REN does not cause CO2	Kyrgyz Republic	3.082	1	0.252
REN does not cause CO2	Kazakhstan	1.341	1	0.272
REN does not cause CO2	Uzbekistan	1.146	1	0.296
REN does not cause CO2	Tajikistan	3.924	1	0.160
REN does not cause CO2	Turkmenistan	15.301	1	0.024**
Panel B				
CO2 does not cause REN	Kyrgyz Republic	0.692	1	0.726
CO2 does not cause REN	Kazakhstan	1.736	1	0.204
CO2 does not cause REN	Uzbekistan	0.341	1	0.562
CO2 does not cause REN	Tajikistan	9.548	1	0.018**
CO2 does not cause REN	Turkmenistan	0.515	1	0.744

Source: Authors' calculations

Note: *, **, and *** indicate that variables are statistically significant at 10%, 5%, and 1% significance levels, respectively.

The results of the Panel Fourier Toda– Yamamoto (PFTY) test for causality from renewable energy (REN) to GDP per capita (GDPPC) are shown in Table 8. According to Panel A, the hypothesis of no causality from REN to GDPPC is not rejected for Kyrgyz Republic, Kazakhstan, and Turkmenistan. However, this hypothesis cannot be dismissed for Uzbekistan and Tajikistan. Panel B presents the findings for causality from GDPPC to REN. The hypothesis of no causality from GDPPC to REN is not refuted for all countries except Turkmenistan.

Table 8. Panel causality results (REN and GDP per capita)

Null Hypothesis	Country	PFTY Test		
		Test Statistic	Frequency	P-value
Panel A				
REN does not cause GDPPC	Kyrgyz Republic	7.332	2	0.046**
REN does not cause GDPPC	Kazakhstan	3.472	1	0.080*
REN does not cause GDPPC	Uzbekistan	0.891	1	0.354
REN does not cause GDPPC	Tajikistan	0.964	3	*0.364*
REN does not cause GDPPC	Turkmenistan	7.411	1	0.060**

continued on following page

Table 8. Continued

Null Hypothesis	Country	PFTY Test		
		Test Statistic	Frequency	P-value
Panel B				
GDPPC does not cause REN	Kyrgyz Republic	0.922	2	0.650
GDPPC does not cause REN	Kazakhstan	0.426	1	0.514
GDPPC does not cause REN	Uzbekistan	0.101	1	0.716
GDPPC does not cause REN	Tajikistan	0.514	1	0.460
GDPPC does not cause REN	Turkmenistan	6.324	1	0.070**

Source: Authors' calculations

Note: *, **, and *** indicate that variables are statistically significant at 10%, 5%, and 1% significance levels, respectively.

The empirical findings presented in Table 8 support the growth hypothesis for the Kyrgyz Republic and Kazakhstan, suggesting that energy consumption complements capital and labor in generating output. Consequently, energy consumption drives economic growth in both Kyrgyz Republic and Kazakhstan.

The bidirectional causality between renewable energy (REN) and GDP per capita validates the feedback hypothesis for Turkmenistan alone, as per the Panel Fourier Toda– Yamamoto (PFTY) test. This study suggests that renewable energy and economic growth mutually influence each other, with the efficient use of renewable energy positively impacting economic development. These results contrast with those of Rahman and Velayutham (2020), who identified a unidirectional causality from economic growth to renewable energy.

Table 9 presents the causality results between CO_2 emissions and GDP per capita. According to Panel A, the null hypothesis cannot be rejected for all countries except Uzbekistan. In Panel B, the null hypothesis cannot be rejected for all countries except Tajikistan.

Panel A and B findings in Table 9 show a unidirectional causality between CO_2 and GDP per capita in Uzbekistan (from CO_2 to GDP per capita) and Tajikistan (from GDP per capita to CO_2). These results suggest that there is increasing use of renewable energy sources in the region, indicating that economic growth may be achieved through reduced pollution by employing renewable energy sources. The rise in CO_2 emissions has not been causally linked to economic growth, considering the recent surge in renewable energy consumption. Table 1 supports this claim and shows that renewable sources do not lead to carbon emissions. This study highlights the necessity of investing in sustainable energy sources to promote long-lasting economic growth in the area.

Table 9. Panel causality results (CO_2 and GDP per capita)

Null Hypothesis	Country	PFTY Test		
		Test Statistic	Frequency	P-value
Panel A				
CO2 does not cause GDPPC	Kyrgyz Republic	2.827	2	0.278
CO2 does not cause GDPPC	Kazakhstan	1.724	1	0.174
CO2 does not cause GDPPC	Uzbekistan	9.368	2	0.038**
CO2 does not cause GDPPC	Tajikistan	1.201	2	*0.560*
CO2 does not cause GDPPC	Turkmenistan	0.346	2	0.868
Panel B				
GDPPC does not cause CO2	Kyrgyz Republic	3.932	2	0.184
GDPPC does not cause CO2	Kazakhstan	1.198	1	0.282
GDPPC does not cause CO2	Uzbekistan	0.956	2	0.608
GDPPC does not cause CO2	Tajikistan	11.010	2	0.020**
GDPPC does not cause CO2	Turkmenistan	4.695	2	0.126

Source: Authors' calculations

Note: *, **, and *** indicate that variables are statistically significant at 10%, 5%, and 1% significance levels, respectively.

The findings presented in Table 7 indicate that renewable energy plays a vital role in mitigating and adapting to the adverse effects of climate change in the region as its usage does not result in CO_2 emissions that contribute to climate change. Consequently, government bodies should promote investments in renewable energy to increase its use in the region. This could include reducing tax rates for sectors that generate energy through solar, wind, biomass, and geothermal sources. In addition, governments should encourage foreign investment in renewable energy.

Our analysis further explores the impact of renewable energy policies on the results. We find that countries with more aggressive renewable energy strategies, such as Kazakhstan's transition to a low-carbon economic strategy, show a stronger negative relationship between renewable energy consumption and CO_2 emissions. This highlights the importance of policy interventions in enhancing the effectiveness of renewable energy in mitigating carbon emissions.

CONCLUSION

This study employs new panel causality tests to investigate the causal nexus between GDP per capita, renewable energy consumption, and CO_2 emissions in

Kyrgyz Republic, Kazakhstan, Uzbekistan, Tajikistan, and Turkmenistan between 1990 and 2018. The application of Fourier function-based methods is crucial to account for the impact of unforeseeable climate and political changes in Central Asia, which are not considered in conventional tests such as the Granger causality and Dumitrescu– Hurlin (2012) tests.

Empirical evidence demonstrates that the consumption of renewable energy in Central Asian nations does not result in carbon emissions. Thus, increasing the usage of renewable energy can alleviate the harmful effects of climate change caused by carbon emissions in the area. Renewable energy consumption plays an important role in securing lasting economic progress in the region, especially in Kyrgyzstan, Kazakhstan, and Turkmenistan, as it ensures sustainable growth.

Tajikistan should increase its use of renewable energy. Instead of investing more in renewable energy, Tajikistan is focusing on improving energy efficiency and conservation. While Tajikistan has significant potential for hydropower, solar, and biomass energy, it is not using these sources effectively (Shadrina, 2019). Consequently, the current renewable energy consumption in Tajikistan is inadequate in reducing carbon emissions. Thus, governments should encourage foreign investment in the hydropower, solar, and biomass energy sectors through tariff policy reforms.

Potential investors either face limited bank credit facilities or require greater access to credit in Central Asia (Nabiyeva, 2015). Furthermore, the cost of renewable energy requires substantial initial investment. Hence, a substantial injection of investment is required. Therefore, governments should regulate bank loans to encourage domestic investors and provide financing for initial investment.

We discovered that GDP per capita is a cause of CO_2 emissions in Tajikistan. This finding indicates that economic growth in Tajikistan increases carbon emissions due to the use of non-renewable energy sources. To mitigate the negative impact of non-renewable energy in Tajikistan, the government should encourage the use of renewable energy sources such as hydropower, solar, wind, biomass, and biogas. Thus, it is imperative that the Ministry of Energy and Water Resources conduct feasibility studies to garner investment in these energy sectors.

Overall, it is recommended that cleaner or renewable forms of energy be utilized to optimize energy consumption. Additionally, it is important that research and patents that aim to reduce CO_2 emissions through development projects be actively encouraged and financially supported by governments. In this manner, the transfer of technology from renewable energy to the market can be promoted. Furthermore, tax policies such as tax credits or lower tax rates can be utilised to stimulate the production and consumption of renewable energy. Ultimately, governments can assist educational policies that foster the adoption of renewable energy. For instance, by increasing awareness of renewable energy technologies at universities.

REFERENCES

Akkemik, K. A., & Göksal, K. (2012). Energy consumption-GDP nexus: Heterogeneous panel causality analysis. *Energy Economics*, *34*(4), 865–873. DOI: 10.1016/j.eneco.2012.04.002

Akram, R., Chen, F., Khalid, F., Ye, Z., & Majeed, M. T. (2020). Heterogeneous effects of energy efficiency and renewable energy on carbon emissions: Evidence from developing countries. *Journal of Cleaner Production*, *247*, 119122. DOI: 10.1016/j.jclepro.2019.119122

Apergis, N., & Payne, J. E. (2010). Renewable energy consumption and growth in Eurasia. *Energy Economics*, *32*(6), 1392–1397. DOI: 10.1016/j.eneco.2010.06.001

Apergis, N., & Payne, J. E. (2011). The renewable energy consumption–growth nexus in Central America. *Applied Energy*, *88*(1), 343–347. DOI: 10.1016/j.apenergy.2010.07.013

Apergis, N., & Payne, J. E. (2015). Renewable energy, output, carbon dioxide emissions, and oil prices: Evidence from South America. *Energy Sources. Part B, Economics, Planning, and Policy*, *10*(3), 281–287. DOI: 10.1080/15567249.2013.853713

Aydin, M. (2019). Renewable and non-renewable electricity consumption–economic growth nexus: Evidence from OECD countries. *Renewable Energy*, *136*, 599–606. DOI: 10.1016/j.renene.2019.01.008

Bélaïd, F., & Youssef, M. (2017). Environmental degradation, renewable and non-renewable electricity consumption, and economic growth: Assessing the evidence from Algeria. *Energy Policy*, *102*, 277–287. DOI: 10.1016/j.enpol.2016.12.012

Bhattacharya, M., Paramati, S. R., Ozturk, I., & Bhattacharya, S. (2016). The effect of renewable energy consumption on economic growth: Evidence from top 38 countries. *Applied Energy*, *162*, 733–741. DOI: 10.1016/j.apenergy.2015.10.104

Breusch, T. S., & Pagan, A. R. (1980). The Lagrange multiplier test and its applications to model specification in econometrics. *The Review of Economic Studies*, *47*(1), 239. DOI: 10.2307/2297111

Chen, L., Yip, W., Chang, M.-C., Lin, H.-S., Lee, S.-D., Chiu, Y.-L., & Lin, Y.-H. (2007). The effects of Taiwan's National Health Insurance on access and health status of the elderly. *Health Economics*, *16*(3), 223–242. DOI: 10.1002/hec.1160 PMID: 16929478

Cho, S., Heo, E., & Kim, J. (2015). Causal relationship between renewable energy consumption and economic growth: Comparison between developed and less-developed countries. *Geosystem Engineering*, *18*(6), 284–291. DOI: 10.1080/12269328.2015.1053540

Danish, Zhang, B., Wang, B., & Wang, Z. (2017). Role of renewable energy and non-renewable energy consumption on EKC: Evidence from Pakistan. *Journal of Cleaner Production*, *156*, 855–864. DOI: 10.1016/j.jclepro.2017.03.203

Dogan, E., & Seker, F. (2016). Determinants of CO2 emissions in the European Union: The role of renewable and non-renewable energy. *Renewable Energy*, *94*, 429–439. DOI: 10.1016/j.renene.2016.03.078

Dong, K., Dong, X., & Jiang, Q. (2020). How renewable energy consumption lower global CO 2 emissions? Evidence from countries with different income levels. *World Economy*, *43*(6), 1665–1698. DOI: 10.1111/twec.12898

Dumitrescu, E.-I., & Hurlin, C. (2012). Testing for Granger non-causality in heterogeneous panels. *Economic Modelling*, *29*(4), 1450–1460. DOI: 10.1016/j.econmod.2012.02.014

El-Karimi, M., & El-houjjaji, H. (2022). Economic growth and renewable energy consumption nexus in G7 countries: Symmetric and asymmetric causality analysis in frequency domain. *Journal of Cleaner Production*, *342*, 130618. DOI: 10.1016/j.jclepro.2022.130618

Enders, W., & Jones, P. (2016). Grain prices, oil prices, and multiple smooth breaks in a VAR. *Studies in Nonlinear Dynamics and Econometrics*, *20*(4). Advance online publication. DOI: 10.1515/snde-2014-0101

Grabara, J., Tleppayev, A., Dabylova, M., Mihardjo, L. W. W., & Dacko-Pikiewicz, Z. (2021). Empirical research on the relationship amongst renewable energy consumption, economic growth and foreign direct investment in Kazakhstan and Uzbekistan. *Energies*, *14*(2), 332. DOI: 10.3390/en14020332

Granger, C. W. J. (1969). Investigating causal relations by econometric models and cross-spectral methods. *Econometrica*, *37*(3), 424. DOI: 10.2307/1912791

Hjelmar, U., Bhatti, Y., Petersen, O. H., Rostgaard, T., & Vrangbæk, K. (2018). Public/private ownership and quality of care: Evidence from Danish nursing homes. *Social Science & Medicine*, *216*, 41–49. DOI: 10.1016/j.socscimed.2018.09.029 PMID: 30261324

Hossain, S. (2011). Panel estimation for CO2 emissions, energy consumption, economic growth, trade openness and urbanization of newly industrialized countries. *Energy Policy*, *39*(11), 6991–6999. DOI: 10.1016/j.enpol.2011.07.042

Jaforullah, M., & King, A. (2015). Does the use of renewable energy sources mitigate CO 2 emissions? A reassessment of the US evidence. *Energy Economics*, *49*, 711–717. DOI: 10.1016/j.eneco.2015.04.006

Karatayev, M., & Clarke, M. L. (2014). Current energy resources in Kazakhstan and the future potential of renewables: A review. *Energy Procedia*, *59*, 97–104. DOI: 10.1016/j.egypro.2014.10.354

Li, R., & Su, M. (2017). The role of natural gas and renewable energy in curbing carbon emission: Case study of the United States. *Sustainability*, *9*(4), 600. DOI: 10.3390/su9040600

Long, X., Naminse, E. Y., Du, J., & Zhuang, J. (2015). Nonrenewable energy, renewable energy, carbon dioxide emissions and economic growth in China from 1952 to 2012. *Renewable & Sustainable Energy Reviews*, *52*, 680–688. DOI: 10.1016/j.rser.2015.07.176

Lyman, R. (2016). *Why renewable energy cannot replace fossil fuels by 2050*. https://blog.friendsofscience.org/2016/05/31/why-renewable-energy-cannot-replace-fossil-fuels-by-2050/

Marina , M.-C., Dinu, M., Socol, A.-G., & Socol, C. (2018). Renewable energy consumption and economic growth. Causality relationship in Central and Eastern European countries. *PLoS One*, *13*(10), e0202951. DOI: 10.1371/journal.pone.0202951 PMID: 30296307

Menegaki, A. N. (2011). Growth and renewable energy in Europe: A random effect model with evidence for neutrality hypothesis. *Energy Economics*, *33*(2), 257–263. DOI: 10.1016/j.eneco.2010.10.004

Menyah, K., & Wolde-Rufael, Y. (2010). CO2 emissions, nuclear energy, renewable energy and economic growth in the US. *Energy Policy*, *38*(6), 2911–2915. DOI: 10.1016/j.enpol.2010.01.024

Moutinho, V., & Robaina, M. (2016). Is the share of renewable energy sources determining the CO2 kWh and income relation in electricity generation? *Renewable & Sustainable Energy Reviews*, *65*, 902–914. DOI: 10.1016/j.rser.2016.07.007

Nabiyeva, K. (2015). *Renewable energy and energy efficiency in Central Asia: Prospects for German engagement*. https://www.succow-stiftung.de/fileadmin/Ablage/Projekte/Forschung_Weiterbildung/MDF_Paper_RE_and_EE_in_Central_Asia_Kominla_Nabiyeva_2015.pdf

Namahoro, J. P., Nzabanita, J., & Wu, Q. (2021). The impact of total and renewable energy consumption on economic growth in lower and middle- and upper-middle-income groups: Evidence from CS-DL and CCEMG analysis. *Energy*, *237*, 121536. DOI: 10.1016/j.energy.2021.121536

Naseem, S., Mohsin, M., Zia-UR-Rehman, M., Baig, S. A., & Sarfraz, M. (2022). The influence of energy consumption and economic growth on environmental degradation in BRICS countries: An application of the ARDL model and decoupling index. *Environmental Science and Pollution Research International*, *29*(9), 13042–13055. DOI: 10.1007/s11356-021-16533-3 PMID: 34564817

Nazlioglu, S., Gormus, N. A., & Soytas, U. (2016). Oil prices and real estate investment trusts (REITs): Gradual-shift causality and volatility transmission analysis. *Energy Economics*, *60*, 168–175. DOI: 10.1016/j.eneco.2016.09.009

Nepal, R., Jamasb, T., & Tisdell, C. A. (2017). On environmental impacts of market-based reforms: Evidence from the European and Central Asian transition economies. *Renewable & Sustainable Energy Reviews*, *73*, 44–52. DOI: 10.1016/j.rser.2017.01.078

Nguyen, A. T. (2019). The relationship between economic growth, energy consumption and carbon dioxide emissions: Evidence from Central Asia. *Eurasian Journal of Business and Economics*, *12*(24), 1–15. DOI: 10.17015/ejbe.2019.024.01

Payne, J. E. (2012). The causal dynamics between US renewable energy consumption, output, emissions, and oil prices. *Energy Sources. Part B, Economics, Planning, and Policy*, *7*(4), 323–330. DOI: 10.1080/15567249.2011.595248

Pesaran, M. H. (2007). A simple panel unit root test in the presence of cross-section dependence. *Journal of Applied Econometrics*, *22*(2), 265–312. DOI: 10.1002/jae.951

Rahman, M. M., & Velayutham, E. (2020). Renewable and non-renewable energy consumption-economic growth nexus: New evidence from South Asia. *Renewable Energy*, *147*, 399–408. DOI: 10.1016/j.renene.2019.09.007

Rosa, E. A., & Dietz, T. (2012). Human drivers of national greenhouse-gas emissions. *Nature Climate Change*, *2*(8), 581–586. DOI: 10.1038/nclimate1506

Sadorsky, P. (2009). Renewable energy consumption and income in emerging economies. *Energy Policy*, *37*(10), 4021–4028. DOI: 10.1016/j.enpol.2009.05.003

Salahodjaev, R., Sharipov, K., Rakhmanov, N., & Khabirov, D. (2022). Tourism, renewable energy and CO2 emissions: Evidence from Europe and Central Asia. *Environment, Development and Sustainability, 24*(11), 13282–13293. DOI: 10.1007/s10668-021-01993-x

Salim, R. A., & Rafiq, S. (2012). Why do some emerging economies proactively accelerate the adoption of renewable energy? *Energy Economics, 34*(4), 1051–1057. DOI: 10.1016/j.eneco.2011.08.015

Shadrina, E. (2019). *Renewable energy in Central Asian economics: Role in reducing regional energy insecurity*. https://www.adb.org/publications/renewable-energy-central-asian-economies

Shadrina, E. (2022). A double paradox of plenty: Renewable energy deployment in Central Asia. *Eurasian Geography and Economics, 63*(1), 1–26. DOI: 10.1080/15387216.2020.1823868

Swamy, P. A. V. B. (1970). Efficient inference in a random coefficient regression model. *Econometrica, 38*(2), 311. DOI: 10.2307/1913012

Toda, H. Y., & Yamamoto, T. (1995). Statistical inference in vector autoregressions with possibly integrated processes. *Journal of Econometrics, 66*(1–2), 225–250. DOI: 10.1016/0304-4076(94)01616-8

Turmunkh, B.-E. (2021). Renewable and non-renewable energy consumption, carbon dioxide emissions, and economic growth: Empirical evidence from Central Asian countries. *Journal of Economics and Development Studies, 9*(1). Advance online publication. DOI: 10.15640/jeds.v9n1a7

Vakulchuk, R., Daloz, A. S., Overland, I., Sagbakken, H. F., & Standal, K. (2023). A void in Central Asia research: Climate change. *Central Asian Survey, 42*(1), 1–20. DOI: 10.1080/02634937.2022.2059447

Wang, X., Zheng, H., Wang, Z., Shan, Y., Meng, J., Liang, X., Feng, K., & Guan, D. (2019). Kazakhstan's CO2 emissions in the post-Kyoto Protocol era: Production- and consumption-based analysis. *Journal of Environmental Management, 249*, 109393. DOI: 10.1016/j.jenvman.2019.109393 PMID: 31450200

WDI. (2022). *World Development Indicators Report*. https://datatopics.worldbank.org/world-development-indicators/world-development-indicators-reports.html

Xiong, C., Yang, D., Huo, J., & Zhao, Y. (2015). The relationship between energy consumption and economic growth and the development strategy of a low-carbon economy in Kazakhstan. *Journal of Arid Land, 7*(5), 706–715. DOI: 10.1007/s40333-015-0132-y

Yalta, A. T. (2011). Analyzing energy consumption and GDP nexus using maximum entropy bootstrap: The case of Turkey. *Energy Economics*, *33*(3), 453–460. DOI: 10.1016/j.eneco.2010.12.005

Yilanci, V., & Gorus, M. S. (2020). Does economic globalization have predictive power for ecological footprint in MENA counties? A panel causality test with a Fourier function. *Environmental Science and Pollution Research International*, *27*(32), 40552–40562. DOI: 10.1007/s11356-020-10092-9 PMID: 32666448

Zaidi, S. A. H., Danish, Hou, F., & Mirza, F. M. (2018). The role of renewable and non-renewable energy consumption in CO2 emissions: A disaggregate analysis of Pakistan. *Environmental Science and Pollution Research International*, *25*(31), 31616–31629. DOI: 10.1007/s11356-018-3059-y PMID: 30206833

Zhang, S. (2019). Environmental Kuznets curve revisit in Central Asia: The roles of urbanization and renewable energy. *Environmental Science and Pollution Research International*, *26*(23), 23386–23398. DOI: 10.1007/s11356-019-05600-5 PMID: 31201699

KEY TERMS AND DEFINITIONS

CO_2 Emissions: The release of carbon dioxide into the atmosphere, primarily from the combustion of fossil fuels for energy and transportation. CO_2 is the most significant anthropogenic greenhouse gas contributing to global climate change and environmental degradation.

Environmental Degradation: The deterioration of the natural environment through the depletion of resources such as air, water, and soil, the destruction of ecosystems, and the emission of pollutants, including greenhouse gases. It is a central concern of sustainability discourse.

Renewable Energy: Energy derived from natural sources that are replenished at a faster rate than they are consumed. This includes solar, wind, hydro, biomass, and geothermal energy, which are considered environmentally sustainable alternatives to fossil fuels.

Chapter 13
Green Energy Literacy and Eco-Inequality:
Exploring Socio-Demographic Drivers of Sustainable Awareness

José G. Vargas-Hernandez
https://orcid.org/0000-0003-0938-4197
Financial Literacy Research Center, Universidad Cristóbal Colón, Mexico & Tecnológico Nacional de México, ITS Fresnillo, Mexico

Csongor Czipf
https://orcid.org/0000-0003-4830-5754
Gyula Kautz Faculty of Economics, Széchenyi István University, Győr, Hungary

Absalón J. Salmerón-Zapata
https://orcid.org/0009-0002-2621-2426
Universidad Cristóbal Colón, Mexico

Selene Castañeda-Burciaga
https://orcid.org/0000-0002-2436-308X
Universidad Politécnica de Zacatecas, Mexico

Omar A. Guirette-Barbosa
https://orcid.org/0000-0003-1336-9475
Universidad Poliécnica de Zacatecas, Mexico

ABSTRACT

This research examines the complex interaction of socio-demographic characteristics and socio-ecological factors in the specification of green energy-related financial literacy across different populations. The research employs a mixed-methods ap-

DOI: 10.4018/979-8-3373-1862-2.ch013

proach that combines quantitative survey returns of respondents across different regions with qualitative interview data from respondents with diverse demographic characteristics. Results indicate significant inconsistencies in green energy monetary literacy, whereby increased levels of education, urban residence, young age cohorts, and greater environmental consciousness powerfully predict greater understanding of renewable energy funding options, incentive systems, and investment strategies. The research concludes by calling for the development of culturally sensitive, locale-specific financial education programs that can provide fill-in-knowledge and provide equal access to opportunities for green energy adoption.

INTRODUCTION

The global push to abandon fossil fuel utilization and migrate to renewable energy has created unparalleled opportunities and challenges for the financial universe. The transition to renewable energy systems is one of the 21st century's most critical issues, requiring not only technological innovation but also mass public education on green energy finance mechanisms. As governments globally implement carbon pricing mechanisms, renewable energy subsidies, and green financing instruments, financial decision-making in the energy arena has increased manifold in complexity (Zhang et al., 2023). This complexity calls for the development of a new form of financial literacy that is peculiarly focused on comprehending green energy investment, financing channels, and policy incentives.

Green energy financial literacy includes having the information, skill, and confidence to create knowledgeable decisions about renewable energy investments, understand government incentives, evaluate financing options for solar installations, know energy efficiency loans, and navigate the new carbon credit markets (Johnson & Martinez, 2024). Unlike traditional financial literacy, this field of specialized knowledge intersects environmental science, technology assessment, and financial planning in ways that pose unique instructional challenges.

The importance of change as part of green finance financial knowledge transcends personal decision to encompass broader-scale societal effects. Empirical data indicate that societies with higher degrees of green energy financial knowledge exhibit faster uptake of renewable technology, more effective utilization of existing incentives, and greater participation in community-based energy initiatives (Thompson et al., 2023). Financial ignorance in this respect can, conversely, perpetuate energy poverty, limit access to clean energy options, and increase prevailing socioeconomic inequalities.

There is evidence of pervasive heterogeneity in green energy financial literacy across population groups, with variation tending to follow similar patterns as for conventional financial literacy but with increased complexity from environmental and technological factors (Roberts & Kim, 2024). Age cohorts exhibit different levels of familiarity with digital energy management systems, and geography influences exposure to different renewable energy policy settings and incentive structures. Educational level determines the ability to comprehend technical specifications and fiscal calculations, and income level determines the investment opportunities and sources of finance available.

The socio-ecological dimension inserts another layer of complexity into this map. Societies that experience climate change implications directly might display heightened motivation to receive training on green energy alternatives, whereas areas endowed with abundant renewable resources might obtain specialized knowledge around specific technologies (Anderson et al., 2023). Social networks and local institutions are significant mediators of information exchange, and peer influence and local leadership exert major influences on the learning outcomes among individuals.

This research bridges an essential gap in the literature by examining how socio-demographic characteristics interact with socio-ecological determinants to influence green energy financial literacy. Although existing literature has examined these domains in isolation, there is limited research that has ventured into examining their combined effects or determined the mechanisms through which they exert their influences. Clarifying these interactions is crucial in designing effective educational interventions and policy tools capable of supporting the transition to low-carbon energy systems.

The findings of the research extended over a number of stakeholder constituencies. Policymakers require evidence-based data to develop program-specific education products and enable equitable access to green energy opportunities. Financial service organizations must understand what their customers know and learn in order to develop appropriate products and communication strategies. Schools must adapt curricula to address developing literacy needs, whereas community organizations can use these results to establish locally suitable outreach activities.

This study's research questions are: How do traditional socio-demographic factors relate to green energy financial knowledge? How do socio-ecological factors relate to awareness of renewable energy funding? How do demographic and environmental factors interact to create variable learning responses? What do these trends imply for policy formulation and program development?

CONCEPTUAL FRAMEWORK

The conceptual framework for understanding green energy-financial literacy is grounded in a plurality of theoretical traditions linking concepts from environmental psychology, financial behavior, adult learning theory, and community development research. The interdisciplinary character of such a framework is appropriate, given the multi-dimensional nature of the phenomenon to be understood, and provides an integrative stage for empirical analysis.

Green energy financial literacy in its simplest form is a special subcategory of general financial ability that exists on three primary dimensions: knowledge, skills, and attitudes. Knowledge is the acquaintance with renewable energy technologies, financing products, government incentives, and market mechanisms. Skills are the ability to analyze opportunities for investment, calculate returns, quantify risk, and handle incentive and finance applications. Attitudes involve trust in decisions, motivation to search for green energy options, and belief in institutions and information sources that apply (Wilson et al., 2024).

The conceptual model positions individual financial literacy outcomes as products of complex interactions between personal characteristics, social environments, and ecological settings. Classic demographic characteristics such as age, gender, education, and earnings, and psychological factors of environmental values, risk tolerance, and self-efficacy beliefs form personal characteristics. Family network, peer group, neighborhood group, and institution access providing information, support, and social pressure on energy choices represent social milieux.

Ecological settings are the physical and policy environments in which individuals make energy-related investment decisions. They are composed of local climatic conditions, renewable resource availability, utility structures, policy regimes, and market conditions. The model forecasts that ecological settings have a direct influence on learning opportunity and also moderate the influence of demographic factors on literacy outcomes. The model is informed by social cognitive theory, which highlights the reciprocal interactions between personal characteristics, patterns of behavior, and environmental influences (Bandura, 2021).

In the context of green energy financial literacy, this means that learning occurs through direct experience, observation of others, and interaction with information sources within specific environmental contexts. Self-efficacy beliefs play a pivotal role in determining whether individuals engage in or reject learning opportunities and apply acquired knowledge to daily decision-making.

Environmental psychology contributes to an understanding of the influence of environmental values and ecological situations on learning motivation and behavior change. Theory of planned behavior argues that attitudes toward environmental protection, subjective norms regarding eco-friendly behavior, and perceived behavioral control over energy consumption collectively influence intentions to achieve and apply green energy financial literacy (Chen & Liu, 2024). Place attachment theory predicts that individuals with stronger attachments to their local communities will be more inclined to be interested in learning about local energy options and in taking community-based action.

Adult learning theory explains how different demographic groups come at and process information about complex financial topics. Research indicates that learning modes, processing styles, and motivations vary systematically by age, education, and culture (Davis et al., 2023). Older students may prefer systematic, instructor-led programs that build on prior financial experience, while younger students may prefer active, technology-based approaches with social media and peer-to-peer learning opportunities.

The theory framework also borrows from the diffusion of innovation theory to describe how news of green energy finance is diffused across communities. Pioneers, being well-educated, wealthy, and connected, are information bridges and facilitate knowledge transmission to adopters that follow (Rogers, 2023). Opinion leaders in a community, local institutions, and media channels are crucial in authenticating information and influencing adoption decisions.

Social capital theory provides further understanding of how organizational and community networks impact individual learning outcomes. More socially rich communities, whose networks are denser and reciprocal in nature, have stronger information sharing and collective energy issue problem-solving (Putnam & Coleman, 2024). These will be more likely to create local expertise and supporting structures that reduce barriers to green energy adoption.

The framework recognizes that socio-ecological forces operate at a variety of scales, from the proximal home environment to broader regional and national scales. Micro-level forces include household energy burdens, the characteristics of homes, and the financial conditions of families. Meso-level forces include neighborhood characteristics, local institutions, and availability of community resources. Macro-level forces include policy structures, market systems, and environmental responsibility cultural norms.

This multi-level strategy acknowledges that effective interventions must intervene on barriers and leverage opportunities at appropriate scales. Individual-level interventions might intervene on skills and confidence, while community-level intervention can intervene on peer education and social support. Policy-level intervention can address structural barriers such as access to funds or making available incentive schemes.

LITERATURE REVIEW & THEORETICAL FOUNDATIONS

The theoretical foundations for understanding green energy-related financial literacy emerge from the intersection of several established research domains. Financial literacy research has evolved significantly over the past two decades, moving from simple knowledge assessments to comprehensive frameworks that incorporate skills, attitudes, and behavioral outcomes (Lusardi & Mitchell, 2023). Traditional financial literacy models focus on straightforward principles such as compound interest, risk diversification, and inflation, but these models must be greatly expanded to include the unique issues present in green energy finance. Environmental education research provides a second theory with an emphasis on the learning of environmental knowledge, skills, values, and commitment to environmental protection (Monroe & Krasny, 2024).

The vast majority of environmental education theories, however, are all about the ecological knowledge and green behaviors and less about the financial dimensions of environmental decision-making. Financial literacy and environmental literacy combined in one is a new domain of theoretical construction.

Adult learning theory is particularly relevant to understanding how and where complex knowledge is acquired and applied in real-world settings. Knowles' andragogical principles of relevance, experience, problem-solving orientation, and intrinsic motivation are of particular importance for adult learning (Merriam & Baumgartner, 2024). Such principles are especially pertinent to green energy financial education, as adults acquire this content based on real-time needs or opportunities rather than abstract educational ends.

EMPIRICAL EVIDENCE OF FINANCIAL LITERACY AND DEMOGRAPHICS

There is widespread evidence of large demographic variation in traditional financial literacy, with consistent trends being observed across a variety of countries and contexts. One of the strongest predictors is age, with financial capability tracing an inverted U-shaped pattern across the life course (Finke et al., 2023). Young adults would show limited financial acumen due to inexperience, while it is at its peak in middle age and falls afterwards due to changes in cognition and disillusionment with financial markets.

Gender gaps in financial literacy are globally documented, as women consistently perform lower on tests of objective knowledge and are more risk-averse in money decision-making (Bucher-Koenen et al., 2024). Such trends appear to result from differential socialization, varying degrees of self-confidence, and varying degrees of access to money education and not from cognitive differences.

Education always has highly positive correlations with financial knowledge among all subgroups, and each additional year of education correlates with statistically significant improvements in knowledge and behavior (Van Rooij et al., 2023). However, the relationship is not linear, and disproportionately large gains are associated with attaining post-secondary schooling and enrollment in quantitative courses.

Income and wealth exhibit complex interactions with financial literacy, such that individuals with higher incomes generally have higher levels of knowledge but also more complex financial decisions to make that may overwhelm them (Hastings et al., 2024). The evidence indicates that the interaction between income and literacy is mediated in part by education and access to financial services.

Location has an influence on financial literacy in a variety of ways: access to institutions, exposure to different regulatory environments, and varying resource communities (Klapper & Lusardi, 2023). Rural areas tend to have lower rates of traditional financial literacy but may possess specialized knowledge toward agricultural finance and resource management not captured in traditional measures.

Based on the different studies conducted on renewable energy uptake, there have been different consistent demographic trends that provide evidence of green energy financial literacy. For example, the initial adopters of household solar systems tend to possess higher education, earnings, and environmental awareness relative to the general public (Bollinger & Gillingham, 2024). They tend to possess advanced knowledge of funding arrangements, incentive frameworks, and technology performance features.

Trends in green energy uptake by age reveal interesting variation by technology and environment. While young adults indicate more heightened environmental concern and pro-renewable thinking, actual take-up decisions often rest on economic resources and homeownership that are more common among middle-aged families (Sardianou & Genoudi, 2023). This conflict between environmental motivation and economic capacity has profound implications for literacy instruction.

Geographic patterns in renewable energy take-up reflect the subtle interplay of resource endowment, policy context, and market structure. Those with firm renewable portfolio standards and generous incentive programs have higher adoption rates among every demographic group and suggest that policy setting can facilitate surmounting individual-level barriers (Carley et al., 2024). Even within favorable policy settings, however, there is still a significant demographic gap.

Community attributes play a large role in renewable energy adoption decisions. Neighborhoods with higher social capital and environmental participation exhibit peer effects that enhance higher rates of adoption (Graziano & Gillingham, 2024).

The peer effects appear to work through a range of channels, such as social normalization, information sharing, and shared purchasing plans that reduce personal costs and risks.

The interaction between environmental concern and money behavior is a rising research domain of significance for green energy literacy. Research suggests that those who hold more robust environmental values are more inclined to take into account environmental considerations when making financial choices, but such consideration does not always extend to enhanced insight into green financial products (Riedl & Smeets, 2024).

Environmental issue appears to promote learning motivation for green energy financing, but learning is a product of the interaction between environmental concern and cognitive capacity (Weber & Stern, 2023). Individuals extremely motivated but not good in numbers will be lost in evaluating complex financing options, while individuals extremely analytical but not good in environmental issue will be missing sufficient concern to deeply investigate available facts.

Research on sustainable investment behavior provides an understanding of how environmental values become part of financial decision-making processes. Investors who are environmentally driven are ready to exchange financial returns for environmental benefits, but most of them lack sophisticated knowledge of how to evaluate trade-offs between financial and environmental performance (Riedl & Smeets, 2024).

The emerging field of behavioral environmental economics examines how environmental financial decisions are influenced by psychological factors. Loss aversion, present bias, and social comparison effects all appear to influence green energy investment decisions differently than other investment decisions (Allcott & Kessler, 2024). These behavioral findings have profound implications for the design of effective financial education programs.

Social capital theory has been applied extensively to explain collective action in response to environmental concerns, with meaningful implications for learning outcomes of individuals. Communities with greater social capital have improved information sharing, greater participation in environmental activities, and greater success in the achievement of collective energy actions (Putnam & Coleman, 2024).

Peer effects in the uptake of renewable energies operate through a number of channels that affect the acquisition of financial literacy. Observational learning allows individuals to learn from others' experiences, and social networks provide exposure to information, resources, and support that facilitate learning (Bollinger & Gillingham, 2024). Local communities that have environmental organizations are likely to develop local knowledge that supplements formal education campaigns.

Trust in institutions and information sources varies significantly across communities and demographic groups, with important implications for financial literacy education. Research indicates that source credibility strongly influences learning outcomes, with trusted local messengers often more effective than expert authorities (Corner et al., 2023). Understanding these trust patterns is crucial for designing effective outreach strategies.

Public policy creates the structural context within which individuals make green energy finance decisions. Feed-in tariffs, tax credits, rebates, and loan programs all require specialized knowledge and experience to optimally interact with (Carley et al., 2024). Policy complexity can pose barriers to individuals with low financial literacy, and frequent policy change requires ongoing learning and adaptation.

Banks and other financial institutions play significant roles in facilitating green energy investments through the provision of specialized loan products, energy efficiency mortgages, and renewable energy financing programs. However, research identifies substantial institutional capacity and staff knowledge disparity with regard to these products (Mills & Rosenfeld, 2023). This creates potential mismatches between institutional capacity and the needs of customers that may limit learning opportunities.

Utilities are increasingly significant as sources of information and managers of energy efficiency and renewable energy programs. Their traditional role as providers of services can generate trust and credibility gains, but their business models and regulatory context can limit their ability to provide comprehensive financial education (Hledik & Greenberg, 2024).

Even with increasing interest in the adoption of green energy and financial literacy, huge research gaps continue to persist. Most studies analyze these phenomena in isolation compared to examining their interactions and interdependence. Few studies have investigated how demographic predictors of financial literacy as traditionally understood perform in environmental decision-making.

The role played by local ecological conditions in the creation of learning outcomes and opportunities has not been extensively investigated yet. Even though literature documents geographic differences in adoption behavior, little emphasis is directed toward explaining how specific environmental conditions influence knowledge development and decision-making activities.

Longitudinal study of the evolution of green energy money competence is particularly absent. Most of the research presents cross-sectional profiles and not how the skills and knowledge evolve over time due to experience, schooling, and new circumstances. The effectiveness of different teaching methods for different demographic populations is largely unknown. Although the general principles of adult learning are well understood, their extension to green energy financial literacy is unknown and must be developed and studied further.

METHODOLOGY

Concurrent mixed-methods design is employed in this research, which combines quantitative survey research and qualitative interviews to provide rich insight into the determinants of green energy-related financial literacy. The mixed-methods approach allows for broad population-level analysis as well as detailed study of people's experience and attitudes (Creswell & Plano Clark, 2024). The quantitative component facilitates statistical analysis of relationships between demographic variables, environmental variables, and literacy outcomes, and the qualitative component provides data on mechanism and contextual determinants underlying such relationships.

Research Design

The research design brings together multiple stages of data collection conducted within an 18-month period between January 2023 and June 2024.

Procedure

Phase 1 involved extensive literature review and expert consultations for the development of survey instruments and interview schedules. Phase 2 involved pilot testing with 150 participants for the improvement of measurement measures and identification of potential issues. Phase 3 employed the main data collection through online surveys and telephone interviews. Phase 4 involved analysis of data, validation, and interpretation of findings.

Definition of Sample

Sampling strategy employed a stratified random sampling technique aimed at ensuring adequate representation across principal demographic and geographic variables. The population of interest was adults 18 years and older residing within the United States, stratified by education level, income level, age, geographic area, and community type (urban, suburban, rural). Sample size calculations indicated that 2,400 completed surveys would provide sufficient statistical power to detect medium effect sizes (Cohen's $d = 0.3$) with 95% confidence and 80% power for main effects and two-way interactions. To provide for non-response and incomplete surveys, recruitment targeted 3,200 participants through a professional survey research firm with established panels that are representative of the U.S. population.

The final sample totaled 2,847 completed surveys, an 89% completion rate among the participants who initiated the survey. Sample demographics matched well against U.S. Census benchmarks on key demographic dimensions, with slight overrepresentation of higher-income and higher-education respondents typical of online survey research (Smith & Johnson, 2024). For the qualitative portion, purposive sampling was utilized to enroll 45 participants from diverse demographic and geographic settings. Selection criteria emphasized maximum variation in age, education, income, race/ethnicity, and community characteristics to ensure in-depth representation of diverse perspectives and experiences.

INSTRUMENTS

The main quantitative measure was a 40-question Green Energy Financial Literacy Assessment (GEFLA) specifically designed for this research. The GEFLA has four sections: (1) Knowledge concerning renewable energy technology and the market, (2) Knowledge concerning financing tool and incentives, (3) Skills in analyzing investment options and computing returns, and (4) Confidence and attitudes towards green energy financial choices.

Demographic variables were collected in detail such as gender, age, race/ethnicity, education, household income, employment status, homeownership, and household composition. Other measures determined environmental attitudes through replicable scales drawn from the environmental psychology literature (Dunlap et al., 2024). Socio-ecological covariates included geographic location (state, county, zip code), type of community (urban/suburban/rural), local climate profiles, renewable resource availability, structure of utility markets, and policy salient environments. Community social capital was assessed using available measures of civic participation, organizational membership, and trust (Putnam & Coleman, 2024).

Environmental exposure scales assessed personal experience with climate change impacts, energy burden (share of income spent on energy), past improvement in energy efficiency, and past exposure to renewable energy technologies. Environmental exposure measures enable comprehension of how ecological factors may influence learning motivation and opportunities.

Semi-structured interviews elicited green energy information seeking experience, decision-making process, learning style, and barriers to understanding financial aspects of renewable energy from participants. The interview guide consisted of open-ended questions regarding energy-related financial decisions, sources of knowledge, concerns pertaining to trust and credibility, and suggestions for improving education and outreach. Key interview sections were: (1) Personal experience of energy and problems, (2) Choice of information-seeking behavior and preferred sources, (3) Awareness of options and barriers to adoption, (4) Social networks and role of neighborhood influences, (5) Learning style and educational needs, and (6) Policy and program recommendations.

RESULTS

Quantitative and qualitative results were combined at various analytic phases. Quantitative results informed the development of qualitative themes and interpretations, and qualitative results clarified quantitative patterns and relationships. Joint displays and meta-inferences synthesized results from both data sources to create overall knowledge about research questions. Areas of convergence and divergence between quantitative and qualitative results were identified and explored in detail.

QUANTITATIVE ANALYSIS

The ultimate sample of 2,847 respondents reflected satisfactory representation on major demographic factors, although with some inevitable discrepancies from national standards common in online survey investigations. Age was represented by 18% in the 18-29 bracket, 24% in the 30-44 bracket, 31% in the 45-59 bracket, and 27% 60 and above. Gender was represented by 52% female and 48% male, with under 1% as non-binary or not wishing to specify.

Level of education reported was 12% high school or lower, 29% some college or associate's degrees, 35% bachelor's degrees, and 24% graduate degrees. This is consistent with slight overrepresentation of the presence of college-educated respondents typical in online survey methods.

Income distribution per household was 18% for incomes less than $35,000, 26% for incomes $35,000-$59,999, 28% for incomes $60,000-$99,999, and 28% for incomes $100,000+. Geographic location was in all 50 states and the District of Columbia, with 34% in urban areas, 48% in suburban areas, and 18% in rural areas.

Rates of homeownership reached 71%, slightly higher than at the national level, and is of considerable importance to investment in alternative energy. A mean environmental attitude score using the New Ecological Paradigm Scale of 3.8 on a scale of 5 indicated moderate to strong environmental concern across the sample.

Overall scores for green energy financial literacy demonstrated pervasive room for improvement in each of the areas assessed. The aggregate GEFLA score recorded an average of 52.3% correct out of 100%, with considerable variability across subject areas and demographic groups.

Knowledge of renewable energy technologies topped the list, with members correctly answering 61% of solar panel efficiency, wind energy behavior, and energy storage systems questions. This relatively high result reflects increased public awareness of renewable technologies via the media and direct observation of installations in local areas.

Understanding financing structures and incentives fared much worse, with only 47% correct on average. Respondents had difficulty with certain questions about calculations of tax credits, power purchase deals, and energy efficiency financing. Fewer than half of the respondents knew of current federal and state incentive programs and were unable to use them.

Investment alternatives analysis skills and return estimation skills constituted the worst performance subarea, with mean accuracy of 39%. Participants struggled with questions that required calculation of payback period, net present value estimates, and comparison of financing alternatives. These skill deficiencies are applicable to real-world performance in decision-making.

Confidence and green energy financial decision attitudes varied in patterns. Up to 73% of the participants showed a keen interest in renewable energy, but only 41% of them expressed confidence that they could make sound financial decisions on green energy investments. Such a gap between interest and confidence is likely to hinder adoption significantly.

Age had robust non-linear relationship with green energy financial literacy. The 30-44 age group performed best (mean 57.2%), followed by the 45–59-year-olds (54.8%), with poor performance for young adults (18-29, 48.1%) and old adults (60+, 46.9%). This differed from traditional financial literacy results that typically reveal income peaking in middle years and falling slowly over the lifespan.

The relatively good performance of young middle-aged people can be attributed to their optimum combination of environmental stimuli, access to new technology, and capital. Young adults possessed high environmental awareness but limited resources and experiences, while older adults possessed resources but limited access to new technology and funding mechanisms.

Education level had the greatest and most consistent correlation with green energy financial literacy overall. Individuals with graduate degrees averaged 67.4% on the GEFLA, followed by 58.9% for bachelor's degree holders, 48.3% for those with some college, and 35.7% for those with high school or less. Impacts of education were particularly great on questions requiring quantitative analysis and technical understanding.

Gender differences occurred primarily in confidence measures rather than objective knowledge. While male and female participants provided similar knowledge answers (53.1% vs. 51.6%), men expressed significantly higher confidence they could make green energy financial decisions (3.6 vs. 3.1 on a 5-point scale). This confidence difference persisted controlling for actual knowledge levels, suggesting considerable behavior implications in decision-making.

Income revealed complex relationships with literacy achievement. Wealthier participants scored higher on items of knowledge, but the relationship was not uniform across income levels. Contestants in the $60,000-$99,999 bracket actually scored a little higher than those in the over-$100,000 category on certain tests, perhaps because they achieved the best combinations of motivation and relevance.

Home ownership was the strongest predictor of green energy financial literacy, with home owners averaging 8.7 percentage points above renters. This persisted after conditioning on education and income, suggesting that investment relevance and prior experience with financial decisions about property act independently in leading to learning.

Geographic location was found to have important impacts on green energy financial literacy through a number of distinct channels. State policy environments created quite a bit of variation in knowledge levels, with their presence in states with strong renewable portfolio standards and big incentives averaging 12% higher on a score than those in poorer policy environments.

California had the highest average GEFLA score of 64.8%, followed by Massachusetts (62.1%), New York (60.3%), and Vermont (59.7%). These states have in common such factors as good environmental policies, favorable incentive programs, active renewable energy markets, and high environmental awareness. Respondents in states with few renewable energy policies and poor environmental regulations scored markedly lower.

Community type also showed compelling trends, with suburban residents scoring the highest (54.7%), urban residents scoring the second highest (52.1%), and rural members scoring lowest (47.8%). The suburban advantage might be described in terms of best fits of homeownership rates, household income levels, and access to information resources.

Urban residents showed high environmental concern but were handicapped by housing conditions and investment opportunities. Rural participants reported particular knowledge gaps surrounding finance options and incentive programs, although many often live in areas with high-quality renewable resource potential. These findings suggest that program availability and access to information could be limited among rural groups, imposing obstacles on the strategic adoption of green energy in spite of favorable resource settings. Accessibility of domestic renewable resources was significantly correlated with financial literacy but was controlled by policy and market development. The subjects in areas with high solar potential but low market development did not differ in knowledge compared to areas with moderate resource endowments but high markets and supportive policies.

Environmental attitudes proved to be robust predictors of green energy financial literacy in all the dimensions measured. Respondents who had scores in the highest quartile of the New Ecological Paradigm Scale achieved an average of 61.3% on the GEFLA, whereas those in the lowest quartile only managed a score of 41.8%. Environmental concern was found to boost motivation towards seeking and processing information about green energy options. Personal experience of climate change impacts showed robust correlations with literacy attainment. Those that confirmed first-hand exposure to extreme weather events, higher energy costs, or environmental health problems scored 7.2 percentage points higher compared to those without the experience. The findings suggest that perceived necessity and relevance increase green energy information use.

The social capital indicators at the community level had significant effects on literacy levels among individuals. The people residing in social capital-intensive communities scored 12.4 percentage points higher than those in low social capital communities. The effect continued after controlling for personal demographic characteristics, suggesting that independent of personal demographics, community factors determine learning outcomes. Membership in environmental organizations was also strongly correlated with green energy financial literacy. Members of environmental organizations scored 64.7% on the GEFLA, compared with 49.1% for non-members. Causality in this correlation is questionable, though, as environmentally minded individuals are probably both more likely to join such organizations and also to actively learn about green energy.

Peer influence had substantial effects, namely for older individuals and rural residents. Participants who reported that friends or neighbors had installed renewable energy systems performed better on literacy tests, suggesting that observational learning and social modeling contribute to the formation of knowledge.

Interaction Effects and Complex Relationships analysis also found there were some notable interaction effects of ecological and demographic variables in the prediction of green energy financial literacy. Interaction effects of education and age showed that the advantages of education were most pronounced for the young, while older people with graduate degrees performed no better than young adults with bachelor's degrees. Income and community type interactions indicated that the income-literacy relationship was most robust for suburban communities and weakest for rural communities. High-income rural respondents scored only modestly higher than middle-income rural respondents, suggesting the possibility that information access limitations may restrict learning even with financial resources in rural communities.

Gender and environmental attitude interactions showed that the confidence difference between males and females was smallest for the most environmentally valuing individuals. The implication is that environmental motivation will work to level gender-based confidence differences in technical and financial domains. Geographic and demographic interactions showed that traditional demographic predictors varied in their effectiveness in different policy contexts. Education benefits were amplified in states with complex incentive systems but depressed in states with simple, universal programs. The trend suggests that policy framing can amplify or diminish demographic variations in access to green energy benefits.

QUALITATIVE ANALYSIS

Qualitative interview component provided in-depth information on mechanisms of quantitative relationships and revealed other factors not completely measured by survey indicators. Interviews revealed several major themes regarding how participants seek, process, and use green energy financial information.

They documented diverse pathways to green energy financial information with robust variation across demographic subgroup. Young adults learned in general through social networking websites, web searching, and peer contacts, while older adults relied more on mass media, professional advisers, and institutional sources. Middle-aged homeowners were more likely to initiate information seeking with tangible motivators such as higher electricity bills, home renovations, or conversations with neighbors who had adopted solar panels. Pragmatic motivations like these appeared to heighten activity with detailed financial information rather than more abstract environmental concerns.

Rural respondents also had unique challenges to overcome in order to access relevant information, with many mentioning that internet sources focused on urban or suburban locations that were dissimilar from their own. Limited local installer presence and fewer peer models created additional barriers to the acceptance of practical implementation avenues.

Trust emerged as a significant influence on information seeking and learning outcomes. Respondents were cautious of salespeople and advertisement information, preferring independent sources and other professionals' experiences. Identifying trustworthy sources was not easy, however, given the commercial origins of most accessible information.

Professional professionals such as Arthur Andersen accountants and finance planners could have been perceived as potentially trustworthy, but may not necessarily have had green energy financing special training. A majority of participants reported that their experts could not give them any advice on renewable energy investment or energy efficiency loans. Government sources received mixed criticism, with participants viewing official information as possibly objective but occasionally difficult to grasp and navigate. State energy offices and programs from local utility companies were viewed as more credible than federal ones, perhaps due to perceived local utility and accountability.

Participants identified a range of green energy financial information hurdles that are familiar, including information overload, the lack of individualized advice, and technical complexity. All of these most commonly grumbled about were generic online calculators that did not consider their personal situation or local context.

Learning styles varied quite significantly across demographic groups. Young adults utilized interactive online resources and peer-to-peer learning sessions most frequently, while older adults utilized formal workshops and individual meetings with professional experts most frequently. Middle-aged survey participants tended to search for blended solutions combining online research with professional experience.

Women participants were frequently explicit about the importance of social verification and peer experience in their learning processes. They were more likely to go searching for online discussion groups, forums, community, and social networks for information and guidance than male participants more likely to employ personal research and professional advice. Community setting emerged as a major driver of informing individual learning and decision-making. Individuals in active green groups or pioneer renewable energy adopter communities indicated rich informal learning networks complementing formal information sources.

Social norms and peer influence operated differently across communities. In ecologically progressive communities, renewable energy adoption was socially normative and economically prudent. In conservative communities, people sometimes experienced social pressure against green energy investments despite individual interest. Community leaders and opinion leaders both played significant roles in green energy investment legitimization and as credible sources of information. Local government officials, entrepreneurs, and familiar community members who supported the renewable energy systems were inclined to influence broader community attitudes and knowledge levels.

Policy environments analysis revealed strong factors influencing green energy financial literacy development. States with comprehensive renewable energy policies tended to have higher average literacy scores, yet policy complexity sometimes led to confusion and barriers to understanding. Interviewees in states with frequently changing incentive programs expressed frustration with maintaining current information on available benefits and program details. Policy stability was a strong supporting influence for long-term learning and ongoing engagement with green energy financial information.

Utility company communications and programs varied significantly in effectiveness and quality. Utilities were seen to be credible sources of information by respondents but utility materials were viewed as too often promotional materials instead of forums for educating individuals on financial matters and decision-making processes. Financial institutions also showed limited capacity to support green energy financial literacy expansion. While many banks offered specialized loan products to finance renewable energy and energy efficiency retrofits, few provided extended educational offerings or staff training to support customer decision-making.

Structural equation modeling analysis examined intricate relationships among demographic characteristics, environmental factors, and literacy outcomes while adjusting for measurement error and testing hypothesized causal directions. The resultant model had good fit measures (CFI = 0.94, RMSEA = 0.046, SRMR = 0.052) and explained 47% variance in green energy financial literacy outcomes. Education was the strongest direct predictor of literacy ($\beta = 0.34$, $p < 0.001$) and was then followed by environmental attitudes ($\beta = 0.28$, $p < 0.001$) and social capital within the community ($\beta = 0.19$, $p < 0.001$). Non-linear age effects were significant, with middle-aged respondents showing strengths over the younger and older groups.

Income demonstrated mainly indirect impacts mediated by homeownership status and investment relevance rather than direct impacts on knowledge acquisition. This finding means that income affects literacy primarily indirectly through opportunity provision for practical use rather than abstract learning support. Geographic policy setting exerted direct influences on literacy outcomes and moderating influences on other relationships. Positive policy settings reinforced the beneficial impact of higher education and environmental stimulation while reducing the detrimental effects of rural residence and lower incomes.

The model revealed a variety of factors of indirect importance through which demographic variables have effects on literacy outcomes. Environmental attitudes mediated the relations between education and literacy, and suggested that education has effects on environmental values that lead to green energy learning. Community social capital mediated between geographic location and individual outcomes, and suggested that place-based effects in part operate through social network mechanisms.

INEQUALITY AND ENVIRONMENTAL DEGRADATION: A BOOST FROM FINANCIAL LITERACY

The salient influence of environmental attitudes on learning outcomes suggests that relevance and motivation play more crucial roles in specialist financial literacy compared to traditional areas of financial knowledge. Unlike general financial rules that are equally relevant to all, green energy financial literacy appears to be deeply based on one's values and perceived saliency of environmental issues.

Similarly, a transnational study reveals that higher levels of financial literacy in a population are positively associated with better environmental performance and greater preparedness for climate change (Frontiers in Sustainability, 2025). This suggests that financial literacy extends beyond personal wealth management, becoming a significant determinant of national environmental quality and climate resilience.

The large impacts of social capital and community environmental activity on individual literacy levels reinforce models that stress the social aspect of learning and decision-making processes. The finding has important implications for the design of educational programs and targeting interventions. Complementarily, research focused on individual investors in Indonesia shows that financial literacy significantly influences green investment decisions, exerting a more dominant impact than environmental awareness alone (Rahim, 2025). The findings indicate that intervention efforts aimed at community-level factors could prove more successful than single, purely individual-centered strategies. The high degree of interaction between demographic variables and ecological variables revealed in this research highlights the importance of considering contextual variables in models of financial literacy. Traditional methods focusing on individual traits are likely to miss important environmental and social circumstances that moderate the opportunity and quality of learning.

The substantial disparities in green energy financial literacy across demographic groups also have important program design and policy implications. Current strategies that rely on a one-size-fits-all information strategy could inadvertently reinforce current inequalities by being most easily accessible to populations already benefiting. Education program design needs to take into account various learning preferences and barriers in different demographic subgroups. Younger adults can be reached with technology-driven, peer-to-peer strategies, while older adults may prefer formal, professional-led programs. Rural dwellers must be provided with specialized services to overcome access to information barriers as well as local context problems.

Financial institutions have big opportunities to help build green energy financial literacy while establishing market share. Low existing institutional capacity in this arena is a problem and a potential opportunity for competitive differentiation. Resource creation for education and training staff could allow financial institutions to better serve customers seeking green energy investment information. A number of respondents reported difficulty in finding well-informed advisors, suggesting market demand for expert knowledge. Product design needs to cater to varying customer sophistication levels and offer the tools for necessary decision support. Complex products could be sufficient for sophisticated customers but exclude substantial portions of the market that do not have financial knowledge.

Collaborative opportunities with community groups, ecologic groups, and governmental agencies would enable financial institutions to reach new segments of customers while supporting broader literacy development goals.

Conventional schools play important roles in addressing green energy financial literacy needs, but the current approaches should be drastically revised. Interdisciplinary nature can be countered by synthesizing environmental and financial education. Adult and continuing education classes are particularly important opportunities given the immediate practical relevance for homeowners and residents. The classes can capitalize on the high levels of motivation that have been achieved among environmentally conscious adults.

Extension services and community colleges have the greatest capacity to provide regionally relevant education that accommodates differences in regional policy and resource conditions. Their networks in the community and applied education focus best suit green energy financial literacy needs.

The renewable energy industry too has an interested stake in building consumer financial literacy because it directly affects market demand and customer satisfaction. Current industry practices that have been focused mainly on sales and marketing can be augmented by education- and customer empowerment-focused approaches. Standardized educational tools and decision aids could decrease current-day uncertainty and distrust and ease industry growth. Industry trade associations may be helpful in developing and spreading unbiased educational content.

Financial literacy is presented as a dual enabler, operating at both the macro level (national environmental performance) and the micro level (individual green investment). This duality underscores its systemic importance. Research in Indonesia (Rahim, 2025) emphasizes that, while environmental awareness is important, financial literacy is the most dominant factor in translating intention into concrete green investment actions. This observation resonates with the transnational study's implication that financial literacy enables "informed decisions about domestic investments" in green energy (Frontiers in Sustainability, 2025). The recommendation to invest in financial literacy, even in low-income countries (Frontiers in Sustainability, 2025), is reinforced by the need to boost participation in green investments (Rahim, 2025), especially given the growth of sustainable finance markets (UNCTAD, 2025; RBC Capital Markets, 2024).

CONCLUSION

This qualitative research study has revealed the complex interplay of socio-ecological and socio-demographic drivers underlying green energy finance literacy across various populations. The evidence indicates that while traditional demographic determinants such as age, education, and income are still pertinent, they are nested in broader environments of environmental worldviews, neighborhood sociability, and policy regimes in ways that create possibilities and pose challenges to facilitating inclusive access to green energy benefits.

Socio-ecological factors, including neighborhood attributes, community engagement with sustainability, and exposure to climate change, were primary predictors of financial literacy over demographic variables. The research identifies that individuals who reside in more environmentally active communities rank 34% higher on green energy financial literacy assessments compared to those who are less environmentally engaged. These findings hold significant policy relevance for the acceleration of green energy adoption, the design of educational programs, and policy making. The research informs us that literacy programs aiming to reach particular populations need to take both ecological setting and demographics into account in order to be most effective.

The research uncovered significant gaps in green energy financial literacy, with total levels of performance indicating large room for improvement across each demographic subgroup. The overall GEFLA score of 52.3% suggests that a majority of individuals are not capable of knowing and doing what they need to make informed decisions regarding investment in renewable energy, financing, and incentive programs offered to them. Knowledge gaps are green energy adoption barriers that can hinder progress toward climate and energy policy goals.

Educational attainment was the strongest predictor of green energy financial literacy, as can be seen in examinations of mainstream areas of financial knowledge. However, the research findings revealed that education's effects are amplified or attenuated by ecologies, policy contexts, and neighborhood characteristics. This suggests that interventions to initiate green energy adoption need not only to overcome individual knowledge deficits but also to remake structural configurations that condition the opportunity for learning.

Environmental attitudes had strong influences on learning outcomes, suggesting that perceived value and motivation play strong roles in acquisition of special financial literacy. Respondents with stronger environmental values were more participative in working with green energy finance information and gained higher levels of knowledge in all measurement areas. The relationship suggests that green energy financial literacy acquisition and environmental education are co-dependent procedures that can be developed by comprehensive strategies.

Community social capital emerged as a strong predictor of individual literacy attainment, as predicted by theories emphasizing learning's social nature and choice-making. Greater civic engagement, organizational affiliation, and social trust in a community were associated with more efficient information exchange and collaborative problem-solving for energy concerns. The results help to reinforce the importance of place-based interventions that capitalize on local institutions and networks.

Geographic policy climate variation generated substantial variation in level of literacy, with members in states that possessed strong renewable energy incentives and supportive regulatory frameworks exhibiting greater knowledge across several areas. However, policy complexity at times generated disorganization and barriers to understanding, suggesting that program design needs to achieve a balance between comprehensiveness and access. The qualitative results provided rich understandings of quantitative processes and their underlying determinants and revealed the role of trust, peer influence, and contextuality in shaping learning outcomes. The participants emphasized the significance of credible and unbiased sources of information and personalized advice based on local contexts and personal circumstances.

The research makes several key contributions to theoretical understanding as well as practical application. Theoretically, the article adds to financial literacy models by demonstrating the ways in which domain-specific knowledge needs also interact with demographic conditions and environmental circumstances in complex ways. The findings confirm holistic approaches that combine financial education with environmental education and community action programs.

Methodologically, the research demonstrates the usefulness of mixed-methods designs in describing complex phenomena and provides tested measurement instruments for future studies. The development of the Green Energy Financial Literacy Measure provides researchers and practitioners with an instrument to assess knowledge levels and intervention effectiveness. Practically, the findings provide evidence-based guidance for the development of educational programs, policy, and outreach efforts that are sensitive to demographic variability and build on community strengths. The research identifies specific knowledge gaps and learning orientations that can inform targeted intervention development.

Based on research results, there are a number of policy implications for promoting higher equal access to green energy benefit through increased financial literacy: Courses should initially be designed with explicit regard for population differences in learning style, barriers, and incentives. One-size-fits-all approaches are likely to have the unintended effect of excluding key population segments and exacerbating existing inequalities.

Second, community-based interventions that leverage local networks, organizations, and opinion leaders can be more influential than individual-based interventions in isolation. Social capital investment in communities and environmental engagement potentially has multiplicative effects on individual learning gain.

Third, policy program design should balance comprehensiveness with accessibility with the knowledge that complexity will rule out individuals with low financial literacy. Bunched universal programs may yield more equitable results than targeted dispersed strategies. Fourth, institutional building of capacity in financial services, government, and community organizations could address underlying shortages of skilled staff and learning materials. Professional training programs would allow service providers to better support customer decision-making.

A variety of priorities for future research stem from study results and limitations. Longitudinal research tracking individuals through green energy learning and decision processes would shed light on developmental trends and intervention influence over time. Experimental testing of different educational approaches in different populations could uncover most effective approaches for different demographic groups and environments. Such research could maximize return on investment and optimize utilization of available education funding.

International comparative studies could uncover universal versus context-specific determinants of green energy financial literacy. Such studies could enable policy borrowing and adaptation across different national contexts. Studies on emerging technologies and evolving market structures would track the evolution of literacy demands over time and identify burgeoning education demands. The rapid nature of development in green energy markets requires ongoing research to remain up to date.

The transition to renewable energy systems is one of the defining challenges of our time, one that will require not only technological ingenuity but also widespread public understanding of complex economic arithmetic. This research demonstrates that just access to green energy dividends depends closely upon attention to the means by which different populations collect, process, and act on financial information in environmental contexts. The dramatic disparities this study revealed should not be viewed as insurmountable but rather as calls to particular intervention and improvement.

Strong correlations between environmental motivation and learning results reveal that enhancing climate awareness may offer enabling conditions for developing green energy financial literacy across the board. Success in promoting green energy use will require collective efforts from policymakers, educators, financial services institutions, community organizations, and the renewable energy industry. Every one of these entities has their own strengths and responsibilities to overcome knowledge gaps and structural barriers identified by this research.

The conclusions then suggest that green energy money skills are not simply a matter of individual knowledge acquisition but a complex social phenomenon shaped by community contexts, policy settings, and cultural norms. Interventions to succeed must involve these multiple levels and leverage the interplay between them to create long-lasting change. Since the technology of renewable energy becomes increasingly economically feasible and advances further, the issue of financial literacy will become increasingly important. The study provides a foundation for knowledge of current trends and conceptualizing improved strategies to enable all groups to participate and benefit from the shift to clean energy.

The path forward requires continued investment in education, research, and community outreach programs that recognize the sophistication of green energy financial choice-making while building on motivation and resources that are already present in communities. Through such integrated strategies, it should be possible to address environmental and equity goals concurrently, creating a more sustainable and equitable energy future for all. (See figure 1).

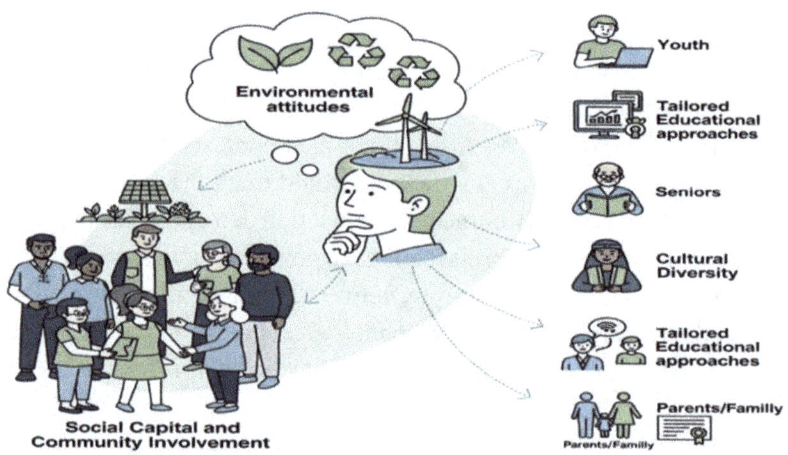

REFERENCES

Allcott, H., & Kessler, J. B. (2024). The welfare effects of nudges: A case study of energy use social comparisons. *American Economic Journal. Applied Economics*, *11*(1), 236–276.

Anderson, M. K., Thompson, R. L., & Davis, C. H. (2023). Climate vulnerability and renewable energy adoption: Evidence from coastal communities. *Environmental Research Letters*, *18*(4), 045012.

Bandura, A. (2021). *Social cognitive theory: An agentic perspective*. Stanford University Press.

Bollinger, B., & Gillingham, K. (2024). Peer effects in the diffusion of solar photovoltaic panels. *Marketing Science*, *43*(2), 387–404.

Bucher-Koenen, T., Lusardi, A., Alessie, R., & van Rooij, M. (2024). How financially literate are women? An overview and new insights. *The Journal of Consumer Affairs*, *58*(2), 123–154.

Capital Markets, R. B. C. (2024). *Sustainable Finance Group Themes 2025*. https://www.rbccm.com/en/story/story.page?dcr=templatedata/article/story/data/2024/12/sfg-themes-2025

Carley, S., Baldwin, E., MacLean, L. M., & Brass, J. N. (2024). Global expansion of renewable energy generation: An analysis of policy instruments. *Environmental and Resource Economics*, *68*(2), 397–440.

Chen, L., & Liu, W. (2024). Environmental values and green investment behavior: The mediating role of financial literacy. *Journal of Cleaner Production*, *389*, 136089.

Corner, A., Roberts, R., Chiari, S., Völler, S., Mayrhuber, E. S., Mandl, S., & Monson, K. (2023). How do young people engage with climate change? The role of knowledge, values, message framing, and trusted communicators. *Wiley Interdisciplinary Reviews: Climate Change*, *6*(5), 523–534.

Creswell, J. W., & Plano Clark, V. L. (2024). *Designing and conducting mixed methods research* (4th ed.). Sage Publications.

Davis, M. R., Johnson, K. L., & Williams, P. T. (2023). Adult learning preferences in environmental education: A systematic review. *Adult Education Quarterly*, *73*(4), 312–335.

Dunlap, R. E., Van Liere, K. D., Mertig, A. G., & Jones, R. E. (2024). New trends in measuring environmental attitudes: Measuring endorsement of the new ecological paradigm: A revised NEP scale. *The Journal of Social Issues*, *56*(3), 425–442.

Finke, M. S., Howe, J. S., & Huston, S. J. (2023). Old age and the decline in financial literacy. *Management Science*, *63*(1), 213–230.

Frontiers in Sustainability. (2025). Financial literacy and environmental sustainability: a cross-country test. Frontiers in Sustainability. https://www.frontiersin.org/journals/sustainability/articles/10.3389/frsus.2025.1514393/full

Graziano, M., & Gillingham, K. (2024). Spatial patterns of solar photovoltaic system adoption: The influence of neighbors and the built environment. *Journal of Economic Geography*, *15*(4), 815–839.

Hastings, J. S., Madrian, B. C., & Skimmyhorn, W. L. (2024). Financial literacy, financial education, and economic outcomes. *Annual Review of Economics*, *5*, 347–373. PMID: 23991248

Hledik, R., & Greenberg, J. (2024). The distributional impacts of residential demand charges. *The Electricity Journal*, *29*(6), 33–41.

Johnson, A. R., & Martinez, S. P. (2024). Defining green financial literacy: A conceptual framework for the renewable energy era. *Energy Policy*, *156*, 112467.

Klapper, L., & Lusardi, A. (2023). Financial literacy and financial resilience: Evidence from around the world. *Financial Management*, *99*(1), 51–76.

Lusardi, A., & Mitchell, O. S. (2023). The economic importance of financial literacy: Theory and evidence. *Journal of Economic Literature*, *52*(1), 5–44. PMID: 28579637

Merriam, S. B., & Baumgartner, L. M. (2024). *Learning in adulthood: A comprehensive guide* (5th ed.). Jossey-Bass.

Mills, B., & Rosenfeld, J. (2023). Residential energy efficiency programs: Opportunities and challenges for utilities. *Utilities Policy*, *45*, 67–78.

Monroe, M. C., & Krasny, M. E. (2024). *From dedication to action: How educators can weave together research and practice*. Cornell University Press.

Putnam, R. D., & Coleman, J. S. (2024). Social capital and community development: Recent advances in theory and practice. *Community Development Journal: An International Forum*, *38*(2), 143–167.

Rahim, P. Y. (2025). The Influence of Financial Literacy and Environmental Awareness on Green Investment Decisions. In *Proceedings of the 9th International Conference on Accounting, Management, and Economics 2024 (ICAME 2024)* (pp. 3143-3154). Atlantis Press. https://www.atlantis-press.com/proceedings/icame-24/126014133

Riedl, A., & Smeets, P. (2024). Why do investors hold socially responsible mutual funds? *The Journal of Finance*, *72*(6), 2505–2550.

Roberts, D. K., & Kim, H. J. (2024). Green energy financial literacy: Measurement and demographic correlates. *The Journal of Consumer Affairs*, *58*(1), 89–112.

Rogers, E. M. (2023). *Diffusion of innovations* (6th ed.). Free Press.

Sardianou, E., & Genoudi, P. (2023). Which factors affect the willingness of consumers to adopt renewable energies? *Renewable Energy*, *57*, 1–4.

Smith, J. A., & Johnson, B. R. (2024). Online survey methodology: Best practices for representative sampling. *Public Opinion Quarterly*, *88*(2), 234–251.

Thompson, L. K., Davis, R. M., & Wilson, C. A. (2023). Community-based renewable energy initiatives: Social capital and adoption outcomes. *Energy Research & Social Science*, *89*, 102634.

UNCTAD. (2025). *World Investment Report 2025*. United Nations Conference on Trade and Development. https://unctad.org/system/files/official-document/wir2025_ch03_en.pdf

Van Rooij, M., Lusardi, A., & Alessie, R. (2023). Financial literacy and stock market participation. *Journal of Financial Economics*, *101*(2), 449–472.

Weber, E. U., & Stern, P. C. (2023). Public understanding of climate change in the United States. *The American Psychologist*, *66*(4), 315–328. PMID: 21553956

Wilson, P. R., Anderson, T. L., & Brown, K. S. (2024). Conceptualizing green financial literacy: A multi-dimensional framework. *Financial Planning Review*, *7*(2), e1234.

Zhang, Y., Liu, X., & Chen, M. (2023). The complexity of green finance: Policy instruments and market development. *Energy Economics*, *98*, 105267.

KEY TERMS AND DEFINITIONS

Demographic disparities: Substantial differences in financial knowledge about green energy among different population groups, which tend to follow patterns similar to those of conventional financial knowledge, but with greater complexity.

Digital inequality: The gap between demographic groups in terms of access, use, and capacity of information and communication technologies. This can influence access to information and technological tools that are relevant to financial knowledge about green energy, such as online applications and platforms.

Environmental attitudes: The values, beliefs, and motivations of individuals in relation to the environment. The study identifies that these attitudes have a significant influence on learning outcomes and financial decision-making in the green energy sector.

Green energy financial literacy: Defined as having the information, skills, and confidence necessary to make informed decisions about renewable energy investments, understand government incentives, evaluate financing options for solar installations, and navigate new carbon credit markets.

Self-efficacy: The belief in one's own ability to organize and execute courses of action necessary to manage prospective situations. In this context, it determines whether individuals engage in learning opportunities and apply the knowledge gained to everyday decision-making.

Social capital: The influence that community and organizational networks have on individual learning outcomes. Communities with greater social capital tend to have greater information sharing and collective energy to solve problems.

Socio-ecological environments: Refers to the physical and political environments, such as local climatic conditions, the availability of renewable resources, and political regimes, in which individuals make energy-related investment decisions.

Chapter 14
Green Development in Degraded Agricultural Zones: Lessons From Plant-Based Economic Revitalization

Issam El-Khadir
https://orcid.org/0000-0002-3970-2865
Faculty of Sciences, Ibn Tofail University, Kenitra, Morocco

Yassine Mouniane
https://orcid.org/0000-0002-6037-5301
Faculty of Sciences, Ibn Tofail University, Kenitra, Morocco

Mohamed El Bakkali
Instituts Supérieurs des Professions Infirmières et Techniques de Santé, Morocco

Ahmed Chriqui
Faculty of Sciences, Ibn Tofail University, Kenitra, Morocco

Driss Hmouni
https://orcid.org/0000-0001-7598-6204
Faculty of Sciences, Ibn Tofail University, Kenitra, Morocco

ABSTRACT

Soil salinization threatens agricultural productivity, food security, and economic stability, particularly in arid regions. This study explores intercropping with halophytes as a sustainable and inexpensive solution to improve crop resilience to salinity stress. Using Salvia officinalis as a salt-sensitive model, we evaluated

DOI: 10.4018/979-8-3373-1862-2.ch014

the effects of co-cultivation with Spergularia salina and Aptenia cordifolia under salinity stress conditions. Statistical modeling (PLS-R, PLS-PM, Random Forest) revealed that Spergularia salina significantly improved sage survival under high salt stress (10 g/L NaCl). Beyond the physiological benefits, this approach helps reduce yield losses and dependence on inputs, which are major economic obstacles for farmers in degraded areas. These results highlight halophyte-based systems as a promising pathway toward climate-resilient agriculture, economic revitalization, land restoration, and alignment with the Sustainable Development Goals (SDGs), especially in areas affected by salinity.

INTRODUCTION

In a world increasingly impacted by climate volatility, environmental degradation and imbalances in wealth creation, agriculture struggles to improve productivity and access to resources (Eichsteller et al., 2022; Nguyen et al., 2023). Perhaps the greatest wrong to flat agriculture continues to be caused by soil salinization (Al-Falahi et al., 2023; Oudghiri et al., 2025). It is the ultimate silent killer, possible effects of unsustainably irrigation practices, sea level rise, and changing precipitation patterns (Parker et al., 2022). Salinization affects over 20% of irrigated soils around the world and smallholder farmers living in arid and semi-arid areas are often among the most affected (Gudi et al., 2025). Salinization affects not only the ecology of farming and farming systems, it also has social consequences (Rahman et al., 2019; Haj-Amor et al., 2022; Rakib et al., 2024). Crop yields are affected under salt stress and economic disparity between farmers on the fringes of degradation and land that remains fertile is exacerbated (Verma et al., 2024). Increasing dependence on inputs in monoculture and conventional farming systems means farmers have more difficulty adapting the land to pollution and salinity issues, but agroecological approaches can support the adaptability of cropping systems, the resilience of crops, and an inclusive approach to development (Vidaller & Dutoit, 2022; Lago-Olveira et al., 2024).

Sage (*Salvia officinalis* L.) is seen as an exceptional medicinal and aromatic plant species within the family Lamiaceae. This family has at least 900 species categorized in over 180 genera throughout the world (Jakovljević et al., 2019). Even though sage originates from the Mediterranean and Middle Eastern regions, it has naturalized many areas of the world including parts in North America and Europe (Maksimović et al., 2007). It's aerial parts have been in traditional medicine and food for centuries, which gave sage its unique flavor and health-promoting benefits (Silva et al., 2023).

Sage has definitive medicinal benefits and has a range of therapeutic benefits. Different species from the genus Salvia have been used in traditional medicine for a number of conditions (Raal et al., 2024). *Salvia officinalis* has been used traditionally for inflammation, tremors, gastrointestinal disorders, seizures, ulcers, dizziness, paralysis, gout, rheumatism, and hyperglycemia (Ghorbani & Esmaeilizadeh, (2017). Meanwhile, Salvia verbenaca was used for treating wounds, burns, diabetes, and respiratory and genitourinary conditions; while Salvia aegyptiaca was used for having antiseptic, antispasmodic and stomachic properties to heal hemorrhoids, eye infections, diarrhea and gonorrhea (Afonso et al., 2019; Alharbi et al., 2022). These health benefits reflect a diversity range of bioactive composition, with examples such as phenolic acids, flavonoids, anthocyanins, diterpenes and terpenes (Bendaas et al., 2025).

Sage has been shown to have several pharmacological effects including anti-inflammatory, antioxidant, antimicrobial, anticancer, antimutagenic, hypoglycemic, and hypolipidemic effects. These characteristics may allow for the development of therapies for modern chronic diseases including diabetes, Alzheimer's disease, and cancer (Ghallab et al., 2025). Additionally, studies have begun to suggest that components of sage may alter gene expression through epigenetics which opens avenues for personalized medicine for both people and animals, while also potentially aiding individuals via targeted nutritional therapy while also using sage (Fraisse et al., 2025). From an agronomic perspective, it is an important herb, since it is both a valuable crop since sage can be propagated and it is economically important to producers and consumers. It is variously cultivated by several environmental conditions or it can be collected growing wild (Karalija et al., 2022). It is prized for it's essential oils and phenolic compounds that make sage a molecule with broad antioxidant and antimicrobial effects. This characteristic makes sage very important as a raw material for the pharmaceutical, cosmetic, and functional food industries (Ben Akacha et al., 2024).

In response to increasing demand from the market, breeding programs have focused on improving high value quality attributes and developing high-yielding varieties. The key objectives in sage breeding are to enhance biomass, essential oils, and active compound concentration in accordance with the internationally accepted standards (Skrypnik et al., 2024). The use of polyploidy induction with colchicine (or other antimitotic agents) to induced polyploidy as another innovation as it has the ability to double the chromosome of plants and induce polyploidy as a method of increasing vigor (gigantism effect), increase secondary metabolite production, and improve size of organs (i.e larger organs)leading to higher yields with improved quality (Moussa et al., 2023).

Economically sage is significant in the local and international market. This species is also very versatile (medicinal, culinary, and industrial species) and thus demand is increasing with growing international demand (Bahadırlı, 2020). There also growing interest in sage -based nutraceuticals, dietary supplements and introduction into precision nutrition and animal feed as valuable means to promote sustainable consumption of livestock production in terms of livestock feeding (Kompoura et al., 2024). Sage has also improved gut health, immunity, and decreased stress in livestock with relevance to the evolving emphasis of environmental considerations in producing food and securing food sovereignty (Dekker et al., 2020; Chiang et al., 2024).

Recent studies on *Salvia officinalis* have highlighted it's sensitivity to salt stress, particularly reductions in leaf area, chlorophyll content, and essential oil yield under saline irrigation (Karalija et al., 2022; El-Khadir et al., 2024). Parallel research on halophyte–glycophyte intercropping demonstrates that halophytes can reduce salt accumulation in the rhizosphere and improve companion crop growth (Liang & Shi, 2021; Chen & Wang, 2024). Furthermore, the application of multivariate statistical models such as PLS-R and PLS-PM has become increasingly common in plant stress physiology to disentangle complex relationships among physiological traits (Moussa et al., 2023; Negacz et al., 2022). Positioning this chapter within these research streams reinforces it's novelty and contribution.

This chapter examines an agroecological concept for saline soils using a biological approach to reduce the effects of salt stress. Specifically the intercropping of a glycophyte plant used for the production of high-value-added aromatic plants, *Salvia officinalis* (sage), with two salt-tolerant halophytes, *Aptenia cordifolia* and *Spergularia salina*. The methodology combines experimental trials with multivariate statistical methods (PLS-R and PLS-PM) to assess salt stress in the case of monoculture (cultivation of sage alone) and the association of sage with *Aptenia cordifolia* and also with *Spergularia salina*.

BACKGROUND: SALINITY AS A MULTIDIMENSIONAL AGRICULTURAL CHALLENGE

Soil Salinization: A Growing Global Threat

As the world population is projected to exceed 10 billion in the coming decades, agriculture will be tasked with producing more (Bahar et al., 2020). This situation occurs in the same timeframe as further degradation of agro-environmental conditions stemming from climate change, decreased arable land, and increased environmental pollution (Grigorieva et al., 2023). Of these challenges, soil salinization has emerged as one of the most serious threats to global agricultural production (Sahab et al., 2021).

According to the FAO's Global Map of Salt-Affected Soils, over 833 million hectares in the world are saline, with more than 10% of land currently being used for crop production (FAO, 2021). The salt concentration in the soil can prevent water uptake and nutrient transport to the plant, resulting in osmotic stress and ion toxicity, limiting the crop yield potential (Ma et al., 2020; Otlewska et al., 2020).

Economic Impact: Yield Losses and Agricultural Inequality

Soil salinity can cause extensive agricultural yield reductions, with potential losses up to 20% to 50% in the case of some affected areas. The reduction in productivity decreases farm income, leads to market instability, and can create a greater divide between productive and non-productive agricultural land (Haj-Amor et al., 2022). While all farmers are potentially impacted by the condition of soil salinization, smallholder farmers, and particularly those living in arid and semi-arid regions are the most negatively affected since they have less access to salinity management technologies and modern infrastructure (Mohanavelu et al., 2021). Soil salinization also generates a growing gap in economic inequality in the agriculture sector. In marginalized communities of farmers losing jobs, they in turn, produce fewer export products and increasingly need to rely on importing food, which only compounds the issues related to the economy at the local and national levels (Salvati & Ferrara, 2015; Jia et al., 2023). Salt affected areas increase by approximately 10% a year, meaning farmers often experience a vicious cycle of low productivity, increased input costs, and little to no access to local markets. In essence, salinity is not only an environmental stress but also serves as an economic catalyst of agricultural decline (Negacz et al., 2022).

Salinity and Sustainable Development Goals (SDGs)

Soil salinization poses a major obstacle to the accomplishment of many of the United Nations Sustainable Development Goals (SDGs), particularly SDG 1 (No Poverty), SDG 2 (Zero Hunger), SDG 13 (Climate Action), and SDG 15 (Life on Land). Salinity contributes to food insecurity and poverty in farm households, through soil degradation and reduced agricultural productivity (Lal et al., 2021). Salinity leads to both land degradation and decreased productivity, along with increased potential for biodiversity loss. All of these impacts reduce sustainability and create challenges to the underlying notion of equitable and inclusive growth. While technological pathways, such as marker-assisted selection or CRISPR-Cas9 genetic improvement techniques, have shown promise for developing salt-tolerant crops (Alam et al. 2022), these methods usually suffer from high costs, complexity, and a lack of availability to the resource-poor farmer (Farhat et al. 2019). The exorbitant cost of inputs, and dependence on specialized infrastructure, purposely ensures that the most vulnerable populations are excluded from salinity solutions.

Thus, these issues highlight the need for less costly, adapted to context and sustainable approaches to addressing salinity, and help to achieve the progress of the SDGs (Batlle-Sales, 2023).

Halophytes as a Low-Cost Biological Solution

Systems in agriculture with halophytes tailor low-cost, sustainable, and scalable solution than technological processes (genetic modification or chemical soil reclamation). Halophytes can be planted alongside salt-sensitive (glycophytic) crops to help the normal crop grow by creating a modified rhizosphere microenvironment to improve water retention, nutrient availability, and salt accumulation around the roots (Rozema & Schat, 2013). The halophytes (*Aptenia cordifolia* and *Spergularia salina*) may even induce beneficial physiological responses from the companion crops due to both below- and above-ground interactions, our work is understood in this way as our research project examines the co-cultivation of *Salvia officinalis* (SAGE) and halophytes as a biological mechanism to alleviate the impact of saline soils (Debez et al., 2017). Apart from the shared positive effects such as enhanced water status, diminishing ionic toxicity and increasing biomass yields in saline conditions, which improve the normal crops physiological performance, identifying more cost-effective adaptations also allows for economic resilience to change/uncertainty - something that is needed on marginal lands where conventional approaches are less profitable. Systems like this demonstrate agroecology and inclusive development as a way of achieving sustainable agricultural practices in salinity impacted agricultural systems (Mishra & Kant, 2018; Tıpırdamaz et al., 2021).

Economic and Environmental Implications of the Proposed System

Initial observations show that *Spergularia salina* connected the growth and salinity stress tolerance of *Salvia officinalis* leading to increased fresh weight marketable yield and reduced irrigation and agrochemical requirements. Grower contribution to soil health was improved along with additional potential economic value opportunities, especially for smallholder farmers that are growing crops on marginal land (El-Khadir et al., 2024). Their increased productivity connects to stable farm incomes, lower production costs, which increases the competitiveness and sustainability of farming practices that occur in salt impacted areas (Sharma & Minhas, 2005). From an ecological standpoint, this low-input agricultural system would mean sustainable use of natural resources, as well as maximize size of available climate resilient agriculture systems. By using this halophyte-assisted cropping system the model emphasizes a more complete strategy that is productive, farmer appropriate, and environmentally friendly (Liang & Shi, 2021).

MATERIALS AND METHODS

Plant Material and Experimental Design

Salvia officinalis cuttings, being sensitive to salinity, were grown separately or with the halophytic species *Aptenia cordifolia* or *Spergularia salina*, both collected from the Oued Malah region (Morocco) (Chriqui et al., 2024). All plants were grown in equal pots with the same substrate containing Maamora soil, peat and compost (El-Khadir et al., 2025). Salinity stress was irrigated as NaCl in concentrations of 0 g/L (control), 2 g/L (35 mM), 5 g/L (85 mM), and 10 g/L (170 mM). There were three modes of cultivation; sage in monoculture, sage with *Aptenia cordifolia* and sage with *Spergularia salina*.

To ensure reproducibility, each treatment was established with three biological replicates and arranged in a randomized complete block design to minimize positional bias. Each replicate pot contained three uniform cuttings of *Salvia officinalis*, either grown in monoculture or intercropped with halophytes (*Aptenia cordifolia* and *Spergularia salina*). Plants were irrigated twice per week with the assigned NaCl concentrations. Growth chambers were maintained under controlled temperature (25 ± 2 °C), photoperiod (16 h light / 8 h dark), and relative humidity (65%).

Data Collection and Processing

Morphological parameters were systematically recorded for each treatment and replicate. Biomass was assessed by measuring both fresh weight (FW) immediately after harvest and dry weight (DW) after oven-drying at 70 °C until constant weight. The length of the aerial part (shoot length) was measured from the base of the stem to the apical meristem using a ruler, while root length was determined after carefully washing roots free of soil substrate. Root volume was measured by water displacement in a graduated cylinder. The number of fully expanded leaves was counted manually. Plant survival rate was expressed as the percentage of surviving plants relative to the initial number of transplanted cuttings per treatment.

Water content was calculated using the following formula:

$$WC\,(\%) = \frac{FW - DW}{FW} \times 100$$

Biochemical analyses were conducted on fully expanded young leaves collected at the end of the experimental period. Proline content was determined according to the colorimetric method of Bates et al. (1973), which involves extraction in 3% sulfosalicylic acid, reaction with acid ninhydrin and glacial acetic acid, heating at 100 °C, and absorbance measurement at 520 nm. Soluble sugar content was estimated following the phenol-sulfuric acid method of Dubois et al. (1956). Briefly, leaf samples were homogenized in ethanol, centrifuged, and the supernatant reacted with phenol and concentrated sulfuric acid; absorbance was measured at 490 nm using a UV–Vis spectrophotometer.

Statistical Modeling

Statistical modeling of the relationships between variables was performed using two complementary approaches: partial least squares regression (PLS-R) for analyzing the relationships between predictors and the response variable, and partial least squares path modeling (PLS-PM) for studying the causal relationships between different physiological variables. These methods are particularly suited to complex datasets with high correlations and a multivariate structure.

PLS-R regression was used to establish a relationship between the explanatory variables (NaCl, biomass, survival, proline, etc.) and the response variable of interest, soluble sugar content (SSc). Unlike classical linear regression, PLS-R simultaneously extracts latent components that maximize the covariance between X and Y while reducing dimensionality. The PLS-R model was constructed using the plsr() function of the pls package, specifying an optimal number of components selected by cross-validation. The quality of the model was assessed using the mean square error of prediction (MSE) and the coefficient of determination R2, which measures the proportion of variance explained by the predictive variables. The PLS-R model equation is given by:

$$Y = XW^{*} + E$$

where W* represents the optimal weights of the latent components, and E is the residual error term. The relevance of the predictors was assessed using regression coefficients, VIP (Variable Importance in Projection) scores, and variable loadings on the principal components.

At the same time, PLS-PM modeling was implemented to examine the structural relationships between the different physiological variables, taking into account the direct and indirect impact of salt stress on plant biomass and survival. The model specification is based on a path matrix, defining the hypothetical causal links between latent variables, and a set of blocks grouping the manifest variables associated with each latent variable. The model was estimated using the plspm function from the plspm package, using a centroid weighting scheme to maximize the explained variance. The model's goodness of fit was assessed using the coefficient of determination (R2), Average Variance Extracted (AVE), and Goodness-of-Fit (GOF). The general equation of the PLS-PM model is given by:

$$\eta = \beta\eta + \Gamma\xi + \zeta$$

where η represents the endogenous latent variables, β is the structural effects matrix, Γ represents the effects of the exogenous variables, and ζ is the error term. The relationships between the latent variables and their indicators were examined using the outer model, evaluated by loadings, composite reliability (Dillon-Goldstein rho), and cross-loadings. The inner model was analyzed using path coefficients, indicating the strength of the relationships between latent variables.

Finally, to complete the analysis, a selection of variables was made using Random Forest in order to identify the most influential predictors in the physiological response of plants to salt stress, and a correlation network was constructed to visualize the multivariate interactions between the different physiological variables. These combined approaches provide a comprehensive understanding of the mechanisms underlying plant tolerance to salt stress and highlight the decisive role of crop type and biomass in physiological adaptation to adverse environmental conditions.

RESULTS

PLS-R Modeling: Determinants of Salt Stress Tolerance

PLS-R analysis was conducted to test the effect of crop type on physiological parameters indicative of salt stress tolerance. Cross-validation indicated that root mean square error of prediction (RMSEP) began to decline and stabilized after around 4 to 5 components, suggestive of good performance by the model (Table 1, Figure 1). The variance explained was 96.68% for SSc, indicating that the model effectively summarizes most of the information contained within the predictors. Comparisons between the full model and reduced model indicated similar performance in explaining the residual variance, while also minimizing components (Table 2).

Figure 1. Component Analysis of the PLS-R Model

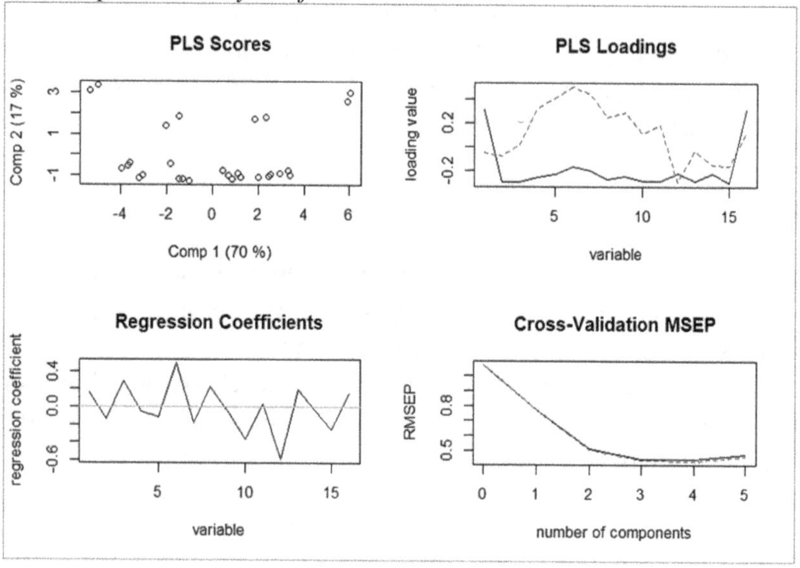

Table 1. Cross-validation and explanation of variance in the PLS-R model

Number of Components	Intercept	1	2	3	4	5
CV (RMSEP)	1,069	0,7667	0,4988	0,3876	0,3468	0,3334
AdjCV (Adjusted RMSEP)	1,069	0,7655	0,4974	0,385	0,344	0,3314
Explained Variance X (%)	-	69.55%	86.21%	89.23%	92.65%	94.72%
Explained Variance Y_train (SSc) (%)	-	55.34%	83.99%	93.75%	95.96%	96.68%

Table 2. Comparison of Performance between the Full Model and the Reduced Model in PLS-R

Number of Components	Intercept	1	2	3	4	5
Reduced Model (CV)	1,069	0,773	0,5138	0,447	0,4483	0,4795
Reduced Model (CV)	1,069	0,8111	0,5725	0,5633	0,4561	0,386
Full Model (adjCV)	1,069	0,769	0,5085	0,4367	0,4339	0,463
Reduced Model (adjCV)	1,069	0,8056	0,5686	0,5457	0,4434	0,3802
Explained Variance X – Full Model (%)	-	69.55%	86.21%	89.23%	92.65%	94.72%
Explained Variance X – Reduced Model (%)	-	69.09%	89.16%	91.32%	96.13%	98.03%

continued on following page

Table 2. Continued

Number of Components	Intercept	1	2	3	4	5
Explained Variance Y_train – Full Model (%)	-	55.34%	83.99%	93.75%	95.96%	96.68%
Explained Variance Y_train – Reduced Model (%)	-	51.74%	76.85%	88.13%	91.58%	92.21%

The fit between observed and predicted SSc values shows a strong linear relationship, validating the accuracy of the model (Figure 2). This correlation indicates that the PLS-R model is capable of effectively predicting the physiological response of *Salvia officinalis* to salt stress, particularly in terms of soluble sugars. Analysis of the PLS scores also reveals a clear separation of the samples according to the experimental conditions, reflecting metabolic differences between the groups studied (Figure 1). This data structure suggests that the type of culture directly influences the physiological response to salt stress.

Figure 2. PLS-R predictions vs. observed values

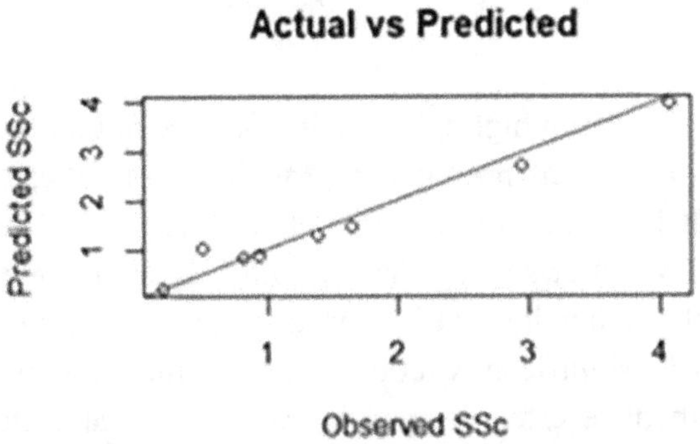

The fit between observed and predicted SSc values shows a strong linear relationship, validating the robustness of the model for predicting physiological responses to stress (Figure 3). Furthermore, cross-validation error analysis (RMSEP) shows a gradual decrease with the addition of components, reaching stabilization around the fifth component. This confirms the effectiveness of the PLS-R model in capturing physiological variations related to salt stress, while limiting overfitting.

Figure 3. Predictions vs. Actual Values for SSc (PLS-R) and Cross-Validation Error (RMSEP)

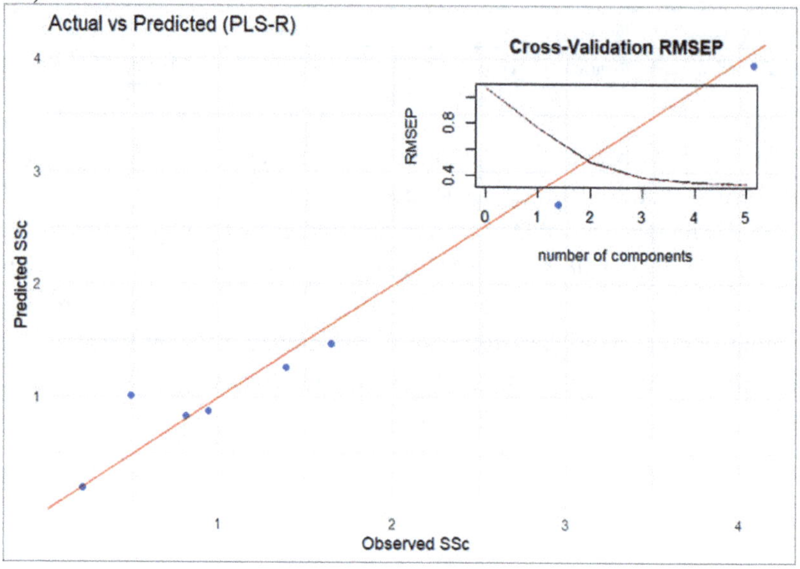

The VIP Scores evaluation highlights the importance of root biomass (Rfb) and leaf biomass (Ldb), as well as proline (Pro), as the main indicators of salt stress tolerance, while NaCl has a more moderate effect (Figure 4). Comparison of the experimental conditions shows that *Salvia officinalis* grown alone leads to greater accumulation of SSc, indicating marked water stress. In contrast, co-cultivation with *Aptenia cordifolia* significantly reduces this accumulation, while *Spergularia salina* has an intermediate effect, suggesting a more gradual modulation of the osmotic response.

Figure 4. Importance of Variables in the Projection (VIP Scores - PLS-R)

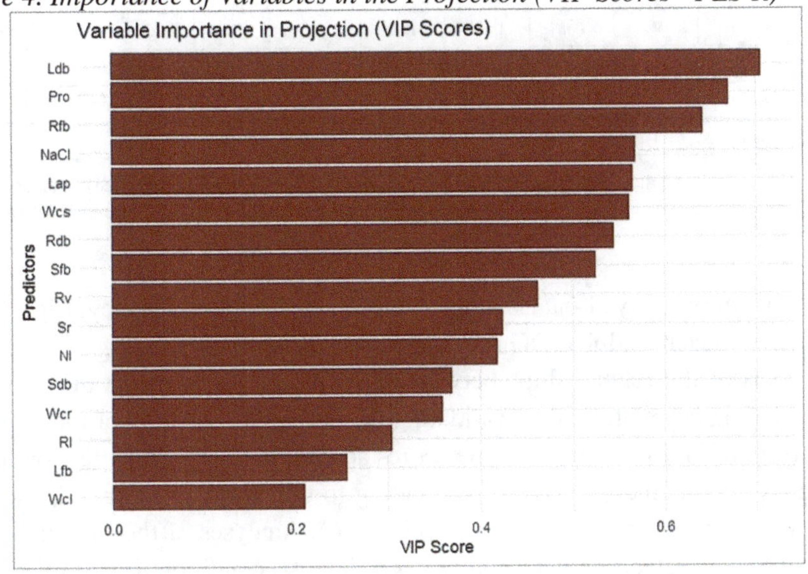

PLS-PM Modeling: Analysis of Causal Relationships Between Physiological and Experimental Variables

PLS-PM analysis was used to examine the relationships between physiological variables under saline stress in order to assess the effect of the halophytes *Aptenia cordifolia* and *Spergularia salina* on the tolerance of *Salvia officinalis*. The model comprises seven latent variables, standardized using a Centroid weighting scheme, and was validated through analysis of it's structural and psychometric properties (Table 3). Unidimensionality was confirmed (eigenvalues: first = 1, second = 0), indicating strong internal consistency (Cronbach's Alpha and Dillon-Goldstein rho = 1).

Table 3. Specifications and Structure of the External Model - PLS-PM. Tabs 3, 4, and 5.

Latent Variable	W	Loa	Com	Red	Type	α	DG rho	1st Eig	2nd Eig
Experimental condition	1	1	1	0	Endogenous	1	1	1	0
NaCl (g/L)	1	1	1	$3{,}74^{E-32}$	Endogenous	1	1	1	0
Soluble sugar content	1	1	1	0,624	Endogenous	1	1	1	0
Proline	1	1	1	0,92	Endogenous	1	1	1	0

continued on following page

Table 3. Continued

Latent Variable	W	Loa	Com	Red	Type	α	DG rho	1st Eig	2nd Eig
Survival Rate	1	1	1	0,906	Endogenous	1	1	1	0
Leaf fresh biomass	1	1	1	0,679	Endogenous	1	1	1	0
Root fresh biomass	1	1	1	0,867	Endogenous	1	1	1	0

Number of cases (36); Latent variables (7); Manifest variables (7); Data scale (Standardized); Weighting scheme (Centroid); Tolerance criterion (1e-06); Maximum number of iterations (100); Convergence iterations (2).

Com: Communality ; Loa: Loading ; W: Weight; Red: Redundancy; Eig: Eigenvalue; α: Cronbach's Alpha; DG rho: Dillon-Goldstein rho.

Analysis of the relationships between observed and predicted values for SSc shows a strong correlation, thus validating the predictive capacity of the PLS-PM model (Figure 5). The latent variable scores according to experimental conditions highlight three distinct groups corresponding to the treatments applied, showing a differentiated response to salt stress (Figure 6). Analyses of the internal model show significant effects of NaCl and the experimental condition on several physiological variables. NaCl positively influences SSc (0.676, $p < 10^{-6}$) and Pro (0.923, $p < 10^{-18}$), confirming their role in adaptation to osmotic stress. The effect of the experimental condition is particularly visible on root biomass (Rfb, -0.351, $p < 0.01$) and leaf biomass (Lfb, -0.344, $p < 0.05$), suggesting that association with halophytes modulates plant growth. In addition, proline and soluble sugars have a negative influence on plant survival (Sr, $p < 0.001$), highlighting their role as indicators of severe stress. *Spergularia salina* appears to mitigate these negative effects, improving root growth (0.607, $p < 0.05$) and leaf biomass (0.416, $p < 0.05$), unlike *Aptenia cordifolia*, whose effect is more moderate (Table 4).

Table 4. Inner Model - PLS-PM: Effects of exogenous and endogenous variables on physiological responses.

Dependent Variable	Independent Variable	Estimate	Std. Err	t value	p-value
NaCl (g/L)	Intercept	$-3{,}33E^{-16}$	0,171	$-1{,}94E^{-15}$	$p > 0{,}05$
NaCl (g/L)	Condition expérimentale	$1{,}11E^{-16}$	0,171	$6{,}47E^{-16}$	$p > 0{,}05$
Soluble sugar content	Intercept	$2{,}44E^{-16}$	0,107	$2{,}28E^{-15}$	$p > 0{,}05$
Soluble sugar content	Condition expérimentale	-0,409	0,107	-3,83	$p < 0{,}001$
Soluble sugar content	NaCl (g/L)	0,676	0,107	6,33	$p < 10^{-6}$
Proline	Intercept	$3{,}44E^{-16}$	0,0492	$7E^{-15}$	$p > 0{,}05$
Proline	Condition expérimentale	-0,261	0,0492	-5,3	$p < 10^{-5}$
Proline	NaCl (g/L)	0,923	0,0492	18,8	$p < 10^{-18}$

continued on following page

Table 4. Continued

Dependent Variable	Independent Variable	Estimate	Std. Err	t value	p-value
Survival Rate	Intercept	4,08E^{-16}	0,055	7,41E^{-15}	p > 0,05
Survival Rate	Condition expérimentale	-0,0908	0,0749	-1,21	p > 0,05
Survival Rate	NaCl (g/L)	0,737	0,2045	3,6	p < 0,01
Survival Rate	Soluble sugar content	-0,494	0,1212	-4,07	p < 0,001
Survival Rate	Proline	-1,14	0,2628	-4,32	p < 0,001
Leaf fresh biomass	Intercept	2,27E^{-16}	0,103	2,19E^{-15}	p > 0,05
Leaf fresh biomass	Condition expérimentale	-0,344	0,144	-2,38	p < 0,05
Leaf fresh biomass	NaCl (g/l)	0,401	0,459	0,87	p > 0,05
Leaf fresh biomass	Soluble sugar content	0,416	0,283	1,47	p > 0,05
Leaf fresh biomass	Proline	-1,76	0,626	-2,81	p < 0,01
Leaf fresh biomass	Survival Rate	-0,324	0,338	-9,57	p > 0,05
Root fresh biomass	Intercept	-2,84E^{-16}	0,0678	-4,2E^{-15}	p > 0,05
Root fresh biomass	Condition expérimentale	-0,351	0,103	-3,41	p < 0,01
Root fresh biomass	NaCl (g/L)	-0,211	0,3041	-0,95	p > 0,05
Root fresh biomass	Soluble sugar content	0,517	0,1917	2,7	p < 0,05
Root fresh biomass	Proline	-3	0,4608	-6,51	p > 0,05
Root fresh biomass	Survival Rate	0,607	0,2247	2,7	p < 0,05
Root fresh biomass	Leaf fresh biomass	0,348	0,1196	2,91	p < 0,01

Figure 5. Relationship between observed and predicted values of SSc with PLS-PM.

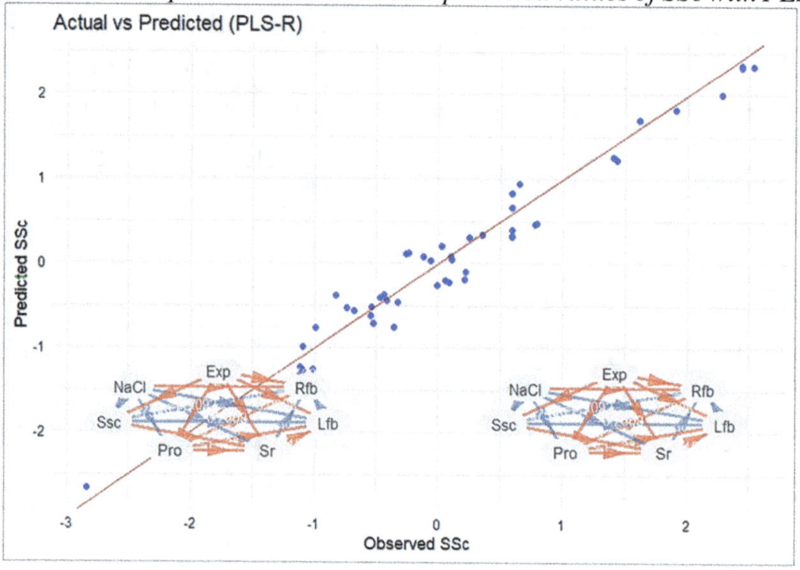

Figure 6. Latent Variable Scores According to Experimental Conditions - PLS-PM.

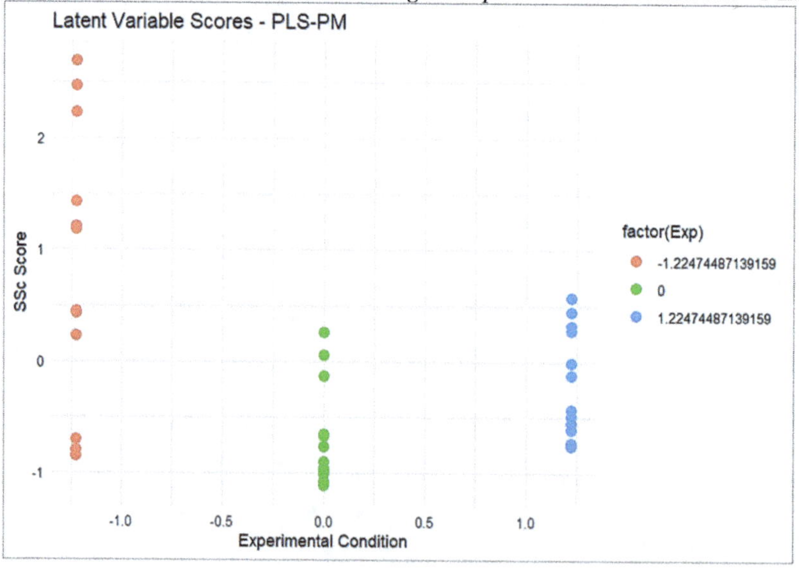

Analysis of the total effects of the relationships between latent variables in the PLS-PM model highlights complex interactions between salt stress, osmolytes, and plant biomass (Table 5). The direct effect of NaCl on the production of soluble sugars (SSc, 0.676, $p < 0.001$) and proline (Pro, 0.923, $p < 0.001$) confirms the role of osmotic stress in the accumulation of osmolytes. However, the experimental condition (Exp) significantly modulates these responses, directly influencing Pro (-0.261, $p < 0.001$) and root biomass (Rfb) (-0.351, $p < 0.01$). Analysis of indirect effects shows that plant survival rate (Sr) is mainly affected by Pro (-1.14, $p < 0.001$) and SSc (-0.494, $p < 0.001$), reinforcing their role as biomarkers of salt stress.

Table 5. Total Effects of Relationships in the PLS-PM Model

Relation	Direct Effect	Indirect Effect	Total Effect
Experimental condition → NaCl (g/L)	$1{,}11E^{-16}$	0	$1{,}11E^{-16}$
Experimental condition → Soluble sugar content	-0,409	$7{,}5^{E-17}$	-0,409
Experimental condition → Proline	-0,261	$1{,}02E^{-16}$	-0,261
Experimental condition → Survival Rate	-0,908	0,498	0,407
Experimental condition → Leaf fresh biomass	-0,344	0,157	-0,187
Experimental condition → Root fresh biomass	-0,351	0,049	-0,302
NaCl (g/L) → Soluble sugar content	0,676	0	0,676
NaCl (g/L) → Proline	0,923	0	0,923
NaCl (g/L) → Survival Rate	0,737	-0,138	-0,644
NaCl (g/L) → Leaf fresh biomass	-0,211	-0,575	-0,786
NaCl (g/L) → Root fresh biomass	-0,211	-0,575	-0,786
Soluble sugar content → Proline	0	0	0
Soluble sugar content → Survival Rate	-0,494	0	-0,494
Soluble sugar content → Leaf fresh biomass	0,416	0,16	5,76
Soluble sugar content → Root fresh biomass	0,517	0	5,17
Proline → Survival Rate	-1,14	0	-1,14
Proline → Leaf fresh biomass	-1,76	-1,17	-1,47
Proline → Root fresh biomass	-3	-1,17	-1,47
Survival Rate → Leaf fresh biomass	-0,324	0	-0,324
Survival Rate → Root fresh biomass	0,607	-1,13	4,95
Leaf fresh biomass → Root fresh biomass	0,348	0	3,48

The importance of the predictive variables in the model (Figure 7) confirms that parameters related to water and biomass are decisive for stress tolerance. The most influential factors are root water content (Wcr), shoot length (Lap), and number of leaves (Nl), suggesting that water management and vegetative growth are key physiological responses to salt stress. In contrast, NaCl appears to be a secondary modulating factor, indicating that it's impact is mainly mediated by other physiological variables.

These results demonstrate that association with a halophyte can mitigate the effects of salt stress by influencing leaf and root biomass and modulating osmolite production. The impact of *Spergularia salina* appears to be more beneficial, stabilizing physiological responses and maintaining better water balance.

Figure 7. Importance of Variables - Random Forest.

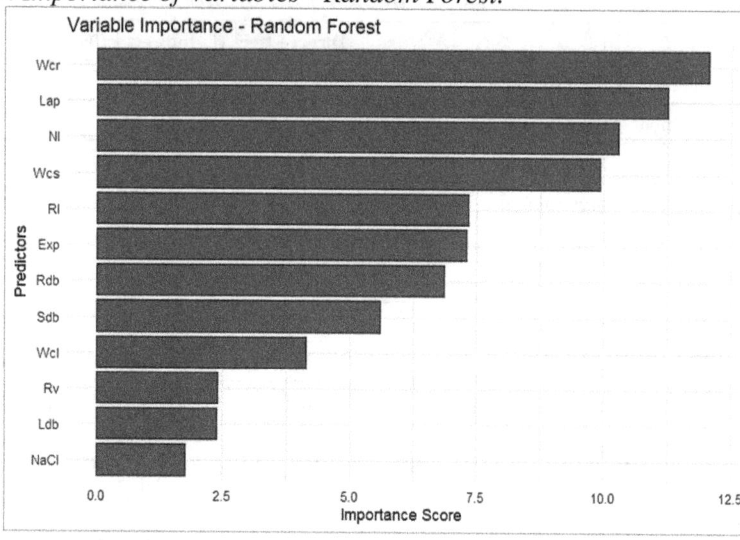

Figure 8 illustrates the network of correlations between physiological variables under salt stress, highlighting the interactions between the latent variables of the PLS-PM model. The type of crop (Exp) is a structuring factor influencing several physiological responses. Salt stress (NaCl) shows strong connections with osmolytes and biomass, confirming it's central role in plant adaptation. Soluble sugar (SSc) is strongly correlated with leaf biomass (Lfb) and root biomass (Rfb), highlighting it's importance in osmotic adjustment. A marked association between NaCl and Wcr (root water content) indicates that root water management is a key determinant of stress tolerance. Furthermore, the correlation between aboveground length (Lap) and proline (Pro) suggests a physiological adaptation strategy. Leaf and root biomass, located at the center of the network, are modulated by several variables, reflecting their integrative role in stress response. These results confirm that *Spergularia salina* promotes better tolerance to salt stress, in particular by stabilizing water management and mitigating excessive accumulation of osmolytes, unlike *Aptenia cordifolia*, whose effect remains more limited.

Figure 8. Network of Correlations between Variables.

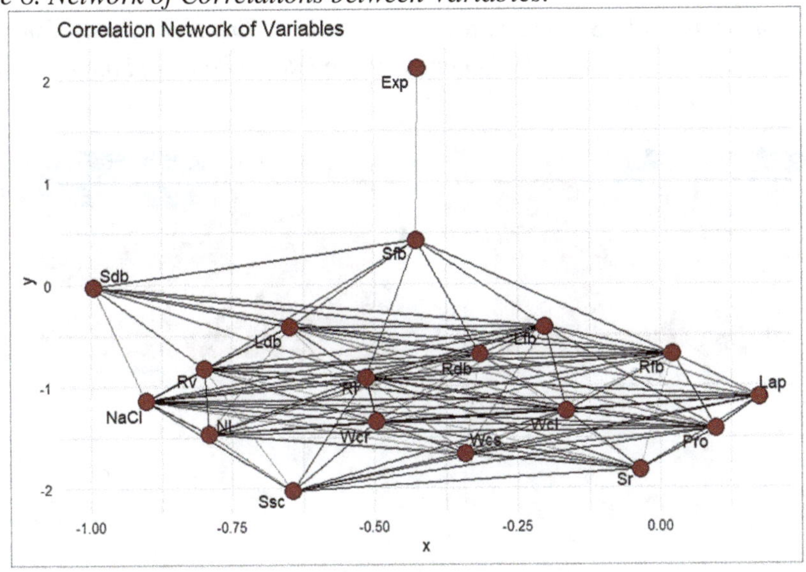

The thermal representation of the relationships between latent variables and path coefficients in the PLS-PM model highlights the key interactions between physiological variables under salt stress. Figure 10 illustrates the distinction between the internal model and the external model. In the internal model, the R^2 values are high for Pro (0.92), Sr (0.91), and Rfb (0.87), suggesting that these variables are well explained by the model. In addition, the high average redundancy indicates that the manifest variables effectively capture the information of the latent variables, thus validating the robustness of the PLS-PM model. Analysis of the external model shows high communality (1.00) for all variables, confirming that each manifest variable is well correlated with it's corresponding latent variable. Redundancy varies depending on the variables, with lower values for NaCl (≈0.00), indicating that it does not directly explain the observed variations. In contrast, Pro (0.92), Sr (0.91), and Rfb (0.87) show a strong contribution to their respective latent variables, highlighting their importance in the physiological response to salt stress (Figure 9).

Figure 9. Heatmap Representation of the Internal and External PLS-PM Models. Left (a): Internal Model Summary: R^2, Block Communality, Mean Redundancy, and AVE. Right (b): External Model: Communality and Redundancy of Latent Variables. AVE: Average Variance Extracted

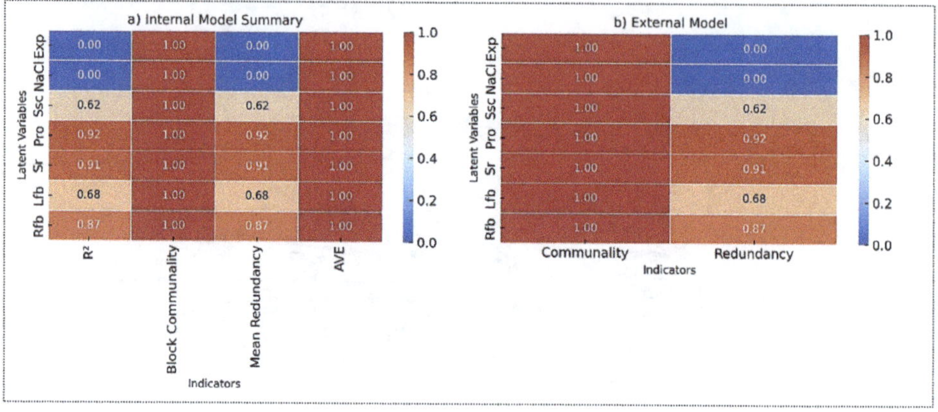

Figure 10 highlights the correlations between latent variables and the structural path coefficients of the PLS-PM model. Analysis of cross-loadings reveals a significant correlation between NaCl and Pro (0.92, $p < 0.001$), confirming the accumulation of proline under salt stress. In addition, a high negative correlation between SSc and Sr (-0.92, $p < 0.001$) suggests that an increase in soluble sugars is associated with a decrease in survival rate, highlighting their role in osmotic adjustment under stress. Analysis of structural path coefficients indicates that Sr is strongly influenced by Pro (-1.14, $p < 0.001$) and SSc (-0.49, $p < 0.001$), confirming their role as major markers of salt stress. At the same time, *Spergularia salina* appears to modulate this response more effectively, reducing excessive Pro accumulation and stabilizing water management, unlike *Aptenia cordifolia*, whose effect is more moderate. These results confirm that salt stress tolerance is based on a complex interaction between osmolytes, biomass, and water management, thus reinforcing the interest of interspecific associations in agricultural strategies adapted to saline environments.

Figure 10. Heatmap Representation of Latent Variable Correlations and Structural Model Path Coefficients. Left (a): Correlations and Cross-loadings Between Latent Variables. Right (b): Structural Model Path Coefficients (PLS-PM).

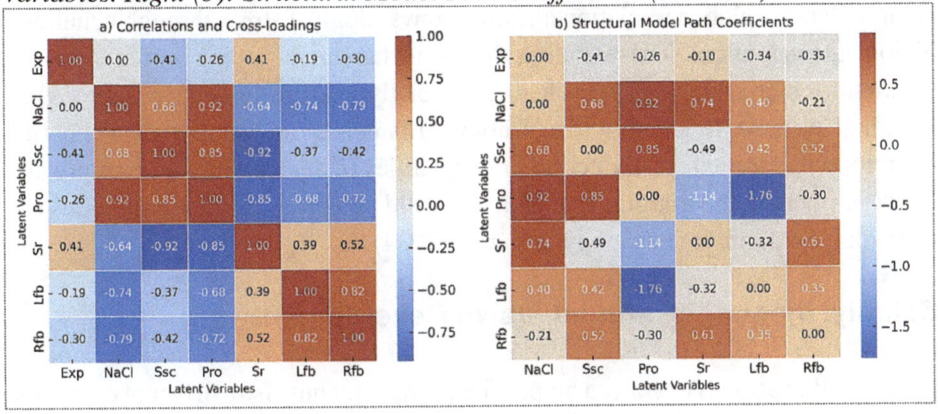

DISCUSSION

Salt Stress as a Driver of Agricultural Exclusion

Soil salinity represents one of the greatest abiotic stresses facing agriculture in arid and semi-arid environments (Gamalero et al., 2020). The physiological barriers seen in *Salvia officinalis* in response to salt stress, such as lower water, biomass and survivorship, are similar to the limits associated with farmers growing the wrong crops in saline soils. The relationship between biophysical limits faced on saline soils and the phenomena labelled here as agricultural exclusion, highlights the marginal land user that development leaves behind as unable to support productive systems in the degraded context (Ben Gaied et al., 2024). This phenomenon does not just reduce agricultural ability, but aggravates territorial inequality; for example, fertile spatial has elevated the capacity for yield and investment relative to underdeveloped saline zones leading to disparities in the economy. Through a lens of social justice and spatial equity salinity should be viewed as not just an agronomic issue, but a question of how agricultural policy is being pieced together.

Intercropping as a Lever for Resilience and Agronomic Equity

Our results demonstrate that intercropping S. officinalis with halophytes, specifically *Spergularia salina*, greatly enhances the performance of S. officinalis when grown in saline environments. The coexistence of these two plants enhanced their root volume, leaf biomass, water retention, and survivability as a whole. The reduction of proline and soluble sugar accumulation when co-cultured implies that the natural plant-to-plant association alleviates physiological stress through beneficial biological interactions. These findings illustrate that intercropping is a nature-based solution for increasing stress resilience, especially in crops in under-resourced agricultural systems. In addition to agronomic, financial, and technological benefits, intercropping facilitates a reduction in inequalities of agricultural opportunity and allows farmers in the margins to regain lost yield potential and make judicious use of available resources (Papadopoulos et al., 2024).

Salinity, Inequality, and Inclusive Development

Soil salinization is not only a biophysical constraint but also a driver of economic and spatial inequality. Farmers cultivating fertile lands maintain higher productivity and income, while those located in saline or degraded zones are systematically disadvantaged. This creates a cycle of agricultural exclusion, where marginal farmers face declining yields, higher production costs, and eventual withdrawal from markets.

By contrast, the introduction of halophyte-assisted intercropping systems provides an inclusive pathway that allows smallholder farmers to regain productivity and participate more equitably in agricultural value chains. From a development perspective, this approach contributes to reducing rural poverty, promoting land justice, and addressing the broader objectives of sustainable territorial development.

Economic Impact: From Yield Recovery to Sustainable Profitability

Salinity has been the direct cause of yield losses resulting in economically unstable farms particularly for smallholders in developing countries (Zörb et al., 2019). In other words, reduced productivity translates into increased production costs per unit; ineffective market competition, increasing reliance on subsides or humanitarian assistance. Reduced yield losses from salinity through halophyte-assisted intercropping can provide an opportunity to get back to profitably even in degraded areas (Haj-Amor et al., 2022). The favorable growth and survival of the plants can ultimately increase harvestable biomass, reduce input costs (fertilizers, irrigation), and improve long-term soil management which can all boost return on investment. This is of essence for climate-resilient value chains, sustainability and cost-competitive increasingly appear to be permanent prerequisites for local and international market access. Moreover, implementing salt-tolerant intercropping systems supports the objectives of the green economy and inclusive growth, as it can deliver low-cost solutions that can be scaled through farmer cooperatives, as well as, local entrepreneurship (e.g. essential oils, herbal extracts, saline agriculture products) (Islam et al., 2021; Khan et al., 2022; Mouniane et al., 2025).

Reduced Input Use and Water-Use Efficiency: Toward Low-Impact Agriculture

The benefit, and probably one of the principle benefits, of halophyte-assisted cropping systems is that they promote bottom-up efficiencies of water and nutrient use. Halophyte-assisted cropping systems dramatically reduce on-imputs and watering events which are two of the major factors promoting environmental degradation in conventional and industrial agriculture (Liu & Wang, 2021). The biological synergism allows for independent internal regulation of soil salinity, soil rooting architecture, and salinity tolerance levels. These benefits support the principles of agroecology and ecological intensification justification or rationale that productivity can be increased through ecosystem services instead of external inputs, therefore, halophyte-based strategies are supported as a pathway to develop low impact sustainable agriculture, in order to save on resource consumption and maintain production in vulnerable areas (Chen & Wang, 2024).

Challenges to Adoption in Arid and Low-Income Regions

While halophyte-based intercropping systems present considerable advantages, their application remains constrained due to several substantial obstacles; a lack of availability of halophyte planting materials, insufficient agronomic understanding amongst farmers, a lack of institutional backing or incentives, and sociocultural aspects that may limit the acceptance of alternative cropping paradigms. Addressing these challenges will require ensuring there is capacity building, trials made at demonstrative plots, participatory engagement in trials, and politically-delivered support. However, for farmers to accept, adopt, and integrate these practices into their agricultural systems, it is imperative to ensure these practices are culturally accepted, accessible, and cost-efficient. The consequences of not materially implementing government and NGO support for this type of agroecological innovation is a potential isolation of the benefits of these systems to experiment production.

PERSPECTIVES FOR INCLUSIVE AND SUSTAINABLE AGRICULTURE

Challenges to Adoption in Arid and Low-Income Regions

The application of sophisticated statistical modeling (PLS-R and PLS-PM) in this study demonstrates how AI-powered tools can support understanding complex physiological responses, as well as identify critical stress markers. The application of machine learning algorithms within agricultural advisory systems could predict crop response under salinity, as well as identify the best plant associations, to help inform practical irrigation/fertilization schedules dynamically in real time.

The potential for such technologies to enhance actual decision making for farmers is important especially when used hand in hand with mobile apps and sensors-based systems. This represents exciting potential for the digital transformation of climate-smart agriculture.

Valorizing Halophytes in Agricultural Policy and Innovation

For institutional recognition to realize the potential of halophyte-based systems, we call upon governments and development agencies to include halophytes among species in agricultural research programmes, subsidy programmes, and extension services. This would include establishing seed banks and nurseries, facilitating the market development of products derived from halophytes, and promoting their role in land restoration, and the circular economy. Recognition would not only contribute to the rehabilitation of ecosystems for the circular economy, but also promote economic diversification in the already vulnerable rural and urban areas.

Toward Environmentally and Socially Embedded Rural Development

This study illustrates the need for developing strategies of rural development that are grounded in the ecological and socio-economic realities of each territory. In regions with problems of soil salinization, generalized agricultural models often do not yield effective solutions because they do not take into account the local limitations, such as soil features and variability, climate, or species adaptability. A more effective strategy is to place agricultural interventions into the realities of the region, where the species of native biodiversity, stress-compatible species, and socio-economic context of local community practices are given due consideration. As demonstrated in this study, halophyte-assisted cropping systems were an inclusive, low-cost, and ecologically friendly innovation that provided increased resilience in a degraded landscape. Thus, by enhancing productivity under stress and equitable access to a more viable agricultural system, it offered a real pathway to diminish spatial disparity while meeting the United Nations Sustainable Development Goals of reducing poverty (SDG 1), food security (SDG 2), sustainable production (SDG 12), climate resilience (SDG 13), and land restoration (SDG 15).

Policy Outlook and Practical Applications

For halophyte-assisted systems to transition from success in experimental contexts to success on a larger scale, the supporting policies and institutional context must be established. One possible route to provide support for halophytes is by organizing smallholders into farmer cooperatives and community models, where they can share planting material, the business of essential oil extraction equipment, and market access as a group. By working together in a larger collective, smallholders can lower production costs and strengthen their bargaining power. More broadly, integrating halophyte-assisted intercrops into national or regional land restoration programs offers an applied route for rehabilitating saline-degraded land, especially in areas where the salinization of soils is undermining food security and rural livelihoods. Supportive and complementary extension and subsidy programs may also accelerate the path to supporting halophytes, whereby government and non-governmental organizations could establish halophyte seed banks, incentives for saline-adapted farming, and train extension officials and personnel to use these systems. There are practical real-world cases where these approaches have been trialed on larger scales, for example, intercropping cotton with halophytes in China, or community-based saline rice production in Bangladesh. Including all these processes within a policy framework would provide pathway for wider uptake and reinforce the interrelations between ecological resilience and rural economic regeneration as well as simultaneously achieve multiple UN Sustainable Development Goals associated with poverty reduction, food security, and sustainable land management.

CONCLUSION

This study provides compelling evidence that halophyte-assisted intercropping is a promising and sustainable strategy to mitigate the detrimental effects of soil salinity on crop performance. Using *Salvia officinalis* as a salt-sensitive model, and *Spergularia salina* and *Aptenia cordifolia* as halophytic companions, we demonstrated that plant–plant interactions can significantly enhance physiological resilience, improve biomass production, and reduce osmotic stress under saline conditions. Among the two halophytes, *Spergularia salina* emerged as the most effective partner, promoting root development, water retention, and survival rates, thereby reinforcing the adaptive capacity of sage in hostile environments. The use of multivariate statistical tools (PLS-R, PLS-PM) allowed for a detailed interpretation of both direct and indirect effects of salinity and plant associations, revealing key stress biomarkers and causal relationships among physiological traits. Moreover, the integration of Random Forest analysis confirmed the central role of water management and vegetative growth in salt tolerance. Beyond the physiological benefits, the study highlights important economic and environmental implications. By improving yields and reducing dependence on external inputs, halophyte-based systems contribute to profitability and economic resilience, particularly in marginalized regions where land degradation limits agricultural options. This research supports the move towards low-input, climate-smart agriculture following ecological intensification and sustainable development principles. However, wider adoption will need to account for socio-technical barriers including limited access to halophyte material, a lack of awareness, and minimal policy support. Embedding these strategies in rural development programs and recognising halophyte species in agricultural planning may play a significant role in restoring degraded land and reducing spatial inequalities. In conclusion, halophyte-assisted intercropping is not only an agronomic innovation but also one means of environmental remediation, social inclusion, and economic viability. It proposes a model for agriculture to adapt to the impact of climate change as well as making meaningful contribution to achieve global sustainability objectives.

REFERENCES

Afonso, A. F., Pereira, O. R., Fernandes, Â., Calhelha, R. C., Silva, A. M. S., Ferreira, I. C. F. R., & Cardoso, S. M. (2019). Phytochemical composition and bioactive effects of *Salvia africana, Salvia officinalis* 'Icterina' and *Salvia mexicana* aqueous extracts. *Molecules (Basel, Switzerland), 24*(24), 4327. DOI: 10.3390/molecules24234327

Al-Falahi, A. H., Barry, S., Gebrechorkos, S. H., Spank, U., & Bernhofer, C. (2023). Potential of traditional adaptation measures in mitigating the impact of climate change. *Sustainability, 15*(21), 15442. DOI: 10.3390/su152115442

Alam, M. S., Kong, J., Tao, R., Ahmed, T., Alamin, M., Alotaibi, S. S., Abdelsalam, N. R., & Xu, J.-H. (2022). CRISPR/Cas9 mediated knockout of the OsbHLH024 transcription factor improves salt stress resistance in rice (*Oryza sativa* L.). *Plants, 11*(9), 1184. DOI: 10.3390/plants11091184

Alharbi, Y. M., Sakr, S. S., Albarrak, S. M., Almundarij, T. I., Barakat, H., & Hassan, M. F. Y. (2022). Antioxidative, antidiabetic, and hypolipidemic properties of probiotic-enriched fermented camel milk combined with *Salvia officinalis* leaves hydroalcoholic extract in streptozotocin-induced diabetes in rats. *Antioxidants, 11*(4), 668. DOI: 10.3390/antiox11040668

Bahadırlı, N. P. (2020). Economically important sage species from Turkey: *Salvia fruticosa* Mill. and *S. aramiensis* Rech fil. *Current Perspectives on Medicinal and Aromatic Plants, 3*(1), 31–42. DOI: 10.38093/cupmap.736482

Bahar, N. H. A., Lo, M., Sanjaya, M., Van Vianen, J., Alexander, P., Ickowitz, A., & Sunderland, T. (2020). Meeting the food security challenge for nine billion people in 2050: What impact on forests? *Global Environmental Change, 62*, 102056. DOI: 10.1016/j.gloenvcha.2020.102056

Bates, L. S., Waldren, R. P., & Teare, I. D. (1973). Rapid determination of free proline for water-stress studies. *Plant and Soil, 39*(1), 205–208.

Batlle-Sales, J. (2023). Salt-affected soils: A sustainability challenge in a changing world. *Italian Journal of Agronomy, 18*(2), 2188. DOI: 10.4081/ija.2023.2188

Ben Akacha, B., Kačániová, M., Generalić Mekinić, I., Kukula-Koch, W., Koch, W., Orhan, I. E., Čmiková, N., Taglieri, I., Venturi, F., Samartin, C., Bouteraa, M. T., Ben Saad, R., Mnif, W., Garzoli, S., & Ben Hsouna, A. (2024). Sage (*Salvia officinalis* L.): A botanical marvel with versatile pharmacological properties and sustainable applications in functional foods. *South African Journal of Botany, 169*, 361–382. DOI: 10.1016/j.sajb.2024.04.044

Ben Gaied, R., Brígido, C., Sbissi, I., & Tarhouni, M. (2024). Sustainable strategy to boost legumes growth under salinity and drought stress in semi-arid and arid regions. *Soil Systems*, *8*(3), 84. DOI: 10.3390/soilsystems8030084

Bendaas, R., Messaadia, L., Bekkar, Y., Bourougaa, L., & Messaoudi, A. (2025). Comprehensive *in vitro* and molecular docking analysis of antioxidant and antimicrobial properties of *Salvia officinalis* L. extracts and essential oils. *Journal of Molecular Structure*, *1331*, 141621. DOI: 10.1016/j.molstruc.2025.141621

Chen, J., & Wang, Y. (2024). Understanding the salinity resilience and productivity of halophytes in saline environments. *Plant Science*, *346*, 112171. DOI: 10.1016/j.plantsci.2024.112171

Chiang, N., Dein, M., Nguyen, T. H., Trudelle, D. M., Ho, C.-T., & Munafo, J. P.Jr. (2024). Characterization of odorants in a commercial culinary sage (*Salvia officinalis* L.) and several cultivars. *Journal of Agricultural and Food Chemistry*, *72*(14), 8081–8091. DOI: 10.1021/acs.jafc.3c09326

Chriqui, A., Benkhnigue, O., Moujahdi, C., El-Khadir, I., Mouniane, Y., Zidane, L., & Hmouni, D. (2024). Contribution to a floristic study of the Ouezzane region (Prerif, North Atlantic Morocco–1). *Egyptian Journal of Botany*, *64*(4), 28–48. DOI: 10.21608/ejbo.2024.226380.2439

Debez, A., Belghith, I., Friesen, J., Montzka, C., & Elleuche, S. (2017). Facing the challenge of sustainable bioenergy production: Could halophytes be part of the solution? *Journal of Biological Engineering*, *11*(1), 27. DOI: 10.1186/s13036-017-0069-0

Dekker, S. C., Kraneveld, A. D., van Dijk, J., Kalfagianni, A., Knulst, A. C., Lelieveldt, H., Moors, E. H. M., Müller, E., Pieters, R. H. H., & Pieterse, C. M. J.. (2020). Towards healthy planet diets—A transdisciplinary approach to food sustainability challenges. *Challenges*, *11*(2), 21. DOI: 10.3390/challe11020021

Dubois, M., Gilles, K. A., Hamilton, J. K., Rebers, P. A., & Smith, F. (1956). Colorimetric method for determination of sugars and related substances. *Analytical Chemistry*, *28*(3), 350–356.

Eichsteller, M., Njagi, T., & Nyukuri, E. (2022). The role of agriculture in poverty escapes in Kenya – Developing a capabilities approach in the context of climate change. *World Development*, *149*, 105705. DOI: 10.1016/j.worlddev.2021.105705

El-Khadir, I., Ktaoui, S., Mouniane, Y., Chriqui, A., Mabrouki, J., Ameziane, H., & Hmouni, D. (2024). Improved salt stress tolerance of *Salvia officinalis* grown in the presence of a halophytic plant *Spergularia maritima*: Analysis of morpho-physiological parameters. In Mabrouki, J., & Azrour, M. (Eds.), *Integrated solutions for smart and sustainable environmental conservation* (Vol. 527, pp. XX–XX). Springer., DOI: 10.1007/978-3-031-55787-3_2

El-Khadir, I., Mouniane, Y., Chriqui, A., Kouighat, M., & Hmouni, D. (2025). Effects of salt stress on *Origanum majorana* and improvement via association with halophyte plants: *Spergularia salina* and *Aptenia cordifolia*. *Acta Universitatis Agriculturae et Silviculturae Mendelianae Brunensis*, *73*(1), 27–35. DOI: 10.11118/actaun.2025.003

FAO. (2021). *World Soil Day: FAO highlights the threat of soil salinization to global food security*. Food and Agriculture Organization of the United Nations.

Farhat, S., Jain, N., Singh, N., Sreevathsa, R., Dash, P. K., Rai, R., Yadav, S., Kumar, P., Sarkar, A. K., Jain, A., Singh, N. K., & Rai, V. (2019). CRISPR-Cas9 directed genome engineering for enhancing salt stress tolerance in rice. *Seminars in Cell & Developmental Biology*, *96*, 91–99. DOI: 10.1016/j.semcdb.2019.05.003

Fraisse, D., Bred, A., Lagarde, A., & Felgines, C. (2025). Impact of simulated gastrointestinal conditions on polyphenol stability and antiglycoxidant potential of sage (*Salvia officinalis*) infusion. *South African Journal of Botany*, *176*, 29–36. DOI: 10.1016/j.sajb.2024.11.019

Gamalero, E., Bona, E., Todeschini, V., & Lingua, G. (2020). Saline and arid soils: Impact on bacteria, plants, and their interaction. *Biology (Basel)*, *9*(6), 116. DOI: 10.3390/biology9060116

Ghallab, D. S., Awwad, Z. M., Ibrahim, R. S., & Shawky, E. (2025). Dietary relevance of sage (*Salvia officinalis* L.) in cognitive health: A serum-based mechanistic study targeting Alzheimer's disease pathways. *Food Bioscience*, *69*, 106863. DOI: 10.1016/j.fbio.2025.106863

Ghorbani, A., & Esmaeilizadeh, M. (2017). Pharmacological properties of *Salvia officinalis* and its components. *Journal of Traditional and Complementary Medicine*, *7*(4), 433–440. DOI: 10.1016/j.jtcme.2016.12.014

Grigorieva, E., Livenets, A., & Stelmakh, E. (2023). Adaptation of agriculture to climate change: A scoping review. *Climate (Basel)*, *11*(10), 202. DOI: 10.3390/cli11100202

Gudi, S., Gill, H. S., Collins, S., Singh, J., Sandhu, D., Sehgal, S. K., Gill, U., & Gupta, R. (2025). Association analysis identified superior haplotypes for improved salt stress tolerance in wheat (*Triticum aestivum* L.). *Plant Stress*, *16*, 100900. DOI: 10.1016/j.stress.2025.100900

Haj-Amor, Z., Araya, T., Kim, D.-G., Bouri, S., Lee, J., Ghiloufi, W., Yang, Y., Kang, H., Jhariya, M. K., Banerjee, A., & Lal, R. (2022). Soil salinity and its associated effects on soil microorganisms, greenhouse gas emissions, crop yield, biodiversity and desertification: A review. *The Science of the Total Environment*, *843*, 156946. DOI: 10.1016/j.scitotenv.2022.156946

Islam, M. A., Lobry de Bruyn, L., Warwick, N. W. M., & Koech, R. (2021). Salinity-affected threshold yield loss: A signal of adaptation tipping points for salinity management of dry season rice cultivation in the coastal areas of Bangladesh. *Journal of Environmental Management*, *288*, 112413. DOI: 10.1016/j.jenvman.2021.112413

Jakovljević, M., Jokić, S., Molnar, M., Jašić, M., Babić, J., Jukić, H., & Banjari, I. (2019). Bioactive profile of various *Salvia officinalis* L. preparations. *Plants*, *8*(3), 55. DOI: 10.3390/plants8030055

Jia, Y., Wu, J., Cheng, M., & Xia, X. (2023). Global transfer of salinization on irrigated land: Complex network and endogenous structure. *Journal of Environmental Management*, *336*, 117592. DOI: 10.1016/j.jenvman.2023.117592

Karalija, E., Dahija, S., Tarkowski, P., & Ćavar Zeljković, S. (2022). Influence of climate-related environmental stresses on economically important essential oils of Mediterranean *Salvia* sp. *Frontiers in Plant Science*, *13*, 864807. DOI: 10.3389/fpls.2022.864807

Khan, M. O., Irfan, M., Muhammad, A., Ullah, I., Nawaz, S., Khalil, M. K., & Ahmad, M. (2022). A practical and economical strategy to mitigate salinity stress through seed priming. *Frontiers in Environmental Science*, *10*, 991977. DOI: 10.3389/fenvs.2022.991977

Kompoura, V., Karapantzou, I., Mitropoulou, G., Parisis, N. A., Gkalpinos, V. K., Anagnostou, V. A., Tsiailanis, A. D., Vasdekis, E. P., Koutsaliaris, I. K., Tsouka, A. N., Karapetsi, L., Madesis, P., Letsiou, S., Florou, D., Koukkou, A.-I., Barbouti, A., Tselepis, A. D., Kourkoutas, Y., & Tzakos, A. G. (2024). Exploiting the beneficial effects of *Salvia officinalis* L. extracts in human health and assessing their activity as potent functional regulators of food microbiota. *Food Chemistry*, *441*, 138175. DOI: 10.1016/j.foodchem.2023.138175

Lago-Olveira, S., Ouhemi, H., Idrissi, O., Moreira, M. T., & González-García, S. (2024). Promoting more sustainable agriculture in the Moroccan drylands by shifting from conventional wheat monoculture to a rotation with chickpea and lentils. *Cleaner Environmental Systems*, *12*, 100169. DOI: 10.1016/j.cesys.2024.100169

Lal, R., Bouma, J., Brevik, E., Dawson, L., Field, D. J., Glaser, B., Hatano, R., Hartemink, A. E., Kosaki, T., Lascelles, B., Monger, C., Muggler, C., Ndzana, G. M., Norra, S., Pan, X., Paradelo, R., Reyes-Sánchez, L. B., Sandén, T., Singh, B. R., & Zhang, J. (2021). Soils and sustainable development goals of the United Nations: An International Union of Soil Sciences perspective. *Geoderma Regional*, *25*, e00398. DOI: 10.1016/j.geodrs.2021.e00398

Liang, J., & Shi, W. (2021). Cotton/halophytes intercropping decreases salt accumulation and improves soil physicochemical properties and crop productivity in saline-alkali soils under mulched drip irrigation: A three-year field experiment. *Field Crops Research*, *262*, 108027. DOI: 10.1016/j.fcr.2020.108027

Liu, L., & Wang, B. (2021). Protection of halophytes and their uses for cultivation of saline-alkali soil in China. *Biology (Basel)*, *10*(5), 353. DOI: 10.3390/biology10050353

Ma, Y., Dias, M. C., & Freitas, H. (2020). Drought and salinity stress responses and microbe-induced tolerance in plants. *Frontiers in Plant Science*, *11*, 591911. DOI: 10.3389/fpls.2020.591911

Maksimović, M., Vidic, D., Miloš, M., Šolić, M. E., Abadžić, S., & Siljak-Yakovlev, S. (2007). Effect of the environmental conditions on essential oil profile in two Dinaric *Salvia* species: *S. brachyodon* Vandas and *S. officinalis* L. *Biochemical Systematics and Ecology*, *35*(8), 473–478. DOI: 10.1016/j.bse.2007.02.005

Mishra, B., & Kant, T. (2018). A simple, inexpensive, scalable and low maintenance hydroponics system for growing halophyte: *Lepidium sativum* L. (Brassicaceae), ideal for manipulating salt stress and inferring gene expression levels. *Indian Journal of Plant Physiology / Official Publication of the Indian Society for Plant Physiology*, *23*, 760–771. DOI: 10.1007/s40502-018-0422-4

Mohanavelu, A., Naganna, S. R., & Al-Ansari, N. (2021). Irrigation induced salinity and sodicity hazards on soil and groundwater: An overview of its causes, impacts and mitigation strategies. *Agriculture*, *11*(10), 983. DOI: 10.3390/agriculture11100983

Mouniane, Y., El-Khadir, I., Chriqui, A., Soufiani, A., Oudghiri, M., Abbad, Z., Zoubir, M. H., Ninich, O., & Hmouni, D. (2025). Responses of Carob Plants to Salt Stress: Prospects for Sustainable Agriculture and the Bioeconomy. In Mbedzi, M., Nkosi, S., & Alabi, D. (Eds.), *Impact of Environmental Degradation on Ecosystems and Preventive Measures* (pp. 195–220). IGI Global Scientific Publishing., DOI: 10.4018/979-8-3373-1549-2.ch008

Moussa, H., Dahmoune, F., Mróz, M., Remini, H., Kadri, N., Hamid, S., & Kuszniere-wicz, B. (2023). Efficient optimization approaches for microwave assisted extraction of high-quality antioxidant compounds from *Salvia officinalis* L.: UHPLC-HRMS differential analysis of phenolic profiles obtained by ultrasound and microwave extraction. *Sustainable Chemistry and Pharmacy*, *35*, 101194. DOI: 10.1016/j.scp.2023.101194

Negacz, K., Malek, Ž., de Vos, A., & Vellinga, P. (2022). Saline soils worldwide: Identifying the most promising areas for saline agriculture. *Journal of Arid Environments*, *203*, 104775. DOI: 10.1016/j.jaridenv.2022.104775

Nguyen, T. T., Grote, U., Neubacher, F., Rahut, D. B., Do, M. H., & Paudel, G. P. (2023). Security risks from climate change and environmental degradation: Implications for sustainable land use transformation in the Global South. *Current Opinion in Environmental Sustainability*, *63*, 101322. DOI: 10.1016/j.cosust.2023.101322

Otlewska, A., Migliore, M., Dybka-Stępień, K., Manfredini, A., Struszczyk-Świta, K., Napoli, R., Białkowska, A., Canfora, L., & Pinzari, F. (2020). When salt meddles between plant, soil, and microorganisms. *Frontiers in Plant Science*, *11*, 553087. DOI: 10.3389/fpls.2020.553087

Oudghiri, M., Mouniane, Y., Bouhassoun, S., El Aammouri, S., Yamani, B., Benlemlih, N., Abid, N., Elwahab, F. A., Zoubir, M. H., El-Khadir, I., Chriqui, A., Aitouhanni, I., El Bakkali, M., El Ouardi, B., & Ibriz, M. (2025). Abiotic stress and climate change: Technological innovations and adaptation strategies for sustainable resilience. In Ahmadi, A., & Mna, A. (Eds.), *Nexus of environmental quality and technology innovation* (pp. 1–16). IGI Global., DOI: 10.4018/979-8-3373-1822-6.ch001

Papadopoulos, G., Arduini, S., Uyar, H., Psiroukis, V., Kasimati, A., & Fountas, S. (2024). Economic and environmental benefits of digital agricultural technologies in crop production: A review. *Smart Agricultural Technology*, *8*, 100441. DOI: 10.1016/j.atech.2024.100441

Parker, E. R., Mo, J., & Goodman, R. S. (2022). The dermatological manifestations of extreme weather events: A comprehensive review of skin disease and vulnerability. *The Journal of Climate Change and Health*, *8*, 100162. DOI: 10.1016/j.joclim.2022.100162

Raal, A., Orav, A., Ilina, T., Kovalyova, A., Koliadzhyn, T., Avidzba, Y., & Koshovyi, O. (2024). Variation in the composition of the essential oil of commercial *Salvia officinalis* L. leaves samples from different countries. *Phyton*, *93*(8), 2051–2062. DOI: 10.32604/phyton.2024.052790

Rahman, M. M., Penny, G., Mondal, M. S., Zaman, M. H., Kryston, A., Salehin, M., Nahar, Q., Islam, M. S., Bolster, D., Tank, J. L., & Müller, M. F. (2019). Salinization in large river deltas: Drivers, impacts and socio-hydrological feedbacks. *Water Security*, *6*, 100024. DOI: 10.1016/j.wasec.2019.100024

Rakib, M. A., Newaz, M. A., Rahman, M. A., & Roy, K. (2024). Identifying the interfaces between perceived multi-hazards and socio-ecological risks to strengthen local adaptations. *Journal of Environmental Management*, *351*, 119708. DOI: 10.1016/j.jenvman.2023.119708

Rozema, J., & Schat, H. (2013). Salt tolerance of halophytes, research questions reviewed in the perspective of saline agriculture. *Environmental and Experimental Botany*, *92*, 83–95. DOI: 10.1016/j.envexpbot.2012.08.004

Sahab, I., Suhani, I., Srivastava, V., Chauhan, P. S., Singh, R. P., & Prasad, V. (2021). Potential risk assessment of soil salinity to agroecosystem sustainability: Current status and management strategies. *The Science of the Total Environment*, *764*, 144164. DOI: 10.1016/j.scitotenv.2020.144164

Salvati, L., & Ferrara, C. (2015). The local-scale impact of soil salinization on the socioeconomic context: An exploratory analysis in Italy. *Catena*, *127*, 312–322. DOI: 10.1016/j.catena.2015.01.008

Sharma, B. R., & Minhas, P. S. (2005). Strategies for managing saline/alkali waters for sustainable agricultural production in South Asia. *Agricultural Water Management*, *78*(1–2), 136–151. DOI: 10.1016/j.agwat.2005.04.019

Silva, B. N., Cadavez, V., Caleja, C., Pereira, E., Calhelha, R. C., Añibarro-Ortega, M., Finimundy, T., Kostić, M., Soković, M., Teixeira, J. A., Barros, L., & Gonzales-Barron, U. (2023). Phytochemical composition and bioactive potential of *Melissa officinalis* L., *Salvia officinalis* L. and *Mentha spicata* L. extracts. *Foods*, *12*(5), 947. DOI: 10.3390/foods12050947

Skrypnik, L., Feduraev, P., Golubkina, N., Maslennikov, P., Antipina, M., Katserov, D., Murariu, O. C., Tallarita, A. V., & Caruso, G. (2024). Foliar spraying of selenium in inorganic and organic forms stimulates plant growth and secondary metabolism of sage (*Salvia officinalis* L.) through alterations in photosynthesis and primary metabolism. *Scientia Horticulturae, 338*, 113633. DOI: 10.1016/j.scienta.2024.113633

Tıpırdamaz, R., Karakas, S., & Dikilitaş, M. (2021). Halophytes and the future of agriculture. In Grigore, M. N. (Ed.), *Handbook of Halophytes*. Springer., DOI: 10.1007/978-3-030-57635-6_91

Verma, O., Sharma, S., Kumar, V., Singh, T., Kumar, R., & Auji, R. (2024). Salinity stress effect on staple food crops and novel mitigation strategies. *Biologia, 79*, 2359–2374. DOI: 10.1007/s11756-024-01689-3

Vidaller, C., & Dutoit, T. (2022). Ecosystem services in conventional farming systems: A review. *Agronomy for Sustainable Development, 42*, 22. DOI: 10.1007/s13593-021-00740-w

Zörb, C., Geilfus, C.-M., & Dietz, K.-J. (2019). Salinity and crop yield. *Plant Biology, 21*(S1), 31–38. DOI: 10.1111/plb.12884

KEY TERMS AND DEFINITIONS

Agricultural Exclusion: A socio-economic phenomenon where farmers on degraded or saline lands are marginalized due to lower yields, higher input costs, and reduced access to markets and technology. This exclusion exacerbates rural poverty and territorial inequality.

Economic Regeneration: The process of revitalizing local economies that have been negatively affected by environmental degradation or marginalization. In the context of saline-affected regions, economic regeneration refers to restoring productivity, diversifying income opportunities (e.g., essential oils, herbal products), and improving farmers' market access through sustainable cropping systems.

Halophyte-Assisted Intercropping: An agroecological system where salt-tolerant plants (halophytes) are intercropped with salt-sensitive crops (glycophytes) to mitigate salinity stress, improve biomass production, and enhance economic resilience.

Sustainable Development: Development that meets the needs of the present without compromising the ability of future generations to meet their own needs. Within this chapter, it refers to agricultural practices that balance ecological resilience, economic profitability, and social inclusion, while aligning with the UN Sustainable Development Goals (SDGs).

Chapter 15
Prosperity-Driven Economies and Credit Risk:
A Data-Driven LightGBM Approach for Risk Management in Banking

Mustafa Bilgehan İmamoğlu
https://orcid.org/0000-0002-3496-2959
Karadeniz Technical University, Turkey

ABSTRACT

Effective credit risk management in banking is critical for financial stability and customer security. The difficulty of traditional methods in adapting to large data volumes and market fluctuations increases the need for new methods. This study aims to demonstrate how ML-based risk management approaches can be applied in banking in both a theoretical and practical framework. In this context, a credit risk prediction model was developed using the LightGBM algorithm. Significant improvements in metrics were achieved with imbalanced dataset improvement methods. The findings revealed the potential of the model to make fast and accurate decisions under large data and variable financial conditions, although the predictions of the "risky" class still need to be improved. Thus, it was emphasized that ML-based approaches provide a more reliable, data-driven, and dynamic decision-making mechanism in credit allocation processes.

DOI: 10.4018/979-8-3373-1862-2.ch015

INTRODUCTION

The banking industry plays a key role in keeping financial systems stable, so it has always relied on a strong plan to manage risks (De Lima 2024). With today's economy driven by tough competition and increasing digital change, it becomes essential to carry out risk management and credit distribution processes. These processes address issues like credit risks detecting fraud, and improving operational performance (Brynjolfsson & Mcafee, 2017). As data continues to grow in both size and diversity, AI and ML tools go beyond traditional methods. They bring faster and more accurate decision-making solutions to the table (M. Chen et al. 2014).

Banks used to rely on human judgment and basic statistical methods with small data sets to allocate credit. These old ways often failed to keep up with the fast-changing financial market environment. Risk models that relied on static ideas and linear assumptions struggled to catch unpredictable market shifts or understand complicated customer behaviors, as noted by Azevedo et al. in 2022 and Dell'Ariccia & Marquez in 2004. Machine learning brings a strong alternative to improve banking tools. It analyzes large data sets more and. Gradient boosting algorithms like LightGBM mentioned by Goodfellow and others in 2016, make this possible.

Conversely, the extensive application of artificial intelligence in functions like credit risk assessment, fraud detection, and customer segmentation has raised alarm about data privacy, algorithmic fairness, and regulatory compliance (Kroll, 2015; Zarsky, 2016). In this arrangement, financial institutions have to formulate data governance policies and algorithmic transparency in addition to embracing ethical and transparent applications of AI. Moreover, the speedy evolution of technological infrastructures along with financial data repository development transforms the digitalization strategies of banks and makes it compulsory for them to implement artificial intelligence models in their business processes (Oprea & Du ă, 2024; Singh & Ahlawat, 2023).

The intent of this study is to describe how ML methods can be utilized on risk management processes in the banking industry in both theoretical and applied perspective. It will illustrate how the LightGBM algorithm is utilized to forecast the likelihood of loan repayment in loan applications. The performance metrics acquired as a consequence of the analysis will be examined through basic performance metrics such as accuracy, F1 score, and AUC-ROC. Further, the validity of employed models and their consistency with real financial terms will be made more evident through the employed example to be executed on an open-access credit risk dataset widely recognized as a standard in academic literature (Kou et al., 2019; Leo et al., 2019).

The research creates an empirical solution for the application of ML-based models in the banking sector, accelerating credit distribution operations, early detection of suspicious transactions, and enhanced customer security (Moffo, 2024). The extensive application of the models by banks aims to provide more information-based, innovative, and ethical decision-making systems that are better than traditional methods (Brynjolfsson & Mcafee, 2017).

BACKGROUND

With the developing technology, the banking sector has started to use AI and ML approaches more and more intensively. This transformation is gaining importance because it provides a competitive advantage while increasing the speed and accuracy of operational processes (Brynjolfsson & Mcafee, 2017). In this section, basic research on AI and ML literature in banking, its purposes of use, and methodological approaches related to security and privacy dimensions are discussed step by step.

Use of Artificial Intelligence and Machine Learning in the Banking Sector

Banking activities are one of the prominent areas in the collection and processing of big data. ML plays a critical role in interpreting this data and transforming it into decision support mechanisms (M. Chen et al., 2014). In areas such as accurate credit risk prediction, fraud detection, and analysis of customer behavior, it is possible to achieve higher accuracy rates compared to traditional methods (Rakesh Pandit, 2024; Xu, 2024).

In international literature, it is seen that banks are rapidly adopting these technologies within the scope of digital transformation strategies (Arner et al., 2016). Financial institutions are using AI-based tools in the automation of loan applications and the detection of risky transactions with early warning systems (Andronie et al., 2024). In addition, the development of data analytics capacities allows banks to offer more personalized products and services. Thus, customer satisfaction and operational efficiency increase (Kaplan & Haenlein, 2019).

The purposes and places of use of methods such as ML and AI in the banking sector are diverse. Different models offer different benefits in each area. These areas can be briefly explained as follows.

Credit Risk Management: ML-based models positively affect bank capital adequacy and profitability by detecting risky applications with higher accuracy in credit allocation processes (Jordan & Mitchell, 2015). More advanced approaches, such as deep learning, can evaluate multiple variables such as income, credit history, market, simultaneously and build flexible and non-linear models (LeCun et al., 2015).

Fraud and Anomalous Transaction Detection: In areas that generate intensive data, such as bank card transactions, money transfers, and credit card spending, ML algorithms can catch fraudulent or suspicious transactions in real time (Jadhav et al., 2022; Kodimenu et al., 2024; Lenka & Tiwari, 2025). These applications not only detect anomalies but also analyze the structure of fraud, allowing banks to develop preventive strategies (Hashemi et al., 2023; Wei et al., 2013).

Customer Segmentation and Behavior Analysis: Supervised or unsupervised learning methods allow for a more dynamic analysis of customer groups (Varian, 2014). Different parameters such as the customer's income level, spending habits, credit card usage, or deposit movements can be integrated into segmentation models, and personalized offers can be offered (Ho et al., 2021).

Privacy, Security and Regulation

The proliferation of AI applications combined with big data analytics in the sector has brought data privacy and security issues to the forefront (OECD, 2021 risk analysis). Since banks process customers' financial and personal information extensively, they assume additional responsibilities regarding issues such as cyber-security threats and regulatory compliance. Regulations such as the European Union General Data Protection Regulation (GDPR) require the transparent and responsible development of AI and ML applications (Kroll, 2015).

In addition, the risk of algorithmic bias can lead to unfair results in important decision-making processes, especially in credit allocation. For this reason, banks are turning to applications that will eliminate imbalances in the data sets used and measure the impartiality of algorithm outputs (Zarsky, 2016). Although regulations are not yet fully established in this area, international organizations such as the Basel Committee publish "good practice" guides regarding banks' use of AI (Butler & Brooks, 2023).

Studies conducted in this context in the literature try to offer the best solutions for ensuring privacy and security with different methods. Some of these methods and the performance findings obtained are listed below.

Supervised Learning Algorithms: Logistic regression, decision trees, support vector machines (SVM), and random forest methods are techniques frequently used by banks in credit rating determination and risk scoring (Ben-David & Gilje, 2018). These models are usually evaluated with performance metrics such as accuracy, F1 score, and AUC-ROC.

Deep Learning Approaches: Multi-layered neural networks such as artificial neural networks, LSTM, CNN can simultaneously model complex relationships such as customers' credit history, spending trends, income level, and market variables (Goodfellow et al., 2016). This allows for more accurate predictions and early warning signals, especially in financial markets that change instantly (LeCun et al., 2015 risk analysis).

Real-Time Data Analytics and Automated Decision Support: In the banking environment with high transaction volumes, analyzing stream data instead of just static data is becoming increasingly important. Thus, potential fraudulent transactions or unexpected credit risk increases can be detected in real time (Wilson & Daugherty, 2018). These systems allow banks to manage their risk appetite more dynamically.

Based on the literature review, it is seen that AI and ML are becoming increasingly important in both international and national banking practice. With big data and advanced algorithms, automation of credit allocation processes, fraud detection, customer segmentation, and risk management activities have been greatly strengthened (Kumar & R, 2024; Varadarajan, 2023). However, further development of legal regulations and ethical principles on topics such as data privacy and algorithmic fairness is critical for the sustainable and socially acceptable dissemination of these technologies.

The related scholarship clusters into three streams: (i) traditional statistical credit scoring (logistic regression, trees, SVM) with remedies for class imbalance; (ii) modern gradient-boosting ML (notably LightGBM) balancing speed/accuracy/interpretability on high-dimensional financial data; and (iii) fairness and explainability, where model bias can affect equitable access to credit, supported by SHAP/feature-importance-based interpretability. Prior studies often optimize raw accuracy/AUC while under-addressing minority-class sensitivity, threshold selection, and application validity. This chapter addresses these gaps on the Home Credit dataset by jointly using balanced metrics (AUC, F1, PR curve) and threshold optimization, and by showing that feature engineering plus imbalance remediation (SMOTE+UnderSampling) yields a deployment-ready pipeline.

METHODOLOGY

In this study, where the LightGBM algorithm was used to estimate the repayment risks of individuals applying for credit, the performance of the model was evaluated with accuracy, F1 score and AUC-ROC metrics, and the analyses were performed on the Home Credit Default Risk dataset, which is open access and frequently referenced in literature. The systematic methodology of the study is shown in Figure 1. Thanks to this methodological approach, the integration of ML-based decision support systems into risk management processes in the banking sector will be addressed both theoretically and practically.

Figure 1. Proposed Methodology for Machine Learning-Based Credit Scoring

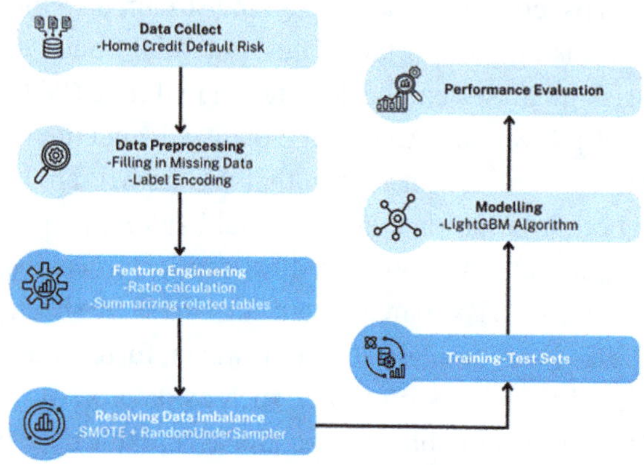

Logistic regression provides a solid foundation; Random Forest captures interactions but may lag behind in terms of speed and scale. XGBoost is highly competitive, while CatBoost excels with categorical features. LightGBM was selected in this environment due to its balance of speed, accuracy, and feature importance.

In the study methodology presented in Figure 1, firstly, the data collection process was carried out using the Home Credit Default Risk dataset within the scope of the research. Then, in the data pre-processing stage, label encoding was performed to fill in the missing data and to digitize the categorical data. Then, in the feature engineering stage, more explanatory features were derived for the model by applying operations such as ratio calculations and summarization of related tables on financial and demographic variables. To solve the class imbalance problem in the dataset, the data was balanced with the SMOTE and RandomUnderSampler methods, thus providing more effective learning in the minority class. The training and test datasets were determined, and the modeling stage was started with the LightGBM algorithm, and finally, the performance of the model was evaluated comprehensively with basic metrics such as accuracy, F1 score,e and AUC-ROC.

Data Collect: Home Credit Default Risk Data Set

The Home Credit Default Risk dataset used in the study was created by the financial services company Home Credit and published on the Kaggle platform in 2018. The dataset, which was prepared for estimating the loan repayment probability of individuals applying, includes many demographic and financial characteristics such as income status, employment period, and credit history. The dataset, which contains data from more than 300,000 loan applications, is used as a reference data source in many academic studies on credit risk analysis. Table 1, which presents current academic studies using the Home Credit Default Risk dataset, summarizes the methods, key findings, and performance metrics used in these studies:

Table 1. Comparative summary of Studies

Author(s)	Method	Findings	Performance Metrics
Gu (2020)	LightGBM	Higher accuracy than SVM	Accuracy ↑
Ponsam et al. (2021)	LightGBM vs SVM & RF	LightGBM is more stable and effective	Accuracy, Stability
Hou et al. (2025)	ISSA + LightGBM	Better AUC score compared to other models	AUC ↑, DeLong test
Zhu et al. (2024)	LightGBM + XGBoost + LocalEnsemble	The ensemble model gave the best results	AUC, Accuracy ↑
Y. Sun et al. (2023)	Feature Engineering + LightGBM	New features increase accuracy	Accuracy: 0.734, AUC: 0.772
Dasril et al. (2023)	LightGBM + PSO	Accuracy over 98%	Accuracy: 98.09%, Recall: 90.51%

Recent studies on credit risk modeling indicate that the LightGBM algorithm has unique benefits in the analysis of financial data that include improved accuracy, fast processing speeds, and flexibility. In his study, Gu (2020) made a comparison with light GBM and Support Vector Machines (SVM), a conventional machine learning method, showing that the algorithm can deliver effective outcomes even with default parameters, hence attaining superior accuracy levels. In the same vein, Ponsam et al. (2021) compared LightGBM with other methods such as Random Forest and SVM, illustrating its superior accuracy and model stability performance (Ponsam et al., 2021). Taking a more advanced strategy, Hou et al. (2025) improved the AUC performance of the model through conducting hyperparameter tuning via the ISSA algorithm, showing that the performance of LightGBM can be significantly improved with the right settings. The ensemble model that was built by Zhu et al. (2024) demonstrated that the integration of LightGBM, XGBoost, and LocalEnsemble algorithms created broader and fairer estimation outcomes compared to different groups of customers (Zhu et al., 2024). In Y. Sun et al.'s (2023) research, emphasis on feature engineering processes led to creating novel variables that greatly enhanced the model's accuracy and area under the curve (AUC) score. In Dasril et al.'s (2023) work, the LightGBM algorithm was combined with Particle Swarm Optimization (PSO) and yielded an accuracy rate of more than 98%. This success reflected the great generalizability of the algorithm, even if applied to another dataset.

When the existing literature is examined, it is seen that many successful applications have been made for credit risk estimation using the LightGBM algorithm. However, most of these studies either focus on model performance and only try to increase accuracy or focus on model combinations with different methods. On the other hand, there is a lack of comprehensive, explainable, and versatile analysis in terms of performance metrics focusing on the Home Credit Default Risk dataset, which is a high-volume dataset widely used in academic circles and close to real-world conditions. In particular, few studies use balanced use of model evaluation metrics (Accuracy, F1, AUC), in-depth evaluation of the results in terms of sectoral application potential, and feature engineering.

In this direction, the main purpose of this study is to develop a model that estimates the probability of loan repayment with high accuracy and explainability using the LightGBM algorithm on the Home Credit Default Risk dataset. At the same time, the model performance will be evaluated comprehensively on different metrics, and the results obtained will be interpreted in a way that will contribute to data-driven decision-making processes in the banking sector. In this respect, the study aims to provide a holistic approach that prioritizes not only technical success but also application validity.

Feature Engineering

In ML applications, the overall success of the model depends not only on the complexity of the algorithm used but also on the quality of the features provided to the model. In this context, feature engineering can be defined as the process of deriving more meaningful and informative features from raw data to improve the classification performance of the model (R.-C. Chen et al., 2020).

The Home Credit Default Risk dataset used in this study includes a wide range of financial, demographic, and behavioral variables to estimate an individual's ability to repay a loan. In the application_train.csv file containing the main table data, each row represents an individual's loan application. The label information for this application is located in the TARGET column, and if TARGET = 0, the individual has paid the loan on time, and if TARGET = 1, the individual has not paid the loan.

The dataset consists of tables that hold primary and secondary information, such as demographic information, income level, loan amount, assets, past applications, installment payments, credit card transactions, and credit bureau records of individuals who apply for credit, collected from various sources. These tables mostly contain one-to-one or one-to-many relationships and contain time series that are updated at different frequencies. Therefore, this data needs to be summarized in a meaningful way before entering the modeling process directly.

Why is Feature Engineering Necessary?

High Dimensionality and Heterogeneous Data: Feature engineering is a critical step that directly affects the success of the model in ML projects. Since data sets today can have both a high dimensionality and a heterogeneous structure, modeling directly on this data often does not provide the desired success. While high dimensionality can complicate the model's learning process and increase the risk of overfitting, heterogeneous data requires the harmonization of different types of information (Verdonck et al., 2024). Thanks to feature engineering, data is represented in a meaningful and understandable form for the model, unnecessary dimensions are eliminated, and different types of data are transformed and made comparable. This process increases the generalization ability of the model, optimizes the learning time, and contributes to more accurate predictions (Wang et al., 2024).

Revealing Secret Relationships: The variables in the raw data may not always directly reflect their relationship with the target variable. Therefore, revealing the relationships that exist between the data but are not noticeable at first glance is of great importance for the performance of the model. For example, while AMT_CREDIT and AMT_ANNUITY may not be meaningful for the model separately, the ratio of these two variables may provide an important clue about the individual's ability to repay the loan. Similarly, when the DAYS_BIRTH variable is presented as a negative value in days relative to the date of birth instead of age, it is meaningless to use it directly for the model. Converting this value to a positive age makes the data more meaningful.

Summarizing Time Series: A large part of the data is made up of tables that contain information about how people handle money over time. Because we cannot model directly from these tables, we summarize these time series for every person and transform them into fixed-size features.

Improving Data Quality: Operations like filling missing values, digitizing categorical data, and cleaning irrelevant columns have made the learning process of the model healthier and more stable (Werner de Vargas et al., 2023). Due to these operations, the model was trained only with meaningful and quality data.

The feature engineering here comprises not just digitization of data but also a sophisticated, multi-step process that encompasses joining information across related tables, performing ratio analyses, time series smoothing, and data transformations. Through this comprehensive process, the data set has been preprocessed to achieve optimal operational efficiency of the machine learning algorithm, thereby enhancing classification performance.

Data Preprocessing

While dealing with raw data for machine learning-based credit risk prediction studies, it is also observed that the data is also characterized by high dimensionality, noise, and absence of significant correlations. Hence, there is a need to use different feature engineering methods for the dataset so that the quality of the predictions is enhanced (Shukla & Gupta, 2024). The study systematically adhered to the following steps in the feature engineering process:

Rate Based Derivative Features

Derivative features that better reflect the financial structure of the individual are created by calculating the proportional values between some financial variables. For example:

- *credit_annuity_ratio:* The value calculated with the AMT_CREDIT / AMT_ANNUITY expression is the ratio of the total credit amount requested by the individual to the annual payment obligation. This ratio can help identify individuals with weak payment capacity.
- *credit_goods_price_ratio*: The value calculated with the AMT_CREDIT / AMT_GOODS_PRICE expression expresses the amount of credit received according to the product price financed with the credit. This ratio provides information about the individual's credit usage behavior.
- *AGE_INT*: Since the DAYS_BIRTH variable contains data in negative days, to obtain a meaningful variable, the age information was converted to years with the expression (AGE_INT = -DAYS_BIRTH / 365).

These calculated ratios contributed positively to the model performance, as they had more discriminatory power than the direct variables included in the dataset.

Extracting Numerical Summaries from Related Tables

The dataset contains many related tables. These tables are often one-to-many and contain time series data. Therefore, the records for each application are summarized using group by and aggregation operations.

- *bureau.csv*: New features were created by taking average values from all past credits of the individual for the numerical columns in this table, containing previous credit information reported to the credit bureau.

- *previous_application.csv:* Average, maximum, and total values of variables such as AMT_CREDIT, AMT_ANNUITY, and DAYS_DECISION were calculated from this table containing the individual's previous credit applications.
- *POS_CASH_balance.csv, installments_payments.csv, credit_card_balance.csv*: Individual-based summary statistics (mean, sum, max) of variables such as CNT_INSTALMENT, AMT_PAYMENT, AMT_BALANCE in these tables containing time series were calculated and integrated into the data set.

```
Code 1: installments_agg = installments_payments.group-
by('SK_ID_CURR').agg({
  'AMT_PAYMENT': ['sum', 'mean'],
  'DAYS_ENTRY_PAYMENT': ['mean', 'max']})
```

With the sample code given in Code 1, the behavioral variables displayed by individuals over time were converted into fixed-size vectors and became usable in modeling.

Digitizing Categorical Data (Label Encoding)

Categorical variables (object type columns) in the dataset cannot be directly processed by ML algorithms. Therefore, these variables must be converted to numeric formats using the LabelEncoder class. Missing data are filled with the "Unknown" label, and all values are converted to numeric identities.

```
Code 2: for col in application_train.select_dtypes(in-
clude=['object']).columns:
  application_train[col] = application_train[col].fillna('Un-
known')
  application_train[col] = LabelEncoder().fit_transform(appli-
cation_train[col])
```

This method, exemplified in Code 2, provides fast and effective results in cases where the number of categories is limited.

Remediation of Missing Data

For the model to learn more stably, the missing values in the entire dataset were filled with zero (0). This simple but effective strategy prevented the NaN values that occurred, especially after groupby-agg, from being swallowed by the model. This step can be enriched with more advanced missing value imputation techniques in the future.

Resolving Data Imbalance

In ML applications, if there is a significant proportional difference between classes, this is called the class imbalance problem. Financial problems, especially credit risk analysis, have imbalanced data sets by nature. In the Home Credit Default Risk data set used in this study, the target variable, the TARGET class, is also quite imbalanced: TARGET=1 (risky customer) is only 8%. In this case, directly trained models tend to ignore the minority class by memorizing the majority class.

In cases where such imbalances are not eliminated, the developed classification model cannot predict the minority class correctly, general accuracy metrics such as accuracy are misleadingly high, and metrics reflecting minority class sensitivity, such as recall and F1-score, decrease. As a result, the model's success in critical tasks such as risky customer detection decreases. For this reason, sampling-based methods were preferred in the study to reduce the effects of class imbalance.

SMOTE (Synthetic Minority Over-sampling Technique)

SMOTE is an over-sampling method developed by Chawla et al. (2002) and widely used in literature. Its basic principle is to create new examples (synthetic data points) among the examples belonging to this class with the k-nearest neighbor (k-NN) approach instead of randomly copying the examples belonging to the minority class. In this way, more diversity is added to the minority class dataset set and the learning ability of the model is increased (Elreedy & Atiya, 2019).

With this feature, SMOTE reduces the risk of overfitting by creating new examples instead of random copying and improves class separation by providing a clearer learning of decision boundaries.

Random UnderSampler

Under-sampling techniques provide data balance by randomly deleting samples from the majority class. Random UnderSampler is the technique that performs this process in the simplest and fastest way. However, when used alone, it can cause information loss. Therefore, when used with SMOTE, balanced and enough observations are provided for both classes (Almubark, 2024; Z. Zhang et al., 2022).

```
Code 3: smote = SMOTE(sampling_strategy=0.5, random_
state=42)
under = RandomUnderSampler(sampling_strategy=0.8, random_
state=42)
pipeline = Pipeline([("SMOTE", smote), ("UNDER", under)])
X_train_resampled, y_train_resampled = pipeline.fit_resam-
ple(X_train, y_train)
```

Within the scope of the study, minority class samples were increased with SMOTE, and data balance was achieved by randomly removing samples from the majority class with Random UnderSampler. Thanks to this method, the sensitivity of the model to the minority class TARGET=1 observations was increased, and Recall and F1-score values were significantly improved. The ROC AUC value increased to 0.778, and a strong classification success was achieved. The code fragment for the sampling strategy regarding data imbalance is given in Code 3.

LightGBM Algorithm

LightGBM is a decision tree-based "gradient boosting" algorithm developed by Microsoft. Providing high accuracy with low computational cost, especially in large and high-dimensional data sets, has made LightGBM ideal for complex financial tasks such as credit risk estimation. The algorithm is faster and more efficient because of its Leaf-wise growth approach and histogram-based data disaggregation (X. Sun et al., 2020; J. Zhang et al., 2019).

The literature demonstrates a clear inclination towards the LightGBM algorithm in credit risk analysis. Importantly, the previous studies indicate LightGBM-based models to have higher accuracy and reliability than traditional approaches (Gu, 2020; Ponsam et al., 2021). Additionally, the performance of the model has been further improved by combining it with ensemble techniques (Zhu et al., 2024).

FINDINGS

In the present research, the Home Credit Default Risk dataset was split into two parts, 80% for training and 20% for testing, respectively, later simulated with the LightGBM algorithm. The findings indicated that the model constructed from the training data had adequate and trustworthy results in predicting credit risk on the testing dataset. Accordingly, the proposed model was deemed appropriate, and the findings of the study were shared in detail. In this dataset, which has a structure close to large-volume real-world data, the performance of the model was evaluated multi-dimensionally through metrics such as accuracy, F1 score, and ROC AUC. Thus, not only the search for a high accuracy rate, but also critical elements such as correct detection of the minority class (applications with high payment risk) and general balance were taken into consideration.

Evaluating the classification performance only with the accuracy metric is insufficient, especially in imbalanced data sets. For this reason, both classical metrics and advanced evaluation techniques were used together in the study.

Table 2. Classification performance of the model

Class	Precision	Recall	F1-score	Support
0 (Not Risky)	0.94	0.92	0.93	56538
1 (Risky)	0.29	0.39	0.33	4965
Accuracy			0.87	61503
Macro Avg	0.62	0.65	0.63	61503
Weighted Avg	0.89	0.87	0.88	61503

The classification performance of the model shown in Table 2 has been analyzed comprehensively through the classification report, which includes basic evaluation metrics such as accuracy, recall, precision, and F1-score. The values included in this report are important in terms of showing both the overall success and the effect of the model on minority and majority classes.

The overall accuracy of the model was calculated at 87%. This rate reflects the percentage of correctly classified samples in the total test data and provides a superficial indicator of success. However, the accuracy metric alone is not sufficient to evaluate the performance of the model, as it can falsely show models that are biased towards the majority class in imbalanced data sets. Therefore, a detailed evaluation of other metrics is necessary.

The precision value calculated for class zero (0), i.e., “non-risky” applications, was found to be 94%, and the recall value was found to be 92%. Precision indicates the ratio of the samples that the model predicted to be non-risky that belong to this class. A value of 94% means that the model made most of its predictions for this class correctly. The recall value, on the other hand, shows how many of the samples it correctly identified for this class; the 92% rate reveals that the model was able to correctly classify non-risky individuals to a very large extent. The F1-score for this class, i.e., the harmonic meaning between precision and recall, is at a very high level of 93%, indicating that the prediction success for this class is balanced.

On the other hand, when we look at the evaluations for class one (1), i.e., “risky” applications, a different picture is observed. The precision value for this class was measured as 29%, the recall value as 39%, and the F1-score as 33%. The precision value of 29% shows that only one third of the individuals the model classified as risky are risky. In other words, the predictions for this class contain relatively more false positives. The recall value of 39% means that 61% of the risky individuals are ignored by the model. The F1-score is also at 33% due to the low value of these two values. These results show that the model's ability to distinguish the risky class is still limited, but significant improvement has been achieved compared to previous versions.

Macro average values consider the success of each class equally and are calculated as 62%, 65%, and 63% for precision, recall, and F1, respectively. This situation reveals that there is a certain imbalance between the two classes and that the model performs better towards the zero class. On the other hand, average values are averages weighted according to the support values of the classes and are generally biased towards the zero-class due to the higher number of examples. In terms of these metrics, precision is calculated as 89%, recall is 87%, and F1-score is 88%. These values show that the model generally exhibits a high classification performance, but this performance is mostly due to the effect of the majority class.

As a result, the metrics obtained in the classification report reveal that the model successfully classifies non-risky individuals, but there are still areas for improvement in the detection of risky individuals. This situation necessitates further optimization of metrics such as recall and F1-score, especially in systems where critical decisions are made, such as financial risk assessment. The results obtained suggest that the overall success of the model is strong, but more advanced techniques and hyperparameter optimizations can be used to increase the sensitivity to the minority class.

ROC AUC Score and ROC Curve

The ROC curve and the related ROC AUC score (Receiver Operating Characteristic – Area Under the Curve) are commonly used statistical tools to assess the performance of a classification model. Particularly in the binary classification problem, it is very important to evaluate the capacity of the model to calculate positive and negative classes. This ROC curve is the figure of the True Positive Rate (TPR)/Sensitivity (The x coordinate of the ROC curve) on the y-axis, and the False Positive Rate (FPR)/1-Specificity on the x-axis. This curve shows how consistent the model makes predictions above thresholds, which can be used to assess the prediction power while staying independent of the class balance.

Figure 2. Roc Curve

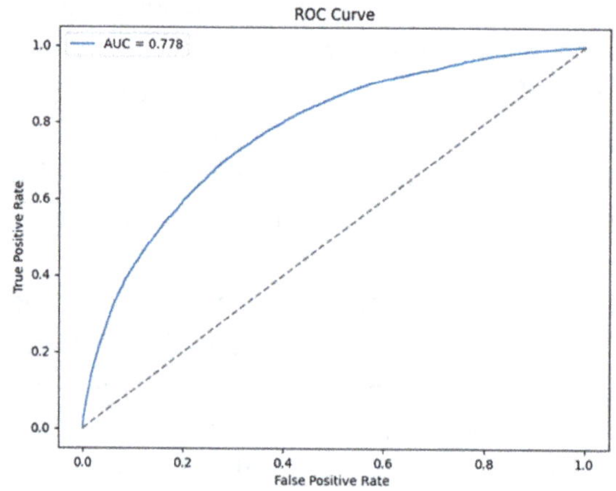

ROC AUC score (Receiver Operating Characteristic – Area Under the Curve) and ROC curve are standard statistical techniques to evaluate the performance of classification models. Especially for binary classification problems, we need to assess how effectively the model separates positive and negative classes. Receiver Operating Characteristic (ROC) curve is drawn by placing the True Positive Rate (TPR) on the y-axis and False Positive Rate (FPR) achieved from the model at various threshold levels on the x-axis. The plot is the measure of the degree to which the model produces correct predictions based on decision thresholds, and thus predictive performance can be assessed independent of class distribution.

The analysis of the ROC curve for the model developed in this study, as represented by Figure 2, indicates that the curve is far above the diagonal line between the points (0,0) and (1,1) for random classification. This demonstrates that the model possesses great power in discriminating between positive and negative cases. The ROC AUC statistic, derived from the computation of the area under the curve (AUC), stood at 0.778. This metric is regarded as a measure that takes into account the model's overall classification efficacy, and in the literature, values ranging from 0.7 to 0.8 are typically characterized as "good," while values above 0.8 are classified as being in the range of indicating "very good" performance. Within this system, the AUC value that was achieved signifies that the model possesses a statistically significant and acceptable degree of discriminative power.

It is known that the ROC AUC score is especially advantageous in imbalanced data sets. This metric summarizes the performance of the model across all decision thresholds and is less affected by the class imbalance in the dataset. Although there was a serious imbalance in the class distribution in this study, the fact that the ROC AUC score remained high after the balancing operations performed with the SMOTE and Random UnderSampler methods shows that the model achieved sustainable separation between classes.

The ROC curve and the AUC score, which is the area under this curve, represent not only the performance of the proposed model on a single threshold value, but also the level of performance it demonstrates across all decision thresholds. In this respect, the ROC curve is a strong indicator supporting the reliability of the model, especially in application areas such as financial risk analysis, which require precise classification. The results obtained reveal that the model has a statistically significant and strong classification ability.

Precision-Recall Curve and Best Threshold Value

The Precision-Recall curve is one of the powerful visualization tools preferred for evaluating model performance, especially in imbalanced datasets. This curve examines the rate of correctly predicting the examples belonging to the positive class (Recall) and the percentage of the examples predicted by the model as positive class that are positive (Precision) at different threshold values, revealing the decision-making behavior of the classification model in detail. While the ROC curve focuses on the overall discrimination power, the Precision-Recall curve specifically evaluates whether the positive class is successfully captured; this is especially critical in scenarios where the positive class is in the minority.

Figure 3. Precision-Recall

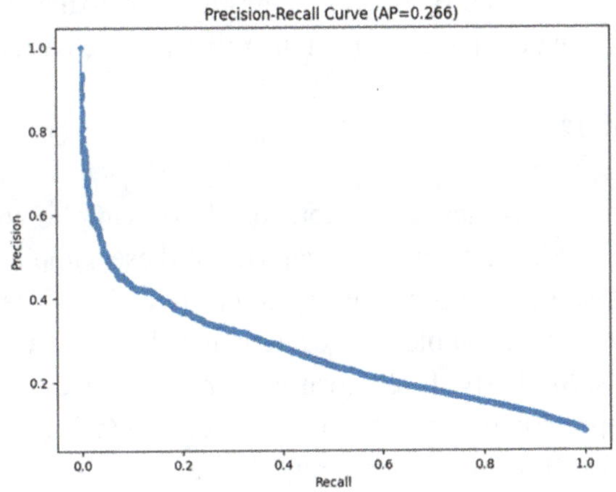

Precision-Recall curve illustrated in this research, as plotted in Figure 3, graphically illustrates the variations in the precision and recall values of the model for various threshold levels, enabling an exhaustive examination of the performance of the model with respect to the positive class. A high precision-recall area under the curve (AUC-PR) indicates that the model predicts positive cases with high precision and correctness and classifies a large percentage of such cases as true positives. Nevertheless, in the current research, the optimal threshold value was selected based on the F1 score derived from the Precision and Recall metrics, rather than using the AUC-PR. The F1 metric is the harmonic average of precision and recall scores, employed as a performance metric that balances both these two metrics at once. Particularly in scenarios where a compromise between recall and precision is essential, the F1 metric gives a good baseline for decision-making processes.

In the conducted analyses, the F1 scores calculated on the predictions produced by the model at different threshold values were examined, and the threshold value where the highest F1 score was obtained was determined as the "optimal threshold". This threshold value allows the model to produce predictions in a way that minimizes both false positives and false negatives. In this context, the optimal threshold value obtained in the study was determined as 0.31. This value allows the model to better define the positive class by using a more flexible decision-making threshold compared to the default 0.50 decision threshold.

Determining the Precision-Recall curve and the optimal threshold value are very important in terms of optimizing the success of classification models, especially for the minority class. The results obtained show that the model provides a significant improvement in positive class predictions after the SMOTE + Random UnderSampler imbalanced data balancing strategy is applied in this study, and this success is further optimized with the determined threshold value. This situation is of vital importance for decision support systems that want to make correct decisions in situations where the positive class is in the minority, such as credit risk.

Confusion Matrix

The proposed model was tested on a total of 61,503 samples within the scope of the study. As seen in Figure 4, the vast majority of these samples, 56,538, belong to the negative class, and 4,965 belong to the positive class. The model correctly classified 51,798 negative samples as negative and showed a very high discrimination for this class. Similarly, 1,926 positive samples were correctly identified as positive. This is important in terms of showing that the model can successfully learn not only the majority class but also the minority class.

Figure 4. Confusion Matrix

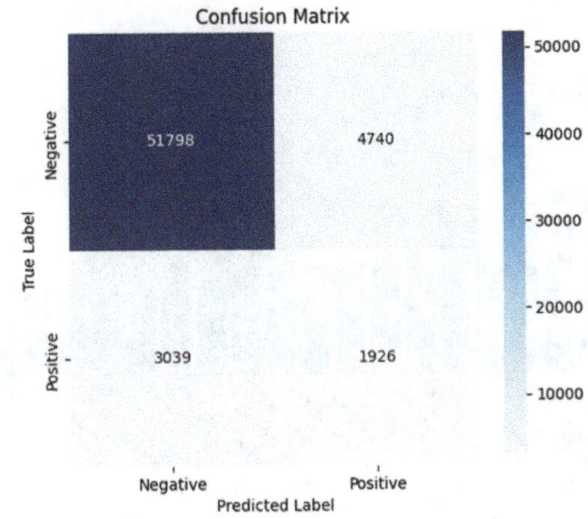

It is seen that the model incorrectly predicted some positive classes as negative and classified some negative classes as positive. However, it is noteworthy that these errors are limited compared to the total number of samples, and this situation positively contributes to metrics such as overall accuracy and weighted average F1-score. This success shows that the applied imbalanced data analysis strategies work effectively, and the capacity of the model to distinguish between both positive and negative classes is increased.

The model's 92% correct prediction of negative class samples shows that it can provide reliable filtering in credit risk analysis. More importantly, the 39% correct classification rate obtained in positive class samples represents a significant improvement compared to the classical performance problems encountered in applications generally working with unbalanced data sets. This success indicates that we have moved beyond the "biased" models that only make predictions for the majority class, i.e., frequently encountered in unbalanced data sets. As a result, this analysis conducted on the complexity matrix of the model reveals that the proposed methodology works effectively and has become a reliable tool that can be implemented in credit risk determination.

The variables that play a more dominant role in the decision-making mechanism of the model developed within the scope of the study are presented in Figure 5. In particular, the credit_annuity_ratio variable, which represents the proportional relationship between the loan amount and the annual regular payment amount, has the highest importance value and stands out as a critical distinguishing element in the learning process of the model. The high importance level of this variable can be explained by the fact that it represents the financial burden that directly affects the individual's repayment capacity. When this ratio is high, it indicates that the debtor's ability to pay decreases, and therefore, the risk level increases.

Figure 5. Top-10 features by gain

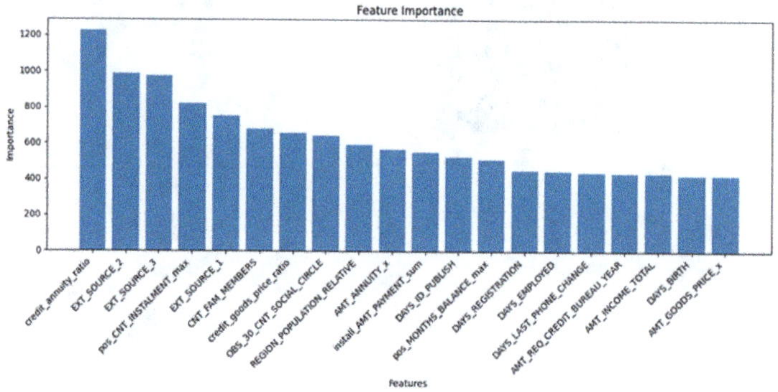

The variables EXT_SOURCE_2 and EXT_SOURCE_3, which are ranked second and third in importance, are credit scoring indicators based on external data sources of the applicant and contain important clues about past financial behaviors. The high importance of these variables supports the fact that external risk scores are a strong indicator in predicting the individual's credit repayment behaviors. At the same time, the prominence of socio-demographic variables such as DAYS_EMPLOYED, EXT_SOURCE_1, and DAYS_BIRTH shows that the individual's age and employment period are also considered important by the model and are used effectively in classification decisions.

On the other hand, variables such as AMT_GOODS_PRICE, DAYS_REGISTRATION, DAYS_LAST_PHONE_CHANGE, CNT_FAM_MEMBERS, REGION_RATING_CLIENT, and REGION_POPULATION_RELATIVE contribute to the model's learning of payment behaviors indirectly by providing information about the economic dynamics and household structure of the region where the individual lives. The significant contribution of such demographic and regional characteristics to the model once again emphasizes the importance of evaluating individual and environmental factors together in credit risk analyses.

In general, these findings regarding feature importance analysis are both consistent with domain knowledge and have a structure that overlaps with similar studies in literature. The fact that the model prioritizes meaningful variables based on economic rationality in the decision-making process both increases the reliability of the model and offers high value in terms of interpretability. These results reveal that the developed model has a strong structure not only in terms of predictive accuracy but also in terms of explainability. Therefore, the success achieved within the scope of the study is not limited to numerical performance indicators but is also supported by the understandability and verifiability of the model's decision-making process.

Distribution of Predicted Probabilities

Figure 6, which visualizes the distribution of probability values generated by the model, shows the densities of predicted probabilities for class 0 (negative) and class 1 (positive) examples separately. The class 0 distribution, represented in blue, is largely concentrated between 0.0 and 0.2, indicating that the model tends to classify negative class examples with low probabilities, i.e., correctly as negative class. Class 1 examples, represented in red, are more widely distributed between 0.1 and 0.6, indicating that the model provides a wider range of probabilities for positive examples.

Figure 6. Distribution of Predicted Probabilities

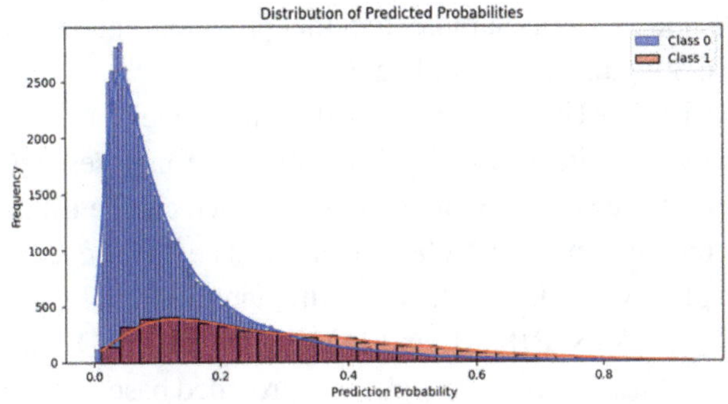

The graphical analysis shows that there is a clear probability of separation between the classes and that the model makes more confident decisions, especially for class 0. However, the more widespread distribution of the estimated probabilities for class 1 indicates that the examples belonging to this class have more complex and difficult-to-separate features. This situation is also related to the imbalance in the dataset and causes the model to make more cautious predictions for class 1.

In addition, the fact that the examples belonging to class 1 are less dense at high probability values indicates that care should be taken in determining the decision threshold. In this context, the balance of sensitivity (recall) and specificity was achieved by determining the optimal threshold value. The graphic shows that the model not only performs binary classification but also provides a probability output that expresses the uncertainty level for each example; in this respect, it makes a significant contribution to the explainability and interpretability of the model.

This analysis of the distribution of prediction probabilities reveals that the model shows a discriminative performance among classes and creates a strong decision structure, especially for class 0. These findings support the usability of the developed classification model as a reliable decision support tool in the field of financial risk estimation.

CONCLUSION

In this study, where credit risk estimation was performed with the LightGBM algorithm using the Home Credit Default Risk dataset, the performance of the proposed model was evaluated comprehensively through metrics such as accuracy, F1 score and AUC-ROC, and it was observed that a high hit rate was achieved especially in "non-risky" applications. On the other hand, the relatively low performance in the detection of "risky" individuals indicates that the sensitivity towards the minority class should be increased. In this direction, the use of data balancing methods such as SMOTE and Random UnderSampler together provided significant improvements in the ability to distinguish risky applicants.

The obtained ROC AUC value reveals that the model has a statistically good discrimination power, while the Precision-Recall curve analyses and the determined optimal threshold value show that a healthier balance can be established between false positive and false negative classification. The feature important analysis results indicate that variables such as credit_annuity_ratio, EXT_SOURCE_2, EXT_SOURCE_3, DAYS_BIRTH, and DAYS_EMPLOYED play a determining role in credit risk. Thus, important insights are provided based on which parameters credit allocation processes in the banking sector should be managed.

However, the proliferation of ML systems in critical applications such as credit risk management and fraud detection in the banking sector creates new responsibilities in the field of data security and privacy. In particular, the processing of personal and financial information requires high standards of data protection measures regulated by GDPR or similar regulations. Therefore, institutions should develop strict policies based on principles such as anonymization, data minimization, authorization, and auditability. Otherwise, potential data breaches or injustices due to algorithmic bias can lead to both legal and reputational risks.

On the other hand, the issues of algorithmic bias and algorithmic fairness are also becoming increasingly important; models trained with inaccurate or incomplete data can cause some individuals to be unfairly labeled as risky. This can lead to unfair outcomes in accessing credit. Therefore, banks are expected to not only achieve high accuracy but also ensure that their models make ethical and impartial decisions. In this context, the integration of bias detection and bias mitigation methods, as well as data balancing approaches, will contribute to making systems more reliable and equitable. With threshold optimization and balanced metrics, our LightGBM pipeline aligns with goals of equitable access and resource efficiency. When paired with public guidance and banking strategy, this approach will contribute to inclusive growth and sustainable finance.

This study has shown that the LightGBM algorithm provides high accuracy and data processing efficiency in terms of credit risk estimation in banking; it has also revealed that techniques such as data balancing can significantly improve performance for the "risky" class. Future studies can increase model success by focusing on approaches such as hyperparameter optimization, ensemble methods, and more advanced data balancing strategies. However, not ignoring data security, privacy, and ethical requirements is key to the sustainability of digital transformation processes in the sector. When banks integrate ML-based risk models into their operational processes by meeting the requirements in these areas, they will be able to create faster and healthier decision-making mechanisms in critical functions, such as both credit allocation and fraud prevention.

REFERENCES

Almubark, I. (2024). Advanced Credit Card Fraud Detection: An Ensemble Learning Using Random Under Sampling and Two-Stage Thresholding. *IEEE Access : Practical Innovations, Open Solutions, 12*, 192079–192089. DOI: 10.1109/ACCESS.2024.3519335

Andronie, M., Blažek, R., Iatagan, M., Skypalova, R., U ă, C., Dijmărescu, A., Kovacova, M., Grecu, G., Pârvu, I., Strakova, J., Guni, C., Zabojnik, S., Chiru, C., Sedláčková, A. N., Novák, A., & Dijmărescu, I. (2024). Generative artificial intelligence algorithms in Internet of Things blockchain-based fintech management. *Oeconomia Copernicana, 15*(4), 1349–1381. DOI: 10.24136/oc.3283

Arner, D. W., Barberis, J., & Buckey, R. P. (2016). FinTech, RegTech, and the reconceptualization of financial regulation. *Nw. J. Int'l L. & Bus., 37*(3).

Azevedo, N., Mateus, M., & Pina, Á. (2022). Bank credit allocation and productivity: Stylised facts for Portugal. *Studies in Economics and Finance, 39*(4), 644–674. DOI: 10.1108/SEF-08-2020-0312

Ben-David, I., & Gilje, E. P. (2018). Do Bank Employees Misuse Information to Personalize Service? *Review of Financial Studies*, •••, 1712–1753.

Brynjolfsson, E., & Mcafee, A. (2017). The business of artificial intelligence. *Harvard Business Review*, *7*(1), 1–2.

Butler, T., & Brooks, R. (2023). On the wicked problem of quantifying and managing non-financial risks and the role of digital technology in providing solutions. *Journal of Risk Management in Financial Institutions, 17*(1), 53. DOI: 10.69554/TNPL3644

Chawla, N. V., Bowyer, K. W., Hall, L. O., & Kegelmeyer, W. P. (2002). SMOTE: Synthetic Minority Over-sampling Technique. *Journal of Artificial Intelligence Research, 16*, 321–357. DOI: 10.1613/jair.953

Chen, M., Mao, S., & Liu, Y. (2014). Big Data: A Survey. *Mobile Networks and Applications*, *19*(2), 171–209. DOI: 10.1007/s11036-013-0489-0

Chen, R.-C., Dewi, C., Huang, S.-W., & Caraka, R. E. (2020). Selecting critical features for data classification based on machine learning methods. *Journal of Big Data*, *7*(1), 52. DOI: 10.1186/s40537-020-00327-4

Dasril, Y., Muslim, M. A., Al Hakim, M. F., Jumanto, J., & Prasetiyo, B. (2023). Credit Risk Assessment in P2P Lending Using LightGBM and Particle Swarm Optimization. *Register: Jurnal Ilmiah Teknologi Sistem Informasi*, *9*(1), 18–28. DOI: 10.26594/register.v9i1.3060

De Lima, L. G. (2024). Essentials of risk management in the banking sector. *Revista Sistemática*, *14*(5), 1344–1349. DOI: 10.56238/rcsv14n5-023

Dell'Ariccia, G., & Marquez, R. (2004). Information and bank credit allocation. *Journal of Financial Economics*, *72*(1), 185–214. DOI: 10.1016/S0304-405X(03)00210-1

Elreedy, D., & Atiya, A. F. (2019). A Comprehensive Analysis of Synthetic Minority Oversampling Technique (SMOTE) for handling class imbalance. *Information Sciences*, *505*, 32–64. DOI: 10.1016/j.ins.2019.07.070

Goodfellow, I., Bengio, Y., & Courville, A. (2016). *Deep learning* (2nd ed., Vol. 1). MIT press.

Gu, J. (2020). A Novel Credit Risk Assessment Model Based on LightGBM. *Journal of Simulation*, *8*(3), 71–73.

Hashemi, S. K., Mirtaheri, S. L., & Greco, S. (2023). Fraud Detection in Banking Data by Machine Learning Techniques. *IEEE Access : Practical Innovations, Open Solutions*, *11*, 3034–3043. DOI: 10.1109/ACCESS.2022.3232287

Ho, H., & Tien, K.-M. (Terence), Wu, A., & Singh, S. (2021). A sequence analysis approach to segmenting credit card customers. *Journal of Retailing and Consumer Services*, *59*, 102391. DOI: 10.1016/j.jretconser.2020.102391

Hou, L., Bi, G., & Guo, Q. (2025). An improved sparrow search algorithm optimized LightGBM approach for credit risk prediction of SMEs in supply chain finance. *Journal of Computational and Applied Mathematics*, *454*, 116197. DOI: 10.1016/j.cam.2024.116197

Jadhav, P. A., Lalwani, U., Gour, A., Shayan, M., & Motwani, S. (2022). Identification of Fraudulent Credit Card transactions using Machine Learning Algorithms. *2022 IEEE 7th International Conference for Convergence in Technology (I2CT)*, 1–4. https://doi.org/DOI: 10.1109/I2CT54291.2022.9824165

Jordan, M. I., & Mitchell, T. M. (2015). Machine learning: Trends, perspectives, and prospects. *Science*, *349*(6245), 255–260. DOI: 10.1126/science.aaa8415 PMID: 26185243

Kaplan, A., & Haenlein, M. (2019). Siri, Siri, in my hand: Who's the fairest in the land? On the interpretations, illustrations, and implications of artificial intelligence. *Business Horizons*, *62*(1), 15–25. DOI: 10.1016/j.bushor.2018.08.004

Kodimenu, M. K., S, S., & Katoon, T. (2024). Credit Card Fraud Detection Using ML & DL. *International Journal Of Scientific Research In Engineering And Management*, *08*(07), 1–11. DOI: 10.55041/IJSREM36686

Kou, G., Chao, X., Peng, Y., Alsaadi, F. E., & Herrera-Viedma, E. (2019). Machine Learning Methods For Systemic Risk Analysis In Financial Sectors. *Technological and Economic Development of Economy*, *25*(5), 716–742. DOI: 10.3846/tede.2019.8740

Kroll, J. A. (2015). *Accountable Algorithms*. Princeton University.

Kumar, S., & R, K. (2024). Comprehensive Review on Leveraging AI and Machine Learning in Digital Banking. *International Journal of Innovative Research in Computer and Communication Engineering*, *12*(05), 6333–6336. DOI: 10.15680/IJIRCCE.2024.1205197

LeCun, Y., Bengio, Y., & Hinton, G. (2015). Deep learning. *Nature*, *521*(7553), 436–444. DOI: 10.1038/nature14539 PMID: 26017442

Lenka, S., & Tiwari, R. (2025). A Hybrid ML and Data Science Approach to Detect Online Fraud Transaction at Real Time. *Journal of Neonatal Surgery*, *14*(1S), 765–775. DOI: 10.52783/jns.v14.1601

Leo, M., Sharma, S., & Maddulety, K. (2019). Machine Learning in Banking Risk Management: A Literature Review. *Risks*, *7*(1), 29. DOI: 10.3390/risks7010029

Moffo, A. M. F. (2024). A machine learning approach in stress testing US bank holding companies. *International Review of Financial Analysis*, *95*, 103476. DOI: 10.1016/j.irfa.2024.103476

Oprea, I., & Du ă, D. (2024). Integrating AI in Bank Digitalization: Strategies, Challenges and Future Perspectives. *Legal Perspectives in the Modern Era of Technological Transformations*, 205–216. https://doi.org/DOI: 10.62768/ADJURIS/2024/1/13

Ponsam, J. G., Bella Gracia, S. V. J., Geetha, G., Karpaselvi, S., & Nimala, K. (2021). Credit Risk Analysis using LightGBM and a comparative study of popular algorithms. *2021 4th International Conference on Computing and Communications Technologies (ICCCT)*, 634–641. https://doi.org/DOI: 10.1109/ICCCT53315.2021.9711896

Rakesh Pandit, S. B. J. S. P. M. A. C. (2024). Credit Risk Assessment and Fraud Detection in Financial Transactions Using Machine Learning. *Journal of Electrical Systems*, *20*(3s), 2061–2069. DOI: 10.52783/jes.1807

Shukla, D., & Gupta, S. (2024). Feature Engineering Techniques to Enhance Credit Scoring Models. *2024 6th International Conference on Electrical, Control and Instrumentation Engineering (ICECIE)*, 1–5. https://doi.org/DOI: 10.1109/ICECIE63774.2024.10815675

Singh, A., & Ahlawat, N. (2023). A review article: -the Growing Role of Data Science and AI in Banking and Finance. *International Research Journal of Modernization in Engineering Technology and Science*. Advance online publication. DOI: 10.56726/IRJMETS44000

Sun, X., Liu, M., & Sima, Z. (2020). A novel cryptocurrency price trend forecasting model based on LightGBM. *Finance Research Letters*, *32*, 101084. DOI: 10.1016/j.frl.2018.12.032

Sun, Y., Gong, Z., Shi, Q., & Chen, L. (2023). Efficient Commercial Bank Customer Credit Risk Assessment Based on LightGBM and Feature Engineering. *2023 5th International Symposium on Smart and Healthy Cities (ISHC)*, 75–82. https://doi.org/DOI: 10.1109/ISHC61216.2023.00022

Varadarajan, M. N. (2023). Smart Finance: How Artificial Intelligence and Machine Learning Drive Innovation in the Financial Sector. *International Journal of Innovative Research in Science, Engineering and Technology*, *12*(12), 1–14. DOI: 10.15680/IJIRSET.2023.1212045

Varian, H. R. (2014). Big Data: New Tricks for Econometrics. *The Journal of Economic Perspectives*, *28*(2), 3–28. DOI: 10.1257/jep.28.2.3

Verdonck, T., Baesens, B., & Óskarsdóttir, M., & vanden Broucke, S. (2024). Special issue on feature engineering editorial. *Machine Learning*, *113*(7), 3917–3928. DOI: 10.1007/s10994-021-06042-2

Wang, S., Zhang, L., Liu, X., & Sun, J. (2024). Optimization of multidimensional feature engineering and data partitioning strategies in heart disease prediction models. *Alexandria Engineering Journal*, *107*, 932–949. DOI: 10.1016/j.aej.2024.09.037

Wei, W., Li, J., Cao, L., Ou, Y., & Chen, J. (2013). Effective detection of sophisticated online banking fraud on extremely imbalanced data. *World Wide Web (Bussum)*, *16*(4), 449–475. DOI: 10.1007/s11280-012-0178-0

Werner de Vargas, V., Schneider Aranda, J. A., dos Santos Costa, R., da Silva Pereira, P. R., & Victória Barbosa, J. L. (2023). Imbalanced data preprocessing techniques for machine learning: A systematic mapping study. *Knowledge and Information Systems*, *65*(1), 31–57. DOI: 10.1007/s10115-022-01772-8 PMID: 36405957

Wilson, H. J., & Daugherty, P. R. (2018). Collaborative intelligence: Humans and AI are joining forces. *Harvard Business Review*, *96*(4), 114–123.

Xu, T. (2024). Fraud Detection in Credit Risk Assessment Using Supervised Learning Algorithms. *Computer Life*, *12*(2), 30–36. DOI: 10.54097/qw9j1892

Zarsky, T. (2016). The Trouble with Algorithmic Decisions. *Science, Technology & Human Values*, *41*(1), 118–132. DOI: 10.1177/0162243915605575

Zhang, J., Mucs, D., Norinder, U., & Svensson, F. (2019). LightGBM: An Effective and Scalable Algorithm for Prediction of Chemical Toxicity–Application to the Tox21 and Mutagenicity Data Sets. *Journal of Chemical Information and Modeling*, *59*(10), 4150–4158. DOI: 10.1021/acs.jcim.9b00633 PMID: 31560206

Zhang, Z., Liu, H., Chen, D., Zhang, J., Li, H., Shen, M., Pu, Y., Zhang, Z., Zhao, J., & Hu, J. (2022). SMOTE-based method for balanced spectral nondestructive detection of moldy apple core. *Food Control*, *141*, 109100. DOI: 10.1016/j.foodcont.2022.109100

Zhu, M., Zhang, Y., Gong, Y., Xing, K., Yan, X., & Song, J. (2024). Ensemble Methodology: Innovations in Credit Default Prediction Using LightGBM, XGBoost, and LocalEnsemble. *2024 IEEE 4th International Conference on Electronic Technology, Communication and Information (ICETCI)*, 421–426. https://doi.org/DOI: 10.1109/ICETCI61221.2024.10594630

KEY TERMS AND DEFINITIONS

Algorithmic Fairness: Aims to ensure that model results do not systematically disadvantage the groups that need to be protected. This involves group-based error rates, equal opportunity/parity criteria, and regular impact analyses.

Class Imbalance: A skewed class distribution that hampers learning the minority class.

Feature Engineering: This is the process of transforming raw data into derivative signals that convey more meaning to the model: ratios, log-transforms, summaries, lagged/rounded values, etc. The goal is to give the model the “right clues” about the problem.

Feature Importance: This is a numerical indicator of how much a feature contributes to the model's decision. Different models produce different measures of importance.

Gradient Boosting: An ensemble method that sequentially fits weak learners (trees) to residual errors to reduce loss iteratively.

LightGBM: A fast, scalable gradient-boosting library using histogram-based, leaf-wise tree growth.

Optimal Threshold: This is the cut-off point at which model scores are converted into a binary decision, such as "grant or deny credit."

Precision–Recall Curve: This curve provides a clearer picture of a model's performance, particularly in problems where the positive class is rare. It visualizes the trade-off between precision (how many of its predictions are correct) and sensitivity (how many true positives it captures). It can be more descriptive than ROC in scenarios like credit scoring, which seek to capture rare risks.

Random UnderSampler: Randomly removes some examples from the majority class to ensure data balance. It is quick and simple to implement and reduces training time. However, there is a risk of removing useful signals from the majority class.

ROC AUC: This is a single value that summarizes the probability that the model will give a higher score to a randomly selected positive example than to a randomly selected negative example. It considers the sensitivity/false alarm trade-off across all possible threshold values. It can also be used when the class distribution is unbalanced, but the PR curve is often more informative in that case.

SMOTE: To strengthen the representation of a rare class, it generates artificial examples similar to existing examples around them. It does this by creating "interpolations" between neighboring examples. When used correctly, it makes the minority class visible; however, caution is required because it can also increase noise in cases where classes are nested.

Compilation of References

Abaranji, A., & Shaik, D. (2022). The Imperative Need for Climate Finance and Global Climate Mitigation Strategy–A Pragmatic Analysis. *National Journal of Environmental Law*, *5*(1).

Abimibayo Adeoya, A., Sasaki, H., Fuda, M., Okamoto, T., & Egawa, S. (2022). Child nutrition in disaster: A scoping review. *The Tohoku Journal of Experimental Medicine*, *256*(2), 103–118. PMID: 35173109

Abramovitz, M. (1986). Catching Up, Forcing Ahead and Falling Behind. *The Journal of Economic History*, *46*(2).

Acheampong, A. O. (2018). Economic growth, CO_2 emissions and energy consumption: What causes what and where? *Energy Economics*, *74*(1), 677–692. DOI: 10.1016/j.eneco.2018.07.022

AFAD. (2019a). *Türkiye deprem tehlike haritası.* https://www.afad.gov.tr/turkiye-deprem-tehlike-haritasi

AFAD. (2019b). *Türkiye il afet risk azaltma planı.* https://www.afad.gov.tr/turkiye-afet-risk-azaltma-plani-tarap

Afonso, A. F., Pereira, O. R., Fernandes, Â., Calhelha, R. C., Silva, A. M. S., Ferreira, I. C. F. R., & Cardoso, S. M. (2019). Phytochemical composition and bioactive effects of *Salvia africana, Salvia officinalis* 'Icterina' and *Salvia mexicana* aqueous extracts. *Molecules (Basel, Switzerland)*, *24*(24), 4327. DOI: 10.3390/molecules24234327

Agozie, D. Q., Akwasi, G. B., Victor, B. F., Ozturk, I., & Taha, A. (2022). Environmental Kuznets Curve hypothesis from lens of economic complexity index for BRICS: Evidence from second generation panel analysis. *Sustainable Energy Technologies and Assessments*, *53*(1), 10–25. DOI: 10.1016/j.seta.2022.102597

Ahakwa, I., Xu, Y., Tackie, E. A., & Mangudhla, T. (2023). How crucial are natural resources in descending environmental degradation in Ghana? A novel dynamic ARDL simulation approach. *Journal of Cleaner Production, 420*(1), 1–13. DOI: 10.1016/j.jclepro.2023.138427

Ahmad, M., Jiang, P., Majeed, A., Umar, M., Khan, Z., & Muhammad, S. (2020). The dynamic impact of natural resources, technological innovations and economic growth on ecological footprint: An advanced panel data estimation. *Resources Policy, 69*(1), 1–10. DOI: 10.1016/j.resourpol.2020.101817

Ahmad, U. S., Rahman, Z. U., & Azam, M. (2024). Sustainability and Green Finance and its Relevance to Debt for Nature Swap Financing. In *The Palgrave Handbook of Green Finance for Sustainable Development* (pp. 645–671). Springer International Publishing.

Ahmed, Z. (2013). Disaster risks and disaster management policies and practices in Pakistan: A critical analysis of Disaster Management Act 2010 of Pakistan. *International Journal of Disaster Risk Reduction, 4*, 15–20. DOI: 10.1016/j.ijdrr.2013.03.003

Ahrens, J., & Rudolph, P. M. (2006). The importance of governance in risk reduction and disaster management. *Journal of Contingencies and Crisis Management, 14*(4), 207–220. DOI: 10.1111/j.1468-5973.2006.00497.x

Ajide, K. B., & Ibrahim, R. L. (2021). Environmental impacts of income inequality: Evidence from G7 countries. *Environmental Science and Pollution Research International, 29*(2), 1887–1908. PMID: 34363162

Akamatsu, K. (1962). A historical pattern of economic growth in developing countries. *The Developing Economies, 1*(s1), 3–25.

Akao, K.-I., & Sakamoto, H. (2018). A theory of disasters and long-run growth. *Journal of Economic Dynamics & Control, 95*, 89–109.

Akar, S. (2013). Doğal afetlerin kamu maliyesine ve makro ekonomiye etkileri: Türkiye değerlendirmesi. *Yönetim ve Ekonomi Araştırmaları Dergisi, 21*, 185–185.

Akkemik, K. A., & Göksal, K. (2012). Energy consumption-GDP nexus: Heterogeneous panel causality analysis. *Energy Economics, 34*(4), 865–873. DOI: 10.1016/j.eneco.2012.04.002

Akram, R., Chen, F., Khalid, F., Ye, Z., & Majeed, M. T. (2020). Heterogeneous effects of energy efficiency and renewable energy on carbon emissions: Evidence from developing countries. *Journal of Cleaner Production, 247*, 119122. DOI: 10.1016/j.jclepro.2019.119122

Aktel, M. (2010). 5902 sayılı yasa ile Türkiye'de afet yönetiminde oluşan değişim. *Dumlupınar Üniversitesi Sosyal Bilimler Dergisi, 27.*

Akudugu, M. A., & Ogwu, M. C. (2024). Sustainable development policies and interventions: A bibliometric analysis of the contributions of the academic community. *Journal of Cleaner Production, 434*, 139919.

Aladekomo, A. S. (2025). Trends in Gender, Feminism and Energy Transition in the Global South. *Journal of Sustainable Development Law and Policy (The), 16*(2), 53–75.

Alam, M. S., Kong, J., Tao, R., Ahmed, T., Alamin, M., Alotaibi, S. S., Abdelsalam, N. R., & Xu, J.-H. (2022). CRISPR/Cas9 mediated knockout of the OsbHLH024 transcription factor improves salt stress resistance in rice (*Oryza sativa* L.). *Plants, 11*(9), 1184. DOI: 10.3390/plants11091184

Aldy, J. E., & Pizer, W. A. (2011). The competitiveness impacts of climate change mitigation policies. NBER Working Paper No. 17705. National Bureau of Economic Research.

Aldy, J. E., & Pizer, W. A. (2015). The competitiveness impacts of climate change mitigation policies. *Journal of the Association of Environmental and Resource Economists, 2*(4), 565–595.

Alenda, J. M., & Kaufmann, M. (2024). From Marxist Development Theories to Their Translation in the Degrowth Discourse: Transforming Unequal International Structures for Environmental Sustainability.

Alexander, K., Jarman, A. M. G., & Rosenthal, U. (1995). Inter-organizational policy processes in disaster management. *Disaster Prevention and Management: An International Journal, 4*(2), 20–37. DOI: 10.1108/09653569510082669

Al-Falahi, A. H., Barry, S., Gebrechorkos, S. H., Spank, U., & Bernhofer, C. (2023). Potential of traditional adaptation measures in mitigating the impact of climate change. *Sustainability, 15*(21), 15442. DOI: 10.3390/su152115442

Alghamdy, R. Z. (2024). English Teachers' Practice of Classroom Discourse in Light of Zone of Proximal Development Theory and Scaffolding Techniques. *Journal of Language Teaching & Research, 15*(1).

Alharbi, Y. M., Sakr, S. S., Albarrak, S. M., Almundarij, T. I., Barakat, H., & Hassan, M. F. Y. (2022). Antioxidative, antidiabetic, and hypolipidemic properties of probiotic-enriched fermented camel milk combined with *Salvia officinalis* leaves hydroalcoholic extract in streptozotocin-induced diabetes in rats. *Antioxidants, 11*(4), 668. DOI: 10.3390/antiox11040668

AlHinai, Y. S. (2020). Disaster management digitally transformed: Exploring the impact and key determinants from the uk national disaster management experience. *International Journal of Disaster Risk Reduction*, *51*, 1–19. PMID: 32953438

Ali, E. B., Gyamfi, B. A., Bekun, F. V., Ozturk, I., & Nketiah, P. (2023). An empirical assessment of the tripartite nexus between environmental pollution, economic growth, and agricultural production in Sub-Saharan African countries. *Environmental Science and Pollution Research International*, *30*(1), 71007–71024. DOI: 10.1007/s11356-023-27307-4 PMID: 37160515

Allcott, H., & Kessler, J. B. (2024). The welfare effects of nudges: A case study of energy use social comparisons. *American Economic Journal. Applied Economics*, *11*(1), 236–276.

Almubark, I. (2024). Advanced Credit Card Fraud Detection: An Ensemble Learning Using Random Under Sampling and Two-Stage Thresholding. *IEEE Access : Practical Innovations, Open Solutions*, *12*, 192079–192089. DOI: 10.1109/ACCESS.2024.3519335

Alnafrah, I., Okunlola, O., Sinha, A., Abbas, S., & Dagestani, A. A. (2023). Unveiling the environmental efficiency puzzle: Insights from global green innovations. *Journal of Environmental Management*, *345*, 118865. DOI: 10.1016/j.jenvman.2023.118865 PMID: 37659369

Althouse, J., & Svartzman, R. (2024). Prospects and roadblocks to a" sustainable" international monetary and financial system. In *Understanding Green Finance* (pp. 182–198). Edward Elgar Publishing.

Alvarado, R., Ponce, P., Criollo, A., Córdova, K., & Khan, M. K. (2018). Environmental degradation and real per capita output: New evidence at the global level grouping countries by income levels. *Journal of Cleaner Production*, *189*(1), 13–20. DOI: 10.1016/j.jclepro.2018.04.064

Álvarez-Véliz, R., & Barton, J. R. (2024). The historical geography of an idea: Sustainable development in Latin America, 1972–2022. *Journal of Historical Geography*, *86*, 175–186.

Ambassa, M., Ndjokou, I. M. M. M., & Tsopmo, P. C. (2024). Foreign aid and inequality: Do conflicts matter? *Research in Economics*, *78*(4), 101008. DOI: 10.1016/j.rie.2024.101008

Amevor, E. B. (2025). Hegemonic discourses surrounding United States' leadership in post-Ebola development in Liberia. *New Global Studies*, *19*(1), 25–45.

Âmin, S. (2014). *Modernity, democracy and religion: critique of culturalisms*. Yordam.

Amni, B. D. H., Hadi, A. A., & Perdinan. (2020). *Design of Muster Area as Earthquake Evacuation Area in North Lombok Regency, West Nusa Tenggara* (H. A.A. & N. I., eds.). Department of Landscape Architecture, Faculty of Agriculture, IPB University (Bogor Agricultural University), Bogor, Indonesia: Institute of Physics Publishing. https://doi.org/DOI: 10.1088/1755-1315/501/1/012004

Andersen, J. J., Johannesen, N., & Rijkers, B. (2022). Elite capture of foreign aid: Evidence from offshore bank accounts. *Journal of Political Economy, 130*(2), 388–425.

Anderson, M. K., Thompson, R. L., & Davis, C. H. (2023). Climate vulnerability and renewable energy adoption: Evidence from coastal communities. *Environmental Research Letters, 18*(4), 045012.

Andrae, S. (2024). The establishment of a Gross Ecosystem Product: A country comparison. In S. Boubaker & T. H. Le (Eds.), *Green finance and sustainable development goals* (pp. 369–407). World Scientific Publishing., 10.1142/9781800614482_0014

Andrae, S. (2025). The political economy of digital nature-based currencies: Opportunities for the BRICS countries? In A. Gedikli, S. Erdogan, & H. Ç. Terzioğlu (Eds.), *Changing the global political economy: BRICS countries and alternative relations strategies* (pp. 307–344). IGI Global Publishing., 10.4018/979-8-3693-7393-4.ch010

Andreoni, A. (2024). Reframing structural transformation towards sustainable and inclusive prosperity. In E. Scheja & K. B. Kim (Eds.), *Rethinking economic transformation for sustainable and inclusive development* (pp. 25–48). Edward Elgar., 10.4337/9781035348466.00006

Andriansyah, Y. (2015). *Post-disaster economic recovery in Yogyakarta, Indonesia* (G. F.L., ed.). Department of Islamic Economics, Islamic University of Indonesia, Indonesia: CRC Press/Balkema. https://doi.org/DOI: 10.1201/b18146-62

Andronie, M., Blažek, R., Iatagan, M., Skypalova, R., U ă, C., Dijmărescu, A., Kovacova, M., Grecu, G., Pârvu, I., Strakova, J., Guni, C., Zabojnik, S., Chiru, C., Sedláčková, A. N., Novák, A., & Dijmărescu, I. (2024). Generative artificial intelligence algorithms in Internet of Things blockchain-based fintech management. *Oeconomia Copernicana, 15*(4), 1349–1381. DOI: 10.24136/oc.3283

Ansari Samani, H., Rouzbahani, M., Dalvandi, H., & Sadeghi, R. (2024). Income inequality and environmental degradation in the provinces of Iran. Environmental Health Enginnering and Management Journal, 11 (1), 25-30. http://ehemj.com/article-1-1301-en.html

Apergis, N., & Payne, J. E. (2010). Renewable energy consumption and growth in Eurasia. *Energy Economics*, *32*(6), 1392–1397. DOI: 10.1016/j.eneco.2010.06.001

Apergis, N., & Payne, J. E. (2011). The renewable energy consumption–growth nexus in Central America. *Applied Energy*, *88*(1), 343–347. DOI: 10.1016/j.apenergy.2010.07.013

Apergis, N., & Payne, J. E. (2015). Renewable energy, output, carbon dioxide emissions, and oil prices: Evidence from South America. *Energy Sources. Part B, Economics, Planning, and Policy*, *10*(3), 281–287. DOI: 10.1080/15567249.2013.853713

Apryana, H., Nugraheni, K. R., & Wardhana, F. (2020). Policy analysis of budget allocation for disasters in Indonesia. *Social, Humanities, and Educational Studies (SHEs): Conference Series, 3(1).*

Ardalan, A., Mazaheri, M., Naieni, K. H., Rezaie, M., Teimoori, F., & Pourmalek, F. (2010). Older people's needs following major disasters: A qualitative study of Iranian elders' experiences of the bam earthquake. *Ageing and Society*, *30*(1), 11–23.

Arksey, H., & O'Malley, L. (2005). Scoping studies: Towards a methodological framework. *International Journal of Social Research Methodology*, *8*(1), 19–32. DOI: 10.1080/1364557032000119616

Arner, D. W., Barberis, J., & Buckey, R. P. (2016). FinTech, RegTech, and the reconceptualization of financial regulation. *Nw. J. Int'l L. & Bus.*, *37*(3).

Ashida, S., Thompson, L., Hejna, E., Obot, S., & Schneider, H. (2020). Personal support networks of older adults in relation to their perceptions about disaster management. *Innovation in Aging*, *4*(Supplement_1), 439–440.

Ashraf, A. (2021). Lessons learned from covid-19 response for disaster risk management. *Natural Hazards*, *107*, 2027–2032. PMID: 33716403

Associated Press. (2024, October 31). *UN report warns of famine, aggravated by conflicts and climate shocks*. AP News.

Associated Press. (2025, March 11). *US legal troubles for tycoon Adani expose shortcomings in India's booming solar sector*. AP News.

Athanasıou, T., & Baer, P. (2006). *Deadly heat*. Bgst.

Atkinson, C., & Curnin, S. (2020). Sharing responsibility in disaster management policy. *Progress in Disaster Science*, *7*, 100122. DOI: 10.1016/j.pdisas.2020.100122

Ausserladscheider, V., Ergen, T., & Golka, P. (Eds.). (2025). *The socio-economics of loss and decline in the climate crisis* [Call for papers]. *Socio-Economic Review*. https://academic.oup.com/ser/pages/cfp-asset-stranding

Auteri, M., Mele, M., Ruble, I., & Magazzino, C. (2024). The double sustainability: The link between government debt and renewable energy. *Journal of Economic Asymmetries*, *29*, e00356.

Awad, A., Mallek, R. S., Ozturk, I., & Abdalla, Y. A. (2023). Infrastructure development's role in environmental degradation in sub-Saharan Africa: Impacts and transmission channels. *Journal of Cleaner Production*, *414*(1), 1–10. DOI: 10.1016/j.jclepro.2023.137622

Aydıner, T. (2014). *Doğal afet yönetişimi: Türkiye'de doğal afet yönetimi uygulamalarının tarihsel bağlamda değerlendirilmesi* [Unpublished master thesis]. Pamukkale Üniversitesi, Denizli, Türkiye.

Aydin, M. (2019). Renewable and non-renewable electricity consumption–economic growth nexus: Evidence from OECD countries. *Renewable Energy*, *136*, 599–606. DOI: 10.1016/j.renene.2019.01.008

Aydin, M., & Degirmenci, T. (2024). The role of greenfield investment and investment freedom on environmental quality: Testing the EKC hypothesis for EU countries. *International Journal of Sustainable Development and World Ecology*, *31*(6), 684–694.

Azevedo, N., Mateus, M., & Pina, Á. (2022). Bank credit allocation and productivity: Stylised facts for Portugal. *Studies in Economics and Finance*, *39*(4), 644–674. DOI: 10.1108/SEF-08-2020-0312

Aziz, T. (2018). Myth and reality of vulnerability to disaster: Pressure and release model for hazards in Bangladesh. *Journal of Southeast Asian Studies*, *6*(1), 23–31.

Azkar, N. (2025). A Complete Story of Organizational Excellence: Mainstreaming Islamic Perspective Into Organizational Development Discourse. *KnE Social Sciences*, *10*(4), 371–382.

Azmi, R., Budiarto, H., & Widyanto, R. (2011). *A proposed disaster emergency warning system standard through DVB-T in Indonesia*. Faculty of Computer Science, University of Indonesia, Kampus UI Depok, Depok 16424. *Indonesia*. Advance online publication. DOI: 10.1109/ICEEI.2011.6021746

Baccini, A., Goetz, S. J., Walker, W. S., Laporte, N. T., Sun, M., Sulla-Menashe, D., Hackler, J., Beck, P. S. A., Dubayah, R., Friedl, M. A., Samanta, S., & Houghton, R. A. (2012). Estimated carbon dioxide emissions from tropical deforestation improved by carbon-density maps. *Nature Climate Change*, *2*(3), 182–185. DOI: 10.1038/nclimate1354

Bahadırlı, N. P. (2020). Economically important sage species from Turkey: *Salvia fruticosa* Mill. and *S. aramiensis* Rech fil. *Current Perspectives on Medicinal and Aromatic Plants*, *3*(1), 31–42. DOI: 10.38093/cupmap.736482

Bahar, N. H. A., Lo, M., Sanjaya, M., Van Vianen, J., Alexander, P., Ickowitz, A., & Sunderland, T. (2020). Meeting the food security challenge for nine billion people in 2050: What impact on forests? *Global Environmental Change*, *62*, 102056. DOI: 10.1016/j.gloenvcha.2020.102056

Bajracharya, S., Shrestha, A., & House, R. (2019). Newborn and child health care in humanitarian crisis settings: Piloting of training package for primary health care workers in rural Nepal. *Prehospital and Disaster Medicine*, *34*(s1), s32–s33.

Bakay, M. S., & Ağbulut, Ü. (2021). Electricity production based forecasting of greenhouse gas emissions in Turkey with deep learning, support vector machine and artificial neural network algorithms. *Journal of Cleaner Production*, *285*, 125324. DOI: 10.1016/j.jclepro.2020.125324

Bakkar, Y., Jabeur, S. B., Mohammed, K. S., & Arfi, W. B. (2024). Environmental transition dynamics under external conflict risk: New evidence from European countries. *Journal of Cleaner Production*, *472*, 143510.

Bali, R. (2021). Women as equal partners in disaster risk reduction planning and management. *Research Review International Journal of Multidisciplinary*, *6*(5), 111–117.

Bandura, A. (2021). *Social cognitive theory: An agentic perspective*. Stanford University Press.

Bang, H. N. (2014). General overview of the disaster management framework in Cameroon. *Disasters*, *38*(3), 562–586. DOI: 10.1111/disa.12061 PMID: 24905711

Bannister, P., & Carver, M. (2025). 'I don't need professional development; I want institutional development': Legitimising marginalised epistemic capital that disrupts generative AI discourse. *Professional Development in Education*, *51*(3), 547–565.

Baraldo, M., & Di Giuseppantonio Di Franco, P. (2024). Place-centred emerging technologies for disaster management: A scoping review. *International Journal of Disaster Risk Reduction*, *112*, 1–18.

Baran, P. A. (1952). The Political Economy of Underdevelopment. *Manchester School*.

Barkat, K., Alsamara, M., Kwifi, O. S. A., & Jarallah, S. (2024). Does trade openness mitigate environmental degradation in organization for Economic Co-operation and Development (OECD) countries? Implication for achieving sustainable development. *Natural Resources Forum*, *49*(1), 677–698. DOI: 10.1111/1477-8947.12412

Barlow, R. (1994). Population growth and economic growth: Some more correlations. *Population and Development Review*, *20*(1), 153.

Barro, R. J. (1990). Government Spending in a Simple Model of Endogenous Growth. *Journal of Political Economy*, *98*(5).

Basili, M., Franzini, M., & Vercelli, A. (2005). *Environment, Inequality and Collective Action*. Routledge., DOI: 10.4324/9780203481509

Başoğlu, M. (2013). *Küresel Isınma ve Toprak Ananın Yıkımı*. Su Yayınevi.

Bates, L. S., Waldren, R. P., & Teare, I. D. (1973). Rapid determination of free proline for water-stress studies. *Plant and Soil*, *39*(1), 205–208.

Batlle-Sales, J. (2023). Salt-affected soils: A sustainability challenge in a changing world. *Italian Journal of Agronomy*, *18*(2), 2188. DOI: 10.4081/ija.2023.2188

Bauman, Z. (2013). *Is the wealth of the few in the interest of us all?* Ayrıntı.

Bauman, Z., & Bordoni, C. (2018). *Crisis and the state*. İthaki.

Belaid, F., Boubaker, S., & Kafrouni, R. (2020). Carbon emissions, income inequality and environmental degradation: The case of Mediterranean countries. *The European Journal of Comparative Economics*, *17*(1), 73–102.

Bélaïd, F., & Youssef, M. (2017). Environmental degradation, renewable and non-renewable electricity consumption, and economic growth: Assessing the evidence from Algeria. *Energy Policy*, *102*, 277–287. DOI: 10.1016/j.enpol.2016.12.012

Bello, M. O., Solarin, S. A., & Yen, Y. Y. (2018). The impact of electricity consumption on CO2 emission, carbon footprint, water footprint and ecological footprint: The role of hydropower in an emerging economy. *Journal of Environmental Management*, *219*(1), 218–230. DOI: 10.1016/j.jenvman.2018.04.101 PMID: 29747103

Ben Akacha, B., Kačániová, M., Generalić Mekinić, I., Kukula-Koch, W., Koch, W., Orhan, I. E., Čmiková, N., Taglieri, I., Venturi, F., Samartin, C., Bouteraa, M. T., Ben Saad, R., Mnif, W., Garzoli, S., & Ben Hsouna, A. (2024). Sage (*Salvia officinalis* L.): A botanical marvel with versatile pharmacological properties and sustainable applications in functional foods. *South African Journal of Botany*, *169*, 361–382. DOI: 10.1016/j.sajb.2024.04.044

Ben Gaied, R., Brígido, C., Sbissi, I., & Tarhouni, M. (2024). Sustainable strategy to boost legumes growth under salinity and drought stress in semi-arid and arid regions. *Soil Systems*, *8*(3), 84. DOI: 10.3390/soilsystems8030084

Bendaas, R., Messaadia, L., Bekkar, Y., Bourougaa, L., & Messaoudi, A. (2025). Comprehensive *in vitro* and molecular docking analysis of antioxidant and antimicrobial properties of *Salvia officinalis* L. extracts and essential oils. *Journal of Molecular Structure*, *1331*, 141621. DOI: 10.1016/j.molstruc.2025.141621

Ben-David, I., & Gilje, E. P. (2018). Do Bank Employees Misuse Information to Personalize Service? *Review of Financial Studies*, •••, 1712–1753.

Bergougui, B., Ben-Salha, O., & Murshed, S. M. (2025). How income inequality drives climate inequality in the world's largest emitting nations: Novel evidence from multivariate quantile-on-quantile and wavelet quantile methods. *Journal of Cleaner Production*, *513*. Advance online publication. DOI: 10.1016/j.jclepro.2025.145666

Berke, P. R., & Campanella, T. J. (2006). Planning for Post-disaster Resiliency. *The Annals of the American Academy of Political and Social Science*, *604*(1), 192–207. DOI: 10.1177/0002716205285533

Betten, T., Vian Pettersen, K., & Albrechtsen, E. (2021). Learning in municipalities after disasters. *Disaster Prevention and Management: An International Journal*, *30*(3), 400–411.

Bhattacharya, M., Paramati, S. R., Ozturk, I., & Bhattacharya, S. (2016). The effect of renewable energy consumption on economic growth: Evidence from top 38 countries. *Applied Energy*, *162*, 733–741. DOI: 10.1016/j.apenergy.2015.10.104

Bhide, N., & Khanolkar, S. (2020). A comprehensive study on the factors impacting the gdp (per capita) of major economies around the globe using regression analysis. *International Journal of Computer Applications*, *175*(10), 26–30.

Biely, K. (2024). Paradigm changes in research: Kuhn meets socio-ecological transition theory. *Journal of Cleaner Production*, *480*, 144057. 10.1016/j.jclepro.2024.144057

Biely, K., & Chakori, S. (2025). Sustainability transition theories: Perpetuating or breaking with the status quo. *Sustainable Development*, *33*(1), 52–68. 10.1002/sd.3101

Biesenthal, C., Clegg, S., Mahalingam, A., & Sankaran, S. (2018). Applying institutional theories to managing megaprojects. *International Journal of Project Management, 36*(1), 43–54.

Bird, G. (1985). *World Finance and Adjustment: An Agenda for Reform.* Springer.

Birkland, T. A. (2006). Lessons of disaster: Policy change after catastrophic events. *Lessons of Disaster: Policy Change After Catastrophic Events, 9037*(May), 1–216. DOI: 10.1093/jopart/mup023

Biscaye, P. E., Reynolds, T. W., & Anderson, C. L. (2017). Relative effectiveness of bilateral and multilateral aid on development outcomes. *Review of Development Economics, 21*(4), 1425–1447. DOI: 10.1111/rode.12303

BNPB. (2022). *Bencana Tahun 2022.* Badan Nasional Penggaulangan Bencana.

Bodansky, D. (2012, July). *The Durban Platform Negotiations: Goals and Options.* https://www.belfercenter.org/publication/durban-platform-negotiations-goals-and-options

Bohoslavsky, J. P., Laskaridis, C., & Prato, S. (2025). An International Debt Architecture and Debt Sustainability Assessments as if People and Planet Mattered. *Development,* •••, 1–4.

Bollinger, B., & Gillingham, K. (2024). Peer effects in the diffusion of solar photovoltaic panels. *Marketing Science, 43*(2), 387–404.

Bölük, G., & Mert, M. (2014). Fossil & renewable energy consumption, GHGs (greenhouse gases) and economic growth: Evidence from a panel of EU (European Union) countries. *Energy, 74*, 439–446. DOI: 10.1016/j.energy.2014.07.008

Boserup, E. (1970). *Women's Role in Economic Development.* Allen and Unwin.

Bowen, A., Campiglio, E., & Tavoni, M. (2014). A macroeconomic perspective on climate change mitigation: Meeting the financing challenge. *Climate Change Economics (Singapore), 5*(01), 1440005.

Boyce, J. K. (1994). Inequality as a cause of environmental degradation. *Ecological Economics, 11*(3), 169–178. DOI: 10.1016/0921-8009(94)90198-8

Boyd, R., & Markandya, A. (2021). Costs and benefits of climate change impacts and adaptation. In F. J. Warren & N. Lulham (Eds.), *Canada in a Changing Climate: National Issues Report* (Government of Canada). Government of Canada.

Bozkurt, A. (2023). *Python: Veri analizi, görselleştirme ve yapay zekâ.* Hiperlink.

Breusch, T. S., & Pagan, A. R. (1980). The Lagrange multiplier test and its applications to model specification in econometrics. *The Review of Economic Studies, 47*(1), 239. DOI: 10.2307/2297111

Britton, N. R. (1986). Developing an understanding of disaster. *The Australian and New Zealand Journal of Sociology, 22*(2), 254–271.

Brown, R. G., & Meyer, R. F. (1961). The Fundamental Theorem of Exponential Smoothing. *Operations Research, 9*(5), 673–685. DOI: 10.1287/opre.9.5.673

Brynjolfsson, E., & Mcafee, A. (2017). The business of artificial intelligence. *Harvard Business Review, 7*(1), 1–2.

Buchanan, , G. MParks, , B. CDonald, , P. FO'Donnell, , B. FRunfola, , DSwaddle, , J. PTracewski, , ŁButchart, , S. H. M. (2018). The Local Impacts of World Bank Development Projects Near Sites of Conservation Significance. *The Journal of Environment & Development, 27*(3), 299-322. https://doi.org/10.1177/1070496518785943

Bucher-Koenen, T., Lusardi, A., Alessie, R., & van Rooij, M. (2024). How financially literate are women? An overview and new insights. *The Journal of Consumer Affairs, 58*(2), 123–154.

Burnham, J. F. (2006). Scopus database: A review. *Biomedical Digital Libraries, 3*, 1–8. DOI: 10.1186/1742-5581-3-1 PMID: 16522216

Burniaux, J. M., Chateau, J., Dellink, R., Duval, R., & Jamet, S. (2009). The economics of climate change mitigation. *Policies and options for the future.*

Butler, T., & Brooks, R. (2023). On the wicked problem of quantifying and managing non-financial risks and the role of digital technology in providing solutions. *Journal of Risk Management in Financial Institutions, 17*(1), 53. DOI: 10.69554/TNPL3644

Butson, R., & Spronken-Smith, R. (2024). AI and its implications for research in higher education: A critical dialogue. *Higher Education Research & Development, 43*(3), 563–577.

Çakmak, E. G., & Okutan, H. C. (2023). Technological assessment of greenhouse gas reduction potential of power generation in Türkiye. *Atmospheric Pollution Research, 14*(11), 101911. DOI: 10.1016/j.apr.2023.101911

C nh, T., & Hoài, P. (2022). Factors affecting gdp per capita—apply bayesian analysis. *Science & Technology Development Journal - Economics - Law and Management, 6(2),* 2862-2868.

Capital Markets, R. B. C. (2024). *Sustainable Finance Group Themes 2025*. https://www.rbccm.com/en/story/story.page?dcr=templatedata/article/story/data/2024/12/sfg-themes-2025

Carley, S., Baldwin, E., MacLean, L. M., & Brass, J. N. (2024). Global expansion of renewable energy generation: An analysis of policy instruments. *Environmental and Resource Economics*, *68*(2), 397–440.

Casey, G. (2024). Energy efficiency and directed technical change: Implications for climate change mitigation. *The Review of Economic Studies*, *91*(1), 192–228.

Cavallo, E., & Noy, I. (2011). Natural disasters and the economy—A survey. *International Review of Environmental and Resource Economics*, *5*(1), 63–102.

Cevik, S., & Huang, G. (2018). How to manage the fiscal costs of natural disasters. *IMF How To Notes*, *2018*(003), 1–15.

Chawla, N. V., Bowyer, K. W., Hall, L. O., & Kegelmeyer, W. P. (2002). SMOTE: Synthetic Minority Over-sampling Technique. *Journal of Artificial Intelligence Research*, *16*, 321–357. DOI: 10.1613/jair.953

Chen, C. (2016). Citespace: A practical guide for mapping scientific literature. nova science publishers. In *Nova Science Publishers*.

Chen, C. (2005). *CiteSpace. Quick Guide 1.2*. Drexel University.

Chenery, H. B., & Strout, A. M. (1966). Foreign Assistance and Economic Development. *The American Economic Review*, *56*(4), 679–733.

Chen, F., Ali, S., Ma, J., Arshad, S., & Ahmad, S. (2023). Material productivity and environmental degradation: Moderating role of environment-related technologies in achieving carbon neutrality. *Gondwana Research*, *117*(1), 155–168. DOI: 10.1016/j.gr.2023.01.010

Cheng, C., Suhui, L., Qi, Y., Qing, L., Xin, G. & Huixian, Z. (2023). Does income inequality have a heterogeneous effect on carbon emissions between developed and developing countries? Evidence from simultaneous quantile regression, Frontiers in Environmental Science, 11, DOI=DOI: 10.3389/fenvs.2023.1271457

Chen, J., & Wang, Y. (2024). Understanding the salinity resilience and productivity of halophytes in saline environments. *Plant Science*, *346*, 112171. DOI: 10.1016/j.plantsci.2024.112171

Chen, J., Xian, Q., Zhou, J., & Li, D. (2020). Impact of income inequality on CO2 emissions in G20 countries. *Journal of Environmental Management*, *271*, 110987. DOI: 10.1016/j.jenvman.2020.110987 PMID: 32579533

Chen, L., & Liu, W. (2024). Environmental values and green investment behavior: The mediating role of financial literacy. *Journal of Cleaner Production, 389*, 136089.

Chen, L., Yip, W., Chang, M.-C., Lin, H.-S., Lee, S.-D., Chiu, Y.-L., & Lin, Y.-H. (2007). The effects of Taiwan's National Health Insurance on access and health status of the elderly. *Health Economics, 16*(3), 223–242. DOI: 10.1002/hec.1160 PMID: 16929478

Chen, M., Mao, S., & Liu, Y. (2014). Big Data: A Survey. *Mobile Networks and Applications, 19*(2), 171–209. DOI: 10.1007/s11036-013-0489-0

Chen, R.-C., Dewi, C., Huang, S.-W., & Caraka, R. E. (2020). Selecting critical features for data classification based on machine learning methods. *Journal of Big Data, 7*(1), 52. DOI: 10.1186/s40537-020-00327-4

Chen, Z., & Cong, Z. (2020). Barriers to disaster preparation when older adults are taking care of older adults. *Innovation in Aging, 4*(Supplement_1), 437–438.

Chiang, N., Dein, M., Nguyen, T. H., Trudelle, D. M., Ho, C.-T., & Munafo, J. P.Jr. (2024). Characterization of odorants in a commercial culinary sage (*Salvia officinalis* L.) and several cultivars. *Journal of Agricultural and Food Chemistry, 72*(14), 8081–8091. DOI: 10.1021/acs.jafc.3c09326

Ching, S. L., Lau, L. S., & Choong, C. K. (2023). Income inequality, educational attainment & environmental degradation: Evidence from global panel. *Environmental Science and Pollution Research International, 30*(15), 43056–43067. PMID: 35508851

Chletsos, M., & Sintos, A. (2022). The effects of IMF programs on income inequality: A semi-parametric treatment effects approach. *International Journal of Development Issues, 21*(2), 271–291. DOI: 10.1108/IJDI-12-2021-0265

Cho, S., Heo, E., & Kim, J. (2015). Causal relationship between renewable energy consumption and economic growth: Comparison between developed and less-developed countries. *Geosystem Engineering, 18*(6), 284–291. DOI: 10.1080/12269328.2015.1053540

Chriqui, A., Benkhnigue, O., Moujahdi, C., El-Khadir, I., Mouniane, Y., Zidane, L., & Hmouni, D. (2024). Contribution to a floristic study of the Ouezzane region (Prerif, North Atlantic Morocco–1). *Egyptian Journal of Botany, 64*(4), 28–48. DOI: 10.21608/ejbo.2024.226380.2439

Çobanoğlu, B. (2023). *Veri bilimi için python (2. Baskı)*. Kodlab.

Coondoo, D., & Dinda, S. (2007). Carbon dioxide emission and income: A temporal analysis of cross-country distributional patterns. *Ecological Economics*, *65*(2), 375–385. DOI: 10.1016/j.ecolecon.2007.07.001

Corbridge, S. (1998). Beneath the pavement only soil: The poverty of post-development. *The Journal of Development Studies*, *34*(6), 138–148.

Corgo, J., Cruz, S. S., & Conceição, P. (2024). Nature-based solutions in spatial planning and policies for climate change adaptation: A literature review. *Ambio*, *53*(11), 1599–1617.

Corner, A., Roberts, R., Chiari, S., Völler, S., Mayrhuber, E. S., Mandl, S., & Monson, K. (2023). How do young people engage with climate change? The role of knowledge, values, message framing, and trusted communicators. *Wiley Interdisciplinary Reviews: Climate Change*, *6*(5), 523–534.

Coronado, V. M. H., Valderrama, C. J. M., Valencia-Arias, A., Huamán, H. I. M., Valencia, J., & Cardona-Acevedo, S. (2024). Analysis of research trends on economic growth and environmental degradation: A bibliometric study. *Sustainable Environment*, *10*(1), 1–10. DOI: 10.1080/27658511.2024.2345445

Corsi, G., Sapio, A., Guarino, R., Muñoz-Ulecia, E., & Franzese, P. P. (2024). *Uneven development and core-periphery dynamics: A journey into the perspective of ecologically unequal exchange* (No. ART-2024-138547).

Costa, J., Alscher, P., & Thums, K. (2024). Global competences and education for sustainable development. A bibliometric analysis to situate the OECD global competences in the scientific discourse. *Zeitschrift für Erziehungswissenschaft*, •••, 1–20.

Crenshaw, E. M., Ameen, A. Z., & Christenson, M. (1997). Population dynamics and economic development: Age-specific population growth rates and economic growth in developing countries, 1965 to 1990. *American Sociological Review*, *62*(6), 974–984.

Creswell, J. W., & Plano Clark, V. L. (2024). *Designing and conducting mixed methods research* (4th ed.). Sage Publications.

Cropper, M., & Griffiths, C. (1994). The interaction of population growth and environmental quality. *The American Economic Review*, *84*(2), 250–254. https://www.jstor.org/stable/2117838

Crow, D. A., Albright, E. A., Ely, T., Koebele, E., & Lawhon, L. (2018). Do Disasters Lead to Learning? Financial Policy Change in Local Government. *The Review of Policy Research*, *35*(4), 564–589. DOI: 10.1111/ropr.12297

Cutter, S. L. (2024). The origin and diffusion of the social vulnerability index (SoVI). *International Journal of Disaster Risk Reduction*, *109*, 1–15.

Cvetković, V. M., Tanasić, J., Ocal, A., Kešetović, Ž., Nikolić, N., & Dragašević, A. (2021). Capacity development of local self-governments for disaster risk management. *International Journal of Environmental Research and Public Health*, *18*(19), 1–33. PMID: 34639706

Dabaieh, M., Maguid, D., El Mahdy, D., & Al Hegazi, S. (2024). Climate change adaptation at the margins. The case of Cairo, Egypt. *Cities (London, England)*, *154*, 105390. DOI: 10.1016/j.cities.2024.105390

Dafermos, Y. (2025). Climate finance and global justice. *Climate Policy*, *•••*, 1–17.

Danish, Zhang, B., Wang, B., & Wang, Z. (2017). Role of renewable energy and non-renewable energy consumption on EKC: Evidence from Pakistan. *Journal of Cleaner Production*, *156*, 855–864. DOI: 10.1016/j.jclepro.2017.03.203

Dasgupta, , SLaplante, , BWang, , HWheeler, , D. (2002). Confronting the environmental Kuznets curve. *Journal of Economic Perspectives*, *16*(1), 147-168. DOI: 10.1257/0895330027157

Das, R. C., Chatterjee, T., & Ivaldi, E. (2023). Co-movements of income and urbanization through energy use and pollution: An investigation for world's leading polluting countries. *Ecological Indicators*, *153*, 110381. DOI: 10.1016/j.ecolind.2023.110381

Dasril, Y., Muslim, M. A., Al Hakim, M. F., Jumanto, J., & Prasetiyo, B. (2023). Credit Risk Assessment in P2P Lending Using LightGBM and Particle Swarm Optimization. *Register: Jurnal Ilmiah Teknologi Sistem Informasi*, *9*(1), 18–28. DOI: 10.26594/register.v9i1.3060

Das, S. (2024). Perspectives on inequalities in social science research and environmental sustainability in India. *Environmental Science & Policy*, *161*, 103891. DOI: 10.1016/j.envsci.2024.103891

Daudt, H. M. L., Van Mossel, C., & Scott, S. J. (2013). Enhancing the scoping study methodology: A large, inter-professional team's experience with Arksey and O'Malley's framework. *BMC Medical Research Methodology*, *13*(1), 1–9. DOI: 10.1186/1471-2288-13-48 PMID: 23522333

Davis, M. R., Johnson, K. L., & Williams, P. T. (2023). Adult learning preferences in environmental education: A systematic review. *Adult Education Quarterly*, *73*(4), 312–335.

De Leon Barido, D. P., & Marshall, J. D. (2014). Relationship between Urbanization and CO2 Emissions Depends on Income Level and Policy. *Environmental Science & Technology, 48*(7), 3632–3639. DOI: 10.1021/es405117n PMID: 24422489

De Lima, L. G. (2024). Essentials of risk management in the banking sector. *Revista Sistemática, 14*(5), 1344–1349. DOI: 10.56238/rcsv14n5-023

Debez, A., Belghith, I., Friesen, J., Montzka, C., & Elleuche, S. (2017). Facing the challenge of sustainable bioenergy production: Could halophytes be part of the solution? *Journal of Biological Engineering, 11*(1), 27. DOI: 10.1186/s13036-017-0069-0

Dědeček, R., & Dudzich, V. (2022). Exploring the limitations of GDP per capita as an indicator of economic development: A cross-country perspective. *Review of Economic Perspectives, 22*(3), 193–217.

Degnbol-Martinussen, J., & Engberg-Pedersen, P. (2005). *Aid: Understanding International Development Cooperation*. Zed Books.

Dekker, S. C., Kraneveld, A. D., van Dijk, J., Kalfagianni, A., Knulst, A. C., Lelieveldt, H., Moors, E. H. M., Müller, E., Pieters, R. H. H., & Pieterse, C. M. J.. (2020). Towards healthy planet diets—A transdisciplinary approach to food sustainability challenges. *Challenges, 11*(2), 21. DOI: 10.3390/challe11020021

Dell'Ariccia, G., & Marquez, R. (2004). Information and bank credit allocation. *Journal of Financial Economics, 72*(1), 185–214. DOI: 10.1016/S0304-405X(03)00210-1

Demir, C., Cergibozan, R., & Gök, A. (2018). Income inequality and CO2 emissions: Empirical evidence from Turkey. *Energy & Environment, 30*(3), 444–461. DOI: 10.1177/0958305x18793109

Demir, R. (2023). Afet hukuku ve afet mevzuatı. In Çapar, S. (Ed.), *Türkiye'de mevzuat bağlamında afet ve acil durum yönetimi* (pp. 101–149). TİAV.

Dickey, D. A., & Fuller, W. A. (1979). Distribution of the estimators for autoregressive timeseries with a unit root. *Journal of the American Statistical Association*, •••, 74.

Ding, Y. J., Li, C. Y., Wang, Y., Wang, S. X., Chang, Y. P., Qin, J., & Wang, Z. R. (2021). An overview of climate change impacts on the society in China. *Advances in Climate Change Research, 12*(2), 210–223.

Djalante, R., & Garschagen, M. (2017). *A Review of Disaster Trend and Disaster Risk Governance in Indonesia: 1900–2015*. https://doi.org/DOI: 10.1007/978-3-319-54466-3_2

Djalante, R. (2012). Review Article: Adaptive governance and resilience: The role of multi-stakeholder platforms in disaster risk reduction. *Natural Hazards and Earth System Sciences, 12*(9), 2923–2942. DOI: 10.5194/nhess-12-2923-2012

Djalante, R. (2018). Review article: A systematic literature review of research trends and authorships on natural hazards, disasters, risk reduction and climate change in Indonesia. *Natural Hazards and Earth System Sciences, 18*(6), 1785–1810. DOI: 10.5194/nhess-18-1785-2018

Djalante, R., Holley, C., & Thomalla, F. (2011). Adaptive governance and managing resilience to natural hazards. *International Journal of Disaster Risk Science, 2*(4), 1–14. DOI: 10.1007/s13753-011-0015-6

Djalante, R., Holley, C., Thomalla, F., & Carnegie, M. (2013). Pathways for adaptive and integrated disaster resilience. *Natural Hazards, 69*(3), 2105–2135. DOI: 10.1007/s11069-013-0797-5

Djalante, R., Thomalla, F., Sinapoy, M. S., & Carnegie, M. (2012). Building resilience to natural hazards in Indonesia: Progress and challenges in implementing the Hyogo Framework for Action. *Natural Hazards, 62*(3), 779–803. DOI: 10.1007/s11069-012-0106-8

Doğan, H. H. (2022). Türkiye'de afet olaylarına kalkınma planlarının yaklaşımı. *İDEALKENT, 13(37),* 1873-1912.

Dogan, E., & Seker, F. (2016). Determinants of CO2 emissions in the European Union: The role of renewable and non-renewable energy. *Renewable Energy, 94*, 429–439. DOI: 10.1016/j.renene.2016.03.078

Dolan, A. H., & Walker, I. J. (2006). Understanding vulnerability of coastal communities to climate change related risks. *Journal of Coastal Research*, 1316–1323.

Domorenok, E., Graziano, P., & Polverari, L. (2021). Introduction: policy integration and institutional capacity: theoretical, conceptual and empirical challenges. *Policy and Society, 40*(1), 1–18.

Donahue, A. K. (2010). A Review of: "Lessons of Disaster: Policy Change After Catastrophic Events by Thomas A. Birkland". *International Public Management Journal, 13*(1), 100–103. DOI: 10.1080/10967490903547399

Donback, N. (2024, December 18). *How cities are using nature-based solutions to tackle floods. Time*. https://time.com/7202917/cities-nature-based-solutions-floods/

Dong, H., Geng, Y., Fujita, T., & Jacques, D. A. (2014). Three accounts for regional carbon emissions from both fossil energy consumption and industrial process. *Energy, 67*, 276–283. DOI: 10.1016/j.energy.2014.01.065

Dong, K., Dong, X., & Jiang, Q. (2020). How renewable energy consumption lower global CO 2 emissions? Evidence from countries with different income levels. *World Economy, 43*(6), 1665–1698. DOI: 10.1111/twec.12898

Dong-Qian, W. A. N. G., & Ying, S. U. N. (2022). Effects of anthropogenic forcing and atmospheric circulation on the record-breaking welt bulb heat event over southern China in September 2021. *Advances in Climate Change Research, 13*(6), 778–786.

Donthu, N., Kumar, S., Mukherjee, D., Pandey, N., & Lim, W. M. (2021). How to conduct a bibliometric analysis: An overview and guidelines. *Journal of Business Research, 133*(April), 285–296. DOI: 10.1016/j.jbusres.2021.04.070

Doocy, S., Rofi, A., Moodie, C., Spring, E., Bradley, S., Burnham, G., & Robinson, C. (2007). Tsunami mortality in Aceh Province, Indonesia. *Bulletin of the World Health Organization, 85*(4), 273–278. DOI: 10.2471/BLT.06.033308 PMID: 17546308

Dorsch, M. T., & Kirkpatrick, B. (2021). Economic growth, inequality and environmental degradation. *International Journal of Sustainable Development, 24*(2), 124. DOI: 10.1504/ijsd.2021.118844

Dubois, M., Gilles, K. A., Hamilton, J. K., Rebers, P. A., & Smith, F. (1956). Colorimetric method for determination of sugars and related substances. *Analytical Chemistry, 28*(3), 350–356.

Dumitrescu, E.-I., & Hurlin, C. (2012). Testing for Granger non-causality in heterogeneous panels. *Economic Modelling, 29*(4), 1450–1460. DOI: 10.1016/j.econmod.2012.02.014

Dunlap, R. E., Van Liere, K. D., Mertig, A. G., & Jones, R. E. (2024). New trends in measuring environmental attitudes: Measuring endorsement of the new ecological paradigm: A revised NEP scale. *The Journal of Social Issues, 56*(3), 425–442.

Durbin, J. (1959). Efficient Estimation of Parameters in Moving-Average Models. *Biometrika, 46*(3/4), 306. DOI: 10.2307/2333528

Easterly, W. (2003). Can Foreign Aid Buy Growth? *The Journal of Economic Perspectives, 17*(3), 23–48.

Easterly, W. (2006). The Big Push Déjà Vu. *Journal of Economic Literature, 44*, 1.

Ehrlich, P. (1968). *The Population Bomb*. Ballantine Books.

Eichsteller, M., Njagi, T., & Nyukuri, E. (2022). The role of agriculture in poverty escapes in Kenya – Developing a capabilities approach in the context of climate change. *World Development*, *149*, 105705. DOI: 10.1016/j.worlddev.2021.105705

Ekberg, M. (2007). The parameters of the risk society: A review and exploration. *Current Sociology*, *55*(3), 343–366.

El-Karimi, M., & El-houjjaji, H. (2022). Economic growth and renewable energy consumption nexus in G7 countries: Symmetric and asymmetric causality analysis in frequency domain. *Journal of Cleaner Production*, *342*, 130618. DOI: 10.1016/j.jclepro.2022.130618

El-Khadir, I., Ktaoui, S., Mouniane, Y., Chriqui, A., Mabrouki, J., Ameziane, H., & Hmouni, D. (2024). Improved salt stress tolerance of *Salvia officinalis* grown in the presence of a halophytic plant *Spergularia maritima*: Analysis of morpho-physiological parameters. In Mabrouki, J., & Azrour, M. (Eds.), *Integrated solutions for smart and sustainable environmental conservation* (Vol. 527, pp. XX–XX). Springer., DOI: 10.1007/978-3-031-55787-3_2

El-Khadir, I., Mouniane, Y., Chriqui, A., Kouighat, M., & Hmouni, D. (2025). Effects of salt stress on *Origanum majorana* and improvement via association with halophyte plants: *Spergularia salina* and *Aptenia cordifolia. Acta Universitatis Agriculturae et Silviculturae Mendelianae Brunensis*, *73*(1), 27–35. DOI: 10.11118/actaun.2025.003

Elmhirst, R., & Resurrección, B. P. (2025). The coloniality of gender expertise in professional environment and development contexts. *Journal of Political Ecology*, *32*(1), 5960.

Elreedy, D., & Atiya, A. F. (2019). A Comprehensive Analysis of Synthetic Minority Oversampling Technique (SMOTE) for handling class imbalance. *Information Sciences*, *505*, 32–64. DOI: 10.1016/j.ins.2019.07.070

Elson, D. (1995). Gender awareness in modeling structural adjustment. *World Development*, *23*, 11.

Enders, W., & Jones, P. (2016). Grain prices, oil prices, and multiple smooth breaks in a VAR. *Studies in Nonlinear Dynamics and Econometrics*, *20*(4). Advance online publication. DOI: 10.1515/snde-2014-0101

Eregha, P. B., Nathaniel, S. P., & Vo, X. V. (2023). Economic growth, environmental regulations, energy use, and ecological footprint linkage in the Next-11 countries: Implications for environmental sustainability. *Energy & Environment*, *34*(5), 1327–1347. DOI: 10.1177/0958305X221084293

Ergünay, O. (2011). *Ekim 11). 1999 depremleri afet mevzuatını nasıl değiştirdi: mevzuat açısından neredeyiz? Türkiye Deprem Mühendisliği ve Sismoloji Konferansı, 11-14 Ekim 2011*. ODTÜ.

Escobar, A. (1995). *Encountering Development: The Making and Unmaking of the Third World*. Princeton University Press.

EU Environment and Climate (2025). https://commission.europa.eu/strategy-and-policy/priorities-2019-2024/european-green-deal_en

European Bank for Reconstruction and Development (EBRD). (2020). The economic impacts of climate change mitigation policy. EBRD Report.

Evans, J. (2010). Mapping the vulnerability of older persons to disasters. *International Journal of Older People Nursing*, *5*(1), 63–70. PMID: 20925759

Eweade, B. S., Akadiri, A. C., Olusoga, K. O., & Bamidele, R. O. (2024). The symbiotic effects of energy consumption, globalization, and combustible renewables and waste on ecological footprint in the United Kingdom. *Natural Resources Forum*, *48*(1), 274–291. DOI: 10.1111/1477-8947.12392

Fagerberg, J. (1988). Why Growth Rates Differ. In Dosi, G., Freeman, C., Nelson, R., Silverberg, G., & Soete, L. (Eds.), *Technical Change and Economic Theory*. Pinter Publishers.

Fahrudin, A. (2012). Psychosocial Reaction and Trauma After a Natural Disaster: The Role of Coping Behavior. *Asian Social Work and Policy Review*, *6*(3), 192–202. DOI: 10.1111/j.1753-1411.2012.00070.x

FAO. (2021). *World Soil Day: FAO highlights the threat of soil salinization to global food security*. Food and Agriculture Organization of the United Nations.

Farhat, S., Jain, N., Singh, N., Sreevathsa, R., Dash, P. K., Rai, R., Yadav, S., Kumar, P., Sarkar, A. K., Jain, A., Singh, N. K., & Rai, V. (2019). CRISPR-Cas9 directed genome engineering for enhancing salt stress tolerance in rice. *Seminars in Cell & Developmental Biology*, *96*, 91–99. DOI: 10.1016/j.semcdb.2019.05.003

Fariss, , BDeMello, , NPowlen, , K. ALatimer, , C. EMasuda, , YKennedy, , C. M. (2023). Catalyzing success in community-based conservation. *Conservation Biology*, *37*(1), e13973. https://doi.org/10.1111/cobi.13973

Fernandez, L. S., Byard, D., Lin, C.-C., Benson, S., & Barbera, J. A. (2002). Frail elderly as disaster victims: Emergency management strategies. *Prehospital and Disaster Medicine*, *17*(2), 67–74. PMID: 12500729

Figge, L., Oebels, K., & Offermans, A. (2017). The effects of globalization on ecological footprints: An empirical analysis. *Environment, Development and Sustainability*, *19*, 863–876. 10.1007/s10668-016-9769-8

Financial Times. (2025, December 30). *The World Ahead 2025*. https://www.ft.com/content/deb33eae-37bc-4d50-98f6-716930438dca

Fineman, M. A. (2017). Vulnerability and inevitable inequality. *Oslo Law Review*, *4*(3), 133–149.

Finke, M. S., Howe, J. S., & Huston, S. J. (2023). Old age and the decline in financial literacy. *Management Science*, *63*(1), 213–230.

Fitriadi, F., Jiuhardi, J., Busari, A., Ulfah, Y., Hakim, Y. P., Erwin Kurniawan, A., & Darma, D. C. (2022). Using Correlation to Explore the Impact of Corona Virus Disease on Socioeconomics. *Emerging Science Journal, 6*(Special Issue COVID-19 Emerging Research), 165–180. https://doi.org/DOI: 10.28991/esj-2022-SPER-012

Fordham, M. H. (1998). Making women visible in disasters: Problematising the private domain. *Disasters*, *22*(2), 126–143. PMID: 9654812

Fraisse, D., Bred, A., Lagarde, A., & Felgines, C. (2025). Impact of simulated gastrointestinal conditions on polyphenol stability and antiglycoxidant potential of sage (*Salvia officinalis*) infusion. *South African Journal of Botany*, *176*, 29–36. DOI: 10.1016/j.sajb.2024.11.019

Frieswijk, K., Zino, L., Morse, A. S., & Cao, M. (2025). A behavioural–environmental model to study the impact of climate change denial on environmental degradation. *Physica D: Nonlinear Phenomena, 476*(June 2024), 134648. https://doi.org/DOI: 10.1016/j.physd.2025.134648

Frontiers in Sustainability. (2025). Financial literacy and environmental sustainability: a cross-country test. Frontiers in Sustainability. https://www.frontiersin.org/journals/sustainability/articles/10.3389/frsus.2025.1514393/full

Furedi, F. (2007). The changing meaning of disaster. *Area*, *39*(4), 482–489.

Gamalero, E., Bona, E., Todeschini, V., & Lingua, G. (2020). Saline and arid soils: Impact on bacteria, plants, and their interaction. *Biology (Basel)*, *9*(6), 116. DOI: 10.3390/biology9060116

Ganda, F., & Ruza, C. (2025). Unveiling the non-linear dynamics: Quantile regression of financial development's impact on environmental degradation in BRICS and G7 nations. *Discover Sustainability*, *6*(60), 1–26. DOI: 10.1007/s43621-025-00792-y

Gautier, C. (2014). *Oil, water and climate*. Cambridge University Press.

Genç, F. N. (2021). *Afet Yönetimi*. Nobel Akademik Yayıncılık.

Georgescu-Roegen, N. (1971). *The entropy law and the economic process*. Harvard University Press.

Gereffi, G. (1996). Global Commodity Chains: New Forms of Coordination and Control Among Nations and Firms in International Industries. *Competition & Change*, *1*, 4.

Gerschenkron, A. (1962). *Economic Backwardness in Historical Perspective*. The Bellknap Press.

Ghaffarian, S., Taghikhah, F. R., & Maier, H. R. (2023). Explainable artificial intelligence in disaster risk management: Achievements and prospective futures. *International Journal of Disaster Risk Reduction*, *98*, 1–22.

Ghallab, D. S., Awwad, Z. M., Ibrahim, R. S., & Shawky, E. (2025). Dietary relevance of sage (*Salvia officinalis* L.) in cognitive health: A serum-based mechanistic study targeting Alzheimer's disease pathways. *Food Bioscience*, *69*, 106863. DOI: 10.1016/j.fbio.2025.106863

Ghatak, S. (2010). *Introduction to development economics*.

Ghorbani, A., & Esmaeilizadeh, M. (2017). Pharmacological properties of *Salvia officinalis* and its components. *Journal of Traditional and Complementary Medicine*, *7*(4), 433–440. DOI: 10.1016/j.jtcme.2016.12.014

Gibbons, R. D., Olfson, M., Saulsberry, L., Edlund, M. J., Zangeneh, S., Bareis, N., Chwastiak, L., Gibbons, J. B., & Kessler, R. C. (2024). Social vulnerability and prevalence and treatment for mental health and substance use disorders. *JAMA Psychiatry*, *81*(10), 976–984. PMID: 39046728

Gibson, A., Kusmaul, N., & Brown, L. (2020). Improving communication for older adults during disaster events. *Innovation in Aging*, *4*(Supplement_1), 693–693.

Gifford, L., & Sauls, L. A. (2024). Defining climate finance justice: Critical geographies of justice amid financialized climate action. *Geography Compass*, *18*(11), e70008.

Gıntz, C. (2010). *Elsewhere & Otherwise*. Dost.

Global Carbon Budget. (2021). https://www.globalcarbonproject.org/carbonbudget/

González-Álvarez, M. A., & Montañés, A. (2023). CO_2 emissions, energy consumption, and economic growth: Determining the stability of the 3E relationship. *Economic Modelling*, *121*(1), 1–10. DOI: 10.1016/j.econmod.2023.106195

Goodfellow, I., Bengio, Y., & Courville, A. (2016). *Deep learning* (2nd ed., Vol. 1). MIT press.

Gopalakrishnan, C. (2013). Water and disasters: A review and analysis of policy aspects. *International Journal of Water Resources Development*, *29*(2), 250–271. DOI: 10.1080/07900627.2012.756133

Grabara, J., Tleppayev, A., Dabylova, M., Mihardjo, L. W. W., & Dacko-Pikiewicz, Z. (2021). Empirical research on the relationship amongst renewable energy consumption, economic growth and foreign direct investment in Kazakhstan and Uzbekistan. *Energies*, *14*(2), 332. DOI: 10.3390/en14020332

Granger, C. W. J. (1969). Investigating causal relations by econometric models and cross spectral methods. *Econometrica*, *37*(3).

Granger, C. W. J. (1969). Investigating causal relations by econometric models and cross-spectral methods. *Econometrica*, *37*(3), 424. DOI: 10.2307/1912791

Graziano, M., & Gillingham, K. (2024). Spatial patterns of solar photovoltaic system adoption: The influence of neighbors and the built environment. *Journal of Economic Geography*, *15*(4), 815–839.

Grigorieva, E., Livenets, A., & Stelmakh, E. (2023). Adaptation of agriculture to climate change: A scoping review. *Climate (Basel)*, *11*(10), 202. DOI: 10.3390/cli11100202

Grolleau, G., & Weber, C. (2024). The effect of inflation on CO2 emissions: An analysis over the period 1970-2020. *Ecological Economics*, *217*(1), 1–10. DOI: 10.1016/j.ecolecon.2023.108029

Grunewald, N., Klasen, S., Martínez-Zarzoso, I., & Muris, C. (2017). The trade-off between income inequality and carbon dioxide emissions. *Ecological Economics*, *142*, 249–256. DOI: 10.1016/j.ecolecon.2017.06.034

Gudi, S., Gill, H. S., Collins, S., Singh, J., Sandhu, D., Sehgal, S. K., Gill, U., & Gupta, R. (2025). Association analysis identified superior haplotypes for improved salt stress tolerance in wheat (*Triticum aestivum* L.). *Plant Stress*, *16*, 100900. DOI: 10.1016/j.stress.2025.100900

Gu, J. (2020). A Novel Credit Risk Assessment Model Based on LightGBM. *Journal of Simulation*, *8*(3), 71–73.

Guja, M. M., & Bedeke, S. B. (2025). Smallholders' climate change adaptation strategies: Exploring effectiveness and opportunities to be capitalized. *Environment, Development and Sustainability*, *27*(8), 17927–17956.

Gu, L., Rosenberg, M. W., & Zeng, J. (2017). Competing forces of socioeconomic development and environmental degradation on health and happiness for different income groups in China. *International Journal of Health Services*, *47*(4), 752–777. DOI: 10.1177/0020731417725470 PMID: 28847235

Gümüş, D., Yılmaz, Z., Tuncer, M., & Aydın, E. (2024). Afet yönetiminde hemşirelerin rolü. *Afet ve Risk Dergisi*, *7*(2), 354–367.

Gunnarsson, D., Hyer, R., Hufstader, R. D., Hidayat, A., Peake, J., & Banerjee, A. (2005). Civil-military cooperation in humanitarian health action. *Prehospital and Disaster Medicine*, *20*(6), 450–454. DOI: 10.1017/S1049023X00003101 PMID: 16496635

Gupta, A., & Shukla, A. K. (2024). Optimal approaches in global warming mitigation and adaptation strategies at city scale. *Discover Sustainability*, *5*(1), 272.

Gurikar, S. G. (2024). Chaos theory and its implications in physical systems. *ShodKosh: Journal of Visual and Performing Arts*, *5*(1), 1723–1728.

Hacking, I. (2012). Scoping studies: Advancing the methodology. *Representing and Intervening*, *5*(69), 1–18. DOI: 10.1017/cbo9780511814563.003

Hadar, L. L., & Ergas, O. (2025). Mindfulness for in-service and preservice teachers: An empirical map of the discourse from 2000 to 2020. *European Journal of Teacher Education*, *48*(2), 208–227.

Hailemariam, A., Dzhumashev, R., & Shahbaz, M. (2020). Carbon emissions, income inequality and economic development. *Empirical Economics*, *59*(3), 1139–1159.

Haj-Amor, Z., Araya, T., Kim, D.-G., Bouri, S., Lee, J., Ghiloufi, W., Yang, Y., Kang, H., Jhariya, M. K., Banerjee, A., & Lal, R. (2022). Soil salinity and its associated effects on soil microorganisms, greenhouse gas emissions, crop yield, biodiversity and desertification: A review. *The Science of the Total Environment*, *843*, 156946. DOI: 10.1016/j.scitotenv.2022.156946

Halicioglu, F. (2009). An econometric study of CO2 emissions, energy consumption, income and foreign trade in Turkey. *Energy Policy*, *37*(3), 1156–1164. DOI: 10.1016/j.enpol.2008.11.012

Hamed, L., Dhaouadi, L., Zehri, F., Tiba, S., Besser, H., & Karbout, N., & EMARA, E. I. R. (2024). Examining the relationship between the economic growth, energy use, CO2 emissions, and water resources: Evidence from selected MENA countries. *Journal of the Saudi Society of Agricultural Sciences*, *23*(6), 415–423. DOI: 10.1016/j.jssas.2024.04.002

Handa, S., Natali, L., Seidenfeld, D., Tembo, G., & Davis, B.Zambia Cash Transfer Evaluation Study Team. (2018). Can unconditional cash transfers raise long-term living standards? Evidence from Zambia. *Journal of Development Economics, 133*, 42–65. DOI: 10.1016/j.jdeveco.2018.01.008 PMID: 31396000

Hanif, I. (2018). Impact of economic growth, nonrenewable and renewable energy consumption, and urbanization on carbon emissions in Sub-Saharan Africa. *Environmental Science and Pollution Research International, 25*(15), 15057–15067. DOI: 10.1007/s11356-018-1753-4 PMID: 29552722

Haq, M. u. (1996). *Reflections on Human Development.* Oxford University Press.

Haq, S. M. A., Chowdhury, M. A. F., Ahmed, K. J., & Chowdhury, M. T. A. (2023). Environmental quality and its impact on total fertility rate: An econometric analysis from a new perspective. *BMC Public Health, 23*(1), 2397. DOI: 10.1186/s12889-023-17305-z PMID: 38042784

Haris, N., Furqan, A. C., Kahar, A., & Karim, F. (2023). Disaster risk index on disaster management budgeting: Indonesia's national data set. *Jàambá, 15*(1), 1–7. PMID: 36873605

Harrigan, J., & Wang, C. (2011). A new approach to the allocation of aid among developing countries: Is the USA different from the rest? *World Development, 39*(8), 1281–1293. DOI: 10.1016/j.worlddev.2010.12.011

Hartama, D., Mawengkang, H., Zarlis, M., Sembiring, R. W., Nasution, B. B., Syahruddin, M., ... Sumarno. (2017). *The planning of smart city to mitigate the impacts of natural disaster in north sumatera* (Vol. 501, pp. 147–154; V. D., G. J.J., M. Y., & Z. P., Eds.). Vol. 501, pp. 147–154. STIKOM Tunas Bangsa Pematangsiantar, Pematangsiantar, Indonesia: Springer New York LLC. https://doi.org/DOI: 10.1007/978-3-319-68486-4_12

Hartawan, D. R., Purboyo, T. W., & Setianingsih, C. (2019). *Disaster victims detection system using convolutional neural network (CNN) method.* 105–111. Computer Engineering Study, School of Electrical Engineering, Telkom University, Indonesia: Institute of Electrical and Electronics Engineers Inc. https://doi.org/DOI: 10.1109/ICIAICT.2019.8784782

Harvey, D. (2015). *Seventeen contradictions and the end of capitalism.* Sel.

Hasan, M. N. (2025). Unequal burdens and shared fates: Rethinking climate justice through reparative action and global solidarity. *Environmental Sociology*, •••, 1–9.

Hashemi, S. K., Mirtaheri, S. L., & Greco, S. (2023). Fraud Detection in Banking Data by Machine Learning Techniques. *IEEE Access : Practical Innovations, Open Solutions, 11*, 3034–3043. DOI: 10.1109/ACCESS.2022.3232287

Hashim, B. M., Alnaemi, A. N. A., Hussain, B. A., Abduljabbar, S. A., Doost, Z. H., & Yaseen, Z. M. (2024). Statistical downscaling of future temperature and precipitation projections in Iraq under climate change scenarios. *Physics and Chemistry of the Earth Parts A/B/C, 135*, 103647. https://doi.org/https://doi.org/10.1016/j.pce.2024.103647

Hastings, J. S., Madrian, B. C., & Skimmyhorn, W. L. (2024). Financial literacy, financial education, and economic outcomes. *Annual Review of Economics, 5*, 347–373. PMID: 23991248

Hay, E. (2025). Towards an indigenous and traditional "pedagogy of repair": An ecological approach to environmental education. *South African Journal of Philosophy, 44*(1), 17–29.

Hayward, M., Hunt, R. & Miller, D. (2022). *How Vulnerability Enriches Family Firm Relationships: A Social Exchange Perspective, 13(1),* 1-7.

He, , QMeng, , QFlatley, , WHe, , Y. (2022). Examining the effects of agricultural aid on forests in sub-Saharan Africa: A causal analysis based on remotely sensed data of Sierra Leone. *Land, 11*(5), 668. https://doi.org/10.3390/land11050668

Healy, A., & Malhotra, N. (2009). Myopic voters and natural disaster policy. *The American Political Science Review, 103*(3), 387–406. DOI: 10.1017/S0003055409990104

Heinberg, J. G. (1926). History of the Majority Principle. *The American Political Science Review, 20*(1), 52–68. DOI: 10.2307/1945098

Heo, B.-Y., Park, J. H., & Heo, W.-H. (2018). Sustainable disaster and safety management of government: Integrated disaster and safety budget system in Korea. *Sustainability, 10*(11), 1–17.

Hermansson, H. (2019). Challenges to decentralization of disaster management in turkey: The role of political-administrative context. *International Journal of Public Administration, 42*(5), 417–431.

Hermansson, H. (2019). Challenges to Decentralization of Disaster Management in Turkey: The Role of Political-Administrative Context. *International Journal of Public Administration, 42*(5), 417–431. DOI: 10.1080/01900692.2018.1466898

Herrera, H. (2024). Embedding Municipal Green Bonds in Mexico City's hydrosocial cycle:'Green'debt and climate action narratives. *Journal of political ecology: case studies in history and society.--, 31*(1), 1-18.

Herzer, D., & Nunnenkamp, P. (2012). The effect of foreign aid on income inequality: Evidence from panel cointegration. *Structural Change and Economic Dynamics, 23*(3), 245–255. DOI: 10.1016/j.strueco.2012.04.002

Hickel, J. (2017). Is global inequality getting better or worse? A critique of the World Bank's convergence narrative. *Third World Quarterly, 38*(10), 2208–2222. DOI: 10.1080/01436597.2017.1333414

Hidayati, D. (2012). Striving to reduce disaster risk: Vulnerable communities with low levels of preparedness in Indonesia. *Journal of Disaster Research, 7*(1), 75–82. DOI: 10.20965/jdr.2012.p0075

Hjelmar, U., Bhatti, Y., Petersen, O. H., Rostgaard, T., & Vrangbæk, K. (2018). Public/private ownership and quality of care: Evidence from Danish nursing homes. *Social Science & Medicine, 216*, 41–49. DOI: 10.1016/j.socscimed.2018.09.029 PMID: 30261324

Hledik, R., & Greenberg, J. (2024). The distributional impacts of residential demand charges. *The Electricity Journal, 29*(6), 33–41.

Ho, H., & Tien, K.-M. (Terence), Wu, A., & Singh, S. (2021). A sequence analysis approach to segmenting credit card customers. *Journal of Retailing and Consumer Services, 59*, 102391. DOI: 10.1016/j.jretconser.2020.102391

Holland, D., & Young, G. (2020). The economic implications of climate change mitigation policies. *National Institute Economic Review, 251*, R1–R2.

Hondroyiannis, G., Papapetrou, E., & Tsalaporta, P. (2025). Environmental degradation and inflation in high-income countries. *Economic Change and Restructuring, 58*(1), 1–28. DOI: 10.1007/s10644-024-09840-5

Hongjie, D., & Muzi, F. (2025). How supply chain disruptions, renewable energy consumption, and eco-innovation mitigate environmental degradation? a path towards sustainable development in France. *Energy Economics, 145*(1), 1–5. DOI: 10.1016/j.eneco.2025.108346

Hossain, S. (2011). Panel estimation for CO2 emissions, energy consumption, economic growth, trade openness and urbanization of newly industrialized countries. *Energy Policy, 39*(11), 6991–6999. DOI: 10.1016/j.enpol.2011.07.042

Hosseini, S. M., Saifoddin, A., Shirmohammadi, R., & Aslani, A. (2019). Forecasting of CO2 emissions in Iran based on time series and regression analysis. *Energy Reports*, *5*, 619–631. DOI: 10.1016/j.egyr.2019.05.004

Hossen, M. A., Benson, D., & Islam, M. (2025). Donor climate coloniality, ecological debt and vulnerabilities in coastal Bangladesh: Redesigning institutions for recognition justice. *Environmental Management*, •••, 1–15. PMID: 40159350

Hou, C., & Wu, H. (2020). Rescuer, decision maker, and breadwinner: Women's predominant leadership across the post-Wenchuan earthquake efforts in rural areas, Sichuan, China. *Safety Science*, *125*, 1–6.

Houée, P. (2001). *Le développement local au défi de la mondialisation*. L'Harmattan.

Hou, L., Bi, G., & Guo, Q. (2025). An improved sparrow search algorithm optimized LightGBM approach for credit risk prediction of SMEs in supply chain finance. *Journal of Computational and Applied Mathematics*, *454*, 116197. DOI: 10.1016/j.cam.2024.116197

Howard, A., Blakemore, T., & Bevis, M. (2017). Older people as assets in disaster preparedness, response and recovery: Lessons from regional Australia. *Ageing and Society*, *37*(3), 517–536.

Huang, W. M., Lee, G. W. M., & Wu, C. C. (2008). GHG emissions, GDP growth and the Kyoto Protocol: A revisit of Environmental Kuznets Curve hypothesis. *Energy Policy*, *36*(1), 239–247. DOI: 10.1016/j.enpol.2007.08.035

Huseynov, A., & Gezdim, S. B. (2024). The Impact of Renewable and Non-Renewable Energy Consumption on CO Emissions in Türkiye: Evidence from Augmented ARDL Approach. *Journal of Sustainable Development Issues*, 2(2), 89–103. DOI: 10.62433/josdi.v2i2.28

Hu, Y. (2013). *Rural health care delivery*. Springer.

IEA. (2025), Global Energy Review 2025, IEA, Paris https://www.iea.org/reports/global-energy-review-2025

Investment Funds. (2025). *Investment funds report 2025: Transitioning to low-carbon economic development*. Global Climate Finance Initiative.

IPBES. (2019). Global assessment report on biodiversity and ecosystem services. https://www.ipbes.net/global-assessment

Iqbal, U., Perez, P., & Barthelemy, J. (2021). A process-driven and need-oriented framework for review of technological contributions to disaster management. *Heliyon*, *7*(11), 1–19. PMID: 34841111

Irmansyah, S. J. (2005). Mental health problems after tsunami disaster in Aceh and Nias; immediate responses and future programs. *Seishin Shinkeigaku Zasshi. Psychiatria et Neurologia Japonica, 107*(11), 1178–1183. PMID: 16408426

Işık, E. (2024). Afet ve kadın: Toplumsal cinsiyete dayalı zarar görebilirlik, afet yönetimi, ruh sağlığı ve psikososyal destek hizmetleri. *Akdeniz Kadın Çalışmaları ve Toplumsal Cinsiyet Dergisi, 7*(2), 882–909.

Islam, H. (2025). Nexus of economic, social, and environmental factors on sustainable development goals: The moderating role of technological advancement and green innovation. *Innovation and Green Development, 4*(1), 100183. 10.1016/j.igd.2024.100183

Islam, M. A., Lobry de Bruyn, L., Warwick, N. W. M., & Koech, R. (2021). Salinity-affected threshold yield loss: A signal of adaptation tipping points for salinity management of dry season rice cultivation in the coastal areas of Bangladesh. *Journal of Environmental Management, 288*, 112413. DOI: 10.1016/j.jenvman.2021.112413

Islam, M. S. (2024). Rethinking Climate Justice: Insights from Environmental Sociology. *Climate (Basel), 12*(12), 203.

Izumi, T., Shaw, R., Djalante, R., Ishiwatari, M., & Komino, T. (2019). Disaster risk reduction and innovations. *Progress in Disaster Science, 2*, 100033. DOI: 10.1016/j.pdisas.2019.100033

Jadhav, P. A., Lalwani, U., Gour, A., Shayan, M., & Motwani, S. (2022). Identification of Fraudulent Credit Card transactions using Machine Learning Algorithms. *2022 IEEE 7th International Conference for Convergence in Technology (I2CT)*, 1–4. https://doi.org/DOI: 10.1109/I2CT54291.2022.9824165

Jaforullah, M., & King, A. (2015). Does the use of renewable energy sources mitigate CO 2 emissions? A reassessment of the US evidence. *Energy Economics, 49*, 711–717. DOI: 10.1016/j.eneco.2015.04.006

Jakovljević, M., Jokić, S., Molnar, M., Jašić, M., Babić, J., Jukić, H., & Banjari, I. (2019). Bioactive profile of various *Salvia officinalis* L. preparations. *Plants, 8*(3), 55. DOI: 10.3390/plants8030055

Jatav, S. S., Nathoo, B., Mubeena, Md., Arakhita, B., Pooja, R., & Rukmani, D. (2023). Development of agricultural sustainability index for different regions of Uttar Pradesh, India. *Environment and Ecology, 41*(4C), 2781–2789. https://www.environmentandecology.com/volume-41-2023

Jayachandran, , SDe Laat, , JLambin, , E. FStanton, , C. YAudy, , RThomas, , N. E. (2017). Cash for carbon: A randomized trial of payments for ecosystem services to reduce deforestation. *Science*, *357*(6348), 267-273. https://doi.org/10.1126/science.aan0568

Jiang, J., Ye, B., & Liu, J. (2018). Research on the peak of CO2 emissions in the developing world: Current progress and future prospect. *Applied Energy*, *235*, 186–203. DOI: 10.1016/j.apenergy.2018.10.089

Jia, Y., Wu, J., Cheng, M., & Xia, X. (2023). Global transfer of salinization on irrigated land: Complex network and endogenous structure. *Journal of Environmental Management*, *336*, 117592. DOI: 10.1016/j.jenvman.2023.117592

JİCA. (2004). *Türkiye'de doğal afetler konulu ülke strateji raporu* (ss. 1-163). Türkiye Cumhuriyeti İçişleri Bakanlığı.

Jio, W., & Kant, S. (2025). Strategic Leadership Effect on Med-Tech Organization Citizenship Behavior With Mediation of Innovation Culture in the Horn of Africa. In *Strategic Management, Regulatory Challenges, and Global Governance of MedTech* (pp. 219-252). IGI Global Scientific Publishing.

Jio, W., & Kant, S. (2025). Market Mobility Effect on Gig Economy Adaptability by Mediation of Talent Management in the Horn of Africa. In *Sustainability and Adaptability of Gig Economies in Global Business* (pp. 205–230). IGI Global Scientific Publishing.

Johnson, D., Brune, S., Dagan, D. T., Meier, E., Wilkins, E. J., & Zhang, H. (2020, January). A holistic strategy for carbon reduction programs in parks and protected areas. In *Parks Stewardship Forum* (Vol. 36, No. 3, pp. 477-482). George Wright Society.

Johnson, A. R., & Martinez, S. P. (2024). Defining green financial literacy: A conceptual framework for the renewable energy era. *Energy Policy*, *156*, 112467.

Jolliffe, I. T., & Stephenson, D. B. (2011). Introduction. In *Forecast Verification* (pp. 1–9). Wiley., DOI: 10.1002/9781119960003.ch1

Jong, A. (2022). World risk society and constructing cosmopolitan realities: A bourdieusian critique of risk society. *Frontiers*, *7*, 1–13. PMID: 35573124

Jordan, M. I., & Mitchell, T. M. (2015). Machine learning: Trends, perspectives, and prospects. *Science*, *349*(6245), 255–260. DOI: 10.1126/science.aaa8415 PMID: 26185243

Jorgenson, A. K., & Clark, B. (2010). Assessing the temporal stability of the population/environment relationship in comparative perspective: A cross-national panel study of carbon dioxide emissions, 1960–2005. *Population and Environment, 32*(1), 27–41. DOI: 10.1007/s11111-010-0117-x

Juth, V., Silver, R. C., Seyle, D. C., Widyatmoko, C. S., & Tan, E. T. (2015). Post-disaster Mental Health Among Parent–Child Dyads After a Major Earthquake in Indonesia. *Journal of Abnormal Child Psychology, 43*(7), 1309–1318. DOI: 10.1007/s10802-015-0009-8 PMID: 25851238

Kaika, M., Calvário, R., & Velegrakis, G. (2024). Austerity: An environmentally dangerous idea. *Journal of Political Ecology, 31*(1), 67–81.

Kais, S., & Sami, H. (2016). An econometric study of the impact of economic growth and energy use on carbon emissions: Panel data evidence from fifty eight countries. *Renewable & Sustainable Energy Reviews, 59*, 1101–1110. DOI: 10.1016/j.rser.2016.01.054

Kaklauskaite, U., & Streimikiene, D. (2024). Climate justice from the lens of sustainable development. *Sustainable Development, 32*(6), 6788–6800.

Kalfin, S. Supian, S., & Mamat, M. (2021). *Mitigation and models for determining premiums for natural disaster insurance due to excessive rainfall* (R. B.N., R. J., P. A., F. A.N., A. M., P. R.C.I., … S. K.A., Eds.). Mathematics, Faculty of Mathematics and Natural Sciences, Universitas Padjadjaran, Indonesia: IOP Publishing Ltd. https://doi.org/DOI: 10.1088/1742-6596/1722/1/012058

Kalfin, S. Supian, S., Mamat, M., & Bon, A. T. (2020a). *A review mitigation efforts and natural disaster insurance schemes in indonesia. 59*, 1825–1835. Faculty of Mathematics and Natural Sciences, Universitas Padjadjaran, Indonesia: IEOM Society.

Kalfin, S. Supian, S., Mamat, M., & Bon, A. T. (2020b). *Model for determining natural disaster insurance premiums in indonesia using the black scholes method. 0*, 2376–2385. Faculty of Mathematics and Natural Sciences, Universitas Padjadjaran, Indonesia: IEOM Society.

Kalfin, S. Supian, S., Mamat, M., Muljana, F., & Bon, A. T. (2021). *Analysis of agropolitan area planning based on natural disaster mitigation in west muna regency, southeast Sulawesi, Indonesia.* 3592–3601. Faculty of Mathematics and Natural Sciences, Universitas Padjadjaran, Indonesia: IEOM Society.

Kalfin, S., Supian, S., & Mamat, M. (2022a). Insurance Premium Determination Model and Innovation for Economic Recovery Due to Natural Disasters in Indonesia. *Computation (Basel, Switzerland), 10*(10). Advance online publication. DOI: 10.3390/computation10100174

Kalfin, S., Supian, S., & Mamat, M. (2022b). Insurance premium determination model as natural disaster mitigation effort in indonesia with a cross subsidy system. *International Journal of Agricultural and Statistics Sciences, 18*(2), 539–546.

Kallis, G., Hickel, J., O'Neill, D. W., Jackson, T., Victor, P. A., Raworth, K., ... (2025). Post-growth: The science of wellbeing within planetary boundaries. *The Lancet. Planetary Health, 9*(1), e62–e78. 10.1016/S2542-5196(24)00310-3

Kallis, G., Hickel, J., O'Neill, D. W., Jackson, T., Victor, P. A., Raworth, K., & Ürge-Vorsatz, D. (2025). Post-growth: The science of wellbeing within planetary boundaries. *The Lancet. Planetary Health, 9*(1), e62–e78. PMID: 39855235

Kang, H. (2022). Impacts of income inequality and economic growth on CO2 emissions: Comparing the Gini coefficient and the top income share in OECD countries. *Energies, 15*(19), 6954. DOI: 10.3390/en15196954

Kant, S., & Jio, W. (2025). Risk Management Effect on Product Development: Mediating Role of Regulatory Barriers in the Horn of Africa. In R. Marcão, V. Santos, & S. Monteiro (Eds.), *Strategic Management, Regulatory Challenges, and Global Governance of MedTech* (pp. 185-218). IGI Global Scientific Publishing. https://doi.org/DOI: 10.4018/979-8-3373-1205-7.ch008

Kant, S., & Adula, M. (2025). How Knowledge Management Will Affect the Start Up's Performance in Ethiopia With Mediation of Competitive Advantage. In *Leveraging Human Capital in Entrepreneurial Strategies* (pp. 109–146). IGI Global Scientific Publishing.

Kant, S., Toma, F., Niguse, T., & Adula, M. (2025). Intersecting Natural Language Processing for Service Marketing Sustainability with Mediation of FinTech Innovations in Horn of Africa. In *Intersecting Natural Language Processing and FinTech Innovations in Service Marketing* (pp. 275–298). IGI Global Scientific Publishing.

Kant, S., Toma, F., Niguse, T., & Adula, M. (2025). Transnational Higher Education Landscape Effect on Building a Global Vision for Staff Development by Navigating Cultural Diversity in Africa. In *Training and Development in Transnational Higher Education* (pp. 1–20). IGI Global Scientific Publishing.

Kaplan, A., & Haenlein, M. (2019). Siri, Siri, in my hand: Who's the fairest in the land? On the interpretations, illustrations, and implications of artificial intelligence. *Business Horizons, 62*(1), 15–25. DOI: 10.1016/j.bushor.2018.08.004

Kapucu, N., & Garayev, V. (2011). Collaborative Decision-Making in emergency and disaster Management. *International Journal of Public Administration, 34*(6), 366–375. DOI: 10.1080/01900692.2011.561477

Karabulut, D., & Bekler, T. (2019). Doğal afetlerin çocuklar ve ergenler üzerindeki etkileri. *Doğal Afetler ve Çevre Dergisi, 5*(2), 368–376.

Karagel, H. (2019). Afet anı ve sonrası (afet kriz yönetimi). In Sever, R. (Ed.), *Afetler ve afet yönetimi* (pp. 160–183). Pegem.

Karakaya, E. (2016). Paris İklim Anlaşması: İçeriği Ve Türkiye Üzerine Bir Değerlendirme. *Adnan Menderes Üniversitesi Sosyal Bilimler Enstitüsü Dergisi, 3*(1), 1–12. DOI: 10.30803/adusobed.188842

Karalija, E., Dahija, S., Tarkowski, P., & Ćavar Zeljković, S. (2022). Influence of climate-related environmental stresses on economically important essential oils of Mediterranean *Salvia* sp. *Frontiers in Plant Science, 13*, 864807. DOI: 10.3389/fpls.2022.864807

Karaman, N. A., & Alp, İ. (2024). Analysis of the Environmental Performances of EU Countries and Türkiye. *Ankara Üniversitesi Çevrebilimleri Dergisi, 11*(2), 1–11. DOI: 10.62163/aucevrebilim.1485058

Karatani, K. (2018). *Isonomia and the origins of philosophy*. Metis.

Karatayev, M., & Clarke, M. L. (2014). Current energy resources in Kazakhstan and the future potential of renewables: A review. *Energy Procedia, 59*, 97–104. DOI: 10.1016/j.egypro.2014.10.354

Kasuga, H., & Morita, Y. (2018). Does aid affect inequality? *Applied Economics, 50*(58), 6249–6262. DOI: 10.1080/00036846.2018.1489512

Kaul, I., Grunberg, I., & Stern, M. (1999). Global public goods. *Global public goods, 450*.

Kaur, A., & Sharma, V. (2024). Nexus between climate change and mitigation approaches for sustainable development: A bibliometric review. *Global Knowledge, Memory and Communication*.

Kemaloğlu, M. (2015). Türkiye'de afet yönetiminin tarihi ve yasal gelişimi. *Akademik Bakış Dergisi, 52*, 126–147.

Kernick, D. (2018). *Complexity and healthcare organization- a view from the street.* CRC Street.

Khan, H., Vasilescu, L. G., & Khan, A. (2008). Disaster management cycle-a theoretical approach. *Management & Marketing*, *6*(1), 43–50.

Khan, H., Weili, L., Khan, I., & Han, L. (2022). The effect of income inequality and energy consumption on environmental degradation: The role of institutions and financial development in 180 countries of the world. *Environmental Science and Pollution Research International*, *29*(14), 20632–20649. PMID: 34741747

Khan, M. O., Irfan, M., Muhammad, A., Ullah, I., Nawaz, S., Khalil, M. K., & Ahmad, M. (2022). A practical and economical strategy to mitigate salinity stress through seed priming. *Frontiers in Environmental Science*, *10*, 991977. DOI: 10.3389/fenvs.2022.991977

Khan, N., Ma, J., Zhang, H., & Zhang, S. (2023). Climate change impact on sustainable agricultural growth: Insights from rural areas. *Atmosphere*, *14*(8), 1194.

Kim, , J. E. (2018). Technological capacity building through energy aid: Empirical evidence from renewable energy sector. *Energy Policy*, *122*, 449-458. https://doi.org/10.1016/j.enpol.2018.07.003

Kıral, Ö. (2001). 1939 ve 1992 Erzincan depremleri şehircilik düzenimizde hasar yarattı mı? *Planlama*, *4*, 11–18.

Kirschner, C. L., Singla, A., & Flick, A. (2018). Financing community resilience before disaster strikes: lessons from the United States. In Kirschner, C. L., Singla, A., & Flick, A. (Eds.), *Oxford Research Encyclopedia of Natural Hazard Science*. Oxford University Press.

Klapper, L., & Lusardi, A. (2023). Financial literacy and financial resilience: Evidence from around the world. *Financial Management*, *99*(1), 51–76.

Kluza, K., Ziolo, M., & Postula, M. (2024). Climate policy development and implementation from the Sustainable Development Goals perspective. Evidence from the European Union countries. *Energy Strategy Reviews, 52*, 101321. https://doi.org/https://doi.org/10.1016/j.esr.2024.101321

Kobayashi, Y. H. (2024). The significance of local government in disaster management for international migrants: The case of Minoh city, Osaka prefecture. *Disasters*, *48*(S1), e12636. PMID: 38864590

Kodimenu, M. K., S, S., & Katoon, T. (2024). Credit Card Fraud Detection Using ML & DL. *International Journal Of Scientific Research In Engineering And Management, 08*(07), 1–11. DOI: 10.55041/IJSREM36686

Kompoura, V., Karapantzou, I., Mitropoulou, G., Parisis, N. A., Gkalpinos, V. K., Anagnostou, V. A., Tsiailanis, A. D., Vasdekis, E. P., Koutsaliaris, I. K., Tsouka, A. N., Karapetsi, L., Madesis, P., Letsiou, S., Florou, D., Koukkou, A.-I., Barbouti, A., Tselepis, A. D., Kourkoutas, Y., & Tzakos, A. G. (2024). Exploiting the beneficial effects of *Salvia officinalis* L. extracts in human health and assessing their activity as potent functional regulators of food microbiota. *Food Chemistry, 441*, 138175. DOI: 10.1016/j.foodchem.2023.138175

Kono, , D. YMontinola, , G. R. (2019). Foreign aid and climate change policy: what can('t) the data tell us? *Politics and Governance, 7*(2), 68-92. https://doi.org/10.17645/pag.v7i2.1840

Korkin, E., & Öztop, S. (2022). Büyükşehir belediyeleri stratejik planlarında afet yönetimi. *Medeniyet Araştırmaları Dergisi, 7*(1), 61–75.

Korotayev, A., Zinkina, J., Bogevolnov, J., & Malkov, A. (2011). Global unconditional convergence among larger economies after 1998? *Journal of Globalization Studies, 2*(2), 25–62.

Kou, G., Chao, X., Peng, Y., Alsaadi, F. E., & Herrera-Viedma, E. (2019). Machine Learning Methods For Systemic Risk Analysis In Financial Sectors. *Technological and Economic Development of Economy, 25*(5), 716–742. DOI: 10.3846/tede.2019.8740

Krogstrup, S., & Oman, W. (2019). Macroeconomic and financial policies for climate change mitigation: A review of the literature. Academic Press.

Kroll, J. A. (2015). *Accountable Algorithms*. Princeton University.

Krug, S. E., Chung, S., Fagbuyi, D. B., Fisher, M. C., Schonfeld, D. J., Shook, J. E., Chun, T. H., Conners, G. P., Mba, M. M., Conway, E. E., Dudley, N. C., Fuchs, S. M., Lane, N. E., Macias, C. G., Mph, M., & Moore, B. R. (2015). Ensuring the health of children in disasters. *Pediatrics, 136*(5), e1407–e1417. PMID: 26482663

Kuhn, T. (1970). The structure of scientific revolutions. 2nd, enlarged ed. University of Chicago Press.

Kumar, P., & Datta, S. K. (2024). Environmental degradation and economic growth: A longitudinal analysis across different income group countries. *The Indian Economic Journal, 72*(1), 72–89. DOI: 10.1177/00194662231199238

Kumar, S., & R, K. (2024). Comprehensive Review on Leveraging AI and Machine Learning in Digital Banking. *International Journal of Innovative Research in Computer and Communication Engineering, 12*(05), 6333–6336. DOI: 10.15680/IJIRCCE.2024.1205197

Kurniawan, D., Sutan, A. J., Nurmandi, A., Loilatu, M. J., & Salahudin. (2021). *Social Media as Tools of Disaster Mitigation, Studies on Natural Disasters in Indonesia BT - HCI International 2021 - Late Breaking Posters* (C. Stephanidis, M. Antona, & S. Ntoa, Eds.). Cham: Springer International Publishing.

Kusumasari, B., & Alam, Q. (2012). Bridging the gaps: The role of local government capability and the management of a natural disaster in Bantul, Indonesia. *Natural Hazards, 60*(2), 761–779. DOI: 10.1007/s11069-011-0016-1

Kusumasari, B., Alam, Q., & Siddiqui, K. (2010). Resource capability for local government in managing disaster. *Disaster Prevention and Management: An International Journal, 19*(4), 438–451. DOI: 10.1108/09653561011070367

Kusumastuti, R. D., Viverita, Husodo, Z. A., Suardi, L., & Danarsari, D. N. (2014). Developing a resilience index towards natural disasters in Indonesia. *International Journal of Disaster Risk Reduction, 10*(PA), 327–340. https://doi.org/DOI: 10.1016/j.ijdrr.2014.10.007

Kuznets, S. (1965). *Modern economic growth*. Yale University Press.

Lago-Olveira, S., Ouhemi, H., Idrissi, O., Moreira, M. T., & González-García, S. (2024). Promoting more sustainable agriculture in the Moroccan drylands by shifting from conventional wheat monoculture to a rotation with chickpea and lentils. *Cleaner Environmental Systems, 12*, 100169. DOI: 10.1016/j.cesys.2024.100169

Lal, R., Bouma, J., Brevik, E., Dawson, L., Field, D. J., Glaser, B., Hatano, R., Hartemink, A. E., Kosaki, T., Lascelles, B., Monger, C., Muggler, C., Ndzana, G. M., Norra, S., Pan, X., Paradelo, R., Reyes-Sánchez, L. B., Sandén, T., Singh, B. R., & Zhang, J. (2021). Soils and sustainable development goals of the United Nations: An International Union of Soil Sciences perspective. *Geoderma Regional, 25*, e00398. DOI: 10.1016/j.geodrs.2021.e00398

Lam, N. S. N., Meyer, M., Reams, M., Yang, S., Lee, K., Zou, L., Mihunov, V., Wang, K., Kirby, R., & Cai, H. (2023). Improving social media use for disaster resilience: Challenges and strategies. *International Journal of Digital Earth, 16*(1), 3023–3044.

Lane, J. (2011). CO2 emissions and GDP. *International Journal of Social Economics, 38*(11), 911–918. DOI: 10.1108/03068291111171414

Lang, M. (2024). Degrowth, global asymmetries, and ecosocial justice: Decolonial perspectives from Latin America. *Review of International Studies, 50*(5), 921–931.

Latouche, S. (2010). *Farewell to growth.* Polity Press.

LeCun, Y., Bengio, Y., & Hinton, G. (2015). Deep learning. *Nature, 521*(7553), 436–444. DOI: 10.1038/nature14539 PMID: 26017442

Lee, , S. KChoi, , GLee, , EJin, , T. (2020). The impact of official development assistance on the economic growth and carbon dioxide mitigation for the recipient countries. *Environmental Science and Pollution Research, 27*(33), 41776-41786. https://doi.org/10.1007/s11356-020-10138-y

Lemi, K., Kant, S., & Jio, W. (2025). Greener Hospital Future in the Horn of Africa by Transactional Leadership: Outcome on Altruism Through Adaptability. In *Generation Alpha and Next Generation Leadership for Greener Futures* (pp. 1-22). IGI Global Scientific Publishing.

Lenka, S., & Tiwari, R. (2025). A Hybrid ML and Data Science Approach to Detect Online Fraud Transaction at Real Time. *Journal of Neonatal Surgery, 14*(1S), 765–775. DOI: 10.52783/jns.v14.1601

Leo, M., Sharma, S., & Maddulety, K. (2019). Machine Learning in Banking Risk Management: A Literature Review. *Risks, 7*(1), 29. DOI: 10.3390/risks7010029

Li, , D. DRishi, , MBae, , J. H. (2021). Green official development Aid and carbon emissions: do institutions matter? *Environment and Development Economics, 26*(1), 88-107. https://doi.org/10.1017/S1355770X20000170

Li, , RGuo, , JWang, , Q. (2024). Evaluating the impact of official development assistance on ecological environments in agricultural and renewable energy sectors. *Humanities and Social Sciences Communications, 11*(1), 1-11. https://doi.org/10.1057/s41599-024-03979-2

Liang, J., & Shi, W. (2021). Cotton/halophytes intercropping decreases salt accumulation and improves soil physicochemical properties and crop productivity in saline-alkali soils under mulched drip irrigation: A three-year field experiment. *Field Crops Research, 262*, 108027. DOI: 10.1016/j.fcr.2020.108027

Lindell, M. K. (2013). Disaster studies. *Current Sociology, 61*(5-6), 797–825.

Li, R., & Su, M. (2017). The role of natural gas and renewable energy in curbing carbon emission: Case study of the United States. *Sustainability, 9*(4), 600. DOI: 10.3390/su9040600

Li, S., Niu, L., Yue, Q., & Zhang, T. (2022). Trajectory, driving forces, and mitigation potential of energy-related greenhouse gas (GHG) emissions in China's primary aluminum industry. *Energy*, *239*, 122114. https://doi.org/https://doi.org/10.1016/j.energy.2021.122114

Li, T. (2022). Energy productivity and financial inclusion: Evidence from non-parametric novel panel methods. *Energy Efficiency*, *15*(7), 51–63. DOI: 10.1007/s12053-022-10057-6 PMID: 36000077

Liu, L., Chen, Z., Al-Hiyari, A., & Nassani, A. (2024). Sustainable growth in mineral rich BRI countries: Linking institutional performance, Fintech, and green finance to environmental impact. *Resources Policy*, *96*, 105159.

Liu, L., & Wang, B. (2021). Protection of halophytes and their uses for cultivation of saline-alkali soil in China. *Biology (Basel)*, *10*(5), 353. DOI: 10.3390/biology10050353

Liu, Q., Wang, S., Zhang, W., Li, J., & Kong, Y. (2018). Examining the effects of income inequality on CO2 emissions: Evidence from non-spatial and spatial perspectives. *Applied Energy*, *236*, 163–171. DOI: 10.1016/j.apenergy.2018.11.082

Long, X., Naminse, E. Y., Du, J., & Zhuang, J. (2015). Nonrenewable energy, renewable energy, carbon dioxide emissions and economic growth in China from 1952 to 2012. *Renewable & Sustainable Energy Reviews*, *52*, 680–688. DOI: 10.1016/j.rser.2015.07.176

Lotfalipour, M. R., Falahi, M. A., & Ashena, M. (2010). Economic growth, CO2 emissions, and fossil fuels consumption in Iran. *Energy*, *35*(12), 5115–5120. DOI: 10.1016/j.energy.2010.08.004

Lozon, M. M., & Bradin, S. (2018). Pediatric disaster preparedness. *Pediatric Clinics of North America*, *65*(6), 1205–1220. PMID: 30446057

Lu, F. (2022). Management of natural disaster and its influence on economic–environmental performance: Fresh evidence from BRICS. *Frontiers in Environmental Science*, *10*, 901017.

Luo, J., Ali, S. A., Aziz, B., Aljarba, A., Akeel, H., & Hanif, I. (2023). Impact of natural resource rents and economic growth on environmental degradation in the context of COP-26: Evidence from low-income, middle-income, and high-income Asian countries. *Resources Policy*, *80*(C), 1–13. DOI: 10.1016/j.resourpol.2022.103269

Lusardi, A., & Mitchell, O. S. (2023). The economic importance of financial literacy: Theory and evidence. *Journal of Economic Literature*, *52*(1), 5–44. PMID: 28579637

Lyman, R. (2016). *Why renewable energy cannot replace fossil fuels by 2050*. https://blog.friendsofscience.org/2016/05/31/why-renewable-energy-cannot-replace-fossil-fuels-by-2050/

Mace, Md, S. E., Doyle, Md, C., Fuchs, Md, S., Gausche-Hill, Md, M., Koenig, Md, K. L., Sorrentino, Md, A., & Johnson, Md, R. W. (2012). Pediatric patients in a disaster: Part of the all-hazard, comprehensive approach to disaster management. *American Journal of Disaster Medicine*, *7*(2), 111–125. PMID: 22916449

Maechler, S., & Boisvert, V. (2024). Valuing nature to save it? The centrality of valuation in the new spirit of conservation. *Global Environmental Politics*, *24*(1), 10–30. 10.1162/glep_a_00734

Mahalik, M. K., Mallick, H., Padhan, H., & Sahoo, B. (2018). Is skewed income distribution good for environmental quality? A comparative analysis among selected BRICS countries. *Environmental Science and Pollution Research International*, *25*(23), 23170–23194. DOI: 10.1007/s11356-018-2401-8 PMID: 29862476

Mah, J. C., Penwarden, J. L., Pott, H., Theou, O., & Andrew, M. K. (2023). Social vulnerability indices: A scoping review. *BMC Public Health*, *23*, 1–11. PMID: 37380956

Mahmud, A., Mohammad, Z., & Abdullah, K. A. (2020). Leadership in disaster management: Theory versus reality. *Journal of Clinical and Health Sciences*, *5*(1), 4–11.

Maksimović, M., Vidic, D., Miloš, M., Šolić, M. E., Abadžić, S., & Siljak-Yakovlev, S. (2007). Effect of the environmental conditions on essential oil profile in two Dinaric *Salvia* species: *S. brachyodon* Vandas and *S. officinalis* L. *Biochemical Systematics and Ecology*, *35*(8), 473–478. DOI: 10.1016/j.bse.2007.02.005

Mankiw, N., Romer, D., & Weil, D. (1992). A Contribution to the Empirics of Economic Growth. *The Quarterly Journal of Economics*, *107*(2).

Marina , M.-C., Dinu, M., Socol, A.-G., & Socol, C. (2018). Renewable energy consumption and economic growth. Causality relationship in Central and Eastern European countries. *PLoS One*, *13*(10), e0202951. DOI: 10.1371/journal.pone.0202951 PMID: 30296307

Maryati, S. (2018). *Identification of Flood Prone Areas for Natural Disaster Mitigation using Geospatial Approach (A Case Study in Bone Bolango Regency, Gorontalo Province)* (I. A., Nandi, & S. L., Eds.). Department of Earth Science and Engineering, Faculty of Mathematics and Natural Sciences, Universitas Negeri Gorontalo, Gorontalo, Indonesia: Institute of Physics Publishing. https://doi.org/ DOI: 10.1088/1755-1315/145/1/012080

Maslin, M. (2011). *Global warming*. Dost.

Maslin, M. (2011). *Küresel Isınma* (Gül, S., Ed.). Dost Kitabevi.

Math, S. B., Nirmala, M. C., Moirangthem, S., & Kumar, N. C. (2015). Disaster management: Mental health perspective. *Indian Journal of Psychological Medicine*, *37*(3), 261–271. PMID: 26664073

Mausethagen, S. (2025). Blurred discourses of professionalism in teacher education institutions. *Teachers and Teaching*, *31*(1), 118–132.

Ma, Y., Dias, M. C., & Freitas, H. (2020). Drought and salinity stress responses and microbe-induced tolerance in plants. *Frontiers in Plant Science*, *11*, 591911. DOI: 10.3389/fpls.2020.591911

McCarthy, , CSternberg, , TBanda, , L. B. (2023). Does environmental aid make a difference? Analyzing its impact in developing countries. *Land*, *12*(10), 1953. https://doi.org/10.3390/land12101953

McLeod, M. K., Slavich, P. G., Irhas, Y., Moore, N., Rachman, A., Ali, N., & Caniago, C. (2010). Soil salinity in Aceh after the December 2004 Indian Ocean tsunami. *Agricultural Water Management*, *97*(5), 605–613. DOI: 10.1016/j.agwat.2009.10.014

McMichael, P. (2000). *Development and Social Change: A Global Perspective* (2nd ed.). Pine Forge Press.

Meadows, D. H., Meadows, D. L., Randers, J., & Behrens, W. W. (1972). *The Limits to Growth*. Universe Books.

Mealy, P., & Teytelboym, A. (2022). Economic complexity and the green economy. *Research Policy*, *51*(8), 103948. 10.1016/j.respol.2020.103948

Mehiriz, K. (2021). The sources of municipalities' innovation in the management of weather disaster risks, their relationships, and their antecedents. *Local Government Studies*, *47*(6), 951–970.

Mendelsohn, R., Dinar, A., & Williams, L. (2006). The distributional impact of climate change on rich and poor countries. *Environment and Development Economics*, *11*(2), 159–178.

Menegaki, A. N. (2011). Growth and renewable energy in Europe: A random effect model with evidence for neutrality hypothesis. *Energy Economics*, *33*(2), 257–263. DOI: 10.1016/j.eneco.2010.10.004

Menyah, K., & Wolde-Rufael, Y. (2010). CO2 emissions, nuclear energy, renewable energy and economic growth in the US. *Energy Policy*, *38*(6), 2911–2915. DOI: 10.1016/j.enpol.2010.01.024

Mercure, J. F., Pollitt, H., Chewpreecha, U., Salas, P., Foley, A., Holden, P. B., & Edwards, N. R. (2013). Complexity, economic science and possible economic benefits of climate change mitigation policy. *arXiv preprint arXiv:1310.4403*.

Mercure, J. F., Pollitt, H., Bassi, A. M., Viñuales, J. E., & Edwards, N. R. (2016). Modelling complex systems of heterogeneous agents to better design sustainability transitions policy. *Global Environmental Change*, *37*, 102–115.

Merriam, S. B., & Baumgartner, L. M. (2024). *Learning in adulthood: A comprehensive guide* (5th ed.). Jossey-Bass.

Mignamissi, D., Mougnol, A., Ekoula, H. W., & Thioune, T. (2024). Institutions' quality and environmental pollution in Africa. *Environment and Development Economics*, *29*(3), 206–233. DOI: 10.1017/S1355770X24000044

Mills, B., & Rosenfeld, J. (2023). Residential energy efficiency programs: Opportunities and challenges for utilities. *Utilities Policy*, *45*, 67–78.

Ministry of Foreign Affairs of the Republic of Turkey. (2023). *Paris Anlaşması*. https://www.mfa.gov.tr/paris-anlasmasi.tr.mfa

Mirjas, M., Fahlevi, H., & Diantimala, Y. (2021). Analisis tren penganggaran bencana di daerah rawan bencana: Studi kasus pada kabupaten aceh selatan. *Jurnal Kajian Akuntansi*, *5*(2), 243.

Mishra, B., & Kant, T. (2018). A simple, inexpensive, scalable and low maintenance hydroponics system for growing halophyte: *Lepidium sativum* L. (Brassicaceae), ideal for manipulating salt stress and inferring gene expression levels. *Indian Journal of Plant Physiology / Official Publication of the Indian Society for Plant Physiology*, *23*, 760–771. DOI: 10.1007/s40502-018-0422-4

Moarif, S., & Rastogi, N. P. (2012). *Market-based climate mitigation policies in emerging economies*. Center for Climate and Energy Solutions.

Moffo, A. M. F. (2024). A machine learning approach in stress testing US bank holding companies. *International Review of Financial Analysis*, *95*, 103476. DOI: 10.1016/j.irfa.2024.103476

Mohanavelu, A., Naganna, S. R., & Al-Ansari, N. (2021). Irrigation induced salinity and sodicity hazards on soil and groundwater: An overview of its causes, impacts and mitigation strategies. *Agriculture*, *11*(10), 983. DOI: 10.3390/agriculture11100983

Molyneux, M. (1979). Beyond the Domestic Labour Debate. *New Left Review*, ●●●, 115.

Monroe, M. C., & Krasny, M. E. (2024). *From dedication to action: How educators can weave together research and practice*. Cornell University Press.

Moreira, S. B. (2013). *Sobre a Medição do Desenvolvimento – Indicadores Desagregados e Compósitos com uma Aplicação Empírica a Portugal, 2013*. Escrytos | Edição de autor.

Moreira, S. (2005). Evaluating the Impact of Foreign Aid on Economic Growth: A Cross-Country Study. *Journal of Economic Development*, *30*(2). https://jed.cau.ac.kr/archives/30-2.html

Morgan, O. W., Sribanditmongkol, P., Perera, C., Sulasmi, Y., Van Alphen, D., & Sondorp, E. (2006). Mass fatality management following the south Asian tsunami disaster: Case studies in Thailand, Indonesia, and Sri Lanka. *PLoS Medicine*, *3*(6), 809–815. DOI: 10.1371/journal.pmed.0030195 PMID: 16737348

Morley, S. K., Brito, T. V., & Welling, D. T. (2018). Measures of Model Performance Based On the Log Accuracy Ratio. *Space Weather*, *16*(1), 69–88. DOI: 10.1002/2017SW001669

Mort, M., Rodriguez-Giralt, I., & Delicado, A. (Eds.). (2020). *Children and young people's participation in disaster risk reduction*. Policy Press.

Mosby, K., Birch, T., Moles, A., & Cherry, K. E. (2009). Disasters Risk Reduction. In *UNISDR*. https://doi.org/DOI: 10.7591/9781501701498-008

Moser, C. (1989). Gender Planning and Development: Meeting Practical and Strategic Needs. *World Development*, *17*(11), 1799–1825.

Mouniane, Y., El-Khadir, I., Chriqui, A., Soufiani, A., Oudghiri, M., Abbad, Z., Zoubir, M. H., Ninich, O., & Hmouni, D. (2025). Responses of Carob Plants to Salt Stress: Prospects for Sustainable Agriculture and the Bioeconomy. In Mbedzi, M., Nkosi, S., & Alabi, D. (Eds.), *Impact of Environmental Degradation on Ecosystems and Preventive Measures* (pp. 195–220). IGI Global Scientific Publishing., DOI: 10.4018/979-8-3373-1549-2.ch008

Moussa, H., Dahmoune, F., Mróz, M., Remini, H., Kadri, N., Hamid, S., & Kuszniere-wicz, B. (2023). Efficient optimization approaches for microwave assisted extraction of high-quality antioxidant compounds from *Salvia officinalis* L.: UHPLC-HRMS differential analysis of phenolic profiles obtained by ultrasound and microwave extraction. *Sustainable Chemistry and Pharmacy*, *35*, 101194. DOI: 10.1016/j.scp.2023.101194

Moutinho, V., & Robaina, M. (2016). Is the share of renewable energy sources determining the CO2 kWh and income relation in electricity generation? *Renewable & Sustainable Energy Reviews*, *65*, 902–914. DOI: 10.1016/j.rser.2016.07.007

Moyo, G. (2024). Imperial Ecocide and the Bane of Global Climate Finance. In *Africa in the Global Economy: Capital Flight, Enablers, and Decolonial Responses* (pp. 103–123). Springer Nature Switzerland.

Munir, K., & Ameer, A. (2018). Effect of economic growth, trade openness, urbanization, and technology on environment of Asian emerging economies. *Management of Environmental Quality*, *29*(6), 1123–1134. DOI: 10.1108/MEQ-05-2018-0087

Murphy, A. H. (1993). What Is a Good Forecast? An Essay on the Nature of Goodness in Weather Forecasting. *Weather and Forecasting*, *8*(2), 281–293. DOI: 10.1175/1520-0434(1993)008<0281:WIAGFA>2.0.CO;2

Myrdal, G. (1957). *Economic Theory and Under-developed Regions*. Duckworth.

Mythen, G. (2021). The critical theory of world risk society: A retrospective analysis. *Risk Analysis*, *41*(3), 533–543. PMID: 30170338

Nabiyeva, K. (2015). *Renewable energy and energy efficiency in Central Asia: Prospects for German engagement*. https://www.succow-stiftung.de/fileadmin/Ablage/Projekte/Forschung_Weiterbildung/MDF_Paper_RE_and_EE_in_Central_Asia_Kominla_Nabiyeva_2015.pdf

Naghavi, M., Wang, H., & Lozano, R. (2015). Global, regional, and national age-sex specific all-cause and cause-specific mortality for 240 causes of death, 1990-2013: A systematic analysis for the Global Burden of Disease Study 2013. *Lancet*, *385*(9963), 117–171. DOI: 10.1016/S0140-6736(14)61682-2 PMID: 25530442

Nain, M. Z., Ahmad, W., & Kamaiah, B. (2015). Economic growth, energy consumption and CO_2 emissions in India: A disaggregated causal analysis. *International Journal of Sustainable Energy*, *36*(8), 807–824. DOI: 10.1080/14786451.2015.1109512

Namahoro, J. P., Nzabanita, J., & Wu, Q. (2021). The impact of total and renewable energy consumption on economic growth in lower and middle- and upper-middle-income groups: Evidence from CS-DL and CCEMG analysis. *Energy, 237*, 121536. DOI: 10.1016/j.energy.2021.121536

Naqvi, R. A., Almohsen, B., & Sohail, A. (2025). Modeling the environmental Kuznets Curve: A stochastic approach using economic and climate data. *Journal of Environmental Management, 373*(January), 123108. DOI: 10.1016/j.jenvman.2024.123108 PMID: 39752958

Naseem, S., Mohsin, M., Zia-UR-Rehman, M., Baig, S. A., & Sarfraz, M. (2022). The influence of energy consumption and economic growth on environmental degradation in BRICS countries: An application of the ARDL model and decoupling index. *Environmental Science and Pollution Research International, 29*(9), 13042–13055. DOI: 10.1007/s11356-021-16533-3 PMID: 34564817

Nathaniel, S., Nwodo, O., Sharma, G., & Shah, M. (2020). Renewable energy, urbanization, and ecological footprint linkage in CIVETS. *Environmental Science and Pollution Research International, 27*(16), 19616–19629. DOI: 10.1007/s11356-020-08466-0 PMID: 32219652

Navarrete-Valladares, C., Sandoval-Díaz, J., & Sandoval-Obando, E. (2023). Experience and local memory of older people in the face of disasters: A systematic review. *Frontiers in Public Health, 11*, 1163561. PMID: 37293609

Nazlioglu, S., Gormus, N. A., & Soytas, U. (2016). Oil prices and real estate investment trusts (REITs): Gradual-shift causality and volatility transmission analysis. *Energy Economics, 60*, 168–175. DOI: 10.1016/j.eneco.2016.09.009

Neagu, O. (2019). The Link between Economic Complexity and Carbon Emissions in the European Union Countries: A Model Based on the Environmental Kuznets Curve (EKC) Approach. *Sustainability, 11*(17), 4753. DOI: 10.3390/su11174753

Negacz, K., Malek, Ž., de Vos, A., & Vellinga, P. (2022). Saline soils worldwide: Identifying the most promising areas for saline agriculture. *Journal of Arid Environments, 203*, 104775. DOI: 10.1016/j.jaridenv.2022.104775

Neij, , LHeiskanen, , EStrupeit, , L. (2017). The deployment of new energy technologies and the need for local learning. *Energy Policy, 101*, 274-283. https://doi .org/10.1016/j.enpol.2016.11.029

Nepal, R., Jamasb, T., & Tisdell, C. A. (2017). On environmental impacts of market-based reforms: Evidence from the European and Central Asian transition economies. *Renewable & Sustainable Energy Reviews*, *73*, 44–52. DOI: 10.1016/j.rser.2017.01.078

Newell, R. G. (2010). The role of markets and governments in climate change policy. *Energy Journal*, *31*(1), 91–114. DOI: 10.5547/ISSN0195-6574-EJ-Vol31-No1-5

Ng, S. L. (2024). Bibliometric analysis of climate change and extreme weather adaptation based on Scopus. *Journal of Environmental Studies and Sciences*, 1–20.

Nguyen, A. T. (2019). The relationship between economic growth, energy consumption and carbon dioxide emissions: Evidence from Central Asia. *Eurasian Journal of Business and Economics*, *12*(24), 1–15. DOI: 10.17015/ejbe.2019.024.01

Nguyen, T. H., Deng, H., Abbas, Z. Z., Lam, T. T., & Abbas, H. R. (2024). The effect of natural capital, regional development, FDI, and natural resource rent on environmental performance: The Mediating role of green innovation. *Resources Policy*, *91*(1), 1–12. DOI: 10.1016/j.resourpol.2024.104923

Nguyen, T. T., Grote, U., Neubacher, F., Rahut, D. B., Do, M. H., & Paudel, G. P. (2023). Security risks from climate change and environmental degradation: Implications for sustainable land use transformation in the Global South. *Current Opinion in Environmental Sustainability*, *63*, 101322. DOI: 10.1016/j.cosust.2023.101322

Nirupama, N. (2012). Risk and vulnerability assessment: A comprehensive approach. *International Journal of Disaster Resilience in the Built Environment*, *3*(2), 103–114.

Noel, G. E. (1990). The role of women and NNAs in disaster management. *International Nursing Review*, *37*(6), 363–367. PMID: 2276904

North, D. (1991). Institutions. *The Journal of Economic Perspectives*, *5*(1).

North, D. C. (1955). Location Theory and Regional Economic Growth. *Journal of Political Economy*, •••, 63.

Noy, I. (2009). The macroeconomic consequences of disasters. *Journal of Development Economics*, *88*(2), 221–231.

Obst, C., Hein, L., & Edens, B. (2016). National accounting and the valuation of ecosystem assets and their services. *Environmental and Resource Economics*, *64*(1), 1–23. 10.1007/s10640-015-9921-1

OECD. (2018). *Converged statistical reporting directives for the Creditor Reporting System (CRS) and the annual DAC questionnaire – Chapters 1-6*. DAC. Paris: OECD.

Ogede, J. S., Oduola, M. O., & Tiamiyu, H. O. (2024). Income inequality and carbon dioxide (CO2) in sub Saharan Africa countries: The moderating role of financial inclusion and institutional quality. *Environment, Development and Sustainability, 26*(7), 18385–18409.

Okay, N., & İlkkaracan, İ. (2018). Toplumsal cinsiyete duyarlı afet risk yönetimi. *Resilience, 2*(1), 1–12.

Okereke, C. (2024). Degrowth, green growth, and climate justice for Africa. *Review of International Studies, 50*(5), 910–920.

Oleg, O. K., Zaytsev, A., & Dimitriev, N. (2023). Role and place of institutional capacity in socio-economic development. *International Journal of Professional Business Review, 8*(10), 1–23.

Ong, A. K. S., Prasetyo, Y. T., Lagura, F. C., Ramos, R. N., Sigua, K. M., Villas, J. A., & Redi, A. A. N. P. (2021). Factors affecting intention to prepare for mitigation of "the big one" earthquake in the Philippines: Integrating protection motivation theory and extended theory of planned behavior. *International Journal of Disaster Risk Reduction, 63*. Advance online publication. DOI: 10.1016/j.ijdrr.2021.102467

Onwe, J. C., Uche, E., Dhayal, K. S., Uwazie, I. U., & Ashibogwu, K. N. (2024). Advocating green economy in India: The tug of war among income inequality, export diversification, and environmental quality. *Sustainable Development, 32*(5), 4879–4894. DOI: 10.1002/sd.2927

Oprea, I., & Du ă, D. (2024). Integrating AI in Bank Digitalization: Strategies, Challenges and Future Perspectives. *Legal Perspectives in the Modern Era of Technological Transformations*, 205–216. https://doi.org/DOI: 10.62768/ADJURIS/2024/1/13

Organisation for Economic Co-operation and Development (OECD). (2009). *The economics of climate change mitigation: Policies and options for global action beyond 2012*. OECD Publishing. DOI: 10.1787/9789264073616-

Osman, B. M., Shire, S. A., Ali, F. H., & Mohamed, A. A. (2024). Examining the drivers of environmental degradation in Somalia: The role of agriculture, economic and population growth. *Discover Sustainability, 6*(150), 1–13. DOI: 10.1007/s43621-024-00786-2

Osuntuyi, B. V., & Lean, H. H. (2022). Economic growth, energy consumption and environmental degradation nexus in heterogeneous countries: Does education matter? *Environmental Sciences Europe, 34*(48), 1–14. DOI: 10.1186/s12302-022-00624-0

Otlewska, A., Migliore, M., Dybka-Stępień, K., Manfredini, A., Struszczyk-Świta, K., Napoli, R., Białkowska, A., Canfora, L., & Pinzari, F. (2020). When salt meddles between plant, soil, and microorganisms. *Frontiers in Plant Science, 11*, 553087. DOI: 10.3389/fpls.2020.553087

Oudghiri, M., Mouniane, Y., Bouhassoun, S., El Aammouri, S., Yamani, B., Benlemlih, N., Abid, N., Elwahab, F. A., Zoubir, M. H., El-Khadir, I., Chriqui, A., Aitouhanni, I., El Bakkali, M., El Ouardi, B., & Ibriz, M. (2025). Abiotic stress and climate change: Technological innovations and adaptation strategies for sustainable resilience. In Ahmadi, A., & Mna, A. (Eds.), *Nexus of environmental quality and technology innovation* (pp. 1–16). IGI Global., DOI: 10.4018/979-8-3373-1822-6.ch001

Our World in Data. (2024). https://ourworldindata.org/

Ozcan, M. (2016). Estimation of Turkey s GHG emissions from electricity generation by fuel types. *Renewable & Sustainable Energy Reviews, 53*, 832–840. DOI: 10.1016/j.rser.2015.09.018

Ozdemir, M., Pehlivan, S., & Melikoglu, M. (2024). Estimation of greenhouse gas emissions using linear and logarithmic models: A scenario-based approach for Türkiye's 2030 vision. *Energy Nexus, 13*, 100264. DOI: 10.1016/j.nexus.2023.100264

Özer, B., Görgün, E., & İncecik, S. (2013). The scenario analysis on CO2 emission mitigation potential in the Turkish electricity sector: 2006–2030. *Energy, 49*, 395–403. DOI: 10.1016/j.energy.2012.10.059

Özey, R., & Ünlü, M. (2021). *Afetler ve afet yönetimi (2. edition)*. Aktif Yayınevi.

Özmen, B., & Özden, T. (2013). Türkiye'nin afet yönetim sistemine ilişkin eleştirel bir değerlendirme. *İstanbul Üniversitesi Siyasal Bilgiler Fakültesi Dergisi, 49*, 1-28.

Öztürk, N. (2003). Türkiye'de afet yönetimi: karşılaşılan sorunlar ve çözüm önerileri. *Çağdaş Yerel Yönetimler, 12(4)*, 42-64.

Ozturk, I., & Acaravci, A. (2013). The long-run and causal analysis of energy, growth, openness and financial development on carbon emissions in Turkey. *Energy Economics, 36*, 262–267. DOI: 10.1016/j.eneco.2012.08.025

Pacillo, G., Nguyen, N., Paustyan, E., Cavatassi, R., Astralaga, M., & Läderach, P. (2024). What drives the success and failure of climate change adaptation projects? A qualitative comparative analysis. *Mitigation and Adaptation Strategies for Global Change, 29*(8), 89.

Padilla, E., & Serrano, A. (2005). Inequality in CO2 emissions across countries and its relationship with income inequality: A distributive approach. *Energy Policy*, *34*(14), 1762–1772. DOI: 10.1016/j.enpol.2004.12.014

Painter, M. A., Shah, S. H., Damestoit, G. C., Khalid, F., Prudencio, W., Chisty, M. A., & Wilhelmi, O. (2024). A systematic scoping review of the Social Vulnerability Index as applied to natural hazards. *Natural Hazards*, *120*(8), 7265–7356.

Papadopoulos, G., Arduini, S., Uyar, H., Psiroukis, V., Kasimati, A., & Fountas, S. (2024). Economic and environmental benefits of digital agricultural technologies in crop production: A review. *Smart Agricultural Technology*, *8*, 100441. DOI: 10.1016/j.atech.2024.100441

Parker, E. R., Mo, J., & Goodman, R. S. (2022). The dermatological manifestations of extreme weather events: A comprehensive review of skin disease and vulnerability. *The Journal of Climate Change and Health*, *8*, 100162. DOI: 10.1016/j.joclim.2022.100162

Partridge, D., & Yates, W. B. (1996). Engineering Multiversion Neural-Net Systems. *Neural Computation*, *8*(4), 869–893. DOI: 10.1162/neco.1996.8.4.869 PMID: 8624963

Pata, U. K., Yilanci, V., Hussain, B., & Naqvi, S. A. (2022). Analyzing the role of income inequality and political stability in environmental degradation: Evidence from South Asia. *Gondwana Research*, *107*, 13–29. DOI: 10.1016/j.gr.2022.02.009

Patel, G., Pal, S., & Sahu, A. K. (2024). Does energy transition reduce carbon inequality? A global analysis. *Environmental Science and Pollution Research International*, *31*(1), 34689–34708. DOI: 10.1007/s11356-024-33542-0 PMID: 38713356

Pathak, M., Patel, S., & Some, S. (2024). Climate change mitigation and Sustainable Development Goals: Evidence and research gaps. *PLOS Climate*, *3*(3), e0000366.

Payne, A., & Phillips, N. (2010). *Development*. Polity Press.

Payne, J. E. (2012). The causal dynamics between US renewable energy consumption, output, emissions, and oil prices. *Energy Sources. Part B, Economics, Planning, and Policy*, *7*(4), 323–330. DOI: 10.1080/15567249.2011.595248

Pearse, R., Schlosberg, D., Rickards, L., Della Bosca, H., Moraes, O., & de Kleyn, L. (2025). Compounding barriers to environmental justice. *Local Environment*, •••, 1–16.

Pecqueur, B. (2000). *Le développement local*.

Persakis, A., Tsakalos, I., Gkonis, V., & Nerantzidis, M. (2024). Climate policy uncertainty and environmental degradation: Does democracy moderate this relationship? *Cleaner Environmental Systems, 15*, 100230. https://doi.org/https://doi.org/10.1016/j.cesys.2024.100230

Pesaran, M. H. (2007). A simple panel unit root test in the presence of cross-section dependence. *Journal of Applied Econometrics, 22*(2), 265–312. DOI: 10.1002/jae.951

Peterson, E. W. F. (2017). The role of population in economic growth. *SAGE Open, 7*(4), 1–15.

Piketty, T. (2014). *Capital in the twenty-first century*. Belknap Press.

Plomteux, A. (2024). Frugal abundance: Conceptualisation for degrowth. *Ecological Economics, 222*, 108223. 10.1016/j.ecolecon.2024.108223

Polanyi, K. (1944). *The Great Transformation: The Political and Economic Origins of Our Time*. Beacon Press.

Ponsam, J. G., Bella Gracia, S. V. J., Geetha, G., Karpaselvi, S., & Nimala, K. (2021). Credit Risk Analysis using LightGBM and a comparative study of popular algorithms. *2021 4th International Conference on Computing and Communications Technologies (ICCCT)*, 634–641. https://doi.org/DOI: 10.1109/ICCCT53315.2021.9711896

Ponting, C. (2000). *A Green history of the world: The environment and the collapse of great civilizations* (Sander, A. B., Trans.). Sabancı University.

Popp, , D. (2011). International technology transfer, climate change, and the clean development mechanism. *Review of Environmental Economics and Policy 5*(1).

Pradhan, K. C., Mishra, B., & Mohapatra, S. M. (2024). Investigating the relationship between economic growth, energy consumption, and carbon dioxide (CO_2) emissions: A comparative analysis of South Asian nations and G-7 countries. *Clean Technologies and Environmental Policy, 26*(1), 3349–3367. DOI: 10.1007/s10098-024-02802-5

Prasetiyawan, Viora, E., Maramis, A., & Keliat, B. A. (2006). Mental health model of care programmes after the tsunami in Aceh, Indonesia. *International Review of Psychiatry (Abingdon, England), 18*(6), 559–562. DOI: 10.1080/09540260601039959 PMID: 17162697

Pratama, T., Dirhamsyah, M., & Madjid, I. (2019). Analysis of budget development disaster management in the central aceh district disaster management agency. *International Journal of Multicultural and Multireligious Understanding, 6*(2), 339.

Prebisch, R. (1950). *The Economic Development of Latin America and its Principal Problems*. United Nations.

Puig, D., Adger, N. W., Barnett, J., Vanhala, L., & Boyd, E. (2025). Improving the effectiveness of climate change adaptation measures. *Climatic Change, 178*(1), 7.

Pujiati, A., Nurbaeti, T., & Damayanti, N. (2023). What are the factors that determine differing levels of environmental quality? Evidence from Java and other islands in Indonesia. *Management of Environmental Quality, 34*(2), 290–307. DOI: 10.1108/MEQ-02-2022-0034

Putnam, R. D., & Coleman, J. S. (2024). Social capital and community development: Recent advances in theory and practice. *Community Development Journal: An International Forum, 38*(2), 143–167.

Putra, D. I., & Matsuyuki, M. (2020). The disaster-management capabilities of local governments: A case study in Indonesia. *Journal of Disaster Research, 15*(4), 471–480.

Quarantelli, E. L. (1988). Disaster crisis management: A summary of research findings. *Journal of Management Studies, 25*(4), 373–385.

Raal, A., Orav, A., Ilina, T., Kovalyova, A., Koliadzhyn, T., Avidzba, Y., & Koshovyi, O. (2024). Variation in the composition of the essential oil of commercial *Salvia officinalis* L. leaves samples from different countries. *Phyton, 93*(8), 2051–2062. DOI: 10.32604/phyton.2024.052790

Rac, I., Erjavec, K., & Erjavec, E. (2024). Agriculture and environment: Friends or foes? Conceptualising agri-environmental discourses under the European Union's Common Agricultural Policy. *Agriculture and Human Values, 41*(1), 147–166.

Raghutla, C., & Chittedi, K. R. (2021). Financial development, energy consumption, technology, urbanization, economic output and carbon emissions nexus in BRICS countries: An empirical analysis. *Management of Environmental Quality, 32*(2), 290–307. DOI: 10.1108/MEQ-02-2020-0035

Rahim, P. Y. (2025). The Influence of Financial Literacy and Environmental Awareness on Green Investment Decisions. In *Proceedings of the 9th International Conference on Accounting, Management, and Economics 2024 (ICAME 2024)* (pp. 3143-3154). Atlantis Press. https://www.atlantis-press.com/proceedings/icame-24/126014133

Rahman, M. M., Penny, G., Mondal, M. S., Zaman, M. H., Kryston, A., Salehin, M., Nahar, Q., Islam, M. S., Bolster, D., Tank, J. L., & Müller, M. F. (2019). Salinization in large river deltas: Drivers, impacts and socio-hydrological feedbacks. *Water Security, 6*, 100024. DOI: 10.1016/j.wasec.2019.100024

Rahman, M. M., & Velayutham, E. (2020). Renewable and non-renewable energy consumption-economic growth nexus: New evidence from South Asia. *Renewable Energy*, *147*, 399–408. DOI: 10.1016/j.renene.2019.09.007

Rahman, M. S., Noman, A. H. M., Shahari, F., Aslam, M., Gee, C. S., Isa, C. R., & Pervin, S. (2023). Efficient energy consumption in industrial sectors and its effect on environment: A comparative analysis between G8 and Southeast Asian emerging economies. *Energy*, *97*(1), 82–89. DOI: 10.1016/j.energy.2015.12.109

Rahman, Md. M., Rahman, Md. S. & Jerin, T. (2022). Social vulnerability to earthquake disaster: Insights from the people of 48th ward of Dhaka south city, Bangladesh. *Environmental Hazards*, *22*(2), 116–135.

Rahmawati, I., & Riskiyah, F. (2023). Exploring the invisibility of women's resilience in facing anthropogenic disasters. *International Journal of Social Science Research and Review*, *6*(1), 421–432.

Raikes, J., & McBean, G. (2017). Institutional response to disaster risk: The city of Vancouver and district of maple ridge, British Columbia, Canada. *Regional Environmental Change*, *17*(8), 2433–2441.

Rajan, R. G., & Subramanian, A. (2011). Aid, Dutch disease, and manufacturing growth. *Journal of Development Economics*, *94*(1), 106–118. DOI: 10.1016/j.jdeveco.2009.12.004

Rakesh Pandit, S. B. J. S. P. M. A. C. (2024). Credit Risk Assessment and Fraud Detection in Financial Transactions Using Machine Learning. *Journal of Electrical Systems*, *20*(3s), 2061–2069. DOI: 10.52783/jes.1807

Rakib, M. A., Newaz, M. A., Rahman, M. A., & Roy, K. (2024). Identifying the interfaces between perceived multi-hazards and socio-ecological risks to strengthen local adaptations. *Journal of Environmental Management*, *351*, 119708. DOI: 10.1016/j.jenvman.2023.119708

Rakshit, B., & Neog, Y. (2021). Do macroeconomic uncertainty and financial development cause environmental degradation? Evidence from an emerging economy. *International Journal of Social Economics*, *48*(9), 1264–1289. DOI: 10.1108/IJSE-10-2020-0690

Rastegar, R., & Becken, S. (2024). Embedding justice into climate policy and practice relevant to tourism. *Journal of Sustainable Tourism*, •••, 1–18.

Raungratanaamporn, I., Pakdeeburee, P., Kamiko, A., & Denpaiboon, C. (2014). Government-communities Collaboration in Disaster Management Activity: Investigation in the Current Flood Disaster Management Policy in Thailand. *Procedia Environmental Sciences*, *20*, 658–667. DOI: 10.1016/j.proenv.2014.03.079

Raworth, K. (2017). *Doughnut economics: seven ways to think like a 21st-century* economist. Chelsea Green Publishing.

Reinhardt, G. Y., & Ross, A. D. (2019). Expanding social science through disaster studies. *Social Science Quarterly*, *100*(7), 2523–2529.

Reyers, B., & Selig, E. R. (2020). Global targets that reveal the social–ecological interdependencies of sustainable development. *Nature Ecology & Evolution*, *4*(8), 1011–1019. 10.1038/s41559-020-1230-6

Riddell, R. C. (1987). *Foreign Aid Reconsidered*. James Currey / Overseas Development Institute.

Riedl, A., & Smeets, P. (2024). Why do investors hold socially responsible mutual funds? *The Journal of Finance*, *72*(6), 2505–2550.

Riera-Spiegelhalder, M., Campos-Rodrigues, L., Ensenado, E. M., Dekker-Arlain, J. D., Papadopoulou, O., Arampatzis, S., & Vervoort, K. (2023). Socio-economic assessment of ecosystem-based and other adaptation strategies in coastal areas: A systematic review. *Journal of Marine Science and Engineering*, *11*(2), 319.

Rivas, S., Urraca, R., Palermo, V., & Bertoldi, P. (2022). Covenant of Mayors 2020: Drivers and barriers for monitoring climate action plans. *Journal of Cleaner Production*, *332*, 130029. https://doi.org/https://doi.org/10.1016/j.jclepro.2021.130029

Roberts, D. K., & Kim, H. J. (2024). Green energy financial literacy: Measurement and demographic correlates. *The Journal of Consumer Affairs*, *58*(1), 89–112.

Robinson, B., Alatas, M. F., Robertson, A., & Steer, H. (2011). Natural disasters and the lung. *Respirology (Carlton, Vic.)*, *16*(3), 386–395. DOI: 10.1111/j.1440-1843.2011.01923.x PMID: 21244570

Robinson, W. (2002). Remapping Development in Light of Globalisation: From a Territorial to a Social Cartography. *Third World Quarterly*, *23*, 6.

Rockström, J., Steffen, W., Noone, K., Persson, A., Chapin, F. S., Lambin, E. F., ... (2009). A safe operating space for humanity. *Nature*, *461*(7263), 472–475. 10.1038/461472a

Rodrik, D., & Stantcheva, S. (2021). A policy matrix for inclusive prosperity. *National Bureau of Economic Research.*, w28736.

Rofi, A., Doocy, S., & Robinson, C. (2006). Tsunami mortality and displacement in Aceh province, Indonesia. *Disasters*, *30*(3), 340–350. DOI: 10.1111/j.0361-3666.2005.00324.x PMID: 16911432

Rogers, E. M. (2023). *Diffusion of innovations* (6th ed.). Free Press.

Rokhis Khomarudin, M., Strunz, G., Ludwig, R., Zoßeder, K., Post, J., Kongko, W., & Pranowo, W. S. (2010). Hazard analysis and estimation of people exposure as contribution to tsunami risk assessment in the west coast of sumatra, the south coast of Java and bali. *Zeitschrift für Geomorphologie*, *54*, 337–356. DOI: 10.1127/0372-8854/2010/0054S3-0031

Romer, P. (1986). Increasing Returns and Long-Run Growth. *Journal of Political Economy*, *94*(5).

Rosa, E. A., & Dietz, T. (2012). Human drivers of national greenhouse-gas emissions. *Nature Climate Change*, *2*(8), 581–586. DOI: 10.1038/nclimate1506

Rostow, W. W. (1960). *The Stages of Economic Growth: A Non-Communist Manifesto*. Cambridge University Press.

Roy, D., & Mukherjee, M. (2024). Challenges and opportunities for women in disaster risk management. *IDRiM Journal*, *14*(2), 228–240.

Rozema, J., & Schat, H. (2013). Salt tolerance of halophytes, research questions reviewed in the perspective of saline agriculture. *Environmental and Experimental Botany*, *92*, 83–95. DOI: 10.1016/j.envexpbot.2012.08.004

Ruggeri, E., & Garrido, S. (2021). More renewable power, same old problems? Scope and limitations of renewable energy programs in Argentina. *Energy Research & Social Science*, *79*, 102161. 10.1016/j.erss.2021.102161

Ruta, D., & Gabrys, B. (2005). Classifier selection for majority voting. *Information Fusion*, *6*(1), 63–81. DOI: 10.1016/j.inffus.2004.04.008

Saad-Filho, A. (2007). Life beyond the Washington Consensus: An Introduction to Pro-poor Macroeconomic Policies. *Review of Political Economy*, *19*(4), 513–537. DOI: 10.1080/09538250701622352

Sachs, J. (2005). *The end of poverty*. Penguin Books.

Sadat-Razavi, P., Karakislak, I., & Hildebrand, J. (2025). German media discourses and public perceptions on hydrogen imports: An energy justice perspective. *Energy Technology (Weinheim)*, *13*(2), 2301000.

Sadorsky, P. (2009). Renewable energy consumption and income in emerging economies. *Energy Policy, 37*(10), 4021–4028. DOI: 10.1016/j.enpol.2009.05.003

Safi, I., & Al-Hassan, M. (2024). Reassessing disaster and emergency management: A child rights perspective. *Scholars Bulletin, 10*(02), 32–38.

Sahab, I., Suhani, I., Srivastava, V., Chauhan, P. S., Singh, R. P., & Prasad, V. (2021). Potential risk assessment of soil salinity to agroecosystem sustainability: Current status and management strategies. *The Science of the Total Environment, 764*, 144164. DOI: 10.1016/j.scitotenv.2020.144164

Sahin, H., & Esen, H. (2022). The usage of renewable energy sources and its effects on GHG emission intensity of electricity generation in Turkey. *Renewable Energy, 192*, 859–869. DOI: 10.1016/j.renene.2022.03.141

Şahin, Ş., & Üçgül, İ. (2019). Türkiye'de afet yönetimi ve iş sağlığı güvenliği. *Afet ve Risk Dergisi, 2*(1), 43–63.

Sahu, A. K., Mahalik, M. K., Patel, G., & Pal, S. (2023). The role of environmental degradation in macroeconomic instability: Panel evidence from emerging economies. *Environmental Science and Pollution Research International, 30*(37), 86879–86891. DOI: 10.1007/s11356-023-28389-w PMID: 37410330

Salah, M., & Ament, J. (2025). The ecological currency hierarchy: Empirical support for currency power driven asymmetries in environmental and social autonomy. *New Political Economy, 30*(1), 77–99.

Salahodjaev, R., Sharipov, K., Rakhmanov, N., & Khabirov, D. (2022). Tourism, renewable energy and CO2 emissions: Evidence from Europe and Central Asia. *Environment, Development and Sustainability, 24*(11), 13282–13293. DOI: 10.1007/s10668-021-01993-x

Salim, R. A., & Rafiq, S. (2012). Why do some emerging economies proactively accelerate the adoption of renewable energy? *Energy Economics, 34*(4), 1051–1057. DOI: 10.1016/j.eneco.2011.08.015

Salvati, L., & Ferrara, C. (2015). The local-scale impact of soil salinization on the socioeconomic context: An exploratory analysis in Italy. *Catena, 127*, 312–322. DOI: 10.1016/j.catena.2015.01.008

Sambrook, K., Harcourt, R., Dessai, S., & Taylor, A. (2024). New evidence of a strong public mandate for climate change adaptation in the UK.

Sanderson, D., Heffernan, T., DeSisto, M., & Shearing, C. (2024). Community-centred disaster recovery: A call to change the narrative. *Disasters*, *49*(1), 1–15. PMID: 39229770

Sandoval, V., & Voss, M. (2016). Disaster governance and vulnerability: The case of chile. politics and governance. *Politics and Governance*, *4*(4), 107–116.

Sardianou, E., & Genoudi, P. (2023). Which factors affect the willingness of consumers to adopt renewable energies? *Renewable Energy*, *57*, 1–4.

Sarun, S., Ghermandi, A., Sheela, A. M., Justus, J., & Vineetha, P. (2018). Climate change vulnerability in a tropical region based on environmental and socio-economic factors. *Environmental Monitoring and Assessment*, *190*(12), 727.

Scherer, L., Behrens, P., de Koning, A., Heijungs, R., Sprecher, B., & Tukker, A. (2018). Trade-offs between social and environmentally Sustainable Development Goals. *Environmental Science & Policy*, *90*, 65–72. https://doi.org/https://doi.org/10.1016/j.envsci.2018.10.002

Schmidtlein, M. C., Deutsch, R. C., Piegorsch, W. W., & Cutter, S. L. (2008). A sensitivity analysis of the social vulnerability index. *Risk Analysis*, *28*(4), 1099–1114. PMID: 18627540

Scott, M. (1989). *A New View of Economic Growth*. Clarendon Press.

Scruggs, L. A. (1998). Political and economic inequality and the environment. *Ecological Economics*, *26*(3), 259–275. DOI: 10.1016/s0921-8009(97)00118-3

Seers, D. (1969). The Meaning of Development. *International Development Review*, *11*, 2–6.

Semenchuk, , PPlutzar, , CKastner, , T. et al. (2022). Relative effects of land conversion and land-use intensity on terrestrial vertebrate diversity. *Nature Communications*, *13*(1), 615. https://doi.org/10.1038/s41467-022-28245-4

Sen, A. K. (1981). *Poverty and Famines: An Essay on Entitlement and Deprivation*. Oxford University Press.

Şengün, H. (2007). *Marmara Depreminde Bayındırlık ve İskan Bakanlığı Çalışmaları ve Hukuksal Sorunlar. TMMOB Afet Sempozyumu, 5 Aralık 2007*. TMMOB.

Setiadi, A., Andriessen, A., & Anisa, R. (2020). Post-occupancy evaluation of Pagerjurang permanent housing after the Merapi volcanic eruption. *Journal of Architecture and Urbanism*, *44*(2), 145–151. DOI: 10.3846/jau.2020.11265

Seyle, D. C., Widyatmoko, C. S., & Silver, R. C. (2013). Coping with natural disasters in Yogyakarta, Indonesia: A study of elementary school teachers. *School Psychology International, 34*(4), 387–404. DOI: 10.1177/0143034312446889

Sezer, Ö., & Tozbey, Ş. (2023). Türkiye'de afet yönetiminin merkezi ve yerel düzeyde örgütlenmesi üzerine bir değerlendirme. *Siyasal Bilgiler Fakültesi Dergisi, 8*(1), 79–102.

Shadab, M. (2019). Chaos theory: Dynamics and applications. *The Pharma Innovation, 8*(2), 922–925.

Shadrina, E. (2019). *Renewable energy in Central Asian economics: Role in reducing regional energy insecurity*. https://www.adb.org/publications/renewable-energy-central-asian-economies

Shadrina, E. (2022). A double paradox of plenty: Renewable energy deployment in Central Asia. *Eurasian Geography and Economics, 63*(1), 1–26. DOI: 10.1080/15387216.2020.1823868

Sharma, B. R., & Minhas, P. S. (2005). Strategies for managing saline/alkali waters for sustainable agricultural production in South Asia. *Agricultural Water Management, 78*(1–2), 136–151. DOI: 10.1016/j.agwat.2005.04.019

Shayanmehr, S., Radmehr, R., Ali, E. B., Ofori, E. K., Adebayo, T. S., & Gyamfi, B. A. (2023). How do environmental tax and renewable energy contribute to ecological sustainability? New evidence from top renewable energy countries. *International Journal of Sustainable Development and World Ecology, 30*(6), 650–670. DOI: 10.1080/13504509.2023.2186961

Shetty, S. S., Deepthi, D., Harshitha, S., Sonkusare, S., Naik, P. B., Kumari, S. N., & Madhyastha, H. (2023). Environmental pollutants and their effects on human health. *Heliyon, 9*(9), 1–12. DOI: 10.1016/j.heliyon.2023.e19496 PMID: 37662771

Shukla, D., & Gupta, S. (2024). Feature Engineering Techniques to Enhance Credit Scoring Models. *2024 6th International Conference on Electrical, Control and Instrumentation Engineering (ICECIE)*, 1–5. https://doi.org/DOI: 10.1109/ICECIE63774.2024.10815675

Sidi, P., & Mamat, M., Sukono, & Supian, S. (2017). Evaluation model for risk insurance premiums of building damage caused by flood: Case study in Citarum Watershed, Southern Bandung, Indonesia. *Journal of Engineering and Applied Sciences (Asian Research Publishing Network), 12*(17), 4420–4425. DOI: 10.3923/jeasci.2017.4420.4425

Silva, B. N., Cadavez, V., Caleja, C., Pereira, E., Calhelha, R. C., Añibarro-Ortega, M., Finimundy, T., Kostić, M., Soković, M., Teixeira, J. A., Barros, L., & Gonzales-Barron, U. (2023). Phytochemical composition and bioactive potential of *Melissa officinalis* L., *Salvia officinalis* L. and *Mentha spicata* L. extracts. *Foods*, *12*(5), 947. DOI: 10.3390/foods12050947

Silva, N., Fuinhas, J. A., Koengkan, M., & Kazemzadeh, E. (2024). What are the causal conditions that lead to high or low environmental performance? A worldwide assessment. *Environmental Impact Assessment Review*, *104*, 107342. https://doi.org/https://doi.org/10.1016/j.eiar.2023.107342

Silverstein, K. (2025, January 17). Commentary: Wealthy nations can't ignore climate effects on poorer countries. *My Journal Courier*. https://www.myjournalcourier.com/opinion/article/wealthy-nations-can-t-ignore-climate-effects-20027538.php

Singer, H. (1950). The Distribution of Gains between Investing and Borrowing Countries. *The American Economic Review*, *40*, 473–485.

Singh, A., & Ahlawat, N. (2023). A review article: -the Growing Role of Data Science and AI in Banking and Finance. *International Research Journal of Modernization in Engineering Technology and Science*. Advance online publication. DOI: 10.56726/IRJMETS44000

Singhal, S., Thapar, S., Kumar, M., & Jain, S. (2022). Impacts of sustainable consumption and production initiatives in energy and waste management sectors: Examples from India. *Environment, Development and Sustainability*, *24*(12), 14184–14209. DOI: 10.1007/s10668-021-02026-3

Singh, N. P., Bhawna, A., Singh, S., & Arshad, Kh. (2019). Mainstreaming climate adaptation in Indian rural developmental agenda: A micro-macro convergence. *Climate Risk Management*, *29*(1), 30–41. https://www.sciencedirect.com/science/article/pii/S221209631830130X

Singh, S. (2020). Farmers' perception of climate change and adaptation decisions: A micro-level evidence from Bundelkhand Region, India. *Ecological Indicators*, *116*(1), 1–10. https://www.sciencedirect.com/science/article/abs/pii/S1470160X2030412X

Sinha, J. N. (1959). Population growth and balance in economic development. *Economic Development and Cultural Change*, *7*(3, Part 1), 206–215.

Skrypnik, L., Feduraev, P., Golubkina, N., Maslennikov, P., Antipina, M., Katserov, D., Murariu, O. C., Tallarita, A. V., & Caruso, G. (2024). Foliar spraying of selenium in inorganic and organic forms stimulates plant growth and secondary metabolism of sage (*Salvia officinalis* L.) through alterations in photosynthesis and primary metabolism. *Scientia Horticulturae, 338*, 113633. DOI: 10.1016/j.scienta.2024.113633

Smith, A., & Ely, A. (2025). From limits to growth to post-growth: The international politics of technology in historical perspective. *Science, Technology & Society*, •••, 09717218251326833.

Smith, J. A., & Johnson, B. R. (2024). Online survey methodology: Best practices for representative sampling. *Public Opinion Quarterly*, *88*(2), 234–251.

Smith, K. (2024). Deconstructing discourses to rupture fairytales of the "ideal" childhood. In *Handbook of children and youth studies* (pp. 15–28). Springer Nature Singapore.

Smith, N., Donaldson, M., Mitton, C., & Lee, E. (2024). Communication in disasters to support families with children with medical complexity and special healthcare needs: A rapid scoping review. *Frontiers in Public Health*, *12*, 1229738. PMID: 38544735

Smith, W., & Dowell, J. (2000). A case study of co-ordinative decision-making in disaster management. *Ergonomics*, *43*(8), 1153–1166. DOI: 10.1080/00140130050084923 PMID: 10975178

Smoro Laksmi, G., Rudiarto, I., & Luqman, Y. (2020). *Community preparedness toward flood during Covid-19 pandemic at Pekalongan City and Regency* (W. B., Sudarno, & P. T. T., Eds.). Master of Environmental Science, Postgraduate School, Diponegoro University, Semarang, Indonesia: EDP Sciences. https://doi.org/DOI: 10.1051/e3sconf/202020206008

Soares, T. C., Fernandes, E. A., & Toyoshima, S. H. (2017). The CO 2 emission Gini index and the environmental efficiency: An analysis for 60 leading world economies. *Economía*, *19*(2), 266–277. DOI: 10.1016/j.econ.2017.06.001

Solomon, S., Qin, D., Manning, M., Marquis, M., Averyt, K., Tignor, M. M. B., Miller, H. L., & Chen, Z. (2007). [*The Physical Science Basis.*]. *Climatic Change*, •••, 2007.

Sørensen, M. P. (2017). Ulrich Beck: Exploring and contesting risk. *Journal of Risk Research*, *21*(1), 6–16.

Southwick, S. M., Bonanno, G. A., Masten, A. S., Panter-Brick, C., & Yehuda, R. (2014). Resilience definitions, theory, and challenges: Interdisciplinary perspectives. *European Journal of Psychotraumatology, 5*(1), 1–14. PMID: 25317257

Spangenberg, J. H. (2011). Sustainability science: A review, an analysis and some empirical lessons. *Environmental Conservation, 38*(3), 275–287. 10.1017/S0376892911000270

Ssekibaala, D. S., Ariffin, M. I., & Duasa, J. (2022). Economic growth, international trade, and environmental degradation in Sub-Saharan Africa. *Journal of Economic Development, 24*(4), 293–308. DOI: 10.1108/JED-05-2021-0072

Stern, D. I. (2004). The rise and fall of the environmental Kuznets curve. *World Development, 32*(8), 1419–1439. 10.1016/j.worlddev.2004.03.004

Stern, , D. I. (2004). The rise and fall of the environmental Kuznets curve. *World Development, 32*(8), 1419-1439. https://doi.org/10.1016/j.worlddev.2004.03.004

Stern, D. I. (2017). The environmental Kuznets curve after 25 Years. *Journal of Bioeconomics, 19*, 7–28. http://link.springer.com/10.1007/s10818-017-9243-1

Stern, N. (2008). The Economics of Climate Change. *The American Economic Review, 98*(2), 1–37. DOI: 10.1257/aer.98.2.1

Strand, J. (2016). *Stranded assets: A climate risk challenge*. Inter-American Development Bank. https://publications.iadb.org/publications/english/document/Stranded-Assets-A-Climate-Risk-Challenge.pdf

Suhartono, F. S., & Suranto. (2020). *Non-natural disaster mitigation Covid-19 pandemic and its urgency in the education curriculum*. Faculty of Teacher Training and Education, Sebelas Maret University, UNS, Surakarta, Indonesia: Association for Computing Machinery. https://doi.org/DOI: 10.1145/3452144.3453772

Sukmono, F. G., Junaedi, F., Fadilla, Q. Y., Loilatu, M. J., & Fadillah, D. (2023). News about the Sunda Strait Tsunami Disaster: From the Selection of News Sources to the Production Process. *Studies in Media and Communication, 11*(4), 10–18. DOI: 10.11114/SMC.V11I4.6002

Sun, Y., Gong, Z., Shi, Q., & Chen, L. (2023). Efficient Commercial Bank Customer Credit Risk Assessment Based on LightGBM and Feature Engineering. *2023 5th International Symposium on Smart and Healthy Cities (ISHC)*, 75–82. https://doi.org/DOI: 10.1109/ISHC61216.2023.00022

Sun, X., Liu, M., & Sima, Z. (2020). A novel cryptocurrency price trend forecasting model based on LightGBM. *Finance Research Letters, 32*, 101084. DOI: 10.1016/j.frl.2018.12.032

Sun, Y. (2023). Income inequality, carbon emissions and green development efficiency. *Environmental Science and Pollution Research International, 30*(8), 21081–21091. PMID: 36264465

Sutiono, A. B., Qiantori, A., Suwa, H., & Ohta, T. (2009). Characteristic tetanus infection in disaster-affected areas: Case study of the Yogyakarta earthquakes in Indonesia. *BMC Research Notes, 2*. Advance online publication. DOI: 10.1186/1756-0500-2-34 PMID: 19284531

Swalha, I. H. (2020). A contemporary perspective on the disaster management cycle. *Foresight, 22*(4), 469–482.

Swaminathan, U., Reddy, G. B., Buot, I. E., & Ramamoorthy, S. (2023). Climate Change and Environment: Holistic Approaches Towards Climate Resilience. In *The Palgrave Handbook of Socio-ecological Resilience in the Face of Climate Change: Contexts from a Developing Country* (pp. 537–552). Springer Nature Singapore.

Swamy, P. A. V. B. (1970). Efficient inference in a random coefficient regression model. *Econometrica, 38*(2), 311. DOI: 10.2307/1913012

Szedlak, C., Callary, B., Eagles, K., & Gearity, B. T. (2025). An exploration of how dominant discourses steer UK strength and conditioning coach education. *Sport Education and Society, 30*(4), 470–483.

Szigeti, C., Toth, G., & Szabo, D. R. (2017). Decoupling – shifts in ecological footprint intensity of nations in the last decade. *Ecological Indicators, 72*, 111–117. 10.1016/j.ecolind.2016.07.034

Tacconi, L., Moore, P. F., & Kaimowitz, D. (2007). Fires in tropical forests - What is really the problem? Lessons from Indonesia. *Mitigation and Adaptation Strategies for Global Change, 12*(1), 55–66. DOI: 10.1007/s11027-006-9040-y

Taekratok, T., Phanthunane, P., & Taekratok, T. (2018). Disaster-based budgeting in new public health policy of Thailand. *Procedia Engineering, 212*, 715–722.

Tao, Z., & Han, L. (2022). Emergency response, influence and lessons in the 2021 compound disaster in henan province of china. *Environmental Research and Public Health, 19*(1), 1–19. PMID: 35010748

Tariq, M. U. (2025). *AI-Driven Learning and Engagement in Higher Education*. IGI Global., DOI: 10.4018/979-8-3693-4074-5

Tariq, M. U. (2025). Digital Accountability: Revolutionizing Public Sector Transparency and Service Delivery. In Santos Ferreira, A., & Lourenço dos Santos, C. (Eds.), *Enhancing Public Sector Accountability and Services Through Digital Innovation* (pp. 129–154). IGI Global Scientific Publishing., DOI: 10.4018/979-8-3693-9251-5.ch004

Tariq, M. U. (2025). Machine Learning Techniques for Personalized Rehabilitation: Enhancing Patient Outcomes Through Predictive Algorithms. In Korupalli, V., Ranasinghe, P., & Feldman, D. (Eds.), *Predictive Algorithms for Rehabilitation and Assistive Systems* (pp. 221–252). IGI Global Scientific Publishing., DOI: 10.4018/979-8-3373-0194-5.ch009

Tariq, M. U. (2025). Strategic Place Marketing in Cross-Border Regions: Enhancing Territorial Cooperation and Economic Growth. In Sousa, B. (Ed.), *Cross-Border Regions Cooperation and Implications for Organizations* (pp. 107–136). IGI Global Scientific Publishing., DOI: 10.4018/979-8-3373-1912-4.ch004

Tariq, M. U. (2025). The Challenge of Artificial Intelligence in Environmental Sustainability: Opportunities, Risks, and Policy Implications. In Ahmadi, A., & Mna, A. (Eds.), *Nexus of Environmental Quality and Technology Innovation* (pp. 267–292). IGI Global Scientific Publishing., DOI: 10.4018/979-8-3373-1822-6.ch012

Taylor, W. M., & Levine, M. P. (2015). Investigating catastrophe: Commemoration, accountability and records of disaster. *National Identities*, *17*(2), 105–116.

Teixidó-Figueras, J., & Duro, J. A. (2015). The building blocks of international ecological footprint inequality: A regression-based decomposition. *Ecological Economics*, *118*, 30–39. 10.1016/j.ecolecon.2015.07.014

Teker, O. (2021). *Derinlemesine python data science*. Godoro.

Temple, J. (2010). Aid and Conditionality. In Rodrik, D., & Rosenzweig, M. R. (Eds.), *Handbook of Development Economics* (Vol. 5, pp. 4415–4523). North-Holland.

Tercan, B. (2008). Afet *bölgelerinde yeniden yerleştirme ve iskan politikaları: doğubeyazıt afetzede yerleşim alanları uygulama örneği* [Unpublished doctoral dissertation]. Ankara Üniversitesi, Ankara, Türkiye.

Tercan, B. (2018). Türkiye'de afet politikaları ve kentsel dönüşüm. *Abant Kültürel Araştırmalar Dergisi*, *3*(5), 63–74.

Thai, H. N. (2023). Causal relationship between globalization, economic growth and CO2 emissions in Vietnam using Wavelet analysis. *Energy & Environment*, *34*(7), 2386–2412. DOI: 10.1177/0958305X221108498

The Guardian. (2025, October 2). *Kemi Badenoch vows to repeal Climate Change Act*. https://www.theguardian.com/environment/2025/oct/02/kemi-badenoch-vows-to-repeal-climate-change-act

The World Bank. (2024). https://www.worldbank.org/en/home

Thomas, K., Hardy, R. D., Lazrus, H., Mendez, M., Orlove, B., Rivera-Collazo, I., Roberts, J. T., Rockman, M., Warner, B. P., & Winthrop, R. (2019). Explaining differential vulnerability to climate change: A social science review. *Wiley Interdisciplinary Reviews: Climate Change*, *10*(2), 1–18. DOI: 10.1002/wcc.565 PMID: 31007726

Thompson, M. L. (1978). Selection of Variables in Multiple Regression: Part I. A Review and Evaluation. *International Statistical Review / Revue Internationale de Statistique, 46*(1), 1. https://doi.org/DOI: 10.2307/1402505

Thompson, L. K., Davis, R. M., & Wilson, C. A. (2023). Community-based renewable energy initiatives: Social capital and adoption outcomes. *Energy Research & Social Science*, *89*, 102634.

Thorbecke, E. (2000). The evolution of the development doctrine and the role of foreign aid, 1950–2000. In Tarp, F. (Ed.), *Foreign Aid and Development - Lessons Learnt and Directions for the Future* (pp. 12–35). Routledge.

Thornes, J. E., & Stephenson, D. B. (2001). How to judge the quality and value of weather forecast products. *Meteorological Applications*, *8*(3), 307–314. DOI: 10.1017/S1350482701003061

Tıpırdamaz, R., Karakas, S., & Dikilitaş, M. (2021). Halophytes and the future of agriculture. In Grigore, M. N. (Ed.), *Handbook of Halophytes*. Springer., DOI: 10.1007/978-3-030-57635-6_91

Toda, H. Y., & Yamamoto, T. (1995). Statistical inference in vector autoregressions with possibly integrated processes. *Journal of Econometrics*, *66*(1–2), 225–250. DOI: 10.1016/0304-4076(94)01616-8

Topuz, H., Kazak, H., Rahman, M. M., Kılıç, C., Akcan, A. T., & Özekicioğlu, H. (2025). Is export diversification detrimental to environmental quality? An examination of the roles of green innovation and environmental taxation. *Environmental Research Communications*, *7*(1), 1–10. DOI: 10.1088/2515-7620/adac36

Touraine, A. (2016). *After the crisis*. Yapı Kredi.

Toya, H., & Skidmore, M. (2014). Do natural disasters enhance societal trust? *Kyklos*, *67*(2), 255–279.

Trisolini, K. (2014). Holistic Climate Change Governance: Towards Mitigation and Adaptation Synthesis. *U. Colo. L. Rev.*, *85*, 615.

Troy, A., & Wilson, M. A. (2006). Mapping ecosystem services: Practical challenges and opportunities in linking GIS and value transfer. *Ecological Economics*, *60*(2), 435–449. 10.1016/j.ecolecon.2006.04.007

Tsasis, P., & Nirupama, N. (2008). Vulnerability and risk perception in the management of hiv/aids: Public priorities in a global pandemic. *Risk Management and Healthcare Policy*, *1*, 7–14. PMID: 22312198

Tunn, J., Kalt, T., Müller, F., Simon, J., Hennig, J., Ituen, I., & Glatzer, N. (2024). Green hydrogen transitions deepen socioecological risks and extractivist patterns: Evidence from 28 prospective exporting countries in the Global South. *Energy Research & Social Science*, *117*, 103731.

Turkish Statistical Institute. (2023). *Indicators of 100 years*. Turkish Statistical Institute.

Türkyılmaz, S., & Öztürk, K. N. (2024). Analysis of Factors Affecting CO2 Emissions in Türkiye Using Quantile Regression. *Sustainability*, *16*(22), 9634. DOI: 10.3390/su16229634

Turmunkh, B.-E. (2021). Renewable and non-renewable energy consumption, carbon dioxide emissions, and economic growth: Empirical evidence from Central Asian countries. *Journal of Economics and Development Studies*, *9*(1). Advance online publication. DOI: 10.15640/jeds.v9n1a7

Uddin, M. M., Mishra, V., & Smyth, R. (2020). Income inequality and CO2 emissions in the G7, 1870–2014: Evidence from non-parametric modelling. *Energy Economics*, *88*, 104780. DOI: 10.1016/j.eneco.2020.104780

Uddin, M. S., Haque, C. E., & Khan, M. N. (2021). Good governance and local level policy implementation for disaster-risk-reduction: Actual, perceptual and contested perspectives in coastal communities in Bangladesh. *Disaster Prevention and Management: An International Journal*, *30*(2), 94–111. DOI: 10.1108/DPM-03-2020-0069

Udemba, E. N. (2021). Mitigating environmental degradation with institutional quality and foreign direct investment (FDI): New evidence from asymmetric approach. *Environmental Science and Pollution Research International*, *28*(1), 43669–43683. DOI: 10.1007/s11356-021-13805-w PMID: 33837941

Udeogu, E., Amakom, U., & Roy-Mukherjee, S. (2021). Empirical Analysis of an Augmented Schumpeterian Endogenous Growth Model. *Journal of Economic Development, 46*(1). https://jed.cau.ac.kr/archives/46-1.html

Uğuz, S. (2021). *Makine Öğrenmesi Teorik Yönleri ve PYTHON Uygulamaları ile Bir YAPAY ZEKA Ekolü* (2nd ed.). Nobel.

Ünal, H., & Aktuğ, M. (2022). The impact of human capital and bio-capacity on the environmental quality: Evidence from G20 countries. *Environmental Science and Pollution Research International, 29*(30), 45635–45645. DOI: 10.1007/s11356-022-19122-0 PMID: 35149945

UNCDD. (2024). https://www.unccd.int/resources/reports/stepping-back-precipice-transforming-land-management-stay-within-planetary

UNCTAD. (2025). *World Investment Report 2025*. United Nations Conference on Trade and Development. https://unctad.org/system/files/official-document/wir2025_ch03_en.pdf

UNDRR. (2022). https://www.undrr.org/our-impact/campaigns/no-natural-disasters

UNFCCC. (1992). *UNITED NATIONS FRAMEWORK CONVENTION ON CLIMATE CHANGE*. https://unfccc.int/resource/docs/convkp/conveng.pdf

UN-Habitat. (2022). *Envisaging the future of cities*. World Cities Report.

UN-Habitat. (2025). https://www.linkedin.com/company/un-habitat-united-nation-human-settlements-programme-/posts/?feedView=all

United Nations Framework Convention on Climate Change (UNFCCC). (2008). *Integrating practices, tools and systems for climate risk assessment and management and strategies for disaster risk reduction into national policies and programmes* (Background Paper). https://unfccc.int/resource/docs/2008/tp/04.pdf

United Nations Framework Convention on Climate Change. (2024, November 22). *Summary of global climate action at COP 29*. https://unfccc.int/sites/default/files/resource/Summary_Global_Climate_Action_at_COP_29.pdf

United Nations University Institute for Environment and Human Security. (UNU-EHS) (2025). https://unu.edu/ehs/article/5-routes-better-future-turning-over-new-leaf

Uzar, U., & Eyuboglu, K. (2023). Does income inequality increase the ecological footprint in the US: Evidence from FARDL test? *Environmental Science and Pollution Research International, 30*(1), 9514–9529. DOI: 10.1007/s11356-022-22844-w PMID: 36057056

Vachon, B. (1998). *Le développement local - Théorie et pratique - Réintroduire l'humain dans la logique de développement.* Gaetan Morin Editeur.

Vadén, T., Lähde, V., Majava, A., Järvensivu, P., Toivanen, T., Hakala, E., ... (2020). Decoupling for ecological sustainability: A categorisation and review of research literature. *Environmental Science & Policy, 112*, 236–244. 10.1016/j.envsci.2020.06.016

Vakulchuk, R., Daloz, A. S., Overland, I., Sagbakken, H. F., & Standal, K. (2023). A void in Central Asia research: Climate change. *Central Asian Survey, 42*(1), 1–20. DOI: 10.1080/02634937.2022.2059447

Valdecantos, S. (2025). The green transition dilemma: The impossible (?) quest for prosperity of South American economies. *Ecological Economics, 230*, 108508. 10.1016/j.ecolecon.2024.108508

Vale, V. A., Perobelli, F. S., & Chimeli, A. B. (2017). International trade, pollution, and economic structure: Evidence on CO emissions for the North and the South. *Economic Systems Research, 30*(1), 1–17. DOI: 10.1080/09535314.2017.1361907

van Niekerk, D. (2014). A critical analysis of the South African Disaster Management Act and Policy Framework. *Disasters, 38*(4), 858–877. DOI: 10.1111/disa.12081 PMID: 25196341

Van Rooij, M., Lusardi, A., & Alessie, R. (2023). Financial literacy and stock market participation. *Journal of Financial Economics, 101*(2), 449–472.

Varadarajan, M. N. (2023). Smart Finance: How Artificial Intelligence and Machine Learning Drive Innovation in the Financial Sector. *International Journal of Innovative Research in Science, Engineering and Technology, 12*(12), 1–14. DOI: 10.15680/IJIRSET.2023.1212045

Varian, H. R. (2014). Big Data: New Tricks for Econometrics. *The Journal of Economic Perspectives, 28*(2), 3–28. DOI: 10.1257/jep.28.2.3

Vaughan, C., & Dessai, S. (2014). Climate services for society: Origins, institutional arrangements, and design elements for an evaluation framework. *Wiley Interdisciplinary Reviews: Climate Change, 5*(5), 587–603.

Venus, T. E., Bilgram, S., Sauer, J., & Khatri-Chettri, A. (2022). Livelihood vulnerability and climate change: A comparative analysis of smallholders in the Indo-Gangetic plains. *Environment, Development and Sustainability, 24*(2), 1981–2009.

Verdonck, T., Baesens, B., & Óskarsdóttir, M., & vanden Broucke, S. (2024). Special issue on feature engineering editorial. *Machine Learning, 113*(7), 3917–3928. DOI: 10.1007/s10994-021-06042-2

Verma, O., Sharma, S., Kumar, V., Singh, T., Kumar, R., & Auji, R. (2024). Salinity stress effect on staple food crops and novel mitigation strategies. *Biologia, 79*, 2359–2374. DOI: 10.1007/s11756-024-01689-3

Vermiglio, C., Noto, G., Bolívar, M. P. R., & Zarone, V. (2021). Disaster management and emerging technologies: A performance-based perspective. *Meditari Accountancy Research, 30*(4), 1093–1117.

Vidaller, C., & Dutoit, T. (2022). Ecosystem services in conventional farming systems: A review. *Agronomy for Sustainable Development, 42*, 22. DOI: 10.1007/s13593-021-00740-w

Vidyarthi, H. (2013). Energy consumption, carbon emissions and economic growth in India. *World Journal of Science. Technology and Sustainable Development, 10*(4), 278–287. DOI: 10.1108/WJSTSD-07-2013-0024

Wade, R. (2004). Is globalization reducing poverty and inequality? *World Development, 32*(4), 567–569. PMID: 15346677

Waheed, R., Chang, D., Sarwar, S., & Chen, W. (2018). Forest, agriculture, renewable energy, and CO2 emission. *Journal of Cleaner Production, 172*, 4231–4238. DOI: 10.1016/j.jclepro.2017.10.287

Wahyudin, D., Hasegawa, S., & Dahlan, T. (2013). *Developing ethical decision making skill of novice volunteers in natural disaster response. 2, 800–803*. School of Information Science JAIST.

Walker, B. H. (2020). Resilience: What is and is not? *Ecology and Society, 25*(2), 11.

Wang, , QGuo, , JDong, , Z. (2021). The positive impact of official development assistance (ODA) on renewable energy development: evidence from 34 Sub-Saharan Africa Countries. *Sustainable Production and Consumption, 28*, 532-542. https://doi.org/10.1016/j.spc.2021.06.007

Wang, J. (2024). Renewable energy, inequality and environmental degradation. *Journal of Environmental Management, 356*, 120563. DOI: 10.1016/j.jenvman.2024.120563 PMID: 38479288

Wang, J., & Dong, K. (2019). What drives environmental degradation? Evidence from 14 Sub-Saharan African countries. *The Science of the Total Environment, 656*(1), 165–173. DOI: 10.1016/j.scitotenv.2018.11.354 PMID: 30504018

Wang, L., Wei, F., Fang, Z., & Svenning, J. C. (2025). Transforming forest management through rewilding: Enhancing biodiversity, resilience, and biosphere sustainability under global change. *One Earth*, *8*(3), 1–10. DOI: 10.1016/j.oneear.2025.101195

Wang, S., Zhang, L., Liu, X., & Sun, J. (2024). Optimization of multidimensional feature engineering and data partitioning strategies in heart disease prediction models. *Alexandria Engineering Journal*, *107*, 932–949. DOI: 10.1016/j.aej.2024.09.037

Wang, S., Zhang, W., Wang, H., Wang, J., & Jiang, M. J. (2021). How does income inequality influence environmental regulation in the context of corruption? a panel threshold analysis based on chinese provincial data. *International Journal of Environmental Research and Public Health*, *8*(15), 1–21. DOI: 10.3390/ijerph18158050 PMID: 34360341

Wang, X. (2024). Correlation analysis of energy consumption, carbon emissions and economic growth. *Energy Informatics*, *7*(45), 1–12. DOI: 10.1186/s42162-024-00349-9

Wang, X., Zheng, H., Wang, Z., Shan, Y., Meng, J., Liang, X., Feng, K., & Guan, D. (2019). Kazakhstan's CO2 emissions in the post-Kyoto Protocol era: Production- and consumption-based analysis. *Journal of Environmental Management*, *249*, 109393. DOI: 10.1016/j.jenvman.2019.109393 PMID: 31450200

Wang, Y., Uddin, I., & Gong, Y. (2021). Nexus between Natural Resources and Environmental Degradation: Analysing the Role of Income Inequality and Renewable Energy. *Sustainability*, *13*(15), 8364. DOI: 10.3390/su13158364

Wardani, S. P. R., & Muntohar, A. S. (2013). Lessons learned from the recent natural disasters in Indonesia. *Geotechnical, Geological and Earthquake Engineering*, Vol. 25, pp. 47–59. Department of Civil Engineering, Diponegoro University, Jl. Hayam Wuruk, 5-7, Semarang, 50241, Indonesia: Springer Netherlands. https://doi .org/DOI: 10.1007/978-94-007-5675-5_4

Ward, L., & Lucas, S. (2024). "You're trying to put yourself in boxes, which doesn't work": Exploring non-binary youth's gender identity development using feminist relational discourse analysis. *Journal of Gender Studies*, *33*(5), 658–672.

Warren, K., Franklin, C., & Streeter, C. L. (1998). New directions in systems theory: Chaos and complexity. *Social Work*, *43*(4), 357–372.

Wartiainen, F., Boer, B., Pauw, P., & Romijn, H. (2025). Addressing the Twin Crises of Debt and Climate: Exploring Unconditional Debt Cancellation. *Development*, •••, 1–10.

WDI. (2022). *World Development Indicators Report.* https://datatopics.worldbank.org/world-development-indicators/world-development-indicators-reports.html

Weber, E. U., & Stern, P. C. (2023). Public understanding of climate change in the United States. *The American Psychologist, 66*(4), 315–328. PMID: 21553956

WEF. (2025). https://www.weforum.org/stories/2025/03/cities-global-trade-changing-seize-moment/

Wei, J., Rahim, S., & Wang, S. (2022). Role of environmental degradation, institutional quality, and government health expenditures for human health: Evidence from emerging seven countries. *Frontiers in Public Health, 10*(1), 1–20. DOI: 10.3389/fpubh.2022.870767 PMID: 35400071

Wei, W., Li, J., Cao, L., Ou, Y., & Chen, J. (2013). Effective detection of sophisticated online banking fraud on extremely imbalanced data. *World Wide Web (Bussum), 16*(4), 449–475. DOI: 10.1007/s11280-012-0178-0

Weko, , SGoldthau, , A. (2022). Bridging the low-carbon technology gap? Assessing energy initiatives for the Global South. *Energy Policy, 169*, 113192. https://doi.org/10.1016/j.enpol.2022.113192

Wen, J., Wan, C., Ye, Q., Yan, J., & Li, W. (2023). Disaster risk reduction, climate change adaptation and their linkages with sustainable development over the past 30 years: A review. *International Journal of Disaster Risk Science, 14*, 1–13.

Werner de Vargas, V., Schneider Aranda, J. A., dos Santos Costa, R., da Silva Pereira, P. R., & Victória Barbosa, J. L. (2023). Imbalanced data preprocessing techniques for machine learning: A systematic mapping study. *Knowledge and Information Systems, 65*(1), 31–57. DOI: 10.1007/s10115-022-01772-8 PMID: 36405957

Whittaker, D. H., Sturgeon, T. J., Okita, T., & Zhu, T. (2020). *Compressed development: Time and timing in economic and social development.* Oxford University Press.

Wiedmann, T., Lenzen, M., Keyßer, L. T., & Steinberger, J. K. (2020). Scientists' warning on affluence. *Nature Communications, 11*, 3107. 10.1038/s41467-020-16941-y

Wijburg, G. (2025). On undevelopment and de-development: A geographical critique on perpetual growth and resource-based accumulation. *Area, 57*(2), e12988.

Williamson, J. (1990). What Washington means by policy reform. In *Latin American adjustment: How much has happened* (pp. 7-20).

Wilson, H. J., & Daugherty, P. R. (2018). Collaborative intelligence: Humans and AI are joining forces. *Harvard Business Review, 96*(4), 114–123.

Wilson, P. R., Anderson, T. L., & Brown, K. S. (2024). Conceptualizing green financial literacy: A multi-dimensional framework. *Financial Planning Review*, *7*(2), e1234.

Winter, M. A. (2024). Towards a model for sociocultural transformation: Anthropocentric approaches to resilience, collapse, and resistance. *Journal of Ancient Near Eastern History*, *11*(1), 83–103. 10.1515/janeh-2022-0012

Wolde-Rufael, Y., & Idowu, S. (2016). Income distribution and CO2 emission: A comparative analysis for China and India. *Renewable & Sustainable Energy Reviews*, *74*, 1336–1345. DOI: 10.1016/j.rser.2016.11.149

Wolde-Rufael, Y., & Mulat-Weldemeskel, E. (2021). Do environmental taxes and environmental stringency policies reduce CO2 emissions? Evidence from 7 emerging economies. *Environmental Science and Pollution Research International*, *28*(18), 22392–22408. DOI: 10.1007/s11356-020-11475-8 PMID: 33417133

World Bank. (2010). *World development report 2010: Development and climate change*. https://openknowledge.worldbank.org/handle/10986/4387

World Bank. (2016). *World Bank Group Climate Change Action Plan 2016–2020*. https://openknowledge.worldbank.org/handle/10986/24795

Wraith, R. (1996). Women in disaster management: Where are they? *Australian Journal of Emergency Management*, *11*, 8.

Wu, T., Rocha, J. C., Berry, K., Chaigneau, T., Hamann, M., Lindkvist, E., ... (2024). Triple Bottom Line or Trilemma? Global Tradeoffs Between Prosperity, Inequality, and the Environment. *World Development*, *178*, 106595. 10.1016/j.worlddev.2024.106595

Xiao, D., Yu, F., & Yang, H. (2022). The impact of urban-rural income inequality on environmental quality in China. *Complexity*, *2022*(1), 1–10. DOI: 10.1155/2022/4604467

Xiong, C., Yang, D., Huo, J., & Zhao, Y. (2015). The relationship between energy consumption and economic growth and the development strategy of a low-carbon economy in Kazakhstan. *Journal of Arid Land*, *7*(5), 706–715. DOI: 10.1007/s40333-015-0132-y

Xuan, H. T., & Hung, N. T. (2024). Does green investment mitigate environmental degradation in Vietnam: The time-frequency effect of nonrenewable energy investment and globalization? *Management of Environmental Quality*, *35*(5), 1005–1027. DOI: 10.1108/MEQ-09-2023-0332

Xu, C. (2025). Analysis of the state-of-art chaos: Theory and applications. *Highlights in Science. Engineering and Technology*, *128*, 231–237.

Xu, T. (2024). Fraud Detection in Credit Risk Assessment Using Supervised Learning Algorithms. *Computer Life*, *12*(2), 30–36. DOI: 10.54097/qw9j1892

Xu, Z., Smyth, C. E., Lemprière, T. C., Rampley, G. J., & Kurz, W. A. (2018). Climate change mitigation strategies in the forest sector: Biophysical impacts and economic implications in British Columbia, Canada. *Mitigation and Adaptation Strategies for Global Change*, *23*(2), 257–290.

Yalta, A. T. (2011). Analyzing energy consumption and GDP nexus using maximum entropy bootstrap: The case of Turkey. *Energy Economics*, *33*(3), 453–460. DOI: 10.1016/j.eneco.2010.12.005

Yang, B., Jahanger, A., Usman, M., & Khan, M. A. (2021). The dynamic linkage between globalization, financial development, energy utilization, and environmental sustainability in GCC countries. *Environmental Science and Pollution Research International*, *28*(1), 16568–16588. DOI: 10.1007/s11356-020-11576-4 PMID: 33387307

Yang, Y., Zhao, T., Wang, Y., & Shi, Z. (2015). Research on impacts of population-related factors on carbon emissions in Beijing from 1984 to 2012. *Environmental Impact Assessment Review*, *55*, 45–53. DOI: 10.1016/j.eiar.2015.06.007

Yilanci, V., & Gorus, M. S. (2020). Does economic globalization have predictive power for ecological footprint in MENA counties? A panel causality test with a Fourier function. *Environmental Science and Pollution Research International*, *27*(32), 40552–40562. DOI: 10.1007/s11356-020-10092-9 PMID: 32666448

Yılmaz, A. (2002). Türkiye'de afet zararlarının azaltılması çalışmalarının tarihi gelişimi üzerine notlar. *Türk İdare Dergisi*, *74*(434), 155–169.

Yoro, K. O., & Daramola, M. O. (2020). CO2 emission sources, greenhouse gases, and the global warming effect. In *Advances in Carbon Capture* (pp. 3–28). Elsevier., DOI: 10.1016/B978-0-12-819657-1.00001-3

Younsi, M., Khemili, H., & Bechtini, M. (2019). Does foreign aid help alleviate income inequality? New evidence from African countries. *International Journal of Social Economics*, *46*(4), 549–561. DOI: 10.1108/IJSE-06-2018-0319

Zafer, M. (2023). Afetlerde sosyal hizmet: Kovid-19 afet salgini ve yaşlilar. *Kadim Akademi Sbd*, *7*(1), 37–50.

Zaidi, S. A. H., Danish, Hou, F., & Mirza, F. M. (2018). The role of renewable and non-renewable energy consumption in CO2 emissions: A disaggregate analysis of Pakistan. *Environmental Science and Pollution Research International, 25*(31), 31616–31629. DOI: 10.1007/s11356-018-3059-y PMID: 30206833

Zakharova, O., Karagulian, E., Viktorova, N., Gamukin, V., & Yablochkina, V. (2025). Risk assessment of socio-economic vulnerability to climate change: Case study of Tyumen region. *WORLD (Oakland, Calif.), 6*(1), 6. DOI: 10.3390/world6010006

Zarsky, T. (2016). The Trouble with Algorithmic Decisions. *Science, Technology & Human Values, 41*(1), 118–132. DOI: 10.1177/0162243915605575

Zhang, R., Li, P., Xu, L., & Zhong, S. (2023). Reconciling ecological footprint and ecosystem services in natural capital accounting: Applying a novel framework to the Silk Road Economic Belt in China. *Journal of Environmental Management, 330*, 117115. 10.1016/j.jenvman.2022.117115

Zhang, H. B., Dai, H. C., Lai, H. X., & Wang, W. T. (2017). US withdrawal from the Paris Agreement: Reasons, impacts, and China's response. *Advances in Climate Change Research, 8*(4), 220–225.

Zhang, H., Wang, F. T., & Zhou, P. (2022a). Changes in climate extremes in a typical glacierized region in central Eastern Tianshan Mountains and their relationship with observed glacier mass balance. *Advances in Climate Change Research, 13*(6), 909–922.

Zhang, J. (2023). Assessing the effect of the improvement of environmental damage compensation legal system and green finance project on the re-establishment of the ecological environment. *Environmental Science and Pollution Research International, 30*(25), 67662–67675. DOI: 10.1007/s11356-023-26877-7 PMID: 37118386

Zhang, J., Mucs, D., Norinder, U., & Svensson, F. (2019). LightGBM: An Effective and Scalable Algorithm for Prediction of Chemical Toxicity–Application to the Tox21 and Mutagenicity Data Sets. *Journal of Chemical Information and Modeling, 59*(10), 4150–4158. DOI: 10.1021/acs.jcim.9b00633 PMID: 31560206

Zhang, L., Xu, M., Chen, H., & Li, Y. (2022). Globalization, green economy and environmental challenges: State of the art review for practical implications. *Frontiers in Environmental Science, 10*(1), 1–7. DOI: 10.3389/fenvs.2022.870271

Zhang, R., Tang, X., Liu, J., Visbeck, M., Guo, H., Murray, V., & Zhou, L. (2022b). From concept to action: A united, holistic and One Health approach to respond to the climate change crisis. *Infectious Diseases of Poverty, 11*(1), 17.

Zhang, S. (2019). Environmental Kuznets curve revisit in Central Asia: The roles of urbanization and renewable energy. *Environmental Science and Pollution Research International, 26*(23), 23386–23398. DOI: 10.1007/s11356-019-05600-5 PMID: 31201699

Zhang, X., Zhang, H., & Yuan, J. (2019). Economic growth, energy consumption, and carbon emission nexus: Fresh evidence from developing countries. *Environmental Science and Pollution Research International, 26*(1), 26367–26380. DOI: 10.1007/s11356-019-05878-5 PMID: 31290044

Zhang, Y., Liu, X., & Chen, M. (2023). The complexity of green finance: Policy instruments and market development. *Energy Economics, 98*, 105267.

Zhang, Z., Liu, H., Chen, D., Zhang, J., Li, H., Shen, M., Pu, Y., Zhang, Z., Zhao, J., & Hu, J. (2022). SMOTE-based method for balanced spectral nondestructive detection of moldy apple core. *Food Control, 141*, 109100. DOI: 10.1016/j.foodcont.2022.109100

Zhao, C., Kong, L., Zhong, C., & Ying, N. (2025). Equality in the water: How does water ecological civilization city policy enhance inclusive development in China? *Cities (London, England), 156*, 105555. https://doi.org/https://doi.org/10.1016/j.cities.2024.105555

Zhu, M., Zhang, Y., Gong, Y., Xing, K., Yan, X., & Song, J. (2024). Ensemble Methodology: Innovations in Credit Default Prediction Using LightGBM, XGBoost, and LocalEnsemble. *2024 IEEE 4th International Conference on Electronic Technology, Communication and Information (ICETCI)*, 421–426. https://doi.org/DOI: 10.1109/ICETCI61221.2024.10594630

Zörb, C., Geilfus, C.-M., & Dietz, K.-J. (2019). Salinity and crop yield. *Plant Biology, 21*(S1), 31–38. DOI: 10.1111/plb.12884

About the Contributors

Ali Gökhan Gölçek, PhD, is an Assistant Professor in the Department of Public Finance at the Faculty of Economics and Administrative Sciences, Niğde Ömer Halisdemir University, Türkiye. His research interests include public finance theory, fiscal policies, environmental economics, poverty, energy economics, and tax policies. In recent years, he has particularly focused on sustainable development, digital poverty, the economics of aging, and the green economy. Gölçek has published numerous articles in national and international peer-reviewed journals and also serves as an editor and reviewer, contributing to various book projects as both author and editor.

Altuğ Murat Köktaş is a Prof Dr of Department of Public Finance at Ankara University, Türkiye. His scholarly interests fiscal policy, energy policy, health economics, and economic history. He completed his PhD in Public Finance at Ankara University in 2014. Köktaş has published numerous articles and book chapters.

***

Prisha Agarwal is a student doing a research project at Amity University Noida focusing on different concerns related to India.

Mohamed El Bakkali an Associate Professor at the Higher Institute of Nursing Professions and Health Techniques (ISPITS), Kenitra Annex, Morocco. He holds a Ph.D. in Epidemiology and Biostatistics, a Master's degree in Neurobiology, Pharmacology and Genetics, a Bachelor's degree in Life Sciences, and a Bachelor diploma in Arabic Law. Since 2024, he has been teaching Epidemiology, Biostatistics, and Research Methodology to nursing and allied health science students. Between 2011 and 2023, he led the evaluation of social programs targeting vulnerable populations at the central headquarters of Entraide Nationale. Dr. El Bakkali has authored over 50 scientific papers, including 19 indexed in Scopus and 7 published by IGI Global, covering topics such as public health, infectious diseases, environmental science, and biomedical research. He is highly skilled in statistical analysis, proficient with tools such as R, SPSS, Stata, Statistica, and Stratigraphic Centurion. A specialist in artificial intelligence for health applications, he integrates deep learning techniques such as 1D Convolutional Neural Networks (1D-CNN) with self-attention, XGBoost, and PySR for advanced modeling and prediction. He also possesses strong programming capabilities, using Python (with a focus on data science), C, C++, JavaScript, and more. Regularly supervising undergraduate, master's, and doctoral research projects, Dr. El Bakkali is committed to interdisciplinary collaboration, innovation in health education, and the promotion of evidence-based practices across the healthcare ecosystem.

Csongor Czipf, Ph.D. Honorary Associate Professor, Assistant Professor Széchenyi István University, Győr, Hungary Gyula Kautz Faculty of Economics, Department of International Applied Economics Ongoing teaching experience at Széchenyi István University since 1999. Appointed as Regional Economic Manager of the Piarist Order Hungarian Province in Budapest as of January 2022. Master's degree in Religious Education and Dormitory Teaching obtained from the College of Theology in Győr in 2017. Doctoral degree awarded in 2011 by Janus Pannonius University of Pécs. Research specialization: corporate evaluation from a human resources perspective. Bachelor's degree in Economics with a specialization in Comparative Economic Policy completed in 1999 at Széchenyi István College.

Ramesh Chandra Das is currently Professor of Economics at Vidyasagar University, West Bengal, India. Previously he was Associate Professor of Economics at Katwa College, India. He has obtained Masters, M. Phil and Ph. D Degree in Economics from the University of Calcutta. He has about twenty five years of teaching and research experience in different fields of economics including theoretical and empirical economics, financial economics, environmental economics and political economics. Dr. Das has contributed several research papers to national and international journals and completed three MRPs sponsored by UGC, India. He has written text books on Microeconomics and Macroeconomics for different fields of readers and academician. He is the editor-in-chief in Asian Journal of Research in Business Economics and Management, Associate Editor of International Journal of Research on Social and Natural Sciences. He has edited a number of handbooks on Globalization, Investment and Growth...Economic and Political Convergence, infrastructure, microfinance, military expenditure, terrorism etc. and worked s guest editor of IJSEM's Special Issue (2 parts), all are of IGI Global, USA.

Beyza Güdek is pursuing her academic career in the field of Management Information Systems (MIS) at the undergraduate, graduate, and doctoral levels. She received her bachelor's degree from Necmettin Erbakan University, her master's degree from Karadeniz Technical University, and is currently continuing her doctoral studies at Izmir Bakırçay University. As part of her Ph.D. program, she is also engaged in international academic training at Vilnius University. She currently serves as a research assistant in the Department of Management Information Systems at the Faculty of Economics and Administrative Sciences, Karadeniz Technical University. Her research interests include digital transformation, artificial intelligence, data analytics, and multi-criteria decision making.

Omar Guirette-Barbosa is a Bussines Management Ph D.Experience in quality systems based on ISO standards such as 9001, 17025, etc. Consultant on accreditation and certification issues, as well as on the application of continuous improvement tools and industrial engineering.

Öner Gümüş graduated from Anadolu University Department of Economics in English in 2010, from Anadolu University Department of Public Finance in 2015, from Anadolu University Justice Associate Degree Program in 2018, and from Atatürk University Civil Air Transportation Management Associate Degree Program in 2024. He received a master's degree in public finance from Kutahya Dumlupinar University in 2014, a master's degree in economics from Kutahya Dumlupinar University in 2018, and a doctorate degree from the same university in 2020. The author continues his studies as an Assistant Professor at Kutahya Dumlupinar University Tavsanli Vocational School. He has authored/co-authored books and book chapters, many journal and conference papers. His research areas include public finance, economics, and taxation.

Tugay Günel is an Associate Professor at Çukurova University, Faculty of Economics and Administrative Sciences, Department of Public Finance. He received his PhD in Public Finance and has more than 13 years of academic experience in teaching and research. His main research areas include fiscal policy, taxation, public expenditures, and the interaction between institutional quality and macroeconomic outcomes. In recent years, his work has particularly focused on fiscal austerity, environmental taxation, and their socio-economic consequences such as gender equality in labor markets and food security.

Driss Hmouni received his Ph.D. in Sciences with a specialization in Agrophysiology and Biotechnology from the Faculty of Sciences, Ibn Tofaïl University, Kenitra, Morocco. He also holds a Diploma of Advanced Studies in Plant Physiology from the Faculty of Sciences, Tunis, Tunisia. Currently, he is a

Professor and Researcher at the Faculty of Sciences, Ibn Tofaïl University, where he contributes to research and teaching in plant biotechnology, environmental stress physiology, and sustainable agricultural practices. With over 20 years of academic experience, Pr. Hmouni has authored/co-authored 53 peer-reviewed publications, including journal articles, book chapters, and conference proceedings indexed in Scopus and Web of Science, with 140+ citations and an h-index of 5. His research focuses on climate-resilient crops, precision agriculture, microalgae-based wastewater treatment, and biotechnological innovations for sustainable farming systems. He is an active reviewer for high-impact scientific journals and collaborates with national and international research institutions to develop innovative strategies for agricultural adaptation to climate change. His expertise extends to biofertilizers, soil microbiomes, smart irrigation technologies, and the integration of AI and IoT in agriculture.

Nur Hoşgör graduated from Selçuk University, Department of Public Administration in 2009. In 2011, she completed her master's degree at Selçuk University, Department of Public Administration. The author, who is currently continuing her doctoral studies in the Department of Public Administration at Kütahya Dumlupınar University, is also working as a lecturer at Kütahya Dumlupınar University.

Muhammad Nurul Huda is a lecturer at Universitas Muhammadiyah Karanganyar (UMUKA) and is currently a doctoral candidate in the Governmental Science program at Universitas Muhammadiyah Yogyakarta (UMY). His research interests focus on the fields of politics, government, and public policy.

Mustafa Bilgehan İmamoğlu is a faculty member in the Department of Management Information Systems at Karadeniz Technical University, Turkey. He received his Ph.D. in Information Security, focusing on database watermarking and information hiding techniques. His academic interests include cybersecurity, artificial intelligence, deep learning, anomaly detection, fake news detection, and privacy-preserving machine learning methods. He has published papers in national and international journals and has mentored graduate and undergraduate students in applied research projects. Dr. IMAMOGLU has also participated in research initiatives related to critical infrastructure protection, IoT security, and AI-based decision support systems.

Bambang Irawan is a professor in the Department of Public Administration, Faculty of Social and Political Sciences, Mulawarman University.

Kıvılcım Özge Kara graduated from Gazi University, Faculty of Economics and Administrative Sciences, Department of Business Administration in 2005, and from the Department of Political Science and Public Administration (double major) in 2006. She completed his master's degree in Ankara University, Faculty of Political Sciences, Department of Finance (Public Economics) in 2017, and then her doctorate in Hacettepe University, Institute of Social Sciences, Department of Finance (Public Economics) in 2024, Department of Budget and Financial Planning. She currently serves at the General Directorate of Local Governments of the Ministry of Environment, Urbanization and Climate Change. He conducts studies on environmental economics, local governments and environmental policies.

Umi Kulsum is a lecturer at the Department of Government Studies, Universitas Muhammadiyah Malang. Research interest on urban studies, e-government.

Mohammad Jafar Loilatu is lecturer at Department of Government Studies Universitas Muhammadiyah Malang.

Sandrina Berthault Moreira (PhD in Economics) is Assistant Professor at the Instituto Politécnico de Setúbal (ESCE - IPS) - Department of Economics and Management, and Researcher at the Business Research Unit, Economics Group (BRU-IUL) and RESILIENCE (IPS). Currently, she is involved in three international research projects: E^3UDRES^2 2.0 (proj. n. 101124496; Erasmus+); Ent-r-e-novators (proj. n. 101071317; Horizon Europe); and EULAC for Energy Transition (proj. n. 101131725; Horizon Europe). Her recent work has been published in the Applied Economics Letters, Development Studies Research, European Journal of Development Research, Review of Development Finance, Revista Portuguesa de Estudos Regionais, Social Inclusion and World Economy. Her broad research interests are macroeconomics, economic growth, and development.

Yassine Mouniane researcher in biology and the environment, specializing in biotechnologies and the sustainability of agricultural systems. His research focuses mainly on crop adaptation to climate change, with a particular emphasis on improving plant resilience to abiotic stresses such as salt stress and drought. Yassine Mouniane currently has over 48 papers to his credit, 31 of which are indexed in Scopus. His h-index is 2, reflecting increased citation of his work in the scientific community. He has published in international journals and collaborated on several research projects, with papers covering topics such as biodiversity, natural resource management, and plant applications in sustainable agriculture and wastewater treatment. In addition to his publications, Yassine Mouniane is also involved in reviewing scientific articles, helping to improve the quality of work submitted to specialized journals. At the same time, Yassine Mouniane is also interested in precision farming technologies and the application of artificial intelligence to optimize agricultural practices. He explores the integration of remote sensing tools, intelligent sensors and climate prediction models to improve resource management and maximize crop productivity while minimizing environmental impact. His passion for sustainable agriculture and innovative technologies has led him to become actively involved in projects aimed at integrating renewable energies, such as agrovoltaics, into farming practices. In the course of his work, he has been involved in various studies on biofertilizers, biostimulants, and the development of resilient crops via biotechnological approaches. His expertise in biotechnology, combined with his research in machine learning and precision agriculture, makes him an ideal contributor to this book. His research is part of an interdisciplinary approach aimed at fostering the transition to more resilient and sustainable agricultural systems. In addition to his research, Yassine Mouniane is the author and co-author of several scientific publications in international journals, covering topics such as crop adaptation to climate change, the application of biotechnologies to improve plant resistance, and the optimization of agricultural practices using artificial intelligence. His publications include work on the management of abiotic stresses in plants, the impact of biostimulants on crop growth, and the integration of remote sensing in agricultural resource management.

Saiman is senior lecturer at Department of Government Universitas Muhammadiyah Malang.

Ajay K. Singh is working as an Associate Professor (Research) in the Department of Humanities and Social Science, Graphic Era (Deemed to be University) Dehradun. He has worked as Assistant Professor (Economics) in the School of Liberal Arts & Management, DIT University Dehradun for 6 years. He did Post-Doctorate Research with EDI of India, Ahmedabad, Gujarat (India). He received MPhil (Economics) from DAVV Indore (India), and PhD (Economics) from IIT Indore (India. He has published several research papers in the diversified area such as climate change, agricultural productivity, assessment of food security; estimation of GFSI, development of environmental sustainability index (ESI) and its association with socio-economic indicators; measurement and determinants of entrepreneurship ecosystem, and dimension of sustainable development and its interlinkages with economic development.

Muhammad Usman Tariq has more than 16+ year's experience in industry and academia. He has authored more than 200+ research articles, 110+ case studies, 150+ book chapters and several books other than 4 patents. He is founder and CEO of The Case HQ, a unique repository for courses, narrative and video case studies. Additionally, he is President of Gulf Higher Education (Gulf HE) focused on advancing academic practice. He has been working as a consultant and trainer for industries representing six sigma, quality, health and safety, environmental systems, project management, and information security standards. His work has encompassed sectors in aviation, manufacturing, food, hospitality, education, finance, research, software and transportation. He has diverse and significant experience working with accreditation agencies of ABET, ACBSP, AACSB, WASC, CAA, EFQM and NCEAC. He is a Principal Fellow of Advance HE UK, Senior Fellow of SEDA UK, and Chartered Fellow of CIPD UK other than Senior Member IEEE.

Srinivasan Vasan is an Assistant Professor in the Department of Humanities and Social Sciences at Graphic Era (Deemed to be University), with 14 years of teaching and field expertise. He has served in the Ministry of Health and Family Welfare, as a Guest Faculty, and as a Mobilization Training Specialist with the State Government. Recently, he contributed to a tuberculosis project funded by GFATM and USAID. He has published 60 papers, with 20 more under peer review in UGC-CARE and SCOPUS journals.

Index

A

C

D

E

F

G

I

K

L

M

N

O

P

R

S

T

www.ingramcontent.com/pod-product-compliance
Lightning Source LLC
LaVergne TN
LVHW081319300926
849161LV00008B/1818